Ecology of Estuaries

Volume II
Biological Aspects

Author

Michael J. Kennish, Ph.D.

Visiting Professor
Institute of Marine and Coastal Sciences
Fisheries and Agriculture TEX Center
Cook College, Rutgers University
New Brunswick, New Jersey

CRC Press
Boca Raton   Ann Arbor   Boston

Library of Congress Cataloging-in-Publication Data
(Revised for vol. 2)

Kennish, Michael J.
  Ecology of estuaries

  Includes bibliographical references.
  Contents: v. 1. Physical and chemical aspects — v. 2. Biological aspects.
  1. Estuarine ecology. I. Title.
QH541.5.E8K46   1986   574.5'26365   85-20937
ISBN 0-8493-5892-2   (vol. 1)
ISBN 0-8493-5893-0   (vol. 2)

   Direct all inquiries to CRC Press, Inc., 2000 Corporate Blvd., N.W., Boca Raton, Florida, 33431.

© 1990 by CRC Press, Inc.

International Standard Book Number 0-8493-5893-0

Library of Congress Card Number 85-20937
Printed in the United States

DEDICATION

TO
My brother, Dr. John M. Kennish, whose dedication
to the pursuit of knowledge in the field of chemistry
has been a great inspiration to me.

and

TO
My sons, Shawn and Michael, who are the promise of
the future and the blessing for today.

PREFACE

The rapid development of estuarine ecology as a field of scientific inquiry reflects a growing awareness of the immense societal importance of these coastal ecosystems. The many, varied, and increasingly complex problems arising from anthropogenic use of estuaries have focused attention on the pressing need to protect estuarine resources and to assess each system's physical, chemical, and biological conditions. Unequivocally, the study of estuaries involves both ecological and societal issues. As a result of the heightened public concern related to the water quality of coastal regions in the U.S., for example, numerous detailed and quantitative research programs were undertaken in the 1960s, 1970s, and 1980s, which provided a basic understanding of the processes operating in many estuaries. The information derived from these programs has also aided decision-making bodies at all levels — that plan for the utilization of natural resources — to make sound decisions dealing with these dynamic waters.

The number of publications on estuaries has increased dramatically during the past two decades concomitant with expanded coverage of the many interdisciplinary topics relevant to estuarine ecology. While the volume of literature on estuaries amassed, scientists deemed it necessary to synthesize the field periodically. Consequently, several books have been produced in recent years which examine various aspects of the discipline.

The principal objective of this book is to review the biological characteristics of estuaries. The volume has been designed as a text for undergraduate and graduate students as well as a reference for scientists conducting research on estuarine systems. However, administrators, managers, decision makers, and other professionals involved in some way with investigations of estuaries can also find value in the publication. I have attempted to integrate the diverse components of estuarine ecology, to assemble citations of the major articles and books treating biological processes in estuaries, and to present the subject matter in an organized framework that should facilitate its use.

I am grateful to the reviewers of this work. In addition, I am indebted to the scientists, colleagues, family members, and friends who have provided encouragement and inspiration during the production of the book. In GPU Nuclear Corporation, I acknowledge the support of D. J. Cafaro and J. J. Vouglitois, and in Rutgers University, I express deep appreciation to J. N. Kraeuter, R. E. Loveland, and R. A. Lutz for innumerable exchanges of ideas on estuarine ecology. The Editorial Department of CRC Press is thanked for its efficiency and guidance during the publication process. Finally, I am most appreciative of my wife, Jo-Ann, and sons, Shawn and Michael, for their unending love during the preparation of the manuscript.

Michael J. Kennish

THE AUTHOR

Michael J. Kennish, Ph.D., is a Visiting Professor in the Institute of Marine and Coastal Sciences at Cook College, Rutgers University, New Brunswick, New Jersey.

He graduated in 1972 from Rutgers University, Camden, New Jersey, with a B.A. degree in geology and obtained his M.S. and Ph.D. degrees in the same discipline from Rutgers University, New Brunswick, New Jersey, in 1974 and 1977, respectively.

Dr. Kennish's professional affiliations include the American Fisheries Society (Mid-Atlantic Chapter), American Geophysical Union, American Institute of Physics, American Society of Limnology and Oceanography, Estuarine and Coastal Sciences Association, Estuarine Research Federation, New England Estuarine Research Society, Atlantic Estuarine Research Society, Southeastern Estuarine Research Society, Gulf Estuarine Research Society, Pacific Estuarine Research Society, National Shellfisheries Association, New Jersey Academy of Science, and Sigma Xi. He is also a member of the Advisory Board of the Fisheries and Aquaculture TEX Center of Rutgers University, overseeing the development of fisheries and shellfisheries in estuarine and marine waters of New Jersey.

Although maintaining research interests in broad areas of marine ecology and marine geology, Dr. Kennish has been most actively involved with investigations of anthropogenic effects on estuarine ecosystems. He is the author of *Ecology of Estuaries* (Volume 1), published by CRC Press, the editor of *Practical Handbook of Marine Science,* published by CRC Press, and the co-editor of *Ecology of Barnegat Bay, New Jersey,* published by Springer-Verlag. In addition to these three books, Dr. Kennish has published articles in scientific journals and presented papers at numerous conferences. Currently, he is the co-editor of the journal, *Reviews in Aquatic Sciences,* and the marine science editor of the journal, *Bulletin of the New Jersey Academy of Science.* His biographical profile appears in *Who's Who in Frontiers of Science and Technology.*

TABLE OF CONTENTS

Introduction 1
I.   Anthropogenic Impacts on Estuaries 1
II.   Plan of This Volume 3
References 7

Chapter 1
Bacteria 9
I.   Introduction 9
II.   Types of Marine Bacteria 10
III. Bacterial Abundance, Biomass, and Production 11
    A.   Abundance 11
    B.   Biomass 18
    C.   Production 18
IV. Microbial Processes 23
    A.   General 23
    B.   Aerobic Environments 24
        1.   Attached and Free-Living Bacteria 24
        2.   Bacteria as Decomposers 25
        3.   Bacterivory 31
    C.   Anaerobic Environments 32
        1.   Fermentation 33
        2.   Dissimilatory Sulfate Reduction 33
        3.   Dissimilatory Nitrogenous Oxide Reduction 35
        4.   Methanogenesis 37
        5.   Iron and Manganese Reduction 38
V.   Summary and Conclusions 38
References 41

Chapter 2
Phytoplankton 51
I.   Introduction 51
II.   Taxonomy 52
    A.   Classes 52
        1.   Diatoms (Class Bacillariophyceae) 54
        2.   Dinoflagellates (Class Dinophyceae) 55
        3.   Coccolithophores and Other Brown-Colored Algae
            (Class Haptophyceae) 56
        4.   Silicoflagellates (Class Chrysophyceae) 56
        5.   Blue-Green Algae (Class Cyanophyceae) 57
        6.   Green-Colored Algae (Class Chlorophyceae) 57
        7.   Euglenoid Flagellates (Class Euglenophyceae) 57
III. Sampling Methods 57
IV. Phytoplankton Biomass and Primary Productivity 58
    A.   Biomass 58
    B.   Productivity 60
V.   Species Composition and Succession 62
    A.   General 62
    B.   Type Estuarine Systems 63
        1.   Chesapeake Bay 63

  2. Barnegat Bay 70
  3. Long Island Sound 74
  4. Narragansett Bay 77
VI. Factors Affecting Primary Production of Phytoplankton 79
 A. General 79
 B. Physical Factors 83
  1. Light 83
  2. Temperature 89
  3. Water Circulation 90
 C. Chemical Factors 91
  1. Salinity 91
  2. Nutrients 92
   a. Nitrogen 93
   b. Phosphorus 96
   c. Silicon 96
   d. Trace Metals 97
   e. Organic Trace Substances 97
 D. Biological Factors 97
  1. Grazing 97
VII. Summary and Conclusions 99
References 102

Chapter 3
Zooplankton **111**
I. Introduction 111
II. Classification 111
 A. Classification By Size 111
  1. Microzooplankton 111
  2. Mesozooplankton 112
  3. Macrozooplankton 113
 B. Classification by Length of Planktonic Life 113
  1. Meroplankton 113
  2. Holoplankton 115
  3. Tychoplankton 117
III. Taxonomy 118
IV. Zooplankton Communities 122
 A. Narragansett Bay 122
 B. Long Island Sound 124
 C. Barnegat Bay 125
 D. Chesapeake Bay 129
 E. North Inlet 132
 F. San Francisco Bay 133
V. Zooplankton Dynamics 134
 A. Biotic and Abiotic Factors 134
  1. Effects of Light 134
  2. Effects of Temperature 135
  3. Effects of Salinity 136
  4. Water Circulation 138
  5. Zooplankton Feeding 139
   a. Copepods 139
   b. Ingestion Rates and Assimilation 140

      6.   Secondary Production 143
VI. Summary and Conclusions 145
References 148

Chapter 4
Benthos **155**
I.   Introduction 155
II.   General Features of the Benthos 156
     A.   Benthic Flora 156
     B.   Benthic Fauna 158
III.   Benthic Flora 159
     A.   Benthic Algae 159
          1.   Benthic Microalgae 159
          2.   Benthic Macroalgae 165
              a.   Zonation 166
              b.   Type Examples 167
                   1.   Grays Harbor, WA 167
                   2.   Barnegat Bay, NJ 167
              c.   Ecological and Economical Value 173
                   1.   Ecological Significance 173
                   2.   Economical Value 173
     B.   Vascular Plants 177
          1.   Salt Marsh Grasses 177
              a.   Species Composition 177
              b.   Salt Marsh Formation 180
              c.   Ecological Importance 182
              d.   Salt Marsh Fauna 185
          2.   Tidal Freshwater Marshes 188
              a.   Benthic Flora 188
          3.   Seagrasses 197
              a.   Species Composition 197
              b.   Benthic Fauna 200
                   1.   Microfauna and Sessile Fauna 200
                   2.   Mobile Epifauna (Creeping, Crawling, or
                       Walking Forms) 201
                   3.   Swimming Fauna 201
               c.   Ecological Significance 201
     C.   Mangroves 202
          1.   Species Composition 202
          2.   Physical and Biological Factors Limiting Distribution 204
              a.   Temperature 204
              b.   Salinity 204
              c.   Tides 204
              d.   Other Factors 205
          3.   Zonation 205
          4.   Succession 206
          5.   Associated Flora and Fauna 206
              a.   Benthic Algae 206
              b.   Benthic Fauna 207
          6.   Ecological Importance 207
              a.   Production 207

          b.   Detritus 208
          c.   Habitat Former 209
          d.   Sediment Stabilizers 209
IV. Benthic Fauna 209
    A.   Microfauna 209
    B.   Meiofauna 210
        1.   Taxonomic Considerations 210
        2.   Abundance 210
        3.   Distribution 215
        4.   Reproduction and Growth 216
        5.   Production 217
    C.   Macrofauna 218
        1.   Distribution 218
           a.   Small-Scale Distribution Patterns 218
           b.   Distribution Patterns on the Scale of the Estuary 222
              1.   Larval Dispersal 224
           c.   Distribution Patterns on a World Scale 228
        2.   Species Composition and Abundance 228
           a.   Long-Term Benthic Macroinvertebrate Studies 229
               1.   Barnegat Bay 229
               2.   Chesapeake Bay 236
               3.   Corpus Christi Bay 238
               4.   Puget Sound 239
               5.   Upper Clyde Estuary 239
               6.   Dutch Wadden Sea 241
        3.   Species Diversity 241
           a.   Causes of Low Species Richness in Estuaries 244
               1.   Stability — Time Hypothesis 245
               2.   Environmental Stress 245
               3.   Resource Stability 246
               4.   Other Hypotheses 246
        4.   Biomass and Productivity 246
        5.   Feeding Strategies 248
        6.   Trophic Interactions 250
           a.   Predator-Prey Experiments 250
           b.   Predation, Competition, Amenalistic Interactions,
              and Community Structure 250
        7.   Animal-Sediment Relationships 251
        8.   Fouling Organisms 253
        9.   Boring Organisms 257
       10. Level-Bottom Community Concept 258
       11. Zonation Concept 259
V.   Summary and Conclusions 260
References 269

Chapter 5
Fishes **291**
I.   Introduction 291
II.   Classification 293
    A.   Classification of Pearcy and Richards 293
    B.   Classification of Perkins 293

C.   Classification of Day 293
D.   Classification of Moyle and Cech 293
E.   Classification of McHugh 295
    1.   Freshwater Fishes 295
    2.   Truly Estuarine Species 295
    3.   Anadromous and Catadromous Species 296
      a.   Anadromous Species 296
      b.   Catadromous Species 296
    4.   Seasonal Marine Migrants 297
    5.   Marine Species Which Use the Estuary Primarily as a
         Nursery Ground 297
    6.   Adventitious Visitors 298
III.   Adaptations to Environmental Conditions 298
A.   Physical-Chemical Factors 298
    1.   Salinity 298
    2.   Temperature 299
    3.   Dissolved Oxygen 300
    4.   Other Factors 302
B.   Biotic Factors 302
    1.   Interspecific Interactions 302
      a.   Predator-Prey Relationships 302
      b.   Competition 302
      c.   Symbiosis 303
        1.   Commensalism 303
        2.   Mutualism 303
        3.   Parasitism 304
    2.   Intraspecific Interactions 304
IV.   Food Habits and Diet 304
V.   Population Dynamics 312
A.   Parameters 312
    1.   Age 312
    2.   Growth 313
    3.   Mortality 315
    4.   Reproduction and Recruitment 316
VI.   Species Abundance and Diversity 319
VII.   Fish Communities 320
A.   Type Examples 320
    1.   Barnegat Bay 320
    2.   Chesapeake Bay 326
    3.   Terminos Lagoon 329
    4.   San Francisco Bay 331
VIII.   Summary and Conclusions 342
References 345

Chapter 6

Trophic Relationships **351**

I.   Introduction 351
II.   Estuarine Food Webs 352
A.   Detritus Food Web 352
    1.   Detritus 352

		2.	Type Detritus Food Webs	356
			a. Georgia Salt Marsh Systems	356
			1. Detritus Production	356
			2. Microbes	357
			3. Macroconsumers	357
			b. Florida Mangroves	359
			1. Detritus Production	359
			2. Consumer Organisms	359
			c. Temperate Seagrasses	360
			1. Detritus Production	360
			2. Barnegat Bay	360
	B.	Grazing Food Web		364
		1.	Phytoplankton and Benthic Algal Production	364
		2.	Type Grazing Food Webs	366
			a. Georgia Salt Marshes	366
			b. Chesapeake Bay	368
			c. Barnegat Bay	368
	C.	Secondary and Tertiary Consumers in Estuaries		370
		1.	Invertebrate Predators	370
		2.	Fishes	370
		3.	Birds	372
III.	Summary and Conclusions			374
References				376

Index **381**

INTRODUCTION

I. ANTHROPOGENIC IMPACTS ON ESTUARIES

Estuaries and nearshore oceanic waters are susceptible to a multitude of human wastes from a burgeoning population in the coastal zone.[1] These highly sensitive ecosystems serve as repositories for dredge spoils, sewage sludge, and industrial and municipal effluents.[2] Areas with the highest levels of pollution border metropolitan centers, where industrial, municipal, and domestic contaminants have accumulated for years. Boston Harbor and Raritan Bay on the East Coast of the U.S. and San Diego Harbor on the West Coast provide examples. Others are San Francisco Bay, contaminated with cadmium, copper, mercury, nickel, and other heavy metals; Elliot Bay near Seattle, WA, contaminated with arsenic, cadmium, copper, lead, zinc, and polychlorinated biphenyls (PCBs); and Commencement Bay, contaminated with pollutants from Tacoma, WA. The U.S. Environmental Protection Agency has designated Commencement Bay as a Superfund site, underscoring the severity of the dilemma in this estuary. Most of the aforementioned problems stem from overpopulation and development in the coastal zone. For example, the U.S. population living within 80 km of the ocean more than doubled between 1940 (42 million) and 1980 (89 million). Recent demographic statistics indicate that the population continued to mount in this region throughout the 1980s, exacerbating the potential environmental dangers in coastal waters. As many as 200 million Americans, representing approximately 80% of the total U.S. population, will reside in coastal areas by the year 2000, according to government projections.[3]

The relative importance of human activity and natural processes as causative agents of estuarine and marine population fluctuations has not been unambiguously established for a large percentage of impacted coastal systems. While man is the source of severe and persistent environmental degradation of estuarine and marine waters, natural processes interacting with anthropogenic wastes may, in some cases, contribute to habitat destruction. Strong currents associated with storm surges, hurricanes, and other meteorological events roil bottom sediments and disperse pollutants from impacted sites to originally unaffected areas; this has been demonstrated for lipophilic organic substances (DDT, PCBs, and other chlorinated hydrocarbons), oil-derived hydrocarbons, xenobiotic (synthetic organic) compounds, trace metals, radionuclides, and pathogens. Although nutrients from sewage effluent are readily biodegraded or assimilated (but in excessive amounts promote eutrophic conditions), the lipophilic organic compounds, xenobiotics, and pathogens often concentrate in biota and sediments, adversely influencing food webs and the functioning of the ecosystems.[4] In this respect, toxic chemicals discharged to estuaries create more insidious problems than some visible pollutants (e.g., oil spills or sewage).[5] Nevertheless, when nutrients overwhelm the capacity of a system to assimilate them, red tides, brown tides, and eutrophic conditions commonly arise which frequently alter its trophic structure through the loss of large numbers of heterotrophs. The uncontrolled growth of certain phytoplankton species (e.g., *Ptychodiscus brevis*) has periodically inflicted mass mortality on fin- and shellfish populations. Neurotoxin released by these dinoflagellate cells accumulates in the siphons and hepatopancreas of shellfish, causing paralytic shellfish poisoning (a neurological disorder) in humans who consume them.[6] Saxitoxin is the toxic agent responsible for this acute illness.

Physical, chemical, and biological processes not only govern the distribution of anthropogenic wastes, but also their fate and overall effect on the environment by altering the concentration, chemical form, bioavailability, or toxicological strength of the contaminants.[7] Capuzzo et al. (p. 15)[4] summarize the following environmental concerns of waste disposal:

1.   The accumulation and transfer of metals and xenobiotic compounds in marine food webs, including accumulation in commercial resources
2.   The toxic effects of such contaminants on the survival and reproduction of marine organisms and the resulting impact on marine ecosystems
3.   The uptake and accumulation of pathogenic organisms in commercially harvested species destined for human consumption
4.   The release of degradable organic matter and nutrients to the ocean, resulting in localized eutrophication and organic enrichment

Biological effects of contaminants discharged into estuarine and marine environments are manifested in impaired feeding, growth, development energetics, and recruitment of populations; they may lead to alterations in reproductive and developmental success of the populations and to changes in the structure and dynamics of biotic communities.[8] The work of Vernberg et al.[9] treats the physiological responses of marine organisms to contamination and pollution of the environment.

Part of the logic of discharging and dumping anthropogenic wastes in estuaries and coastal marine waters is an expected dilution response, with transport and dispersion of contaminants to the offshore presumably mitigating impacts on nearshore habitats. However, hydrographic processes in coastal zones produce more complex circulation patterns than in offshore waters,[10] being characterized by coastal boundary layer effects, broad spectra of turbulent eddies, and flow modulated by local bathymetry and shoreline configurations.[11] These factors make accurate prediction of pollutant dispersal extremely tenuous. Estuarine circulation appears to be even more problematical as disclosed by recent publications covering this topic.[12-15] New monitoring techniques and dispersion models have been introduced to study mixing and transport mechanisms which are essential in evaluating pollutant dispersal. Hydraulic models have been employed for many years, and numerical models since at least the early 1960s, to examine the transport of contaminants. The work of Fischer[12] deals with mathematical models of flow and pollutant spreading in estuarine and coastal ecosystems.

Short- and long-term monitoring of biotic communities is necessary for decision makers to devise sound solutions to acute and insidious aquatic pollution. Ideally, both pre- and post-impact surveys of biota should be conducted to accurately predict pollution effects and to formulate essential remedial action. An ultimate goal of estuarine science is to develop quantitative and predictive models (relying on field and laboratory data) not only of physical, chemical, and biological processes, but also of the anthropogenic stresses imposed on natural systems. Capuzzo and Kester (p. 5)[8] contend that the most complete surveys incorporate "the integration of chemical and biological monitoring; the collection of physiological and biochemical data pertinent to the growth, reproduction, and recruitment of individuals within a population; and modelling the impact both of short-term and of long-term changes in population effort and stability".

The literature on the biological aspects of estuarine ecology is replete with references to pollution impacts on populations attributable to organochlorine insecticides,[16] PCBs,[17] organotin compounds,[18,19] heavy metals,[19] toxic nonmetals,[20] oil spills,[21] and nutrient loadings.[22,23] Although nearly all phyla have been subjected at some time to contaminants in coastal ecosystems, the most adversely affected forms are benthic populations with limited or no mobility; consequently, they have been the focal point of environmental impact studies.[24] Since the early 1980s, the National Oceanic and Atmospheric Administration has performed benthic surveys at approximately 200 locations nationwide to determine the impact of pollutants in estuarine and coastal marine waters.

Biotic monitoring programs have been carried out in estuaries for reasons other than strictly environmental impact assessment. For instance, surveys are commonly executed for purposes of management of amenity, for exploitation of natural resources, for identification of sites for conservation priority, and for basic scientific research.[25] In nearly all cases, the surveys collect data on the abundance and distribution of organisms, together with information on physical and chemical parameters.[26]

It is evident that much ecological research on estuaries requires at least a fundamental knowledge of the biology of the constituent populations. Considering the pervasive and enigmatic anthropogenic problems plaguing these coastal ecotones, it behooves aquatic scientists from all disciplines to make a concerted effort to elucidate the population dynamics of the biota and the effects of pollutants on their life history. Unequivocally, in order to ameliorate or solve these ecological problems and to rehabilitate or protect estuaries, it will be necessary to compile additional facts on the structure of estuarine communities and the biotic interactions and responses of the organisms within the systems under investigation. Perhaps more importantly, an integral part of evaluating the health of these waterways entails the delineation of organism-environment interactions which, in many respects, is much more problematical. *In situ* observations on the effects of abiotic and biotic factors on estuarine organisms must proceed if these prized coastal habitats are to continue to be a source of aesthetic, commercial, and recreational resources for mankind into the 21st century.

II. PLAN OF THIS VOLUME

Chapter 1 focuses on the bacteria of estuaries. The ecology of these microbes has been, until recently, one of the least explored areas of estuarine and marine biology. However, with the advent of the "microbial loop" and the recognition that a substantial fraction of the energy and nutrients in estuaries flows through a food web based on particulate and dissolved organic matter which includes bacteria — and indeed all pro- and eukaryotic unicellular organisms, both autotrophic and heterotrophic — the role of bacteria in the overall functioning of the estuarine ecosystem is being reassessed.[27] The microbial food web provides the ultimate food resources for metazooplankton.[28,29] Bacteria act as key elements in the energy and nutrient budgets of estuaries through their secondary production and decomposition of organic matter, their conversion and transformation of compounds, and their alteration of physical and chemical properties of the environment.

Bacteria may be free floating (bacterioplankton), saprophytic on dying organisms and detritus, or attached to natural or man-made substrates (epibiotic forms). Their abundance in the water column typically ranges from 10^6 to 10^7 cells per milliliter. Peak numbers of the microbes occur in bottom sediments, often exceeding 10^9 cells per cubic centimeter in salt marsh and mudflat substrates. Within the top 20 cm of the sediment profile, cell counts decline with increasing depth below the sediment-water interface. The density of the bacteria rises with correspondingly higher quantities of organic matter in the sediments. In addition, a larger microbial population attaches to finer sedimentary grains (clay and silt) than to coarser particles (sand) due to the greater volume-specific surface area of the smaller grains. By concentrating in cracks, crevices, and depressions of the sedimentary grains, the bacteria are protected from mechanical damage. Research on the growth and production of bacteria in sediments, as well as in the water column, has been greatly aided by the use of a new generation of techniques, such as epifluorescence microscopy, stains (e.g., acridine orange), and radioisotopic applications.

Bacteria are the principal processors of organic matter in aerobic and anaerobic estuarine environments. In aerobic zones, bacteria convert particulate organic matter into dissolved organic matter to meet their nutritional requirements while serving as a food source for microfauna (e.g., ciliates and flagellates) and macrofauna (e.g., detritivores). Anaerobic forms, namely, fermenting bacteria, dissimilatory sulfate-reducing bacteria, dissimilatory nitrogen oxide-reducing bacteria, and methanogenic bacteria, have more specific roles in the transformation of organic matter. Because of the absence or rare occurrence of fungi in anaerobic zones, bacteria maintain a position of preeminence as transformers of organic matter in anoxic estuarine environments.

Chapter 2 details the phytoplankton of estuaries, providing coverage of taxonomy, species composition, succession, biomass, and production of this important autotrophic group. The chapter commences with a taxonomic description of the major classes of phytoplankton encountered in these coastal systems, the most significant ones being diatoms (class Bacillariophyceae) and dinoflagellates (class Dinophyceae). Together, diatoms and dinoflagellates are responsible for a substantial fraction of the primary production generated in the water column of estuaries.

The total production of phytoplankton in estuaries ranges from about 6.8 to 530 g C per square meter per year, with the lowest values recorded in turbid waters and the highest readings in clearer systems. In temperate and boreal regions, peak productivity takes place during the warmer months of the year, remaining relatively high from spring to early fall. Maximum production may arise concomitantly with phytoplankton blooms which pulse periodically from spring through fall. In tropical and subtropical latitudes, where seasonality is much less pronounced, phytoplankton production persists more uniformly year-round.

Physical, chemical, and biological factors — especially light, temperature, water circulation, salinity, nutrients, and zooplankton grazing — control the species composition, abundance, distribution, and production of phytoplankton in estuaries. The effects of these factors are demonstrated for the phytoplankton communities of Chesapeake Bay, Barnegat Bay, Long Island Sound, and Narragansett Bay. All four of these systems display conspicuous patterns in species composition, production, and seasonal cycles of abundance and occurrence.

Chapter 3 discusses the classification, taxonomy, and ecology of estuarine zooplankton. These diminutive organisms are critical links in the food web of estuaries, with most converting plant and detritus to animal matter and comprising the principal herbivorous component of the ecosystems. They exert predation pressure on phytoplankton populations and can regulate the standing crop of the autotrophs during certain times of the year.

Zooplankton are differentiated according to size into three broad groups, specifically, micro-, meso-, and macrozooplankton. Those individuals passing through a plankton net with a mesh size of 202 µm comprise the microzooplankton. Still larger forms collected with plankton nets having a mesh size of 505 µm are defined as macrozooplankton. Protozoans, tintinnids, meroplankton of benthic invertebrates, and copepod nauplii account for the bulk of the microzooplankton in estuaries. The primary members of the mesozooplankton are cladocerans, copepods, rotifers, and larger meroplankton. The macrozooplankton embody four faunal groups, notably, the jellyfish group (hydromedusae, comb jellies, and true jellyfish), crustaceans (amphipods, isopods, mysid shrimp, and true shrimp), polychaete worms, and insect larvae.

Alternatively, zooplankton may be grouped into three classes based on their duration of planktonic life, that is, mero-, holo-, and tychoplankton. Meroplankton (principally planktonic larvae of the benthic invertebrates, benthic chordates, and nekton) spend only a portion of their life cycle in the plankton. Holoplankton (predominantly copepods, cladocerans, and rotifers), in contrast, remain planktonic for their entire life. Tychoplankton are largely small benthic fauna periodically and temporarily inoculated into the water column by wave action, currents, animal-sediment interactions (i.e., bioturbation), or behavioral activity.

Abiotic factors most important in regulating the behavior and population dynamics of zooplankton are light, temperature, salinity, and water circulation. Light governs diel vertical migration patterns of zooplankton, serving as an environmental cue triggering the faunal movements. Temperature strongly influences zooplankton physiology and ecology. Metabolic rates are a function of temperature, with fecundity, duration of life, adult size, and other parameters modulated by temperature levels. Salinity affects the overall distribution of zooplankton in an estuary. The effects of these factors are apparent in Chapter 3 which describes the zooplankton communities of Narragansett Bay (RI), Long Island Sound (NY), Barnegat Bay (NJ), Chesapeake Bay (MD), North Inlet (SC), and San Francisco Bay (CA).

Chapter 4 recounts the benthos of estuaries, that is, the flora and fauna inhabiting bottom sediments. The dominant benthic flora are emergent and submergent vascular plants of salt marsh, seagrass, and mangrove biotopes. These plants play a major ecological role as habitat formers for a broad spectrum of faunal populations. Benthic microalgae are most extensively developed in intertidal zones, where diatoms, blue-green algae, and dinoflagellates commonly grow in dense patches that discolor sediments, rocks, and man-made structures. Green and blue-green algae form mats that bind sediments, especially in warm-water systems, such as those along the Gulf of Mexico. Benthic macroalgae (seaweeds) generally are poorly represented in estuaries, occasionally blanketing small sections of the seafloor, but attaining maximum densities on hard substrates (e.g., rocks, bulkheads, docks, and piers).

The aforementioned plant subsystems exhibit high production. For example, above-ground production of *Spartina* salt marshes ranges from about 200 to 3000 g C per square meter per year, with below-ground plant growth possibly yielding even greater amounts of carbon. Estimates of seagrass production fall between 58 and 1500 g C per square meter per year, and those of mangroves, between 350 and 500 g C per square meter per year. Benthic microalgal production, in turn, typically varies between 25 and 250 g C per square meter per year. Most of the production of vascular plants passes to the detritus food web, being minimally grazed by herbivores, and it supplies a stable energy source for detritivores and other benthic fauna.

Benthic faunal populations are subdivided into micro-, meio-, and macrofaunal components on the basis of their size. The microfaunal populations consist essentially of protozoans. The meiofauna are comprised of temporary and permanent members; the temporary meiofauna embody juvenile stages of the meiofauna, whereas the permanent meiofauna incorporate a host of taxa (i.e., gastrotrichs, kinorhynchs, nematodes, rotifers, archiannelids, halacarines, harpacticoid copepods, ostracods, mystacocarids, and tardigrades, as well as representatives of the bryozoans, gastropods, holothurians, hydrozoans, oligochaetes, polychaetes, turbellarians, nemertines, and tunicates). Weighing less than 10^{-4} g (wet weight), meiofauna are retained on sieves of 0.04 to 0.1 mm mesh, while the microfauna pass through those sieves. The macrofauna refer to larger animals captured on sieves of 0.5 to 2.0 mm mesh. Although far less abundant than the micro- and meiofauna, the benthic macrofauna of estuaries have greater biomasses.

The species composition and abundance of benthic fauna undergo wide temporal and spatial variations due to fluctuations in the abiotic conditions of the environment in addition to competition, predation, and natural periodicities of reproduction, recruitment, and mortality of the biota. Stochastic environmental perturbations, induced in some cases by anthropogenic factors, can precipitate large aperiodic changes in the benthic fauna. Alterations in the benthic community oftentimes are most conspicuous among the opportunistic species (e.g., *Capitella capitata*, *Cistenides* (=*Pectinaria*) *gouldii*, and *Polydora ligni*).

In Chapter 4, the spatial distribution of benthic macroinvertebrates is addressed on three levels: (1) small-scale distributions within the estuary, (2) distribution patterns on the scale of the estuary, and (3) distribution patterns on a world scale. Physical factors (e.g., waves, currents, and sediment type) and biological factors (e.g., competition and predation) strongly affect local distributions of the benthos. Larger-scale distributional patterns, from the head to the mouth of estuaries, manifest from population responses to gradients in environmental conditions. On a world scale, few estuarine benthic macroinvertebrates have a worldwide distribution. Adult migration, movement by rafting, and transport via human activities can broaden their distribution on a global scale.

The benthic macrofauna of estuaries are ecologically important as food for higher-trophic-level consumers, as processors of the large pool of detritus in bottom sediments, and as modifiers of substrates. They alter seafloor sediments by their feeding, burrowing, and physiological activities. Bioturbation influences interparticle adhesion, water content of sediments, bed roughness, and geochemistry of interstitial waters. Animal tubes, pits and depressions, excavation and fecal mounds, crawling trails, and burrows impact fluid motion, sediment erosion, and transport in the benthic boundary layer.

Recent advances in acoustic and photographic techniques have allayed difficulties in benthic sampling. For example, the application of side-scan sonography has facilitated data acquisition on bed roughness, sediment type, and bottom morphology. The REMOTS™ sediment profile camera has proven to be invaluable in investigating the species composition of the benthic community and the geochemistry, bottom micromorphology, and microstratigraphy of the upper part of the sediment column. Electromagnetic current meters, acoustic current meters, thermistors, heat transfer-type current meters, impellors, and laser-doppler current meters are particularly useful in deciphering current profiles and turbulence in the benthic boundary layer.

Chapter 5 reviews the ichthyofauna of estuaries — the largest and most motile component of the estuarine nekton of actively swimming pelagic organisms within the size range of 20 mm to 20 m. Estuarine fishes have an advantage over many other faunal populations; their mobility enables them to avoid unfavorable environmental conditions. In addition to having the physical means of escaping harmful conditions, they display some degree of temperature tolerance and osmoregulatory ability, helping them to adapt to potentially stressful habitats.

Estuarine fish faunas are numerically dominated by a few widespread species having broad tolerances and wide ranges of adaptations. Juveniles, which use the estuary as a nursery, show the greatest abundance. Many of these juveniles are short-lived, euryhaline members of coastal populations that utilize the shelter and rich food supply afforded by the estuarine environment to grow rapidly. While some species occupying estuaries are permanent residents, most consist of seasonal migrants entering these shallow systems from nearshore oceanic areas. Anadromous and catadromous forms strictly employ estuarine waters as a migratory pathway between feeding and spawning grounds. Estuaries are exceptional havens for recreational and commercial fisheries; a large fraction of the total U.S. commercial and recreational catches in marine waters is comprised of species inhabiting these coastal ecosystems at least part of their lives.

Classification schemes published in the scientific literature group estuarine fishes into about five general assemblages. These are freshwater species, residents, diadromous forms, marine migrants, and adventitious visitors. At any given time, individuals from several, or all, of these assemblages may compose the fish community. The species composition, abundance, and distribution of fishes in the community are a function of abiotic and biotic factors. Among biotic factors, interspecific and intraspecific interactions have a profound effect on the community. Predator-prey responses, competition, and symbiotic relationships (i.e., commensalism, mutualism, and parasitism) are interspecific interactions of utmost significance. Research on population dynamics, involving data acquisition on growth, mortality, and recruitment of fish, has advanced man's understanding of the characteristics of estuarine ichthyofauna.

Chapter 6 examines the trophic relationships of estuarine organisms and the flow of energy among producers, consumers, and decomposers. Plants and animals living in estuaries are not isolated entities, but part of a complex food web. Two major interlocking components of the food web exist: (1) the detritus food web and (2) the grazing food web. Differences between these two pathways of energy flow occur principally at the primary producer-primary consumer levels. Hence, in the detritus food web, simply stated, saprophagic bacteria and fungi assimilate dead organic matter and yield inorganic and organic solutes for photosynthetic producers. In the grazing food web, live phytoplankton, and to a lesser degree live microalgae, macroalgae, and vascular plants, contribute to the structural components at the base of the food web. The separation between detritus and grazing food webs becomes obscured among upper-trophic-level consumers, because at secondary consumer levels and above, an intricate network of estuarine organisms derive energy and nutrients from both primary energy pathways. In recent years, the concepts of the microbial loop and microbial food web have gained appeal, although they have fostered additional controversy in trophic analysis.

An estuary usually is dominated by one of these two food webs. Detritus food webs predominate in shallow coastal bays with high vascular plant production. Larger estuaries, such as Chesapeake Bay and Long Island Sound, are typified by grazing food webs. Despite differences in the pathway of energy transfer in estuaries, they all terminate in the same manner with similar higher-order consumers (i.e., carnivorous invertebrates, fishes, birds, and man).

Changes in seasonal occurrence and abundance of producer and consumer species account for seasonal variations in food web constituents. Ontogenetic changes in feeding habits of component fauna further obfuscate the food web structure. These difficulties aside, it is often possible to generate a generalized structure of the trophic relationships in many estuarine systems.

REFERENCES

1. **Biggs, R. B., De Moss, T. B., Carter, M. M., and Beasley, E. L.,** Susceptibility of U.S. estuaries to pollution, *Rev. Aquat. Sci.*, 1, 189, 1989.
2. **Ketchum, B. H., Capuzzo, J. M., Burt, W. V., Duedell, I. W., Park, P. K., and Kester, D. R., Eds.,** *Wastes in the Ocean*, Vol. 6, John Wiley & Sons, New York, 1985.
3. **Hall, C. A. S., Howarth, R., Moore, B., III, and Vörösmarty, C. J.,** Environmental impacts of industrial energy systems in the coastal zone, *Annu. Rev. Energy*, 3, 395, 1978.
4. **Capuzzo, J. M., Burt, W. V., Duedall, I. W., Park, P. K., and Kester, D. R.,** The impact of waste disposal in nearshore environments, in *Wastes in the Ocean*, Vol. 6, Ketchum, B. H., Capuzzo, J. M., Burt, W. V., Duedall, I. W., Park, P. K., and Kester, D. R., Eds., John Wiley & Sons, New York, 1985, 3.
5. **Nybakken, J. W.,** *Marine Biology: An Ecological Approach*, Harper & Row, New York, 1982.
6. **Kennish, M. J., Ed.,** Section 6: phytoplankton, in *Practical Handbook of Marine Science*, CRC Press, Boca Raton, FL, 1989.
7. **Capuzzo, J. M. and Kester, D. R., Eds.,** *Oceanic Processes in Marine Pollution*, Vol. 1, Robert E. Krieger Publishing, Malabar, FL, 1987.
8. **Capuzzo, J. M. and Kester, D. R.,** Biological effects of waste disposal: experimental results and predictive assessments, in *Oceanic Processes in Marine Pollution*, Vol. 1, Robert E. Krieger Publishing, Malabar, FL, 1987, 3.
9. **Vernberg, F. J., Thurberg, F. P., Calabrese, A., and Vernberg, W. B., Eds.,** *Marine Pollution and Physiology: Recent Advances*, University of South Carolina Press, Columbia, 1985.
10. **Dyke, P. P. G., Moscardini, A. O., and Robson, E. H., Eds.,** *Offshore and Coastal Modelling*, Springer-Verlag, New York, 1985.
11. **Lam, D. C. L., Murthy, C. R., and Simpson, R. B.,** *Effluent Transport and Diffusion Models for the Coastal Zone*, Springer-Verlag, New York, 1984.
12. **Fischer, H. B., Ed.,** *Transport Models for Inland and Coastal Waters*, Academic Press, New York, 1981.
13. **van der Kreeke, J., Ed.,** *Physics of Shallow Estuaries and Bays*, Springer-Verlag, New York, 1986.
14. **Kjerfve, B.,** *Hydrodynamics of Estuaries*, Vol. 1, CRC Press, Boca Raton, FL, 1988.
15. **Kjerfve, B.,** *Hydrodynamics of Estuaries*, Vol. 2, CRC Press, Boca Raton, FL, 1988.
16. **Lunsford, C. A. and Blem, C. R.,** Annual cycle of Kepone residue and lipid content of the estuarine clam, *Rangia cuneata, Estuaries*, 5, 121, 1982.
17. **O'Connor, J. M. and Pizza, J. C.,** Dynamics of polychlorinated biphenyls in striped bass from the Hudson River. III. Tissue disposition and routes for elimination, *Estuaries*, 10, 68, 1987.
18. **Weis, J. S. and Perlmutter, J.,** Effects of tributyltin on activity and burrowing behavior of the fiddler crab, *Uca pugilator, Estuaries*, 10, 342, 1987.
19. **Barnett, B., Forbes, S., and Ashcroft, C.,** Heavy metals on the south bank of the Humber estuary, *Mar. Pollut. Bull.*, 20, 17, 1989.
20. **Johns, C., Luoma, S. N., and Elrod, V.,** Selenium accumulation in benthic bivalves and fine sediments of San Francisco Bay, the Sacramento-San Joaquin delta, and selected tributaries, *Estuarine Coastal Shelf Sci.*, 27, 381, 1988.
21. **Kemp, P. F., Swartz, R. C., and Lamberson, J. O.,** Response of the phorocephalid amphipod, *Rhepoxynius abronius*, to a small oil spill in Yaquina Bay, Oregon, *Estuaries*, 9, 340, 1986.
22. **Lee, V. and Olsen, S.,** Eutrophication and management initiatives for the control of nutrient inputs to Rhode Island coastal lagoons, *Estuaries*, 8, 191, 1985.

23. **Meybeck, M., Cauwet, G., Dessery, S., Somville, M., Gouleau, D., and Billen, G.,** Nutrients (organic C, P, N, Si) in the eutrophic river Loire (France) and its estuary, *Estuarine Coastal Shelf Sci.*, 27, 595, 1988.

24. **Loveland, R. E. and Vouglitois, J. J.,** Benthic fauna, in *Ecology of Barnegat Bay, New Jersey,* Kennish, M. J. and Lutz, R. A., Eds., Springer-Verlag, New York, 1984, 135.

25. **Baker, J. M., Hartley, J. P., and Dicks, B.,** Planning biological surveys, in *Biological Surveys of Estuaries and Coasts*, Baker, J. M. and Wolff, W. J., Eds., Cambridge University Press, Cambridge, 1987, 1.

26. **Baker, J. M. and Wolff, W. J., Eds.,** *Biological Surveys of Estuaries and Coasts*, Cambridge University Press, Cambridge, 1987.

27. **Floodgate, G. D. and Jones, E. B. G.,** Bacteria and fungi, in *Biological Surveys of Estuaries and Coasts*, Baker, J. M. and Wolff, W. J., Eds., Cambridge University Press, Cambridge, 1987, 238.

28. **Sherr, E. B. and Sherr, B. F.,** Role of microbes in pelagic food webs: a revised concept, *Limnol. Oceanogr.*, 33, 1225, 1988.

29. **Turner, J. T., Tester, P. A., and Ferguson, R. L.,** The marine cladoceran *Penilia avirostris* and the "microbial loop" of pelagic food webs, *Limnol. Oceanogr.*, 33, 245, 1988.

Chapter 1

BACTERIA

I. INTRODUCTION

The microbiology of estuarine ecosystems is a rapidly evolving area of scientific inquiry. Recent interdisciplinary investigations in marine microbial ecology have prompted reassessment of the views concerning trophodynamic pathways[1] and have inspired additional research into the transformation of organic matter and biogeochemical cycling of elements.[2] In the microheterotrophic community of the sea, bacteria perform a number of significant functions, notably the secondary production and decomposition of organic matter, the conversion and transformation of compounds, and the alteration of physical and chemical properties of the environment (e.g., oxygen levels and pH).[3-6] Perhaps most significantly, bacteria modulate the carbon flux in estuarine and marine systems.[7-9]

The concept of the "microbial loop" in pelagic marine systems is rapidly revolutionizing details of the traditional food chain pathway. In pelagic food webs, for example, the major trophic pathway has historically involved energy flow principally from phytoplankton through zooplankton to higher-order consumers (e.g., fishes). However, marine microheterotrophic activity now appears to be tightly coupled to that of primary producers, with bacteria utilizing dissolved organic matter (DOM) from living phytoplankton as well as dead phytoplankton remains.[10] As much as 10 to 50% of phytoplankton production is converted to DOM and assimilated by bacterioplankton;[1] phagotrophic protists consume the bacterioplankton and, in turn, are ingested by microzooplankton or larger zooplankton.[11-14] Hence, a substantial amount of primary production may cycle through bacteria, much of the DOM being recovered through a "microbial loop" rather than lost through remineralization as occurs in the traditional food chain. The structure and function of the microbial links in pelagic food webs are subjects of ongoing investigations, but remain uncertain. Clearly, opinions differ in terms of the relative importance of bacterioplankton as primarily remineralizers of the nutrients fixed by phytoplankton or as a biomass source for the macroplankton food chain (i.e., zooplankton and phytoplankton).[15]

A benthic microbial loop, analogous to that proposed by Azam et al.,[16] could be even more profound in seafloor sediments, where bacteria consume as much as 80% of the organic inputs.[15] While larger zooplankton ineffectively graze on bacterioplankton in pelagic waters, metazoan bacterivores (e.g., meiofauna and benthic macrofauna) as well as protozoans efficiently ingest the microbes in bottom habitats. Trophic interactions among macrofaunal food webs in seafloor sediments, however, often are poorly understood.[17] Indeed, sediment-ingesting benthic macrofauna eat bacteria, protozoans, and meiofauna, potentially obfuscating energy flow from lower trophic levels. Protozoans, especially microflagellates,[18] remove most of the bacterial production in the sediments, macrofaunal grazing on protozoan bacterivores indirectly accounting for only slight cropping of the bacterial production. Less than 10% of bacterial production is consumed directly by benthic macrofauna in most marine sediments, the highest recorded values (about 40% of production) restricted to intense feeding in surface sediments. Bacterial production in sediments may serve principally as a sink for energy and nutrient flow.[15]

Bacteria play a vital function not only in estuarine and marine food webs, but also in the elemental and mineral cycles of these environments. Dawes[3] discloses that the density of bacteria on detrital organic material usually equals about 2 to 15 cells per 100 μm of surface area. Bacterial colonization of organic detrital material and its subsequent decomposition and mineralization releases inorganic compounds, some of which are assimilated by autotrophic organisms. Mineralization processes can be considered the counterpart of photosynthesis and chemosynthesis.[19]

These microorganisms represent a source of nitrogen for the ecosystem. In the nitrogen cycle, bacteria are involved in nitrogen fixation, nitrification, ammonification, and denitrification. Nitrogen fixation in estuaries is a major process. Dawes (p. 601)[3] states "A large percentage (~80%) of all nitrogen fixation occurs in shallow bays, estuaries, tidal marshes, and shallow neritic sediments. Such habitats appear to be the primary sources of combined nitrogen in oceanic waters."

The role of marine bacteria in the sulfur cycle of estuaries is also prominent. For example, Howarth and Teal[20] discerned that sulfate reduction in sediments of a New England salt marsh amounts to about 1800 g C per square meter per year, approximating the rate of carbon fixation. Other studies focus on the potential value of bacteria in the phosphorus cycle of estuaries. Phosphate-solubilizing bacteria, in addition to the enzyme phosphatase, have been isolated from the bottom sediments of estuaries.[21] These bacteria liberate phosphates, thereby increasing their chemical concentration in the sediments.

In the past decade, new methods of study have finally enabled microbiologists to determine the total numbers, the activity, and the growth rates of planktonic bacteria in nature.[22] Unequivocally, the application of recently developed techniques and instruments has greatly improved our overall knowledge of estuarine and marine bacteria. Chief among these techniques is the determination of adenosine triphosphate and related nucleotides to assess the presence of the microbes.[23,24] Epifluorescence microscopy, together with various stains such as acridine orange and diamidinophenylindole, has provided reliable estimates of bacterial production.[15,25,26] The use of perforated polycarbonate membranes with well-controlled sizes[24,25,27] and radioisotopic methods[28] has been worthwhile in numerous marine bacterial investigations. Tritiated thymidine incorporation into DNA has gained wide acceptance as a preferred method of estimating bacterial production.[29] Measurements of nucleic acid synthesis utilizing ^{3}H-adenine and ^{3}H-thymidine have facilitated research on microbial growth.[30] As new laboratory and field techniques arise, a revision of some of the current concepts of bacteria in estuarine and marine ecosystems is to be expected in the ever-changing microbiological arena.

II. TYPES OF MARINE BACTERIA

Marine bacteria can be subdivided on the basis of habitat, physiology, and ecological roles. In respect to habitat, planktonic, neustonic, epibiotic, benthic, and endobiotic bacteria have been identified.[31] The free-floating bacterioplankton assimilate soluble organic matter in the water column while providing autotrophs with vitamins (e.g., cobalamine) and nitrogen and phosphorus compounds. These bacteria, although attaining high counts in the waters of some estuaries, may be less abundant than bacterial neuston populations that are responsive to the greater concentrations of fixed carbon and nutrients accumulating at the air-seawater interface.[32] Of ecological as well as economical importance, epibiotic bacteria colonize the surfaces of marine substrates, the colonies of bacteria serving as a food source for protozoans and other heterotrophs. Larger numbers of bacteria grow in bottom sediments of estuaries and in sediments of wetland regions bordering these systems. The amassing of organic matter and nutrients in these sediments fosters the proliferation of benthic bacteria.[19] Some marine bacteria enter various biotic relationships — commensalistic, mutualistic, and parasitic. The symbiotic activity of certain endobacteria, that is, the parasitic types, has been well chronicled, primarily because these microbes often cause extensive financial losses via the diseases they inflict on marine plants and animals, particularly commercially and recreationally important finfish and shellfish.

Another grouping of marine bacteria involves their subdivision into autotrophic and heterotrophic forms. Autotrophic bacteria derive energy through photosynthesis (phototrophs) or through the oxidation of nitrogen and sulfur compounds (chemolithotrophs). Included in this category are those bacteria that use hydrogen in water as electron donors, releasing oxygen, and those bacteria that utilize reduced substances (e.g., sulfides, molecular hydrogen, or carbon compounds) as electron donors in photoassimilation of carbon dioxide.[3] Heterotrophic bacteria, saprophytes and parasites, obtain energy from other organic compounds. Photosynthetic bacteria, phototrophs, encompass marine forms such as anoxyphotobacteria and oxyphoto-bacteria. In estuaries, phototrophs have limited significance as primary producers and remain unimportant in biotic transformations.[19] Chemosynthetic bacteria (chemolithotrophs) are integral components in several geochemical cycles (e.g., nitrogen and sulfur cycles). Examples include nitrifying bacteria (family Nitrobacteriaceae), which convert ammonia to nitrate (e.g., *Nitrosomonas*) and nitrite to nitrate (e.g., *Nitrobacter*), and sulfur bacteria (i.e., sulfur-oxidizing and sulfur-granule-containing forms), which oxidize either sulfide, sulfur, or thiosulfate to sulfate. Several species of sulfur bacteria of the genus *Thiobacillus* inhabit the estuarine environment, concentrating in the top few centimeters of bottom sediment. Recently, chemosynthetic bacteria that oxidize sulfur compounds in hydrothermal fluids at oceanic vents have received much attention because the energy derived from their oxidation supports lavish biotic communities.[33]

The heterotrophic forms are the largest group of bacteria occurring in estuaries. Sieburth[31] distinguished two groups of heterotrophic bacteria in the ocean, small bacterioplankton approximately 0.2 to 0.6 μm in diameter and larger epibacteria living on and in organic matter and sediments. These bacteria are decomposers of organic matter and a food source for zooplankton.

As innovative microbiological methods have materialized, new taxonomic information has been compiled on marine bacteria. Detailed treatment of the taxonomy of these microbes can be found elsewhere.[31,34-39] *Bergey's Manual of Determinative Bacteriology*[38] and *Bergey's Manual of Systematic Bacteriology*[39] are considered to be the definitive works on this topic.

III. BACTERIAL ABUNDANCE, BIOMASS, AND PRODUCTION

A. ABUNDANCE

Kemp[15] remarks that abundance estimates of marine bacteria in the literature probably are underestimates. In the past, difficulties in separating attached bacteria from particles have contributed to this shortcoming.[40] Recently, however, improved methodology has ameliorated some of these problems, enabling more accurate estimates of bacterial abundance to be recorded.

Various estimates of bacterial abundance and biomass have been published for estuarine and marine systems. In the euphotic zone of the ocean, Sieburth[41] and Morita[42] reported bacterial abundance values in the range of 10^4 to 10^5 cells per milliliter compared to Fenchel's[43] estimate of 2×10^5 to 10^6 cells per milliliter for offshore waters. The number of bacteria increases landward as the amount of organic matter rises. Hence, the quantity of bacteria in the more productive coastal oceanic waters ranges from 1 to 3×10^6 cells per milliliter.[43,44] In estuarine waters, the concentration of bacteria is somewhat greater, falling between 10^6 to 10^7 cells per milliliter.[45]

Bacterial biomass is likewise positively correlated with organic content as demonstrated clearly by Rublee[46] in marine sediment. Williams[47] showed that the bacterial biomass in oligotrophic and nonoligotrophic oceanic waters equaled approximately 1 to 3 and 2 to 5 μg C per liter, respectively, compared to 5 to 10 μg C per liter in coastal waters. However, Fenchel[43] suggested, for the most productive waters, higher bacterial biomasses of 50 to 100 μg C per liter. The bacterial biomass in estuarine waters can exceed 10 μg C per liter.

Bacteria in the ocean are most often small in size (less than 0.4 μm in diameter) and unattached,[48-50] the larger cells mainly occupying enriched waters at physical interfaces or being attached to particles.[51] Kemp[15] specifies that bacteria inhabiting sediments generally average 0.6 μm in diameter. Wiebe and Pomeroy[52] detected few attached bacteria on particles in most coastal and open oceanic environments. Much (greater than 90%) of the heterotrophic activity in oceanic samples processed by Azam and Hodson[48] was apparent in the water fraction passing through 1-μm Nucleopore filters. This finding differed from that of Hanson and Wiebe[53] who, investigating bacterial processes in estuarine waters, observed the bulk of heterotrophic activity associated with large particles, as did Goulder[54] for an estuary in England.

The abundance of free-living bacteria relative to attached bacteria increases when progressing from the estuary to the open ocean.[55] For example, Bell and Albright[56] documented declining attachment of bacteria downestuary in the Frazer River estuary in British Columbia, Canada. Similarly, Sieburth and Davis[57] confirmed diminishing numbers of attached bacteria downestuary in Narragansett Bay, RI.

Early studies revealed large numbers of microbes coating detrital particles in estuaries,[58,59] but later research employing direct microscopy has not corroborated this phenomenon.[60-62] Sieburth[60] alluded to a paucity of bacteria attached to particles in the water column, although attached bacteria may be more abundant in estuarine settings than elsewhere.[63] Christian et al.[64] counted peak bacterial numbers and microbial ATP concentrations in the water column and not associated with particulate detritus; however, bacteria attached to particles displayed greater heterotrophic activity than free-living forms. Particle attachment, according to Christian et al.[64] appears to enhance microheterotrophic activity. Goulder et al.[63] describe three possible advantages of the attached habit: (1) attached bacteria may develop greater heterotrophic potential per cell, (2) attached bacteria may have a nutritional advantage because of adsorption of dissolved organic compounds onto suspended solids, and (3) attached bacteria may be shielded from grazing zooplankton.

Fenchel,[65] scrutinizing seagrass decomposition, noted that bacteria cover only 2 to 10% of detrital surfaces. Bacterial enumeration on detrital particles from *Zostera*, *Thalassia*, and other vascular plant sources always yields approximately two to ten bacterial cells per 100 μm detrital surface area, which corresponds to 10^9 to 10^{10} bacteria per gram dry weight of detritus (depending on the detrital particle size).[65] In a survey of bacterial colonization of small (5 to 40 μm diameter) detrital particles in salt marsh creeks, Marsh and Odum[66] unveiled uniformly low areas of bacterial coverage, averaging 1.54 ± 1.83% of the particle surface area. Wiebe and Pomeroy[52] rarely ascertained bacterial coverage exceeding 20% of the surface of particulate matter.

Although particles suspended in estuarine waters are not uncommonly devoid of bacteria,[62] the numerical abundance of microbes in estuarine bottom sediments is well established. For instance, direct counts of bacteria in salt marsh sediments of the Newport River estuary range from 8.36 to 10.90×10^9 cells per cubic centimeter of sediment at the surface to 2.19 to 2.58×10^9 cells per cubic centimeter of sediment at a depth of 20 cm.[67] The quantity of bacteria in subtidal sediments of this estuary approximates that recovered in salt marsh sediments. Similar counts have been proffered for mudflat sediments of the Bay of Fundy (i.e., 2 to 5×10^9 cells per cubic centimeter).[68] Data collected by Dale[69] also indicate large numbers of bacteria in intertidal sediments (i.e., 1.17×10^8 to 9.97×10^9 cells per gram of dry sediment). These figures translate into biomass estimates of 5.5 to 26.8 g C per square meter (to 10 cm depth) based on a conversion factor of 2.2×10^{-13} g/cell. Christian et al.[70] divulge comparable microbial biomass figures for salt marsh sediments at Sapelo Island, GA.

In a review of bacterial distribution in estuarine sediments, Rublee[46] lists abundance of total bacteria ranging from 4 to 17×10^9 cells per cubic centimeter in the surface sediments of intertidal and subtidal zones (Table 1). Cell counts decrease with increasing depth of sediment. Rublee,[46] for example, gives bacterial numbers of 1 to 20×10^9 cells per cubic centimeter for surface sediments of salt marshes which decline to 1 to 3×10^9 cells per cubic centimeter at 20 cm depth. A strong correlation exists between bacterial abundance and the organic content of the sediments (Figure 1). Most of the cells are in direct contact with the surfaces of the particles of sediment.

When passing from intertidal to subtidal habitats, the number of bacteria in sediments diminishes, reflecting once again a drop in the organic concentration away from the shoreline.

TABLE 1
Bacterial Abundance in Some Intertidal and Subtidal Sediments[a]

Location/site description	Number of samples	Depth (cm)	Cell numbers × 10⁹		Ref. (sources from Ref. 46)
			Per g dry weight	Per cm³	
Marsh Sediments					
Newport River Estuary,	13	0—1	13.8 (8.5—22.0)	8.5 (5.3—13.7)	This report
NC; *Spartina alterni-*	13	5—6	5.7 (3.2—7.7)	5.4 (3.3—8.1)	
flora marsh, yearly	13	10—11	2.9 (1.8—4.4)	3.0 (1.7—4.2)	
mean	13	20—21	1.8 (1.1—30)	1.9 (1.0—2.9)	
Newport River Estuary,	4	0—1	19.7 (12.0—34.1)	9.4 (8.4—10.9)	Rublee and Dorn-
NC; transect across	4	5—6	7.2 (4.6—9.8)	6.1 (5.1—7.1)	seif (1978)
Spartina marsh	4	10—11	3.2 (2.3—5.0)	3.5 (2.4—4.4)	
	4	20—21	2.1 (1.6—2.3)	2.3 (2.2—2.6)	
Rhode River Estuary,	3	0—1	66.3 (57.2—84.6)	16.0 (13.8—20.4)	This report
MD; *Typha angusti-*	3	5—6	35.6 (24.2—48.0)	9.7 (6.6—13.1)	
folia/Scirpus spp. low	3	10—11	29.9 (27.9—31.9)	3.0 (7.6—8.1)	
marsh	2	20—21	8.8 (7.4—10.3)	3.0 (2.5—3.5)	
Rhode River Estuary,	3	0—1	60.1 (37.5—76.6)	12.0 (11.3—12.4)	This report
MD; three high marsh	3	5—6	51.2 (36.9—69.1)	9.5 (7.2—11.0)	
sites, *Spartina cynuso-*	3	10—11	37.0 (30.8—43.2)	6.1 (4.3—9.3)	
roides, S. patens, Dis-	3	20—21	23.8 (19.9—25.9)	3.9 (2.8—5.1)	
tichlis spicata, Scirpus	1	30—31	31.7	3.6	
spp., *Hibiscus* spp.					
Great Sippewissett	12	0—1	38.0 (10.0—64.0)		J. E. Hobbie and
marsh, MA; tall *Spar-*	12	2—3	28.0		J. Helfrich (per-
tina alterniflora, annual	12	5—6	35.0		sonal communi-
mean	12	9—10	42.0		cation)
	12	19—20	24.0		
	12	29—30	16.0		
Great Sippewissett	12	0—1	54.0 (24.0—80.0)		J. E. Hobbie and
marsh, MA; short *S. al-*	12	2—3	29.0		J. Helfrich (per-
terniflora, annual mean	12	5—6	29.0		sonal communi-
	12	9—10	28.0		cation)
	12	19—20	18.0		
	12	29—30	12.0		
Great Sippewissett	12	0—1	49.0		J. E. Hobbie and
marsh, MA; High	12	2—3		(16.0—120.0)	J. Helfrich (per-
marsh, *S. patens,* an-	12	5—6	35.0		sonal communi-
nual mean	12	9—10	30.0		cation)
	12	19—20	29.0		
	12	29—30	18.0		
			19.0		
Intertidal/Subtidal Mud and Sand Flats					
Sapelo Island, GA; inter-	3	0—1		1.0 (0.6—1.2)	S. Y. Newell (per-
tidal sand flat adjacent					sonal communi-
to *Spartina* marsh, win-					cation)
ter samples					
Newport River Estuary,	18	0—1	3.3 (1.6—5.9)	2.7 (1.2—4.8)	Shelton (1979)
NC; subtidal sand,	18	5—6	2.4 (1.0—5.9)	1.9 (0.8—4.8)	
June—November					
Newport River Estuary,	18	0—1	8.8 (3.1—22.9)	4.2 (2.0—8.3)	Shelton (1979)
NC; subtidal mud,	18	5—6	5.8 (1.3—20.5)	2.7 (1.1—7.4)	
June—November —					
"sulfuretum"					

TABLE 1 (continued)
Bacterial Abundance in Some Intertidal and Subtidal Sediments[a]

Location/site description	Number of samples	Depth (cm)	Cell numbers × 10⁹		Ref. (sources from Ref. 46)
			Per g dry weight	Per cm³	
Intertidal/Subtidal Mud and Sand Flats (continued)					
Rhode River, MD; sub-tidal mud flat	3	0—1	17.4 (9.6—26.8)	9.1 (5.0—14.0)	This report
	2	2—3	16.0 (11.7—19.7)	8.4 (6.1—10.3)	
	3	5—6	13.0 (11.8—14.2)	8.7 (7.9—9.5)	
	3	10—11	8.0 (6.7—9.1)	6.2 (5.2—7.1)	
	2	20—21	4.7 (3.3—6.2)	3.6 (2.5—4.7)	
Lowe's Cove Damaris-cotta Estuary, ME; March—August	6	sfc	2.8 (1.2—4.7)		M. DeFlaun (personal communication)
Halifax, Nova Scotia; high intertidal mud flats, April—November	13	sfc	6.0 (2.2—8.6)	3.5 (2.2—4.8)	L. M. Cammen (personal communication)
	13	1	5.7 (3.7—7.9)	4.5 (3.5—5.5)	
	13	5	3.5 (2.3—5.2)	3.6 (2.9—4.3)	
Halifax, Nova Scotia; low intertidal mudflat, April—November	13	sfc	4.2 (1.5—7.0)	2.8 (1.4—4.3)	L. M. Cammen (personal communication)
	13	1	4.0 (2.8—5.2)	3.5 (2.1—4.1)	
	13	5	4.0 (2.7—5.2)	3.8 (2.2—5.1)	
Petpeswick Inlet, Nova Scotia, seven intertidal mud and sandflat sta-tions, May—September	17	sfc	2.8 (0.3—10.0)		Dale (1974)
	2	1	2.9 (2.1—3.6)		
	1	2	6.3		
	16	5	2.5 (0.1—7.0)		
	14	10	1.8 (0.1—4.7)		

[a]   Values given as mean (and range) of observations. sfc = surface.

From Rublee, P. A., *Estuarine Comparisons*, Kennedy, V. S., Ed., Academic Press, New York, 1982, 159. With permission.

The richness of bacteria in intertidal sediments corresponds to areas of high organic carbon content (i.e., 10 to 50%). Bacterial counts rise appreciably in deposits incorporating detritus; here, small colonies may prevail.[71]

The number of bacteria is about 1000 times greater in bottom sediments than in the water column, being on the order of 10⁹ to 10¹⁰ cells per milliliter.[15,72] According to Meyer-Reil,[71] different types of sediment from diverse habitats contain comparable bacterial abundances in the range of 1 to 507 × 10⁸ cells per dry weight of sediment (Table 2). Data from Table 2 infer a positive correlation with decreasing grain size of sediment. The strong inverse correlation between sediment grain size and bacterial abundance has been mentioned in earlier publications.[67,69,73]

While many workers have theorized on the relation between particle surface area and bacterial abundance, Yamamoto and Lopez[74] provided the first direct confirmation, based on measurements of surface sediments collected from an intertidal salt marsh (Flax Pond, NY), that the relationship between bacterial abundance and specific surface area of sediment holds over a wide range of sediment types. Hence, finer sediments (i.e., clay and silt) offer a greater volume-specific surface area for bacteria than coarser sediments (e.g., sand); consequently, they harbor a larger microbial population. DeFlaun and Mayer,[75] making measurements of both surface area and bacterial abundance, corroborated the aforementioned correlation between bacterial numbers and sediment surface area, but emphasized seasonal variability in the abundance of the microbes. Rublee[72] also delineated seasonal variablity of bacteria in a North Carolina salt marsh,

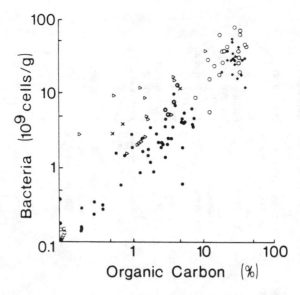

FIGURE 1. Relationship between bacteria and organic carbon in intertidal and shallow subtidal sediments. Closed circle — intertidal, Nova Scotia;[69] open circle — marsh intertidal, Maryland;[46] open star — subtidal, Maryland; triangle — marsh intertidal, North Carolina;[67] X — subtidal, North Carolina; square — subtidal, North Sea coast; closed star — marsh intertidal, Massachusetts. (From Rublee, P. A., *Estuarine Comparisons*, Kennedy, V. S., Ed., Academic Press, New York, 1982, 159. With permission.)

denoting maximum cell counts of 14×10^9 cells per cubic centimeter in late fall and minimum cell counts of approximately 5×10^9 cells per cubic centimeter in late summer.

The greater abundance of microbes in fine-grained sediments may be related to the larger concentration of particulate organic matter accumulating among the smaller particles.[76] Alternatively, nutrient enrichment of these particles can foster proliferation of the microbial populations.[77] Finally, the synergistic effect of nutrient adsorption onto these sedimentary grains and the associated organic matter creates more ideal conditions for microbial growth.

Bacteria colonize only a small portion of the total available area of sedimentary particles.[66,75] They tend to proliferate in crevices, cracks, and depressions of the particles which afford them protection against mechanical damage, such as incurred from abrasion.[78,79] The density of bacteria on the surface of marine sediment, quoted by Dale,[69] equals one cell per 411 μm². Sampling the surface of salt marsh sediments, Rublee and Dornseif[67] registered an average density of bacteria amounting to one cell per 5.31 μm², with a range of 0.31 to 13.73 μm². The organic content[46,80] and temperature[75] of the sediments purportedly influence the density of the microbes.

Most of the bacterial cells are either in direct contact with the particle surface or in indirect contact via detrital material.[71] Hobbie and Lee[81] speculated that significant concentrations of extracellular polysaccharides could result from microbial attachment to sediment surfaces; the polysaccharides, in turn, would serve as a food source for deposit-feeding organisms. The extracellular production of bacteria may be manifested in the slime connections frequently seen between individual cells.[71] Bacterial exudates have been implicated in the trapping of clay in intertidal sediments, a process varying with bacterial activity that is temperature related.[75] Thus, in this case, bacterial activity has been hypothesized to control sediment type (and consequently the surface area available for colonization) rather than the reverse.[74] In addition to temperature and organic content, other factors potentially affect the activity of bacteria, including anions,

TABLE 2
Comparison of Mean Bacterial Numbers in Relationship to Organic Matter and Grain Size in Sediments

Location	Date	Temperature (°C)	Sediment Type	Depth (cm)	Bacterial number (10⁸ g⁻¹)	Organic matter (mg g⁻¹)	Grain size (μm)	Remarks	Ref. (sources from Ref. 71)
Petpeswick Inlet, Nova Scotia	6/3/72 8/30/72	—	Intertidal flats Intertidal flats	0 5	99.70 1.17	3.8[a] 0.1[a]	19.1 132.5	7 stations, different horizons, maximum/minimum values	Dale (1974)
Beaufort Sea, AK	Summer 1975/1976	1.2/−0.1	Arctic sediment	Surface	6.6/106	—	—	33/11 stations	Griffiths et al. (1978)
Kampinge, Baltic Sea	Winter 1976 6/1—5/77	−1.9 15	Arctic sediment Sandy beach	Surface 0	10 14.2	— 2.7	— 277	13 stations 4 stations	Meyer-Reil et al. (1978)
Kiel Bight, Baltic Sea	11/21/74	—	Sandy (14 m)	0	14.02	5.4	410	1 station	Weise and Rheinheimer (1978)
Halifax Harbor, Nova Scotia	—	2	Muddy, aerobic/anaerobic (24 m)	0—100	5—30	2—3[a]	—	1 station, different horizons, different sampling dates	Kepkay et al. (1979)
Windermere, South Basin	4/79	—	Littoral sediment Profundal sediment	2.9 2.1	290 340	— —	— —	Organic matter and particle size given for different sediment fractions	Jones (1980)
Blelham Tarn	4/79	—	Littoral sediment Profundal sediment	3.6 1.2	280 119	— —	— —		
Kiel Fjord/Kiel Bight, Baltic Sea	7/4—12/77 3/6—11/78 11/6—11/78	21.0 3.0 9.0	Sandy beach Sandy beach Sandy beach	0 0 0	10.3 4.5 15.3	5.9 5.1 11.7	293 355 282	12 stations, 3 sampling dates	Meyer-Reil et al. (1980) and Meyer-Reil (unpublished)
Mactan, Philippines	4/2/80	27	Coral beach	0	14.7	—	—	3 stations	Meyer-Reil (1981)
Kiel Bight, Baltic Sea	5/30/80	8	Sandy (10 m)	0—2 2—4 4—6	8.8 6.7 0.7	12.7 8.1 7.2	— — —		Meyer-Reil (unpublished)

							1 station, different horizons	Meyer-Reil (unpublished)
Antarctica	5/28/80		Muddy, anaerobic (28 m)	0—2	84.7	164.4	—	
				2—4	69.0	157.7	—	
				4—6	86.8	160.3	—	
	1/26/81	2	Muddy (1951 m)	0	14.2	109.4	—	
				100	28.6	105.8	—	
				400	6.5	39.8	—	
				800	14.6	98.6	—	
				1100	11.1	82.5	—	

a   Organic carbon (% by weight).

From Meyer-Reil, L.-A., *Heterotrophic Activity in the Sea*, Hobbie, J. E. and Williams, P. J. le B., Eds., Plenum Press, New York, 1984, 523. With permission.

cations, nitrogen concentrations, other inorganic nutrients, pressure, primary producers, and predators.[24,82]

The presence or absence of oxygen largely regulates the overall distribution of microbial populations in the estuarine sediments.[83] The presence of oxygen at the surface supports populations of obligate aerobes which are successively displaced at greater depths by populations of facultative and then strictly anaerobic bacteria. Keith et al.[83] determined the spatial distribution of different physiological groups of bacteria (i.e., nitrifying bacteria, nitrate-respiring bacteria, clostridia, phototrophic bacteria, sulfate-reducing bacteria, and methanogens) in sediments of the Tay estuary. They obtained maximum rates of nitrification at depths of 0 to 10 mm, maximum nitrate respiration rates at depths of 10 to 20 mm, maximum dissimilatory sulfate reduction at depths of 20 to 30 mm, and maximum methanogenesis at depths of 40 to 50 mm. Clearly, bacteria in anoxic marine sediments are key components in the biogeochemical processing of the estuarine seafloor.[84]

Attempts to convert bacterial abundance to biomass have been hindered by several complications. First, accurate measurements of cell diameters are by no means guaranteed.[85] Second, the conversion of cell volume to carbon characteristically varies by a factor of 2, but possibly by as much as a factor of 5;[15,86] the two most commonly used conversion factors are 0.1 and 0.22 g C per cubic centimeter bacterial biovolume.[87,88] Therefore, Kemp[15] advises that bacterial biomass estimates in sediment should be treated with caution, because they are potentially inaccurate by at least a factor of 2.

B. BIOMASS

The microbial biomass of marine and estuarine sediments has been the focus of several surveys. Bacterial biomass accounts for only about 0.2 to 4% of the total organic carbon mass in marine sediments.[15,69,71,85] Nonetheless, Moriarty[28] discerned higher bacterial biomasses in a seagrass biotope of Moreton Bay, Queensland, that is 10 to 14% of the total organic matter content of the sediment (Table 3). When compared with macrofaunal biomass (about 1 to 10 g C per square meter), bacterial biomass is similar, if integrated over the depth range typical of burrowing macrofauna (approximately 10 to 15 cm depth).[15] As organic carbon input rises, so does the microbial biomass, which can surpass that of the macrofaunal biomass in organic-rich sediment.[15,89] Using an average cell volume for sediment bacteria of 0.2 μm^3 (range less than 0.05 to greater than 2.0 μm^3), Rublee[46] computed bacterial biomass in the upper 20 cm of salt marsh sediments, yielding figures in the range of 10 to 20 g C per square meter, which represents a significant proportion of the total microbial biomass. On the basis of ATP assays, Rublee[46] found that the bacterial biomass constituted more than 50% of the ATP estimates of microbial biomass. The bacteria in surface sediments comprised a small portion of the ATP-determined microbial biomass (approximately 15%), but with increasing depth, composed a progressively larger fraction.[46]

C. PRODUCTION

Prior to the use of direct counting by epifluorescence microscopy in the mid-1970s, estimates of bacterial abundance in marine sediments were obtained primarily via a variety of indirect most-probable-number techniques.[90] The inefficiency of these techniques, however, resulted in severe underestimates of bacterial density.[91] With the advent of measuring rates of nucleic acid synthesis, especially DNA, microbiologists have gathered much more meaningful *in situ* productivity values on natural bacterial assemblages.

Moriarty[28] and Karl[91,92] explore the strengths and weaknesses of the nucleic acid synthesis techniques. The basic methodology for calculating growth rates by means of nucleic acid synthesis is, in practice, relatively simple, involving the measurement of the rate of incorporation of radioactive precursor into a macromolecule. Moriarty and Pollard[93] developed the use of tritiated thymidine incorporation into DNA for measuring the growth rate of bacteria in marine sediments, whereas Fuhrman and Azam[94] perfected the technique for application in seawater samples. While some microbiologists advocate radioactive thymidine for labeling DNA, others employ adenine as a labeling agent.[95] Adenine labels RNA as well as DNA; hence, estimates of microbial growth rates can also be derived from rates of RNA synthesis.[96] Table 4 compares the suitability of these methods for measuring bacterial growth rates in seawater. Current radioisotopic methods for determining bacterial production, although holding great promise in microbial investigations of estuaries and oceans, are expected to undergo refinement as their applications accelerate. This refinement, in turn, should enable improvements to be made in the interpretation of results.

Kemp[15] elaborates on potential difficulties in using thymidine incorporation to measure bacterial production. First, a large fraction of bacterial cells, in some cases, may not take up thymidine even though heterotrophically active. Second, one or more assumptions required by the method may not hold, leading to inaccurate production estimates; for example: (1) label specificity for bacteria; and (2) accurate conversion factors to convert from thymidine uptake to biomass production. Drawbacks of applying tritiated thymidine incorporation for measuring bacterial production in sediments could include potential procedural problems, such as the effects of slurrying sediment to disperse added thymidine, rapid abiotic adsorption of label onto sediment grains, greater difficulty in extracting radiolabeled DNA, and the catabolism of thymidine by certain sediment bacterial populations.[15] Despite these possible shortcomings, proponents of the thymidine incorporation method believe that it indeed yields accurate bacterial production estimates[97] as substantiated by independent estimates based on direct counts of bacterial cell numbers,[98-100] ^{32}P incorporation into phospholipids,[101] frequency of dividing cells,[102-105] and oxygen consumption.[106]

Estimates of bacterial production in the water column and sediments of estuaries have been amassed from the rates of DNA synthesis using thymidine and adenine. Ducklow[107] and Moriarty and Pollard,[108] for instance, delineated the bacterial production in the water column of Chesapeake Bay and Moreton Bay (East Australia), respectively. Bacterial production in Chesapeake Bay waters during the spring ranged from 292 to 3130 ng C per liter per hour[107] compared to values of 90 to 300 ng C per liter per hour in Moreton Bay waters during the fall.[108]

A close coupling occurs between bacterial growth and phytoplankton production in the water column. In marine waters, in general, bacterial production as a percent of primary production ranges from 5 to 45%.[28] Malone et al.,[109] however, referenced higher readings for the surface-mixed layer of Chesapeake Bay, where bacteria metabolized approximately 60% of the summer photosynthetic production in 1984. Much of the carbon flow in Chesapeake Bay seems to proceed through the microbial loop, with a substantial fraction of the phytoplankton production supporting microheterotrophic metabolism and high standing stocks and production of aerobic bacterial populations.[110] In seasons and regions of dense phytoplankton assemblages, high bacterial growth rates also are apparent.[28]

Bacterial production in bottom sediments peaks near the sediment-water interface within the upper 2 cm of sediment, and diminishes with increasing depth. At 10 cm depth, bacterial production may be an order of magnitude less than that near the sediment surface. The most productive sediments in estuaries in terms of bacterial growth are those rich in particulate organic matter or enhanced by high dissolved organic concentrations (e.g., mangrove, salt marsh, and seagrass biotopes).[15] In these environments, Moriarty[28] cited bacterial production measurements of 10.5 to 230 mg C per square meter per day. Somewhat higher values were conveyed by Kemp (up to 800 mg C per square meter per day) for these systems (Table 5). Sandy sediments with low organic content display the lowest production figures. Bacterial production in estuarine sediments constitutes a major fraction of the benthic secondary production, equaling or exceeding production of all the macrofaunal assemblages, typically 2.5 to 70 mg C per square meter per day.[15,111]

Alongi[112] gathered seasonal bacterial production data on mangrove sediments of the Cape

TABLE 3
Comparison of Mean and Range of Bacterial Numbers, Biomass, and Biomass/Number Ratio in Relationship to the Organic Matter Content in Sediments

Location	Date	Temperature (°C)	Sediment type	Bacterial number (N) (10^8 g^-1)	Bacterial biomass (B) (10^-3 mg C g^-1)	B/N ratio (10^-11 mg C cell^-1)	Organic matter (OM) (mg g^-1)	Bacterial biomass (% OM)	Ref. (sources from Ref. 71)
Kampinge, Baltic Sea	6/1—5/77	15	Sandy beach	14.2 (7.5—27.2)	26.3 (17.6—45.5)	1.9 (1.7—2.4)	2.7 (2.3—3.2)	1.9	Meyer-Reil et al. (1978)
Kiel Fjord/ Kiel Bight, Baltic Sea	7/4—12/77	21.0	Sandy beach	10.3 (5.7—20.2)	20.4 (11.0—36.7)	2.1 (1.6—2.8)	5.9 (3.3—9.6)	0.69	Meyer-Reil et al. (1980)
	3/6—11/78	3.0	Sandy beach	4.5 (1.4—11.8)	5.6 (1.6—14.1)	1.2 (1.0—1.4)	5.1 (2.0—7.5)	0.22	Meyer-Reil (unpublished)
Kiel Bight, Baltic Sea	11/6—11/78	9.0	Sandy beach	15.3 (4.5—32.1)	12.9 (4.9—33.3)	0.9 (0.6—1.1)	11.7 (3.2—26.1)	0.22	Meyer-Reil (1981)
	3/3—5/8/80	0.8—4.9	Sandy (10 m)	2.0 (1.4—3.5)	5.8 (3.9—10.3)	3.0 (2.7—3.2)	8.1 (7.1—10.3)	0.14	
Mactan	4/2/80	27	Coral beach	14.7 (11.4—17.4)	50.0 (39.0—59.0)	3.4	—	—	Meyer-Reil et al. (1981)
Moreton Bay, Queensland	—	—	Subtidal seagrass sediment (0, 1, 20 cm)	7 (6—8)[a]	240 (110—370)	—	2.3 (2.1—2.5)[b]	10.3[c]	Moriarty (1980)
	—	—	Mid-intertidal seagrass sediment (0, 1, 20 cm)	10 (6—15)[a]	203 (100—320)	—	1.6 (1.1—2.0)[b]	12.3[c]	

			Intertidal sand flat (0, 2, 20 cm)	9 (6—10)[a]	50 (10—80)	—	0.35 (0.29—0.43)[b]	13.7[c]	
	—	—							
Elbe Estuary	2/2/80	3	Mud	31.1	52.4	1.7	—	—	Rheinheimer (unpublished)
Kiel Bight	8/8/80	12	Mud	507	452	0.9	—	—	Rheinheimer (unpublished)

[a] Cells × 10^8 μg^{-1} muramic acid.
[b] Mg organic C g^{-1} sediment.
[c] % organic carbon.

From Meyer-Reil, L.-A., *Heterotrophic Activity in the Sea*, Hobbie, J. E. and Williams, P. J. le B., Eds., Plenum Press, New York, 1984, 523. With permission.

TABLE 4
Comparison of the Suitability of Some Current Methods for Measuring Bacterial Growth Rates in Seawater

Criterion of suitability[a]	1	2	3	4
FDC	+ +[b]	+	+ +	+ +
^{35}S-SO_4 assimilation	− −	+	+	+
^3H-adenine incorporation into RNA	− −	− −	+	+ +
^3H-thymidine incorporation into RNA	+ +	+	+	+ +
Increase in cell number	+ +	+ +	−	−
Increase in cell ATP	−	+	−	+

[a] Criteria:
   1. Method should be specific for bacteria.
   2. Conversion factor should not be growth-rate dependent.
   3. Experimental manipulation should not change growth rate.
   4. Should be sensitive enough to allow short incubations (i.e., min-h).

[b] + +, Meets criterion;
   +, probably meets criterion;
   −, probably does not meet criterion;
   − −, does not meet criterion.

From Azam, F. and Fuhrman, J. A., *Heterotrophic Activity in the Sea*, Hobbie, J. E. and Williams, P. J. le B., Eds., Plenum Press, New York, 1984, 179. With permission.

TABLE 5
Bacterial Production[a] in Sediments of Mangrove, Salt Marsh, and Seagrass Systems to a Depth of 1 cm as Measured by Tritiated Thymidine Incorporation

Habitat	Depth	Production	Ref. (sources from Ref. 15)
Mangrove intertidal	2	800	Alongi (1987)
Mangrove creek bank	1	200—700	Kemp (in press)
	1	230	Moriarty (1986)
Salt marsh	20	20.1	Fallon et al. (1983)
	0.5	450	Kemp (in press)
Seagrass	2	85	Moriarty (1986)
	0.3	40	Moriarty et al. (1982)
	0.5	60	Moriarty et al. (1985)

[a] mg $C/m^2/d$.

From Kemp, P. F., *Rev. Aquat. Sci.*, Vol. 2, Issue 1, 1990. With permission.

York peninsula of Australia (Table 6). Production rates in the peninsula are twice as great and growth rates four times as great during the summer wet season than during the winter season. On Hinchinbrook Island in the dry tropics, Alongi[112] discovered highest production in the spring (4.5 mg C per square meter per day) and lowest production in the fall (0.2 mg C per square meter per day). Stanley et al.[113] followed declining bacterial production in a sediment column of the low and high intertidal zones of the mangroves of Hinchinbrook Island. In low intertidal sediments, production progressively decreased from 1.6 mg C per square meter per day at depths of 0 to 2 cm to 0.6 mg C per square meter per day at depths of 8 to 10 cm. Likewise in the high

TABLE 6
**Seasonal Bacterial Production[a] in Tropical Mangrove Sediments via Rates of
DNA Synthesis**

Location	Habitat and season	Depth (cm)	Production	Ref. (sources from Ref. 90)
Cape York peninsula,	Mangroves, winter	0—2	1.2 (0.6—24)	Alongi (1987)
Qld., Australia	Mangroves, summer	0—2	2.0 (0.2—5.1)	
Hinchinbrook Island,	Mangroves, fall	0—2	0.2 (0.1—0.4)	Alongi (1987)
Australia	Mangroves, winter	0—2	1.4 (0.6—1.9)	
	Mangroves, spring	0—2	4.5 (3.0—6.1)	
	Mangroves, summer	0—2	2.2 (1.6—3.2)	

[a] mg C/m^2/d.

From Alongi, D. M., *Rev. Aquat. Sci.,* 1, 243, 1989. With permission.

intertidal, production gradually dropped from 1.0 mg C per square meter per day at 0 to 2 cm depth to 0.2 mg C per square meter per day at 8 to 10 cm depth.

Bacterial production in surface mangrove sediments of Australia (0.05 to 1.1 g C per square meter per day) generally surpasses that in coral reef sands (0.02 to 0.37 g C per square meter per day).[112,114] Specific growth rates in the mangroves (0.2 to 7.9 per day) also exceed those in the coral reef sediments (0.05 to 0.83 per day). These differences reflect the higher organic concentration and primary productivity encountered in the mangrove habitat.

As the data base on bacterial production in estuarine and marine waters expands, it is becoming evident that production estimates seem to be much higher than deemed reasonable in the recent past.[90] What is the fate of this production?[115,116] Certainly, protozoans and meiofauna consume vast quantities of bacteria in sediment. Does this grazing control the abundance and activities of the benthic microbial populations? Not to be forgotten are the production and growth of anaerobic bacteria in bottom sediments. Of equal significance concerns the question of the fate of bacterial production and the complex heterotrophic processes in the water column. That a strong positive correlation has been established in the past decade between phytoplankton production and bacterial abundance and specific growth rates is unequivocal. The next section will examine microbial processes operating in the estuarine environment.

IV. MICROBIAL PROCESSES

A. GENERAL

Many aspects of the role of microbes in the biological and chemical processes of estuaries remain enigmatic despite years of research. Bacteria, in particular, perform a myriad of tasks critical to the total energy flow of these systems.[90,117,118] Mann (p. 125)[119] encapsulates the chief processes involving marine bacteria: "... (1) release of much of the primary production of plankton and macrophytes as dissolved organic matter (DOM) or as dead particulate organic matter (POM) and colonization of the latter by microbes; (2) uptake of DOM by bacteria with production of living and dead POM; (3) decomposition of the less refractory portions of POM of plant origin by microorganisms; (4) consumption of bacteria in large quantities by protozoa, planktonic filter feeders and benthic detritivores; (5) conversion of some DOM to particulate form by physicochemical processes, and subsequent colonization by microorganisms; (6) sedimentation of the POM to the sea floor and subsequent incorporation into aerobic food chains on the sediment surface, or decomposition by anaerobic organisms in the deeper layers of sediment".

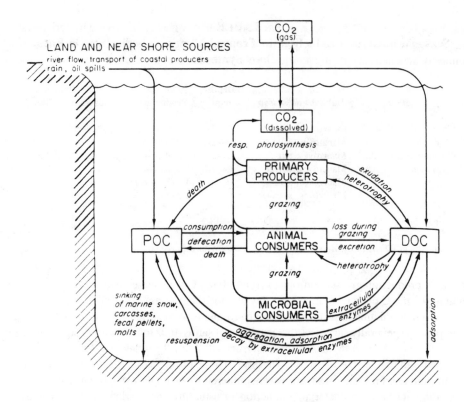

FIGURE 2.   Carbon transfer in aerobic marine environments. The boxes represent pools and the arrows, processes. The inorganic parts of the cycle are simplified. Organic aggregates and debris comprise marine snow. Some resuspension of dissolved organic carbon from sediments into the overlying water is not shown in the diagram. (From Valiela, I., *Marine Ecological Processes*, Springer-Verlag, New York, 1984. With permission.)

B. AEROBIC ENVIRONMENTS
1. Attached and Free-Living Bacteria

Classical energy flow diagrams for estuaries depict heterotrophic bacteria as the principal processors of organic matter, responsible for the breakdown and mineralization of a substantial fraction of this material. Especially in detritus-based systems, such as mangroves, salt marshes, and seagrasses, decomposition of organic material in sediments is viewed essentially as a microbially mediated process, although some larger organisms (e.g., crabs and other crustaceans) may assist in the initial stages of decomposition.[90,120] Meanwhile, the link between bacteria and phytoplankton in the water column occurs through the release of dissolved organic carbon from the autotrophs.[7,110,121,122] Hence, as remarked above, 25 to 30% and up to nearly 70% of the primary production of phytoplankton can enter the microbial food chain to support microheterotrophic metabolism.[121,123-125] An estimated 80 to 90% of the carbon entering this food chain from natural detrital sources may be oxidized by bacteria,[126] the exact amount governed by the carbon conversion efficiency of the bacterial community and the availability of inorganic nutrients.[125,127]

As defined in Hobbie and Williams,[128] marine heterotrophy refers to the process by which carbon autotrophically fixed into organic compounds is transformed and respired. Figure 2 illustrates carbon transfer pathways in aerobic marine environments. Heterotrophic aerobic microbial metabolism transforms DOM and POM to carbon dioxide and biomass.[129] Aerobic heterotrophic bacteria occupy three main habitats — the water column either attached to detrital particles or free floating, the top layer of bottom sediments, and living and dead tissues of plants and animals. Bacteria attached to particles utilize them both as habitat and substrate while also using dissolved organic material as a nutritional source, whereas free-living bacteria must rely on dissolved substrates as an energy base, having no other source of nutrition.[2] The activity and biomass of attached bacteria in estuaries are considered to be significant, those in the open ocean, negligible.[49,130,131]

Microbial activity in estuaries accelerates in the turbidity maximum zone.[132-134] Plummer et al.,[135] conducting research on bacterial distribution in relation to particle concentration and type over a spring-neap tidal cycle in the Tamar estuary, recognized three groups of bacteria: (1) free-living, (2) those attached to permanently suspended particles, and (3) those attached to particles which undergo tidally controlled resuspension and sedimentation. All three bacterial groups increase their activity in the region of the turbidity maximum, with bacteria attached to permanently suspended particles primarily responsible for the increased activity. Other workers have refuted this finding, suggesting that the increase in total bacterial activity at the turbidity maximum is due largely to bacteria associated with resuspended sediment.[136,137]

Fukami et al.[138] demonstrated differences between attached and free-living bacterial communities. In addition to dissimilarities in cell size[52,139] and in rates of assimilation of organic substrates,[140,141] the two types of microbes differ in metabolic respects as well. Thus, Paerl[142] called attention to differences in the metabolic activities of attached and free-living forms, including nutrient uptake rates, utilization, and release. Metabolism of the bacteria reputedly changes in the presence of detritus.[142]

Various microbiologists have argued that free-living bacteria are metabolically active and growing at a rate of one to four divisions per day;[29,97,128,249] others consider these bacteria to be dormant.[143,144] Dissolved organic matter may be too dilute in seawater to support significant growth of free-living forms. Ammerman et al.[145] assume bacterial growth takes place principally on particles. The growth of free-living bacteria could be tightly coupled to the input of useable nutrients.[145]

Fukami et al.[146,147] ascertained changes in abundance of attached bacteria relative to free-living forms in seawater. During decomposition of a phytoplankton bloom, Fukami et al.[146] stressed that the number of free-living bacteria remained almost constant, but the number of attached bacteria increased. As the concentration of suspended solids or particulate organic carbon fluctuates, so does the density of attached bacteria,[148,149] while the density of free-living forms varies little.[150,151]

In open marine waters, most bacteria in the water column are free-living, subsisting on DOM.[128] Attached bacteria, however, probably dominate in certain habitats; they have multiple adaptations for attaching to and degrading particles. Microflagellates appear to be the chief predators of bacterioplankton, capable of removing vast quantities of the microbes from the water column. The trophic significance of this relationship, however, has not been placed into proper perspective. Specifically, it remains uncertain whether bacterioplankton mainly regenerate nutrients back to phytoplankton or function as pathways of carbon flow to zooplankton.

2. Bacteria as Decomposers

Because of the relatively shallow depths and high productivity of estuaries, bottom sediments serve as an ideal site for the accumulation of organic matter and its mineralization via microbial attack.[152] The sediment bacteria, in turn, are a food source for consumer organisms, notably, protozoans and meiofauna.[90,153] The aerobic microheterotrophs inhabiting bottom sediments of estuaries degrade a wide range of natural substrates,[154] but the anaerobic microorganisms — for example, fermenting and dissimilatory nitrogenous oxide-reducing bacteria, dissimilatory sulfate-reducing bacteria, and methanogenic bacteria — exhibit more specific roles in the transformation of organic material.[155]

Only recently have bacteria been viewed as potentially significant competitors for nutrients and food of other organisms.[2] Many estuarine scientists emphasize the role of the microbes in detritus-based systems, elaborating upon the enrichment they bring to particulate debris and the complexity they add to food webs. The traditional concept of bacteria as decomposers in estuaries is that they colonize detrital particles, assimilate nutrients, and convert POM into DOM to meet their energy requirements.[156] According to this paradigm, detritivores subsequently consume the detritus-microbe aggregate, remove the attached microbes as their energy source, but digest little of the detritus. The quantity of detritus digested by primary consumers depends on the extent of modification of this material by autolysis, leaching, microbial decay, and mechanical breakdown.[157] Factors such as nutritional composition of the detritus, its source, and particle size must also be considered.[158] However, detritivores may be capable of directly assimilating more readily decomposable types of detritus (e.g., benthic macroalgae) with little or no microbial enrichment,[126] the more refractory detritus (e.g., vascular plant debris) requiring microbial degradation or enrichment prior to assimilation.[158] By mineralizing plant carbon and assimilating nitrogen, the bacteria decrease the C/N ratio, thereby improving the nutritive value of detritus through time.[159] Although detritivores may directly digest some detritus, the bulk is refractory and excreted free of attached microbes. The value of microbes as a source of food for detritivores may be less important than previously thought, however.[126,160-162] Lopez and Levinton[163] have recently contended that many deposit feeders attain more nutrition from detritus itself than once deemed possible because microbial stripping alone yields an insufficient biomass of bacteria to meet their energy demands (see Chapter 6, Section II. A).

Valiela[164] recounts the multiple processes activated during the decay of organic matter, concentrating on the leaching of soluble compounds, microbial degradation, and consumption by other heterotrophs. Regardless of whether the organic matter originates from phytoplankton, macroalgae, vascular plants, or some other source, a succession of decay phases can be elucidated, all of which influence its depletion. An initial, short-lived leaching phase of nonliving organic matter removes either soluble or autolyzed materials (Figure 3); microbial heterotrophs take up the dissolved materials (e.g., dissolved proteins, amino acids, carbohydrates) and mineralize them into CO_2 and inorganic salts. The leaching phase may persist for several minutes to a few weeks, its longevity being controlled mostly by the kind of detritus under decay.

Subsequent to the leaching phase, a microbially mediated decomposition phase predominates primarily via hydrolysis by enzymes released from the cells of the microbes. Bacteria remove a portion of the soluble products during this phase of degradation, and some of the dissolved organic carbon is lost by leaching. As the decomposition stage proceeds, the more easily degraded compounds (e.g., sugars and various proteins composing detritus) are depleted rather quickly, while the more decay-resistant compounds (e.g, cellulose, waxes, and certain phenolic compounds such as lignins) remain recalcitrant to breakdown. Through time, therefore, the detritus changes in chemical composition; the percent of cellulose and lignin rises (Figure 3). Ultimately, the detritus that remains contains large concentrations of fulvic and humic acids consisting of refractory phenolic polymers and complexes forming marine humus or "Gelbstoff". This final refractory phase may last for years in detritus originating from vascular plants (e.g., *Spartina alterniflora* and *Zostera marina*), the detritus being principally confined to estuarine bottom sediments. In laboratory experiments, for instance, *Z. marina* lost only 35% of its dry weight after 100 d at 29°C.[164] Many detritivores, although ingesting the detritus and attached microbes at any stage during the decomposition process, assimilate little, if any, of the detritus itself while stripping the microbes. These larger heterotrophs do not have the necessary carbohydrates to hydrolyze cellulose and other structural carbohydrates.[165,166] Mechanical fragmentation induced by wave and current action, as well as feeding and burrowing by benthic macrofauna, exposes new surface areas of the detritus to microbial attack and reduces its size.[76] However, in the case of infaunal meiobenthos-microbial food webs at least, small-scale disturbances of surface sediments have been shown to be unimportant to their structure and function.[167]

The decomposability of detritus, as inferred above, is closely linked to the composition of the

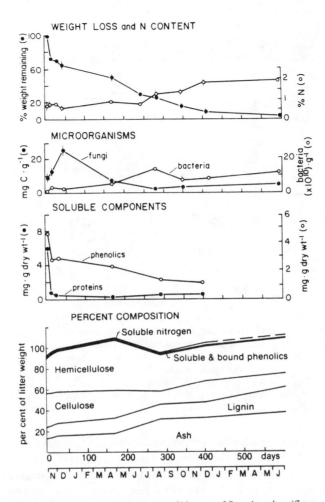

FIGURE 3. Decomposition history of biomass of *Spartina alterniflora* in a Massachusetts salt marsh. Data obtained by means of litter bags set out in November and collected at intervals. The two soluble components are small relative to other chemical components, so the pattern of the third graph does not show in the bottom graph. Cellulose is a straight-chain polymer of glucose, and hemicellulose is a base-soluble cell wall polysaccharide closely associated with cellulose. Lignin is an amorphous phenolic heteropolymer. Data of I. Valiela, John Teal, John Hobbie, Tony Swain, John Wilson, and Robert Buchsbaum. (From Valiela, I., *Marine Ecological Processes*, Springer-Verlag, New York, 1984. With permission.)

source material. Algal detritus high in organic nitrogen, soluble ash, and hydrolyzable constituents decays more rapidly than vascular plant detritus high in cross-linked celluloses and lignins.[157,168] Thus, Rice and Tenore,[157] in a laboratory experiment, reported more rapid degradation of algal detritus compared to vascular plant detritus, even after the readily soluble fraction of the algal detritus was leached. Benner et al.[169] demonstrated that a bacterial assemblage mineralized the polysaccharide component of *S. alterniflora* twice as rapidly as the lignin component (Figure 4). They also surmised that bacteria rather than fungi acted as the primary agents in the degradation of lignocellulosic detritus in salt marsh sediments. As lignocellulosic detritus decomposes, the decay-resistant lignin and its humified derivatives accumulate, forming a long-term pool of organic carbon that delivers a degree of stability to otherwise widely variable benthic food supplies.

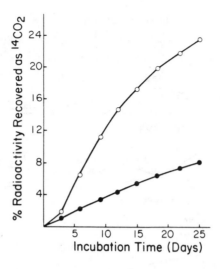

FIGURE 4. Mineralization of *Spartina alterniflora* [^{14}C-lignin] lignocellulose (closed circle) and [^{14}C-polysaccharide] lignocellulose (open circle) by unfractionated salt marsh sediment microflora. The polysaccharide component of the lignocellulose was mineralized 2.9 times faster than the lignin component. (From Benner, R., Newell, S. Y., Maccubbin, A. E., and Hodson, R. E., *Appl. Environ. Microbiol.*, 48, 36, 1984. With permission.)

Lignocellulose, a macromolecular complex of the structural polysaccharides, cellulose and hemicellulose, and the aromatic heteropolymer lignin, is indigestible by most animals.[170] The lignocellulosic polymers are physically intertwined and chemically bound requiring a variety of enzymes for depolymerization.[169] The more recalcitrant lignin component can actually mitigate the biodegradational process of the polysaccharides. Consequently, it represents the rate-limiting step in lignocellulose biodegradation.[170,171]

A subject of controversy over the years centers on the relative importance of bacteria and fungi in the decomposition of organic detritus.[172] Early research emphasized the role of bacteria as the principal decomposers of salt marsh detritus,[59,173] a stance supported by more recent findings.[174,175] Other studies, however, implicate fungi as major decomposers of salt marsh grass.[176-178] Padgett et al.[172] attributed as much as 70% of the oxygen consumption in decomposing *S. alterniflora* detritus to nonbacterial components (probably fungal).

A successional sequence of heterotrophic organisms has been detected colonizing detrital particles.[179,180] Initial colonizers are bacteria followed by fungi, algae, ciliates, flagellates, and larger grazers. Appearance of the ciliates and flagellates signifies the presence of bacteria as a food source. One valuable indicator of this succession is the muramic acid/ATP ratio.[76] Other fauna commonly associated with the detritus — small crustaceans, nematodes, rotifers, and turbellarians — ingest the bacteria in addition to other organisms. The successional sequence of microbes observed on detritus in estuarine bottom sediments parallels that on detrital particles in the water column. In a laboratory setting designed to simulate conditions in the water column, Newell et al.[124] tracked the succession of microorganisms on salt marsh debris derived from *S. alterniflora* and *Juncus roemerianus* and on aged detrital material incubated in estuarine water at temperatures approximating those in the natural habitat. They detailed an initially rapid logarithmic growth of free-living bacteria superseded by a diminution in bacterial numbers concomitant with a rise in the abundance of bacterivorous microflagellates. A mixed population of ciliates, choanoflagellates, amoeboid forms, and attached bacteria replaced these earlier forms to create a complex microbial community associated with the particulate debris. Results of other investigations[44,90,181] corroborate the significance of bacterivores (e.g., microflagellates and ciliates) as grazers of bacteria on detrital particles.

The rate of decomposition of organic detritus increases as microfauna (protozoans) graze on the attached bacteria. Not only does grazing enhance detrital mineralization, but it also stimulates bacterial growth presumably by raising oxygen, nitrogen, and phosphorus levels or by promoting the proliferation of faster growing forms.[76] Moreover, as revealed above, when the detritus ages, its composition changes to reflect greater amounts of nitrogen due presumably to the higher protein content of the microbial biomass.[158] Bacteria, fungi, and extracellular microbial excretion products, particularly mucopolysaccharides,[182] appear to be responsible for this protein enrichment which elevates the nutritive value of the detritus. The rate of nitrogen enrichment varies with detrital source. Detritus derived from phytoplankton, for example, becomes nitrogen-rich during the early stages of decomposition because of bacterial growth, whereas other carbon sources (e.g., zooplankton) may be more resistant to microbial attack and less likely to yield nitrogen-rich detritus quickly.[147] The enrichment of nitrogen into detritus drops the carbon/nitrogen ratio; when this ratio declines over time to a value less than 17:1, it is considered to be an excellent nutritive source for detritivores.[183] In the process of decomposing organic matter, bacteria play a vital role as a net consumer of nutrients, and may even compete with autotrophs for essential nutrients.[184]

Detritus in estuaries has multiple sources — for example, dead plants and animals, fecal pellets of zooplankton and benthic fauna, and amorphous organic matter derived from DOM. No consensus exists as to how bacteria transform POM to living biomass, although it is generally acknowledged that these microbes must first transform POM to DOM to obtain nutritional value from it. In the decomposition of detritus, polymeric compounds must be hydrolyzed extracellularly.[64] Utilizing the dissolved organic carbon, marine bacteria grow rapidly.[145,185,186]

Bacteria are the most efficient transformers of organic substrates of all organisms as evidenced by laboratory investigations of assimilation efficiency.[187,188] While results of early work suggested that bacteria grew rather inefficiently, their main function being mineralizers of organic matter, more recent findings have uncovered higher bacterial growth yields. Conversion efficiencies typically range from 50 to 80% in pure and mixed cultures of bacteria growing on simple and complex substrates.[47] However, such high conversion efficiencies, which are elevated compared to conversion efficiencies of metazoans, do not extend to refractory substrates. Furthermore, relatively low conversion efficiencies may be apparent on less recalcitrant materials as well. For instance, Newell et al.[124,189] stipulated lower conversion efficiencies (approximately 10%) of bacteria on phytoplankton and kelp debris. Variations in conversion efficiencies of bacteria can be ascribed in some cases to differences in the biochemical nature of the substrate or to experimental conditions.[47] Radiochemical techniques employing ^{14}C-labeled substrates have advanced the understanding of bacterial conversion efficiencies in marine systems.[190-193]

The success of marine bacteria as decomposers of organic matter is a function in large part of extracellular enzymes whose efficiency has been tied to various processes (e.g., adsorption and denaturation). The availability of the product of the enzyme reaction, which may be reduced considerably by competition with other heterotrophs, affects the activity of the bacteria. Proteinase, agarase, alginase, esterase, B-glucosidase, phosphatase, succinate dehydrogenase, and amylase are extracellular bacterial enzymes presumed to be important in the decomposition of POM.[71,194-196] Bacteria probably do not excrete these enzymes continuously.[197] When released, the enzymes sorb to organic particles and possibly react with the substrate to form a product subsequently assimilated by bacterial cells. Extracellular polymer fibers enable the microbes to attach to solid surfaces.[198,199] Sediments in intertidal zones appear to elicit greatest extracellular enzyme activity.[200] Bacteria, when attached to a particle surface and in contact with estuarine water, are especially adept at removing DOM and transforming it into POM.[182] Hobbie and Lee[182] hypothesized that the extracellular mucopolysaccharides produced by microbes are more abundant than the microbes themselves and supply the main fraction of food for many benthic fauna. Two lines of evidence support this hypothesis: (1) only a small portion of the total carbon required to feed detritivores is supplied by microbial biomass; and (2) only a small amount of the nitrogen in aged detritus or sediments is accounted for by the microbes. Thus, microorganisms in the estuarine environment must be viewed as essential components in the transformation of organic detritus into microbial protein — either cellular or extracellular excretions — critical to detritivore secondary production.

Hanson,[201] examining the methods of study (isotope and nonisotope tracers) available to measure specific microbial processes, concludes that not all of them are suitable for research on benthic systems. For microbiological research on decomposition rates and utilization of detritus, Hanson[201] advises adopting the least specific measurement to estimate microbial activity or productivity. He, therefore, recommends oxygen consumption to monitor oxidative metabolism, total adenylates to quantify microbial biomass, and DNA synthesis to determine production rates. According to Hanson,[201] other methods of analyzing microbial activities (i.e., changes in adenosine triphosphate and lipid synthesis) do not yield adequate data on total microbial activity, growth rate, and production.

The pool of dissolved organic carbon in estuaries, like that of particulate organic carbon, represents contributions from both allochthonous and autochthonous carbon sources. Allochthonous input of dissolved organic carbon principally occurs via riverine and atmospheric systems, with minor input from the ocean.[202] Autochthonous sources include: (1) exudation and abrasion of macrobenthic algae, (2) exudation of salt marsh macrophytes, (3) excretion or exudation by phytoplankton, (4) release of DOM due to zooplankton feeding and from zooplankton fecal material, and (5) death, lysis of cells, and decomposition.[203] To these five sources can be added release of DOM from seagrass and mangrove systems which supplies a major quantity of estuarine bacteria in temperate and tropical regions, respectively.

Bacterioplankton, important components of the plankton,[204] remove DOM in microgram- or even nanogram-per-liter concentrations.[2] In addition, attached bacteria take up DOM in estuaries.[205] However, in these shallow coastal ecosystems, attached bacteria dominate microheterotrophic processes because of their greater heterotrophic activity relative to free-living forms.[64]

Considerable interest surrounds the uptake of DOM by bacteria in estuarine sediments. Various approaches have been devised to assess the incorporation and mineralization of DOM by the microbes. The kinetics of uptake of ^{14}C-labeled substrates by natural assemblages of bacteria have often been applied to estimate total incorporation.[126] Additional information on the use of ^{14}C-labeled substrates can be derived from Fenchel and Jørgensen[190] and Joint and Morris.[191]

Experimental work on bacterial uptake rates and turnover times in bottom sediments has most frequently employed ^{14}C-glucose (partially or uniformly labeled).[71] A few studies have utilized the uptake of acetate and glutamic acid as well as alanine, aspartic acid, glycine, lactic acid, and urea. Meyer-Reil,[71] in a summary of the literature, compared the turnover times and respiration of dissolved organic carbon by bacteria in sediments and overlying waters of a number of environments, using the aforementioned substrates. Based on this summary, the turnover times are highly variable, ranging from a few minutes to several hundred hours; overlying waters exhibit elevated turnover times, one to two orders of magnitude higher than those of bottom sediments. The respiration also varies greatly (between 0.4 and 100%), although most values fall in the lower end of the range near 20%.

While the concentration of DOM in the open ocean ranges from 0.5 to 1.5 mg/l, substantially higher concentrations (1 to 5 mg/l) characterize estuaries and coastal oceanic waters.[156,206] Wright[203] reviews three basic approaches for evaluating natural dissolved organic carbon compounds in seawater. The first approach identifies and concentrates specific organic solutes.

The second approach analyzes the class of compounds (e.g., dissolved free amino acids or total dissolved carbohydrates). The third approach determines the gross organic fraction associated with dissolved organic carbon, dissolved organic nitrogen, and dissolved organic phosphorus.

Bacteria are the principal metabolizers of DOM.[19] When concentrations of DOM remain high, unicellular algae may also subsist heterotrophically on it. The algae, however, are outcompeted by bacteria for the dissolved material.[207-209]

The flux of dissolved organic carbon has been related to bacterivorous microzooplankton which release or stimulate the release of DOM. Bacterivorous ciliates and microzooplankton contribute both qualitatively and quantitatively to the dissolved organic pool, particularly in nearshore planktonic systems. Furthermore, bacterivory by nanozooplankton (2 to 20 μm in diameter), mainly heterotrophic mastigophoran and ciliated protozoans, and microzooplankton (20 to 200 μm in diameter), mostly ciliated protozoans and micrometazoans, may control bacterioplanktonic metabolic activity, as well as govern species composition and biomass of the bacterioplankton, thereby indirectly influencing dissolved organic carbon levels.[9,16,210,211]

Another factor affecting the flux of DOM is the physical-chemical conversion of DOM to particulate form, a conversion perhaps responsible for much of the organic particles found in the sea.[143] An example of this conversion involves the collection of surface-active material at a gas-liquid interface (e.g., bubbles) and compression of the interface (e.g., breaking waves).[119] Riley[212,213] documented the formation of amorphous organic particles in the sea. He differentiated four categories of particles: (1) aggregated particles that are primarily carbohydrates, (2) flakes that are chiefly protein, (3) fragments that are entirely carbohydrate, and (4) particles that are unclassified.[119,214] The DOM can adsorb directly onto particle surfaces; in general, the adsorption of organic compounds onto surfaces tends to mitigate biodegradation rates.[215]

3. Bacterivory

Kemp[15] and Alongi[90] have discussed grazing of bacteria in estuarine and marine systems. In bottom sediments, protozoans and meiofauna consume substantial quantities of surface bacteria.[114,216] Benthic macrofauna ingest significant amounts of the microbes, too,[115,163,217] but bacterial productivity in the benthos probably outpaces consumption by these larger forms, especially in sediments where bacterial turnover is rapid. Only when macrofaunal populations are exceptionally dense and their feeding confined to a thin layer of surface sediment has grazing by these larger organisms reduced or controlled bacterial abundance. This effect has not been observed among subsurface-feeding macrofauna. Recently, deposit-feeding holothurians and callianassid shrimp have been shown to be unrelenting microbial grazers.[90]

The effect of meiofaunal grazing on sediment bacterial abundance or production has not been resolved. Most of the inferences drawn on meiofaunal grazing are based on correlative data. Some publications, for example, stress negative correlations between abundances of meiofauna and bacteria,[218,219] and others, positive correlations or no correlation between the two groups.[153,220,221] No correlations between bacterial and meiofaunal abundance suggest the lack of meiofaunal control on bacterial numbers.[15] Meiofauna, like benthic macrofauna, possibly affect bacterial dynamics only when their densities are increased to abnormally high levels. Of all the meiofauna, nematodes form the most important bacterivorous element in many bottom sediments.[222]

Protozoans, although a major grazer of bacteria, are generally less abundant in sediment than in the water column. Protozoan bacterivory, notably by microflagellates, can remove a significant fraction of bacterial production when bacterial turnover times are low. Kemp[15] informs microbial ecologists that benthic ciliates and microflagellate protozoans have the potential to be key bacterivores when they graze at maximal rates or inhabit areas in sufficiently large numbers.

In pelagic ecosystems, bacteria are primarily grazed by protozoa. Tightly coupled oscillations in biomasses of bacteria and flagellates have been described in estuarine waters. Bjørnsen et al. (p. 409)[223] underscore the importance of nanoflagellates as bacterial consumers in the water column. They state, "In marine pelagic environments, heterotrophic nanoflagellates are the most probable consumers of bacterial cell production (Fenchel 1984, 1986). Laboratory experiments with cultured flagellates have demonstrated that they are adapted to predation on bacteria-sized particles (Fenchel, 1982), and they occur in seawater in concentrations sufficient to control bacterial biomass (Sherr and Sherr, 1984)."

C. ANAEROBIC ENVIRONMENTS

Areas of estuaries devoid of oxygen are sites of anaerobic microbial mineralization of organic matter. Such areas not only characterize the deeper sediment layers of estuaries, but also the anoxic waters of heavily polluted systems, as well as regions of fjord-type estuaries and coastal lagoons having very poor water circulation due to physical restrictions or stratification.[19,164] Although both bacteria and fungi actively decompose organic matter in aerobic environments, fungi are rare or inactive under anaerobic conditions.[164] Consequently, bacteria serve as the critical agents in the transformation of organic matter in anaerobic habitats. Aside from bacteria, several faunal groups (e.g., micro- and meiofauna; Protozoa, Turbellaria, Nematoda, and Gastrotricha) have representatives which may also be capable of living as anaerobes.[19,224,225]

The depth to the anaerobic zone in estuarine bottom sediments is variable, but usually ranges from a few millimeters below the sediment-water interface in fine-grained, carbon rich sediments to many centimeters below this interface in coarse-grained sediments. The exact position of the anaerobic zone depends on grain size, which regulates, in part, the penetration of dissolved oxygen, and the physical (e.g., waves and currents) and biological (e.g., bioturbation) activity on and in the seafloor. Chemical-reducing conditions take place within the anaerobic zone.[224,226] Despite observing changes in the oxygen concentration when proceeding from the aerobic to the anaerobic zone, Dale[69] found to equivalent numbers of bacteria and no correlation between the Eh of the sediments and bacterial abundance across the zones.

Factors influencing the thickness of the aerobic zone in estuarine sediments include the degree of water turbulence, light intensity at the sediment surface, sediment permeability, and concentration of organic matter.[227] The balance between the downward diffusion of oxygen in the sediments and its consumption determines the depth of the transition zone separating the oxidized and reduced layers. The sources of the oxygen are the overlying water in the estuary, the sediment surface where photosynthesis by autotrophs produces oxygen, and interstitial waters of the bottom sediments.[227]

Fermentation, dissimilatory nitrogenous oxide reduction, dissimilatory sulfate reduction, and methanogenesis comprise the major anaerobic processes in estuarine and marine sediments.[228] These microbally mediated processes can be distinguished on the basis of the specialized biochemistry of the microbes. As clarified by Pomeroy et al.,[129] "... Fermenting (Baker, 1961; Wood, 1961) and dissimilatory nitrogenous oxide reducing (Payne, 1973) bacteria often are generalized with respect to their carbon sources. Their metabolic products are biomass, carbon dioxide, and low molecular weight fermentation products. Dissimilatory sulfate reducing bacteria require specific low molecular weight compounds, such as lactate (Postgate, 1965). They produce biomass, carbon dioxide, and soluble end products, such as acetate. Methanogenic bacteria also use a restricted group of compounds, that is, carbon dioxide, formate, acetate, and methanol (Wolfe, 1971). Their unique end product is methane."

Initially, fermenters act on organic matter in anoxic zones, generating low-molecular-weight organic compounds which are further mineralized by other anaerobic respiratory processes. A major pathway of anaerobic decomposition in shallow marine ecosystems is sulfate reduction.[20,227,229-234] Indeed, it has been deemed the dominant process whereby sediment bacteria oxidize organic carbon.[90] In salt marsh systems, for instance, sulfate-reducing bacteria, together with fermentative forms, constitute the dominant decomposers of organic matter in sediments below 1 cm, with a minor fraction of the total sediment metabolism in anoxic zones ascribable

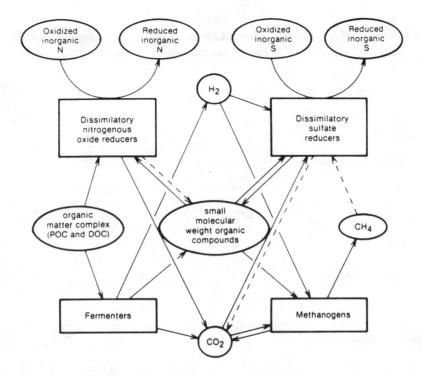

FIGURE 5. Interactions of anaerobic microbial processes in salt marsh sediments. (From Wiebe, W. J., Christian, R. R., Hansen, J. A., King, G., Sherr, B., and Skyring, G., *The Ecology of a Salt Marsh*, Pomeroy, L. R. and Wiegert, R. G., Eds., Springer-Verlag, New York, 1981, 137. With permission.)

to other anaerobic processes.[20,235,236] Hence, oxidation of organic matter via aerobic respiration and sulfate reduction accounts for the largest fraction of carbon mineralization in salt marsh sediments as well as in subtidal estuarine sediments.[232]

Anaerobic microbial metabolism is much more complex than aerobic microbial metabolism, which principally transforms particulate organic carbon and dissolved organic carbon to biomass and carbon dioxide.[129] Figure 5 shows the interactions of anaerobic microbial processes in sediments. Anaerobic reactions have a lower energy yield than aerobic respiration (Table 7).

1. Fermentation

Through processes of fermentation and nitrogenous oxide reduction, organic matter originating from autotrophic and heterotrophic production enters the anaerobic cycle.[228] Fermenters, which may be facultative or obligate anaerobes, use a variety of substrates (e.g., alcohols, amino acids, cellulose, pectin, purines, and sugars). The organic substrates themselves act as electron acceptors in fermentation.[19] End products of this anaerobic metabolism are hydrogen, carbon dioxide (as mentioned above), ammonia, and organic compounds (e.g., alcohols and short-chain fatty acids).[228] In anoxic environments, fermentation yields lower-molecular-weight organic compounds that can be utilized by other microbes (Figure 6).[65,164] Fermentation, along with photosynthesis and respiration, is a major process whereby organisms obtain energy.[164]

2. Dissimilatory Sulfate Reduction

Sulfate-reducing bacteria use SO_4^{2-} as a terminal electron acceptor during the decomposition of organic matter in anoxic sediments, anoxic waters, and anoxic microzones within aerobic sediments.[164] Reduction of sulfate results in the generation of hydrogen sulfide which may concentrate in interstitial water (up to 20 mmol) or precipitate as iron sulfides, creating a black

TABLE 7
Some Representative Reactions Illustrating Pathways of Microbial Metabolism and Their Energy Yields[a]

		Energy yield (kcal)
Aerobic respiration	$C_6H_{12}O_6$ (glucose) $+ 6O_2 = 4CO_2 + 4H_2O$	686
Fermentation	$C_6H_{12}O_6 = 2CH_3CHOCOOH$ (lactic acid)	58
	$C_6H_{12}O_6 = 2CH_3CH_2OH$ (ethanol) $+ 2CO_2$	57
Nitrate reduction and denitrification	$C_6H_{12}O_6 + 24/6\ NO_3^- + 24/5\ H^+ = 6CO_2 + 12/5\ N_2 + 42/5\ H_2O$	649
Sulfate reduction	$CH_3CHOHCOO^-$ (lactate) $+ \frac{1}{2} SO_4 + \frac{3}{2} H^+ = CH_3COO^-$ (acetate) $+ CO_2 + H_2O + \frac{1}{2} HS^-$	8.9
	$CH_3COO^- + SO_4^= = 2CO_2 + 2H_2O + HS^-$	9.7
Methanogenesis	$H_2 + \frac{1}{4} CO_2 = \frac{1}{4} CH_4 + \frac{1}{2} H_2O$	8.3
	$CH_3COO + 4H_2 = 2CH_4 + 2H_2O$	39
	$CH_3COO = CH_4 + CO_2$[b]	6.6
Methane oxidation	$CH_4 + SO_4^= + 2H^+ = CO_2 + 2H_2O + HS^-$	3.1
	$CH_4 + 2O_2 = CO_2 + 2H_2O$	193.5
Sulfide oxidation	$HS^- + 2O_2 = SO_4^= + H^+$	190.4
	$HS^- + 8/5\ NO_3^- + 3/5\ H^+ = SO_4^= + 4/5\ N_2 + 4/5\ H_2O$	177.9

[a] Energy yields vary depending on the conditions, so different measurements may be found in different references. The values reported here are representative.
[b] This reaction is sometimes considered fermentation.

From Valiela, I., *Marine Ecological Processes,* Springer-Verlag, New York, 1984. With permission.

coloration in anoxic sediments. Because of the sulfide formation, the redox potential, Eh, of the interstitial water may vary from −100 to −220 mV.[227] Sulfide, diffusing upward into aerobic layers, can be oxidized spontaneously, or by chemolithotrophic bacteria and by purple or green sulfur bacteria.[227] When iron is reduced from the ferric to the ferrous state, adsorbed phosphates are liberated, enabling the regeneration of important nutrient elements.[119] Sulfate reducers (e.g., the genera *Desulfovibrio*, *Desulfuromonas*, and *Desulfomaculum*) use relatively few substrates provided by fermenters. The following reaction gives an example of dissimilatory sulfate reduction; in it, lactate is an energy source for *Desulfovibrio*:[228]

$$2 \text{ lactate} + SO_4^{-2} \rightarrow 2 \text{ acetate} + 2CO_2 + S^{2-} \tag{1}$$

Dissimilatory sulfate reduction yields end products of value to aerobic heterotrophic processes.

Sulfate-reducing bacteria potentially oxidize as much organic matter to CO_2 as aerobic microbes.[233,237] In some salt marsh and estuarine sediments, sulfate reduction accounts for more than one half of the oxidation of carbon.[232] For instance, Jørgensen[226] attributed 53% of the mineralization of organic matter in the Limfjorden, Denmark, to sulfate reduction. The supply of organic compounds appears to regulate the activity of sulfate reducers.[19] Hence, the significance of sulfate reduction in sediments gradually decreases when progressing from estuarine to offshore waters.[238] Martens[233] formulated an idealized reaction of organic matter decomposition during sulfate reduction:

$$(CH_2O)_x (NH_3)_y (H_3PO_4)_z + \frac{1}{2}xSO_4^{-2} \rightarrow xHCO_3^- + \frac{1}{2}xH_2S + yNH^3 + zH_3PO_4 \tag{2}$$

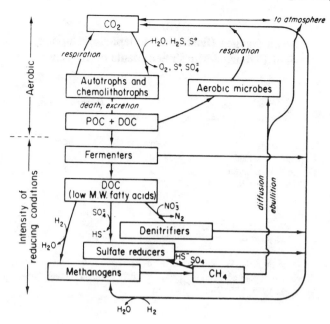

FIGURE 6. Carbon transformations in the transition from aerobic to anaerobic situations. The gradient from aerobic to anaerobic can be thought of as representing a sediment profile, with increased reduction and different microbial processes deeper in the sediment. Boxes represent pools or operations that carry out processes; arrows are processes that can be biochemical transformations or physical transport. Elements other than carbon are shown, where relevant, to indicate the couplings to other nutrient cycles. Some arrows indicate oxidizing and some reducing pathways. (From Valiela, I., *Marine Ecological Processes*, Springer-Verlag, New York, 1984. With permission.)

Salt marsh sediments are excellent sites for assessing sulfate reduction.[239] Typically rich in H_2S and organic matter with large numbers of anaerobic and aerobic bacteria, they have low pH values, and Eh readings generally less than -100 mV.[240] The upward diffusion of sulfide from the anoxic zone and its subsequent oxidation in the aerobic, surficial layer are responsible, in some cases, for more than one half of the total oxygen demand of a salt marsh sediment.[227] Most salt marsh sediments have a high potential for oxygen demand (Table 8).

3. Dissimilatory Nitrogenous Oxide Reduction

Three specific pathways can be distinguished under this category (Figure 7). In dissimilatory nitrogenous oxide reduction, nitrogenous oxides act as terminal electron acceptors. Denitrification produces end products of N_2O and N_2, with nitrate (NO_3^-), nitrite (NO_2^-), nitric oxide (NO), or nitrous oxide (N_2O) being readily reduced in anoxic environments. Sometimes denitrification terminates with the accumulation of nitrous oxide, whereas at other times it proceeds to the end product of dinitrogen (N_2).

A biochemical alternative to denitrification, dissimilatory nitrate reduction to ammonium, can take place in estuarine sediments at rates comparable to that of denitrification.[241,242] In Narragansett Bay, RI, Seitzinger et al.[243] calculated that about 35% of the organic nitrogen remineralized in the sediments was due to denitrification. This process removes nitrogen as N_2 in an amount equal to one half of the inorganic nitrogen loading from urban sewage into the estuary over an annual period.[244,245] Along the South Island, West Coast, New Zealand, denitrification accounted for 82 to 100% of the total nitrate reduction and 3 to 75% of the net

TABLE 8
Mean Oxygen Demand and Depletion Capacity
of Estuarine-Marsh Sediments of Wassaw
Sound, GA

Environment	Oxygen demand[a]	Oxygen depletion capacity[b]
Salt marsh	3.25	766
Tidal creek bank	2.10	515
Tidal creek bottom	3.73	987
Tidal river bottom	2.08	477
Sound bottom	0.89	154

[a]   mg O_2 per cc sediment per day.
[b]   cc water depleted/cc sediment suspended/day.

From Frey, R. W. and Basan, P. B., *Coastal Sedimentary Environments*, 2nd ed., Davis, R. A., Jr., Ed., Springer-Verlag, New York, 1985, 225. With permission.

nitrogen mineralization.[242] Denitrification clearly is an important sink for fixed nitrogen.[245] The conversion of NO_3^- to N_2 in estuarine bottom sediments via bacterial denitrification affects nitrogen cycling in sediments and overlying waters, and at times possibly places limitations on autotrophic growth.[152] As such, denitrification represents a significant pathway that must be considered in the formulation of nitrogen budgets of estuaries.[246,247]

Both denitrification and nitrification undergo diurnal as well as seasonal variations. These two processes ideally operate on opposite sides of an oxic-anoxic boundary, and in the case of shallow estuaries with adequate light penetration to the bottom, can influence benthic autotrophic production. Because of changes in the oxygen profile of these systems, diurnal cycles of denitrification and nitrification may be manifested. Seasonal variations in denitrification, most often evident in mid-latitude estuaries, seem to be more strongly controlled by nitrate availability. Factors other than oxygen and nitrate availability, such as temperature and rate of sedimentation, also influence denitrification in estuarine sediments.[152]

A number of workers have disclosed the small quantitative effect of denitrification on the overall mineralization of organic matter. Valiela[164] found that less than 1% of readily available carbon in a New England salt marsh was consumed by denitrification. Only 2% of the carbon in sediments of the coastal waters of Denmark, according to Jørgensen,[248] was oxidized by this process. The concentration of oxidized nitrogen tends to govern dissimilatory nitrogenous oxide reduction in anaerobic sediments,[164] and since the concentration of oxidized nitrogen may be small, this process is commonly less significant than the mineralization of organic matter via sulfate reduction.

Research on the relative magnitude of the rates of oxygen uptake, denitrification, nitrate reduction to ammonia, and sulfate reduction in Norsminde Fjord, Denmark illustrates a rather large quantitative effect of oxygen uptake and sulfate reduction on the overall mineralization of organic matter.[152] Near the estuarine mouth, oxygen uptake and sulfate reduction predominate in bottom sediments, accounting for 65 and 27% of the total electron flow, respectively, with denitrification and nitrate reduction being responsible for only 8% of the total electron flow. Approaching the head of the estuary, however, nitrate reduction comprises a greater portion of the relative contributions, equaling 33% of the total electron flow compared to 44% for oxygen uptake, 19% for sulfate reduction, and 4% for denitrification.

The concentration of inorganic nutrients can be limiting to dissimilatory processes for certain

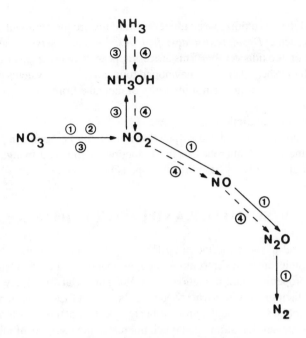

FIGURE 7. Pathways of dissimilatory nitrogenous oxide reduction. (1) Denitrification; (2) dissimilatory reduction (terminates at NO_2); (3) dissimilatory ammonia production; (4) "nitrification" N_2O pathway: ammonia to nitrous oxide. (From Wiebe, W. J., Christian, R. R., Hansen, J. A., King, G., Sherr, B., and Skyring, G., *The Ecology of a Salt Marsh*, Pomeroy, L. R. and Wiegert, R. G., Eds., Springer-Verlag, New York, 1981, 137. With permission.)

anaerobic heterotrophs. As indicated above, dissimilatory sulfate reducers require sulfate, and dissimilatory nitrogenous oxide reducers need oxidized nitrogen species (e.g., nitrite and nitrate) as electron acceptors. When the concentrations of these nutrients decline, dissimilatory processes are directly affected, which is unlike the case for aerobic microheterotrophs.[249]

4. Methanogenesis

Another potentially important pathway of anaerobic decomposition in marine sediments is methanogenesis, whereby strict anaerobes use carbon dioxide or a methyl group as an electron acceptor while producing methane.[228] Methanogenic bacteria utilize only a few substrates which typically form as metabolic end products of other anaerobes. Methanogenic reactions yield low energy (Table 7).

Sulfate reduction usually predominates over methanogenesis in sediments, possibly because sulfate reducers outcompete methanogens for substrates.[250-254] Alternatively, the dominance of sulfate reduction may be ascribable to sulfide poisoning[255] or thermodynamic reasons.[256,257] When sulfate levels are high, methane concentrations characteristically remain low. Hence, sulfate respiration generally provides for 100 to 1000 times more carbon oxidation than methanogenesis when sulfate is present.[258] In sediments having low sulfate levels, however, methanogenesis may be favorable and can become quantitatively important.[254,258] Methane production in lake sediments is inhibited by sulfate levels as low as 0.2 mM.[228,255]

Anaerobic muds rich in organic content (e.g., salt marsh sediments and estuarine mudflats) represent excellent locations for methanogenesis.[164,257] Some of the methane generated in anoxic sediments escapes to the atmosphere by means of diffusion through the sulfate-reducing zone and anaerobic layer[258] or via ebullition.[259] Nonetheless, much of the methane produced in the reduced zone of sediment is oxidized as it moves upward toward the sediment surface — some by sulfate reducers such as *Desulfovibrio desulfuricans*[260] — and it serves as an energy source for bacteria capable of its oxidation.[164,258] This process appears to be especially true in seagrasses, mangroves, and salt marshes where well-developed root and rhizome systems promote aerobic respiration of methane, resulting in a minimal loss of methane from the sediment.[164]

5. Iron and Manganese Reduction

Organic matter can also be oxidized by reactions involving Fe^{3+} and Mn^{4+}.[164] Uncertainty exists concerning the role of microbes in the reduction of iron and manganese. Evidence indicates that bacterial activity may be indirectly responsible for the reduction of Fe^{3+} to Fe^{2+} and Mn^{4+} to Mn^{2+}.

V. SUMMARY AND CONCLUSIONS

Bacteria are critical to the estuarine environment as decomposers of organic matter, as transformers of organic substrates into inorganic compounds, and as agents influencing the physical-chemical properties of these shallow coastal systems. Only relatively recently has the role of bacteria and their grazers in marine food webs become a focal point of research efforts. The concept of microbial loops and the view of bacteria as competitors for food and nutrients of other estuarine organisms has stimulated much interest in the interplay of microbial groups, material cycles, and the factors affecting them. The overwhelming significance of bacteria and protozoa in the heterotrophic activities of estuarine and marine ecosystems has precipitated investigations assessing the controls on microheterotrophic processes. Workers continue to evaluate microbial production from different habitats and the significance of this carbon as an energy source that passes up the food chain to microheterotrophs (e.g., flagellates and ciliates) and detritivores.

Marine bacteria may be subdivided according to their habitat, physiology, and ecological roles. Microbial habitats include planktonic, neustonic, epibiotic, benthic, and endobiotic types. Based on physiology, these microbes have been grouped into autotrophic and heterotrophic forms. In terms of ecological roles, marine bacteria are essential to food chains and elemental cycles (e.g., nitrogen and sulfur cycles).

The abundance of bacteria peaks in estuaries and coastal waters and gradually declines into offshore areas and the open ocean. Estimates of bacterial numbers in estuarine waters fall in the range of 10^6 to 10^7 cells per milliliter; in the coastal ocean, in the range of 1 to 3×10^6 cells per milliliter; and in neritic waters of the ocean, in the range of 10^4 to 10^6 cells per milliliter. Similar spatial trends have been reported for bacterial biomass, with highest values occurring in estuarine waters and exceeding 10 µg C per liter. In coastal oceanic waters, biomass measurements commonly are between 5 and 10 µg C per liter and in offshore waters between 1 and 5 µg C per liter. Bacterial biomass may be as high as 50 to 100 µg C per liter in localized areas.

The greatest number of microbes inhabit bottom sediments. Abundances of bacteria in salt marsh and mudflat sediments often surpass 10^9 cells per cubic centimeter. Bacterial counts drop when proceeding from intertidal to subtidal sediments in concert with the diminution in the concentration of organic matter. Cell counts typically decrease with increasing depth below the sediment-water interface within the top 20 cm of the sediment profile. A strong positive correlation exists here between the amount of organic matter and the abundance of bacteria; for example, the density of bacteria rises in deposits containing detritus. A correlation also holds between the number of bacteria and the size of sedimentary grains. Finer particles (clay and silt) have a larger microbial population attached to them than coarser particles (sand), as a consequence of the greater volume-specific surface area among the smaller grain sizes. Only a small portion of a sedimentary particle is actually colonized by the bacteria. Estimates of the density of bacteria on sediment surfaces range from one cell per 0.3 µm² to one cell per 400 µm².

The microbes concentrate in cracks, crevices, and depressions of the grains where they are protected from mechanical damage. Extracellular polysaccharides produced by bacteria in contact with sediment surfaces serve as a food source for deposit feeders.

Various techniques of study have facilitated research on bacterial growth and production. Chief among these methods are the use of epifluorescence microscopy, stains (e.g., acridine orange), and radioisotopic applications. A major breakthrough has been the use of tritiated thymidine incorporation for determining bacterial production in estuarine and marine environments.

Bacteria are the main processors of organic matter in aerobic and anaerobic environments. Together with fungi, bacteria decompose substantial quantities of organic matter in aerobic zones of estuarine sediments. Perhaps equally important are bacteria inhabiting anaerobic zones. These organisms exhibit more specific roles in the transformation of organic matter; while aerobic forms utilize a diversity of natural substrates, the anaerobes use a more restricted group of compounds. On the basis of their specialized biochemistry, anaerobic forms have been differentiated into four broad types: (1) fermenting bacteria, (2) dissimilatory sulfate-reducing bacteria, (3) dissimilatory nitrogenous oxide-reducing bacteria, and (4) methanogenic bacteria. Anaerobic microbial metabolism is, in some respects, enigmatic and far more complex than heterotrophic aerobic microbial metabolism which principally involves the transformation of dissolved or particulate organic matter to biomass and carbon dioxide.

Aerobic heterotrophic bacteria can be found in three distinct habitats, that is, in the water column either suspended or attached to detrital particles, in the top layer of seafloor sediments, and in living and dead tissues of plants and animals. Free-living bacteria rely on dissolved organic carbon as an energy source whereas attached bacteria not only use dissolved organic carbon as an energy source, but also the attached particle as a substrate and habitat. Fundamental differences between free-living and attached bacteria, such as the inequality in cell size and rate of assimilation of organic substrates, have been established. Metabolic activities — for example, nutrient uptake rates, utilization, and release — probably differ as well. The abundance of the attached microbes fluctuates markedly with the amount of particulate organic carbon; however, the number of free-living bacteria is more constant. As the density of attached bacteria rises in the presence of detritus, microbial metabolism seems to be altered.

Bacterioplankton play a vital role in the trophodynamic pathways of pelagic food webs. Of the total phytoplankton production converted to DOM, 10 to 50% is consumed by bacterioplankton. Ingestion of the bacteria by phagotrophic protists which are consumed, in turn, by microzooplankton or larger zooplankton creates a microbial loop that salvages much of the DOM typically lost through remineralization. Microflagellates constitute the principal grazers of bacteria in the water column. The importance of this trophic link in the food chains of estuaries remains uncertain.

Bacteria are essential links in detritus-based food webs, too, converting POM into DOM to meet their nutritional requirements and serving as a source of food for microfauna (e.g., ciliates and flagellates) and macrofauna (e.g., detritivores). Detritivores depend on microbial enrichment of the more refractory detritus prior to assimilation, but may be capable of directly assimilating the more readily decomposable detritus (e.g., macroalgae) with little or no microbial enrichment. Most detritus ingested by detritus feeders in estuaries is voided relatively unchanged, albeit free of attached bacteria. The detritus may be consumed repeatedly by detritivores, therefore, with the attached bacteria being continually removed from the detritus for nutrition. Although multiple investigations of detritus feeders in estuaries corroborate the significance of attached bacteria as a food source, recent evidence indicates that many deposit feeders obtain at least part of their energy from the detritus itself because the biomass of the bacteria consumed is thought to be insufficient to support the larger bacterivores.

A successional sequence of organisms colonizes detrital particles; bacteria probably act as the initial colonizers followed by fungi, algae, protozoans, other microfauna, and larger grazers.

Grazing on bacteria serves to stimulate their growth, promoting detrital mineralization. The chemical composition of the detritus also changes through time, becoming higher in nitrogen content with the increase in microbial biomass. Extracellular excretion products (mucopolysaccharides) are responsible for much of the protein enrichment of detritus as it ages.

During the decomposition of POM in estuaries, three processes are discernible: (1) leaching of soluble compounds, (2) microbial decay, and (3) heterotrophic consumption. An initial leaching phase during decomposition removes either soluble or autolyzed substances, and it is rather short-lived, lasting for several minutes to a few weeks depending on the kind of detritus. A microbial decay phase, mediated via hydrolysis by enzymes released from microbial cells, occurs subsequent to the leaching phase. Compounds more resistant to decay (e.g., cellulose and lignin) prevail as decomposition proceeds, while the labile components, such as certain proteins and sugars, are depleted. The detritus eventually becomes more recalcitrant to decay, being composed of fulvic and humic acids consisting of refractory phenolic polymers and complexes that form marine humus which concentrates in estuarine bottom sediments. Heterotrophic consumption of bacteria takes place at any time during the process. Mechanical fragmentation of the detritus by physical and biological activity reduces the size of the particulate material, exposing new surface area to microbial attack.

Similar to bacterioplankton, attached bacteria utilize DOM in the water column that originates from allochthonous (e.g., atmospheric and tributary systems) and autochthonous (e.g., exudation products of plants and animals and microbial decomposition products in the estuary) sources. Research on the uptake of DOM by bacteria has often employed ^{14}C-labeled substrates (e.g., ^{14}C-glucose). Bacteria have been shown to outcompete all other organisms for DOM in marine and estuarine environments. Thus, the flux of DOM in estuaries is closely tied to microbial metabolism. However, other factors contribute to the flux of DOM in these systems as well, such as the conversion of DOM to particulate form during the formation of amorphous organic particles.

Because fungi are rare or inactive in anaerobic environments, bacteria represent the principal transformers of organic matter in anoxic areas. Anaerobic conditions characteristically exist in deeper sediment layers, in systems having very poor water circulation resulting from physical restrictions of stratification (e.g., bottom waters of fjord-type estuaries), and some polluted regions. Much microbial decomposition and mineralization of organic matter occurs in salt marsh and estuarine bottom sediments only a few millimeters or centimeters below the sediment-water interface. The depth to the anaerobic zone in seafloor sediments is a function of physical-chemical properties and biological processes. For example, light intensity at the sediment surface, the degree of turbulence on the seafloor, permeability of the sediment, bioturbation, and content of organic matter affect the thickness of the anaerobic zone.

Anaerobic microbial metabolism yields a number of important substances — biomass, carbon dioxide, methane, soluble low-molecule-weight compounds — potentially assimilable by aerobic organisms. Two pathways of anaerobic decomposition in estuaries include fermentation and dissimilatory sulfate reduction. Fermenting bacteria utilize various substrates, for instance, alcohols, amino acids, cellulose, pectin, purines, and sugars. Hydrogen, carbon dioxide, ammonia, and a group of organic compounds (e.g., alcohols and short-chain fatty acids) are derived from this type of anaerobic metabolism. Sulfate-reducing bacteria use SO_4^{2-} as a terminal electron acceptor during the decomposition of organic matter, with hydrogen sulfide generated in the process. The production of sulfides creates a black coloration in anoxic sediments. Sulfate reduction usually accounts for a significant amount of oxidation of organic matter in estuarine sediments; in some systems, greater than 50% of the oxidation of carbon is attributable to this pathway.

Dissimilatory nitrogenous oxide reduction occurs via several specific pathways: (1) denitrification, (2) dissimilatory reduction (terminating at NO_2), (3) dissimilatory ammonia production, and (4) nitrification N_2O pathway: ammonia to nitrous oxide. Nitrogenous oxides act as terminal electron acceptors. These pathways influence the flux of nitrogen in bottom sediments. They also govern, in part, nitrogen cycling in these complex environments. Dissimilatory nitrogenous oxide reduction pathways generally are responsible for less mineralization of organic matter than oxygen uptake and sulfate reduction.

Methanogenic bacteria differ from other anaerobes by using carbon dioxide or a methyl group as an electron acceptor while supplying methane. The oxidation of organic matter by means of methanogenesis can be quantitatively significant in areas where sulfate concentrations remain low. Anaerobic zones of estuarine mudflats and salt marsh sediments that have high organic concentrations are likely sites for methanogenesis.

REFERENCES

1. **Turner, J. T., Tester, P. A., and Ferguson, R. L.,** The marine cladoceran *Penilia avirostris* and the "microbial loop" of pelagic food webs, *Limnol. Oceanogr.*, 33, 245, 1988.
2. **Pomeroy, L. R.,** Microbial processes in the sea: diversity in nature and science, in *Heterotrophic Activity in the Sea,* Hobbie, J. E. and Williams, P. J. le B., Eds., Plenum Press, New York, 1984, 1.
3. **Dawes, C. J.,** *Marine Botany*, John Wiley & Sons, New York, 1981.
4. **Coffin, R. B. and Sharp, J. H.,** Microbial trophodynamics in the Delaware estuary, *Mar. Ecol. Prog. Ser.*, 41, 253, 1987.
5. **Herndl, G. J., Faganeli, J., Fanuko, N., Peduzzi, P., and Turk, V.,** Role of bacteria in the carbon and nitrogen flow between water column and sediment in a shallow marine bay (Bay of Piran, Northern Adriatic Sea), *Mar. Ecol.*, 8, 221, 1987.
6. **Kiene, R. P. and Capone, D. G.,** Microbial transformations of methylated sulfur compounds in anoxic salt marsh sediments, *Microb. Ecol.*, 15, 275, 1988.
7. **Pomeroy, L. R.,** The ocean's food web, a changing paradigm, *BioScience*, 24, 499, 1974.
8. **Jannasch, H. W. and Williams, P. J. le B., Eds.,** *Advances in Aquatic Microbiology*, Vol. 3, Academic Press, Orlando, FL, 1985.
9. **Taylor, G. T., Iturriaga, R., and Sullivan, C. W.,** Interactions of bacterivorous grazers and heterotrophic bacteria with dissolved organic matter, *Mar. Ecol. Prog. Ser.*, 23, 129, 1985.
10. **Biddanda, B. A. and Pomeroy, L. R.,** Microbial aggregation and degradation of phytoplankton-derived detritus in seawater. I. Microbial succession, *Mar. Ecol. Prog. Ser.*, 42, 79, 1988.
11. **Fenchel, T.,** The ecology of heterotrophic microflagellates, *Adv. Microb. Ecol.*, 9, 57, 1986.
12. **Sherr, E. B., Sherr, B. F., Fallon, R. D., and Newell, S. Y.,** Small, aloricate ciliates as a major component of the marine heterotrophic nanoplankton, *Limnol. Oceanogr.*, 31, 177, 1986.
13. **Sherr, E. B., Sherr, B. F., and Paffenhöfer, G.-A.,** Phagotrophic protozoa as food for metazoans: a "missing" trophic link in marine pelagic food webs?, *Mar. Microb. Ecol.*, 1, 61, 1986.
14. **Michaels, A. F. and Silver, M. W.,** Primary production, sinking fluxes and the microbial food web, *Deep Sea Res. Part A*, 35, 473, 1988.
15. **Kemp, P. F.,** The fate of benthic bacterial production, *Rev. Aquat. Sci.*, Vol 2, Issue 1, 1990.
16. **Azam, F., Fenchel, T., Field, J. G., Gray, J. S., Meyer-Reil, L.-A., and Thingstad, T. F.,** The ecological role of water column microbes in the sea, *Mar. Ecol. Prog. Ser.*, 19, 257, 1983.
17. **Alongi, D. M.,** Microbial-meiofaunal interrelationships in some tropical intertidal sediments, *J. Mar. Res.*, 46, 349, 1988.
18. **Nygaard, K., Borsheim, K. Y., and Thingstad, T. F.,** Grazing rates on bacteria by marine heterotrophic microflagellates compared to uptake rates of bacterial-sized monodisperse fluorescent latex beads, *Mar. Ecol. Prog. Ser.*, 44, 159, 1988.
19. **Wolff, W. J.,** Biotic aspects of the chemistry of estuaries, in *Chemistry and Biogeochemistry of Estuaries*, Olausson, E. and Cato, I., Eds., John Wiley & Sons, Chichester, 1980, 263.
20. **Howarth, R. W. and Teal, J. M.,** Sulfate reduction in a New England salt marsh, *Limnol. Oceanogr.*, 24, 999, 1979.
21. **Ayyakkannu, K. and Chandramohan, D.,** Occurrence and distribution of phosphate and solubilizing bacteria and phosphatase in marine sediments at Porto Novo, *Mar. Biol.*, 2, 201, 1971.
22. **Holm-Hansen, O. and Booth, C. R.,** The measurement of adenosine triphosphate in the ocean and its ecological significance, *Limnol. Oceanogr.*, 11, 510, 1966.

23. **Holm-Hansen, O.,** Determination of total microbial biomass by measurement of adenosine triphosphate, in *Estuarine Microbial Ecology*, Stevenson, H. L. and Colwell, R. R., Eds., University of South Carolina Press, Columbia, 1973, 73.

24. **Stevenson, L. H. and Chrzanowski, T. H.,** Marine bacteria, in *Functional Adaptations of Marine Organisms*, Vernberg, F. J. and Vernberg, W. B., Eds., Academic Press, New York, 1981, 71.

25. **Hobbie, J. E., Daley, R. J., and Jasper, S.,** Use of nucleopore filters for counting bacteria by fluorescence microscopy, *Appl. Environ. Microbiol.*, 33, 1225, 1977.

26. **Porter, K. G. and Feig, Y. S.,** The use of DAP1 for identifying and counting aquatic microflora, *Limnol. Oceanogr.*, 25, 943, 1980.

27. **Christian, R. R. and Hall, J. R.,** Experimental trends in sediment microbial heterotrophy: radioisotopic techniques and analysis, in *Ecology of Marine Benthos*, Coull, B. C., Ed., University of South Carolina Press, Columbia, 1977, 67.

28. **Moriarty, D. J. W.,** Measurement of microbial growth rates in aquatic systems using rates of nucleic acid synthesis, *Adv. Microb. Ecol.*, 9, 245, 1986.

29. **Azam, F. and Fuhrman, J. A.,** Measurement of bacterioplankton growth in the sea and its regulation by environmental conditions, in *Heterotrophic Activity in the Sea*, Hobbie, J. E. and Williams, P. J. le B., Eds., Plenum Press, New York, 1984, 179.

30. **Karl, D. M. and Winn, C. D.,** Adenine metabolism and nucleic acid synthesis: applications to microbiological oceanography, in *Heterotrophic Activity in the Sea*, Hobbie, J. E. and Williams, P. J. le B., Eds., Plenum Press, New York, 1984, 197.

31. **Sieburth, J. M.,** *Sea Microbes*, Oxford University Press, New York, 1979.

32. **Dietz, A. S., Albright, L. J., and Tuominen, T.,** Heterotrophic activites of bacterioneuston and bacterioplankton, *Can. J. Microbiol.*, 22, 1699, 1976.

33. **Haymon, R. M. and Macdonald, K. C.,** The geology of deep-sea hot springs, *Am. Sci.*, 73, 441, 1985.

34. **Wood, E. J. F.,** *Microbiology of Oceans and Estuaries*, Elsevier, Amsterdam, 1967.

35. **Baumann, P., Baumann, L., and Mandel, M.,** Taxonomy of marine bacteria: the genus *Beneckea*, *J. Bacteriol.*, 107, 268, 1971.

36. **Baumann, L., Baumann, P., Mandel, M., and Allen, R. D.,** Taxonomy of aerobic marine *Eubacteria*, *J. Bacteriol.*, 110, 402, 1972.

37. **Reicholt, J. L. and Baumann, P.,** Taxonomy of the marine luminous bacteria, *Arch. Mikrobiol.*, 94, 283, 1973.

38. **Buchannan, R. E. and Gibbons, N. E., Eds.,** *Bergey's Manual of Determinative Bacteriology*, 8th ed., Williams & Wilkins, Baltimore, 1974.

39. **Stanley, J. T., Ed.,** *Bergey's Manual of Systemic Bacteriology*, Williams & Wilkins, Baltimore, 1989.

40. **Velji, M. I. and Albright, L. J.,** Microscopic enumeration of attached bacteria of seawater, marine sediment, fecal matter, and kelp blade samples following pyrophosphate and ultrasound treatments, *Can. J. Microbiol.*, 32, 121, 1986.

41. **Sieburth, J. M.,** Bacterial substrates and productivity in marine ecosystems, *Annu. Rev. Ecol. Syst.*, 7, 259, 1976.

42. **Morita, R. Y.,** The role of microorganisms in the environment, in *Oceanic Sound Scattering Prediction*, Anderson, N. R. and Zahuranec, B., Eds., Plenum Press, New York, 1977, 445.

43. **Fenchel, T.,** Suspended marine bacteria as a food source, in *Flows of Energy and Materials in Marine Ecosystems: Theory and Practice*, Fasham, M. J. R., Ed., Plenum Press, New York, 1984, 301.

44. **Fenchel, T.,** Ecology of heterotrophic microflagellates. IV. Quantitative occurrence and importance as consumers of bacteria, *Mar. Ecol. Prog. Ser.*, 9, 35, 1982.

45. **Palumbo, A. V. and Ferguson, R. L.,** Distribution of suspended bacteria in the Newport River estuary, North Carolina, *Estuarine Coastal Mar. Sci.*, 7, 521, 1978.

46. **Rublee, P. A.,** Bacteria and microbial distribution in estuarine sediments, in *Estuarine Comparisons*, Kennedy, V. S., Ed., Academic Press, New York, 1982, 159.

47. **Williams, P. J. le B.,** Bacterial production in the marine food chain: the emperor's new suit of clothes?, in *Flows of Energy and Materials in Marine Ecosystems: Theory and Practice*, Fasham, M. J. R., Ed., Plenum Press, New York, 1984, 271.

48. **Azam, F. and Hodson, R. E.,** Size distribution and activity of marine microheterotrophs, *Limnol. Oceanogr.*, 22, 492, 1977.

49. **Watson, S. W. and Hobbie, J. E.,** Measurement of bacterial biomass as lipopolysaccharide, in *Native Aquatic Bacteria: Enumeration, Activity and Ecology*, Costerton, J. W. and Colwell, R. R., Eds., American Society for Testing Materials, Philadelphia, 1979, 82.

50. **Sleigh, M. A., Ed.,** *Microbes in the Sea*, Ellis Horwood, Chichester, 1987.

51. **Wiebe, W. J.,** Physiological and biochemical aspects of marine bacteria, in *Heterotrophic Activity in the Sea*, Hobbie, J. E. and Williams, P. J. le B., Eds., Plenum Press, New York, 1984, 55.

52. **Wiebe, W. J. and Pomeroy, L. R.,** Microorganisms and their association with aggregates and detritus in the sea: a microscopic study, *Mem. Inst. Ital. Idrobiol.*, 29(Suppl.), 325, 1972.

53. **Hanson, R. B. and Wiebe, W. J.,** Heterotrophic activity associated with particulate size fractions in a *Spartina alterniflora* salt-marsh estuary, Sapelo Island, Georgia, U.S.A. and the continental shelf waters, Mar. Biol., 42, 321, 1977.

54. **Goulder, R.,** Attached and free bacteria in an estuary with abundant suspended solids, *J. Appl. Bacteriol.*, 43, 399, 1977.

55. **Sieburth, J. M.,** Protozoa bacterivory in pelagic marine waters, in *Heterotrophic Activity in the Sea*, Hobbie, J. E. and Williams, P. J. le B., Eds., Plenum Press, New York, 1984, 405.

56. **Bell, C. R. and Albright, L. J.,** Attached and free-floating bacteria in the Frazer River estuary, British Columbia, Canada, *Mar. Ecol. Prog. Ser.*, 6, 317, 1981.

57. **Sieburth, J. M. and Davis, P. G.,** The role of heterotrophic nanoplankton in the grazing and nurturing of planktonic bacteria in the Sargasso and Caribbean Seas, *Ann. Inst. Oceanogr.*, (Paris), 58, 285, 1982.

58. **Darnell, R. M.,** Organic detritus in relation to the estuarine ecosystem, in *Estuaries*, Lauff, G. H., Ed., Publ. 83, American Association for the Advancement of Science, Washington, D.C., 1967, 376.

59. **Odum, E. P. and de la Cruz, A. A.,** Particulate organic detritus in a Georgia salt marsh estuarine ecosystem, in *Estuaries*, Lauff, G. H., Ed., Publ. 83, American Association for the Advancement of Science, Washington, D.C., 1967, 383.

60. **Sieburth, J. M.,** *Microbial Seascapes*, University Park Press, Baltimore, 1975.

61. **Christian, R. R. and Wetzel, R. L.,** Interactions between substrate, microbes, and consumers of *Spartina* detritus in estuaries, in *Estuarine Interactions*, Wiley, M. L., Ed., Academic Press, New York, 1978, 93.

62. **Wetzel, R. L. and Christian, R. R.,** Model studies on the interactions among carbon substrates, bacteria and consumers in a salt marsh estuary, *Bull. Mar. Sci.*, 35, 601, 1984.

63. **Goulder, R., Bent, E. J., and Boak, A. C.,** Attachment to suspended solids as a strategy of estuarine bacteria, in *Feeding and Survival Strategies of Estuarine Organisms*, Jones, N. V. and Wolff, W. J., Eds., Plenum Press, New York, 1981, 1.

64. **Christian, R. R., Hanson, R. B., Hall, J. R., and Wiebe, W. J.,** Aerobic microbes and meiofauna, in *The Ecology of a Salt Marsh*, Pomeroy, L. R. and Wiegert, R. G., Eds., Springer-Verlag, New York, 1981, 113.

65. **Fenchel, T.,** Aspects of the decomposition of seagrasses, in *Seagrass Ecosystems: A Scientific Perspective*, McRoy, C. P. and Helfferich, C., Eds., Marcel Dekker, New York, 1977, 123.

66. **Marsh, D. H. and Odum, W. E.,** Effect of suspension and sedimentation on the amount of microbial colonization of salt marsh microdetritus, *Estuaries*, 2, 184, 1979.

67. **Rublee, P. and Dornseif, B. E.,** Direct counts of bacteria in the sediments of a North Carolina salt marsh, *Estuaries*, 1, 188, 1978.

68. **Cammen, L. M. and Walker, J. A.,** The relationship between bacteria and micro-algae in the sediment of a Bay of Fundy mudflat, *Estuarine Coastal Shelf Sci.*, 22, 91, 1986.

69. **Dale, N.,** Bacteria in intertidal sediments: factors related to their distribution, *Limnol. Oceanogr.*, 19, 509, 1974.

70. **Christian, R. R., Bancroft, K., and Wiebe, W. J.,** Distribution of adenosine triphosphate in salt marsh sediments at Sapelo Island, Georgia, *Soil Sci.*, 119, 89, 1975.

71. **Meyer-Reil, L.-A.,** Bacterial biomass and heterotrophic activity in sediments and overlying waters, in *Heterotrophic Activity in the Sea*, Hobbie, J. E. and Williams, P. J. le B., Eds., Plenum Press, New York, 1984, 523.

72. **Rublee, P. A.,** Seasonal distribution of bacteria in salt marsh sediments in North Carolina, *Estuarine Coastal Shelf Sci.*, 15, 67, 1982.

73. **Hargrave, B. T.,** Aerobic decomposition of sediment and detritus as a function of particle surface area and organic content, *Limnol. Oceanogr.*, 17, 583, 1972.

74. **Yamamoto, N. and Lopez, G.,** Bacterial abundance in relation to surface area and organic content of marine sediments, *J. Exp. Mar. Biol. Ecol.*, 90, 209, 1985.

75. **DeFlaun, M. B. and Mayer, L. M.,** Relationships between bacteria and grain surfaces in intertidal sediments, *Limnol. Oceanogr.*, 28, 873, 1983.

76. **Levinton, J. S.,** *Marine Ecology*, Prentice-Hall, Englewood Cliffs, NJ, 1982.

77. **Jannasch, H. W. and Pritchard, P. H.,** The role of inert particulate matter in the activity of aquatic microorganisms, *Mem. Inst. Ital. Idrobiol.*, 29(Suppl), 289, 1972.

78. **Meadows, P. S. and Anderson, J. G.,** Microorganisms attached to marine and freshwater sand grains, *Nature*, 212, 1059, 1966.

79. **Weise, W. and Rheinheimer, G.,** Scanning electron microscopy and epifluorescence investigation of bacterial colonization of marine sand sediments, *Microb. Ecol.*, 4, 175, 1978.

80. **Cammen, L. M.,** Effect of particle size on organic content and microbial abundance within four marine sediments, *Mar. Ecol. Prog. Ser.*, 9, 273, 1982.

81. **Hobbie, J. E. and Lee, C.,** Microbial production of extracellular material: importance in benthic ecology, in *Marine Benthic Dynamics*, Tenore, K. R. and Coull, B. C., Eds., University of South Carolina Press, Columbia, 1980, 341.

82. **Colwell, R. R. and Morita, R. Y., Eds.,** *Effects of the Ocean Environment on Microbial Activities*, University Park Press, Baltimore, 1974.

83. **Keith, S. M., Russ, M. A., Macfarlane, G. T., and Herbert, R. A.,** The ecology and physiology of anaerobic isolated from Tay estuary sediments, *Proc. R. Soc. Edinburgh Sect. B*, 92, 323, 1987.

84. **Aller, R. C. and Rude, P. D.,** Complete oxidation of solid phase studies by manganese and bacteria in anoxic marine sediments, *Geochim. Cosmochim. Acta*, 52, 751, 1988.

85. **Meyer-Reil, L.-A., Bölter, M., Dawson, R., Liebezeit, G., Szwerinski, H., and Wolter, K.,** Interrelationships between microbiological and chemical parameters of sandy beach sediments, a summer aspect, *Environ. Microbiol.*, 39, 797, 1985.

86. **Bratbak, G.,** Bacterial biovolume and biomass estimations, *Appl. Environ. Microbiol.*, 49, 1488, 1985.

87. **Nagata, T.,** Carbon and nitrogen content of natural planktonic bacteria, *Appl. Environ. Microbiol.*, 52, 28, 1986.

88. **Bratbak, G. and Dundas, I.,** Bacterial dry matter content and biomass estimations, *Appl. Environ. Micrbiol.*, 48, 755, 1984.

89. **McLachlan, A.,** The biomass of macro- and interstitial fauna on clean and wrack-covered beaches in western Australia, *Estuarine Coastal Shelf Sci.*, 21, 587, 1985.

90. **Alongi, D. M.,** The role of soft-bottom benthic communities in tropical mangrove and coral reef ecosystems, *Rev. Aquat. Sci.*, 1, 243, 1989.

91. **Karl, D. M.,** Determination in *in situ* microbial biomass, viability, metabolism and growth, in *Bacteria in Nature*, Poindexter, J. S. and Leadbetter, E. R., Eds., Plenum Press, New York, 1986, chap. 3.

92. **Karl, D. M.,** Selected nucleic acid precursors in studies of aquatic microbial ecology, *Appl. Environ. Microbiol.*, 44, 891, 1982.

93. **Moriarty, D. J. W. and Pollard, P. C.,** DNA synthesis as a measure of bacterial productivity in seagrass sediments, *Mar. Ecol. Prog. Ser.*, 5, 151, 1981.

94. **Fuhrman, J. A. and Azam, F.,** Bacterioplankton secondary production estimates for coastal waters of British Columbia, Antarctica, and California, *Appl. Environ. Microbiol.*, 39, 1085, 1980.

95. **Karl, D. M.,** Simultaneous rates of RNA and DNA synthesis for estimating growth and cell division of aquatic microbial communitites, *Appl. Environ. Microbiol.*, 42, 802, 1981.

96. **Karl, D. M., Winn, C. D., and Wong, D. C. L.,** RNA synthesis as a measure of microbial growth in aquatic environments. I. Evaluation, verification and optimization of methods, *Mar. Biol.*, 64, 1, 1981.

97. **Kirchman, D. L. and Hoch, M. P.,** Bacterial production in the Delaware Bay estuary estimated from thymidine and leucine incorporation rates, *Mar. Ecol. Prog. Ser.*, 45, 169, 1988.

98. **Fuhrman, J. A. and Azam, F.,** Thymidine incorporation as a measure of heterotrophic bacterioplankton production in marine surface waters: evaluation and field results, *Mar. Biol.*, 66, 109, 1982.

99. **Pollard, P. C. and Moriarty, D. J. W.,** Validity of the tritiated thymidine method of estimating bacterial growth rates: measurement of isotope dilution during DNA synthesis, *Appl. Environ. Microbiol.*, 48, 1076, 1984.

100. **Riemann, B.,** Determining growth rates of natural assemblages of freshwater bacteria by means of H-thymidine incorporation into DNA: comments on methodology, *Arch. Hydrobiol. Beih. Ergebn. Limnol.*, 19, 67, 1984.

101. **Moriarty, D. J. W., Pollard, P. C., Hunt, W. G., Moriarty, C. M., and Wassenberg, T. J.,** Productivity of bacteria and microalgae and the effect of grazing by holothurians in sediments on a coral reef flat, *Mar. Biol.*, 85, 293, 1985.

102. **Newell, S. Y. and Fallon, R. D.,** Bacterial productivity in the water column and sediments of the Georgia (USA) coastal zone: estimates via direct counting and parallel measurement of thymidine incorporation, *Microb. Ecol.*, 8, 33, 1982.

103. **Hagström, Å.,** Aquatic bacteria: measurements and growth, in *Current Perspectives in Microbial Ecology*, Klug, M. J. and Reddy, C. A., Eds., American Society of Microbiology, Washington, D.C., 1984, 95.

104. **Riemann, B. and Sondergaard, M.,** Measurements of diel rates of bacterial production in aquatic environments, *Appl. Environ. Microbiol.*, 47, 632, 1984.

105. **Riemann, B., Nielsen, P., Jeppesen, M., Marcussen, B., and Fuhrman, J. A.,** Diel changes in bacterial biomass and growth rates in coastal environments, determined by means of thymidine incorporation into DNA, frequency of dividing cells (FDC), and microautoradiography, *Mar. Ecol. Prog. Ser.*, 17, 227, 1984.

106. **Fallon, R. D., Newell, S. Y., and Hopkinson, C. S.,** Bacterial production in marine sediments: will cell-specific measures agree with whole-system metabolism?, *Mar. Ecol. Prog. Ser.*, 11, 119, 1983.

107. **Ducklow, H. W.,** Chesapeake Bay nutrient and plankton dynamics. I. Bacterial biomass and production during spring tidal destratification in the York River, Virginia, estuary, *Limnol. Oceanogr.*, 27, 651, 1982.

108. **Moriarty, D. J. W. and Pollard, P. C.,** Diel variation of bacterial productivity in seagrass (*Zostera capricorni*) beds measured by rate of thymidine incorporation into DNA, *Mar. Biol.*, 72, 165, 1982.

109. **Malone, T. C., Kemp, W. M., Ducklow, H. W., Boynton, W. R., Tuttle, J. H., and Jonas, R. B.,** Lateral variation in the production and fate of phytoplankton in a partially stratified estuary, *Mar. Ecol. Prog. Ser.*, 32, 149, 1986.

110. **Sellner, K. G.,** Phytoplankton in Chesapeake Bay: role in carbon, oxygen and nutrient dynamics, in *Contaminant Problems and Management of Living Chesapeake Bay Resources*, Majumdar, S. K., Hall, L. W., Jr., and Austin, H. M., Eds., Pennsylvania Academy of Science, Easton, 1987, 134.

111. **Wolff, W. J.,** Estuarine benthos, in *Estuaries and Enclosed Seas*, Ketchum, B. H., Ed., Elsevier, Amsterdam, 1983, 151.
112. **Alongi, D. M.,** Bacterial productivity and microbial biomass in tropical mangrove sediments, *Microb. Ecol.*, 13, 15, 1987.
113. **Stanley, S. O., Boto, K. G., Alongi, D. M., and Gillan, F. T.,** Composition and bacterial utilization of free amino acids in tropical mangrove sediments, *Mar. Chem.*, 22, 13, 1987.
114. **Alongi, D. M.,** Microbial-meiofaunal interrelationships in some tropical intertidal sediments, *J. Mar. Res.*, 46, 349, 1988.
115. **Kemp, P. F.,** Potential impact on bacteria of grazing by a macrofaunal deposit-feeder, and the fate of bacterial production, *Mar. Ecol. Prog. Ser.*, 36, 151, 1987.
116. **Alongi, D. M.,** Detritus in coral reef ecosystems: composition, fluxes and fates, in Proc. 6th Int. Coral Reef Symp., 1988, 1.
117. **Pomeroy, L. R. and Wiebe, W. J.,** Energetics of microbial food webs, *Hydrobiologia*, 159, 7, 1988.
118. **Sherr, B. F., Sherr, E. B., and Hopkinson, C. S.,** Trophic interactions within pelagic microbial communities: interactions of feedback regulation of carbon flow, *Hydrobiologia*, 159, 19, 1988.
119. **Mann, K. H.,** *Ecology of Coastal Waters: A Systems Approach*, University of California Press, Berkeley, 1982.
120. **Tenore, K. R. and Rice, D. L.,** A review of trophic factors affecting secondary production of deposit-feeders, in *Marine Benthic Dynamics*, Tenore, K. R. and Coull, B. C., Eds., University of South Carolina Press, Columbia, 1980, 325.
121. **Williams, P. J. le B.,** Incorporation of microheterotrophic processes into the classical paradigm of the plankton, *Kiel. Meeresforsch.*, 5, 1, 1981.
122. **Biddanda, B. A.,** Microbial aggregation and degradation of phytoplankton-derived detritus in seawater. II. Microbial metabolism, *Mar. Ecol. Prog. Ser.,* 42, 89, 1988.
123. **Larsson, U. and Hagström, Å.,** Phytoplankton extracellular release as an energy source for bacterial growth in a pelagic ecosystem, *Mar. Biol.*, 52, 199, 1979.
124. **Newell, R. C., Linley, E. A. S., and Lucas, M. I.,** Bacterial production and carbon conversion based on saltmarsh plant debris, *Estuarine Coastal Shelf Sci.*, 17, 405, 1983.
125. **Lancelot, C. and Billen, G.,** Activity of heterotrophic bacteria and its coupling to primary production during the spring phytoplankton bloom in the southern bight of the North Sea, *Limnol. Oceanogr.*, 29, 721, 1984.
126. **Newell, R. C.,** The energetics of detritus utilization in coastal lagoons and nearshore waters, *Oceanol. Acta*, 32, 347, 1982.
127. **Fenchel, T. and Blackburn, T. H.,** *Bacteria and Mineral Cycling*, Academic Press, London, 1979.
128. **Hobbie, J. E. and Williams, P. J. le B., Eds.,** *Heterotrophic Activity in the Sea*, Plenum Press, New York, 1984.
129. **Pomeroy, L. R., Bancroft, K., Breed, J., Christian, R. R., Frankberg, D., Hall, J. R., Maurer, L. G., Wiebe, W. J., Wiegert, R. G., and Wetzel, R. L.,** Flux of organic matter through a salt marsh, in *Estuarine Processes*, Vol. 2, Wiley, M., Ed., Academic Press, New York, 1977, 270.
130. **Pedros-Alio, C. and Brock, T. D.,** The importance of attachment to particles for planktonic bacteria, *Arch. Hydrobiol.*, 98, 354, 1983.
131. **Schoenberg, S. A. and Maccubbin, A. E.,** Relative feeding rates on free and particle-bound bacteria by freshwater macrozooplankton, *Limnol. Oceanogr.*, 30, 1084, 1985.
132. **Joint, J. R. and Pomroy, A. J.,** Aspects of microbial heterotrophic production in a highly turbid estuary, *J. Exp. Mar. Biol. Ecol.*, 58, 33, 1982.
133. **Owen, N. J. P.,** Variations in the natural abundance of ^{15}N in estuarine suspended particulate matter: a specific indicator of biological processing, *Estuarine Coastal Shelf Sci.*, 20, 505, 1985.
134. **Owens, N. J. P.,** Estuarine nitrification: a naturally occurring fluidised bed reaction?, *Estuarine Coastal Shelf Sci.*, 22, 31, 1985.
135. **Plummer, D. H., Owens, N. J. P., and Herbert, R. A.,** Bacteria-particle interactions in turbid estuarine environments, *Cont. Shelf Res.*, 7, 1429, 1987.
136. **Bent, E. J. and Goulder, R.,** Planktonic bacteria in the Humber estuary: seasonal variations in population density and heterotrophic activity, *Mar. Biol.*, 62, 35, 1981.
137. **Clarke, K. R. and Joint, I. R.,** Methodology for estimating numbers of free-living and attached bacteria in estuarine water, *Appl. Environ. Microbiol.*, 51, 1110, 1986.
138. **Fukami, K., Simidu, U., and Taga, N.,** Fluctuation of the communities of heterotrophic bacteria during the decomposition process of phytoplankton, *J. Exp. Mar. Biol. Ecol.*, 55, 171, 1981.
139. **Ferguson, R. L. and Rublee, P.,** Contribution of bacteria to standing crop of coastal plankton, *Limnol. Oceanogr.*, 21, 141, 1976.
140. **Hodson, R. E., Maccubbin, A. E., and Pomeroy, L. R.,** Dissolved adenosine triphosphate utilization by free-living and attached bacterioplankton, *Mar. Biol.*, 64, 43, 1981.
141. **Paerl, H. W. and Merkel, S. M.,** Differential phosphorus assimilation in attached vs. unattached microorganisms, *Arch. Hydrobiol.*, 93, 125, 1982.
142. **Paerl, H. W.,** Alteration of microbial metabolic activities in association with detritus, *Bull. Mar. Sci.*, 35, 393, 1984.

143. **Wangersky, P. J.,** The role of particulate matter in the productivity of surface waters, *Helgol. Wiss. Meeresunters.*, 30, 546, 1977.

144. **Stevenson, L. H.,** A case for bacterial dormancy in aquatic systems, *Microb. Ecol.*, 14, 127, 1978.

145. **Ammerman, J. W., Fuhrman, J. A., Hagström, Å., and Azam, F.,** Bacterioplankton growth in seawater. I. Growth kinetics and cellular characteristics in seawater cultures, *Mar. Ecol. Prog. Ser.*, 18, 31, 1984.

146. **Fukami, K., Simidu, U., and Taga, N.,** Change in a bacterial population during the process of degradation of a phytoplankton bloom in a brackish lake, *Mar. Biol.*, 76, 253, 1983.

147. **Fukami, K., Simidu, U., and Taga, N.,** Microbial decomposition of phyto- and zooplankton in seawater. I. Changes in organic matter, *Mar. Ecol. Prog. Ser.*, 21, 1, 1985.

148. **Goulder, R.,** Relationships between suspended solids and standing crops and activites of bacteria in an estuary during a neap-spring-neap tidal cycle, *Oecologia*, 24, 83, 1976.

149. **Fukami, K., Simidu, U., and Taga, N.,** Distribution of heterotrophic bacteria in relation to the concentration of particulate organic matter in seawater, *Can. J. Microbiol.*, 29, 570, 1983.

150. **Wilson, C. A. and Stevenson, L. H.,** The dynamics of the bacterial population associated with a salt marsh, *J. Exp. Mar. Biol. Ecol.*, 48, 123, 1980.

151. **Fukami, K., Simidu, U., and Taga, N.,** Microbial decomposition of phyto- and zooplankton in seawater. II. Changes in the bacterial community, *Mar. Ecol. Prog. Ser.*, 21, 7, 1985.

152. **Jørgensen, B. B. and Sørensen, J.,** Seasonal cycles of O_2, NO_3^- and SO_4^{2-} reduction in estuarine sediments: the significance of an NO_3^- reduction maximum in spring, *Mar. Ecol. Prog. Ser.*, 24, 65, 1985.

153. **Alongi, D. M. and Hanson, R. B.,** Effect of detritus supply on trophic relationships within experimental benthic food webs. II. Microbial responses, fate and composition of decomposing detritus, *J. Exp. Mar. Biol. Ecol.*, 88, 167, 1985.

154. **Wiebe, W. J. and Liston, J.,** Studies of the aerobic, nonexacting, heterotrophic bacteria of the benthos, in *The Columbia River Estuary and Adjacent Ocean Waters: Bioenvironmental Studies*, Pruter, A. T. and Alverson D. L., Eds., University of Washington Press, Seattle, 1972, 281.

155. **Austin, B.,** *Methods in Acquatic Bacteriology,* John Wiley & Sons, 1988.

156. **Kennish, M. J.,** *Ecology of Estuaries*, Vol. 1, CRC Press, Boca Raton, FL, 1986.

157. **Rice, D. L. and Tenore K. R.,** Dynamics of carbon and nitrogen during the decomposition of detritus derived from estuarine macrophytes, *Estuarine Coastal Shelf Sci.*, 13, 681, 1981.

158. **Tenore, K. R., Cammen, L., Findlay, S. E. G., and Phillips, N.,** Perspectives of research on detritus: do factors controlling the availability of detritus to macroconsumers depend on its source?, *J. Mar. Res.*, 40, 473, 1982.

159. **Blum, L. K., Mills, A. L., Zieman, J. C., and Zieman, R. T.,** Abundance of bacteria and fungi in seagrass and mangrove detritus, *Mar. Ecol. Prog. Ser.*, 42, 73, 1988.

160. **Tenore K. R. and Hanson, R. B.,** Availability of detritus of different types and ages to a polychaete macroconsumer, *Capitella capitata, Limnol. Oceanogr.*, 25, 553, 1980.

161. **Newell, R. C. and Field, J.,** The contribution of bacteria and detritus to carbon and nitrogen flow in a benthic community, *Mar. Biol. Lett.*, 4, 23, 1983.

162. **Bowen, S. H.,** Evidence of a detritus food chain based on consumption of organic precipitates, *Bull. Mar. Sci.*, 35, 440, 1984.

163. **Lopez, G. R. and Levinton, J. S.,** Ecology of deposit-feeding animals in marine sediments, *Q. Rev. Biol.*, 62, 235, 1987.

164. **Valiela, I.,** *Marine Ecological Processes*, Springer-Verlag, New York, 1984.

165. **Hylleberg, J.,** Resource partitioning of basis of hydrolytic enzymes in deposit-feeding mud snails (Hydrobiidae), *Oecologia*, 23, 115, 1976.

166. **Bianchi, T. S. and Levinton, J. S.,** The importance of microalgae, bacteria and particulate organic matter in the somatic growth of *Hydrobia totteni, J. Mar. Res.*, 42, 431, 1984.

167. **Alongi, D. M.,** Effect of physical disturbance on population dynamics and trophic interactions among microbes and meiofauna, *J. Mar. Res.*, 43, 351, 1985.

168. **Singh, J. S. and Gupta, S. R.,** Plant decomposition and soil respiration in terrestrial ecosystems, *Bot. Rev.*, 43, 449, 1977.

169. **Benner, R., Newell, S. Y., Maccubbin, A. E., and Hodson, R. E.,** Relative contributions of bacteria and fungi to rates of degradation of lignocellulosic detritus in salt-marsh sediments, *Appl. Environ. Microbiol.*, 48, 36, 1984.

170. **Benner, R. and Hodson, R. E.,** Microbial degradation of the leachable and lignocellulosic components of leaves and wood from *Rhizophora mangle* in a tropical mangrove swamp, *Mar. Ecol. Prog. Ser.*, 23, 221, 1985.

171. **Crawford, R. L.,** *Lignin Biodegradation and Transformation*, Wiley-Interscience, New York, 1981.

172. **Padgett, D. E., Hackney, C. T., and Sizemore, R. K.,** A technique for distinguishing between bacterial and non-bacterial respiration in decomposing *Spartina alterniflora, Hydrobiologia*, 122, 113, 1985.

173. **Burkholder, P. R. and Bornside, G. H.,** Decomposition of marsh grass by aerobic marine bacteria, *Bull. Torrey Bot. Club*, 84, 366, 1957.

174. **Rublee, P., Cammen, L., and Hobbie, J.,** Bacteria in a North Carolina Salt Marsh: Standing Crop and Importance in the Decomposition of *Spartina alterniflora*, University of North Carolina Sea Grant Publ. UNC-SG-78-11, Raleigh, 1978.

175. **Montagna, P. A. and Ruber, E.,** Decomposition of *Spartina alterniflora* in different seasons and habitats of a northern Massachusetts salt marsh, and a comparison with other Atlantic regions, *Estuaries*, 3, 61, 1980.

176. **May, M. S.,** Probable agents for the formation of detritus from the halophyte *Spartina alterniflora*, in *Ecology of Halophytes*, Reimold, R. J. and Queen, W. H., Eds., Academic Press, New York, 1974, 429.

177. **Lee, C., Howarth, R. W., and Howes, B. L.,** Sterols in decomposing *Spartina alterniflora* and the use of ergosterol in estimating the contribution of fungi to detrital nitrogen, *Limnol. Oceanogr.*, 25, 290, 1980.

178. **Newell, S. Y. and Hicks, R. E.,** Direct count estimates of fungal and bacterial biovolume in dead leaves of smooth cordgrass (*Spartina alterniflora* Loisel), *Estuaries*, 5, 246, 1982.

179. **Lopez, G. R., Levinton, J. S., and Slobodkin, L. B.,** The effect of grazing by the detritivore, *Orchestia grillus* or *Spartina* litter and its associated microbial community, *Oecologia*, 30, 111, 1977.

180. **Morrison, S. J., King, J. D., Bobbie, R. J., Bechtold, R. E., and White, D. C.,** Evidence for microfloral succession on allochthonous plant litter in Apalachicola Bay, Florida, U.S.A., *Mar. Biol.*, 41, 229, 1977.

181. **Day, J. W., Jr., Hall, C. A. S., Kemp, W. M., and Yáñez-Arancibia, A.,** *Estuarine Ecology*, John Wiley & Sons, New York, 1989.

182. **Cagle, G. C.,** Fine structure and distribution of extracellular polymer surrounding selected aerobic bacteria, *Can. J. Microbiol.*, 21, 395, 1975.

183. **Russell-Hunter, W. D.,** *Aquatic Productivity*, Macmillan, London, 1970.

184. **Parker, R. R., Sibert, J., and Brown, T. J.,** Inhibition of primary productivity through heterotrophic competition for nitrate in a stratified estuary, *J. Fish Res. Board Can.*, 32, 72, 1975.

185. **Hagström, Å., Ammerman, J. W., Henrichs, S., and Azam, F.,** Bacterioplankton growth in seawater. II. Organic matter utilization during steady-state growth in seawater cultures, *Mar. Ecol. Prog. Ser.*, 18, 41, 1984.

186. **Andersson, A., Lee, C., Azam, F., and Hagström, Å.,** Release of amino acids and inorganic nutrients by heterotrophic marine microflagellates, *Mar. Ecol. Prog. Ser.*, 23, 99, 1985.

187. **Payne, W. J.,** Energy yields and growth of heterotrophs, *Annu. Rev. Microbiol.*, 24, 17, 1970.

188. **Ho, K. P. and Payne, W. J.,** Assimilation efficiency and energy contents of prototrophic bacteria, *Biotechnol. Bioeng.*, 21, 787, 1979.

189. **Newell, R. C., Lucas, M. I., and Linley, E. A. S.,** Rate of degradation and efficiency of conversion of phytoplankton debris by marine microorganisms, *Mar. Ecol. Prog. Ser.*, 6, 123, 1981.

190. **Fenchel, T. M. and Jørgensen, B. B.,** Detritus food chains of aquatic ecosystems and the role of bacteria, *Adv. Microb. Ecol.*, 1, 1, 1976.

191. **Joint, I. R. and Morris, R. J.,** The role of bacteria in the turnover of organic matter in the sea, *Oceanogr. Mar. Biol. Annu. Rev.*, 20, 65, 1982.

192. **Wolter, R.,** Bacterial incorporation of organic substances released by natural phytoplanton populations, *Mar. Ecol. Prog. Ser.*, 7, 287, 1982.

193. **Newell, R. C.,** The biological role of detritus in the marine environment, in *Flow of Energy and Materials in Marine Ecosystems: Theory and Practice*, Fasham, M. J. R., Ed., Plenum Press, New York, 1984, 317.

194. **Corpe, W. A. and Winters, H.,** Hydrolytic enzymes of some periphytic bacteria, *Can. J. Microbiol.*, 18, 1483, 1972.

195. **Corpe, W. A.,** Periphytic marine bacteria and the formation of microbial films on solid surfaces, in *Effect of the Ocean Environment on Microbial Activites*, Colwell, R. R. and Morita, R. Y., Eds., University Park Press, Baltimore, 1974, 397.

196. **Kim, J. and Zobell, C. E.,** Occurrence and activities of cell-free enzymes in oceanic environments, in *Effect of the Ocean Environment on Microbial Activites*, Colwell, R. R. and Morita, R. Y., Eds., University Park Press, Baltimore, 1974, 368.

197. **Burns, R. G.,** Microbial adhesion to soil surfaces: consequences for growth and enzyme activities, in *Microbial Adhesion to Surfaces*, Berkeley, R. C. W., Lynch, J. M., Melling, J., Rutter, P. R., and Vincent, B., Eds., Ellis Horwood, London, 1980, 249.

198. **Marshall, K. C.,** Mechanism of adhesion of marine bacteria to surfaces, in *Proc. 34th Int. Cong. on Marine Corrosion and Fouling*, Acker, R. R., Brown, B. R., De Palma, J. R., and Iverson, W. P., Eds., Northwestern University Press, Evanston, IL, 1973, 625.

199. **Costerton, J. W., Geesey, G. C., and Cheng, K.-J.,** How bacteria stick, *Sci. Am.*, 238, 86, 1978.

200. **King, G. M. and Klug, M. J.,** Sulfhydrolase activity in sediments of Wintergreen Lake, Kalamazoo County, Michigan, *Appl. Environ. Microbiol.*, 39, 950, 1980.

201. **Hanson, R. B.,** Measuring microbial activity to assess detrital decay and utilization, in *Marine Benthic Dynamics*, Tenore, K. R. and Coull, B. C., Eds., University of South Carolina Press, Columbia, 1980, 347.

202. **Williams, P. J. le B.,** Biological and chemical dissolved organic material in sea water, in *Chemical Oceanography*, Vol. 2, 2nd ed., Riley, J. P. and Skirrow, G., Eds., Academic Press, New York, 1975, 301.

203. **Wright, R. T.,** Dynamics of pools of dissolved organic carbon, in *Heterotrophic Activity in the Sea,* Hobbie, J. E. and Williams, P. J. le B., Eds., Plenum Press, New York, 1984, 121.

204. **Pedrosalio, C.,** Toward and autecology of bacterioplankton, in *Plankton Ecology: Succession in Plankton Communities,* Sommer, U., Ed., Springer-Verlag, Berlin, 1989, 297.

205. **Wangersky, P. J.,** Organic particles and bacteria in the ocean, in *Heterotrophic Activity in the Sea,* Hobbie, J. E. and Williams, P. J. le B., Eds., Plenum Press, New York, 1984, 263.

206. **Head, P. C.,** Organic processes in estuaries, in *Estuarine Chemistry,* Burton, J. D. and Liss, P. S., Eds., Academic Press, London, 1976, 53.

207. **Hoppe, H. G.,** Determination and properties of actively metabolizing heterotrophic bacteria in the sea, investigated by means of microautoradiography, *Mar. Biol.,* 36, 291, 1976.

208. **Meyer-Reil, L.-A.,** Autoradiography and epifluorescence microscopy combined for the determination of number and spectrum of actively metabolizing bacteria in natural waters, *Appl. Environ. Microbiol.,* 36, 506, 1978.

209. **Kirchman, D. L., Mazella, L., Alberte, R. S., and Mitchell, R.,** Epiphytic bacterial production on *Zostera marina, Mar. Ecol. Prog. Ser.,* 15, 17, 1984.

210. **Ducklow, H. W.,** Production and fate of bacteria in the oceans, *BioScience,* 33, 494, 1983.

211. **Sieburth, J. M.,** Protozoan bacterivory in pelagic marine waters, in *Heterotrophic Activity in the Sea,* Hobbie, J. E. and Williams, P. J. le B., Eds., Plenum Press, New York, 1984, 405.

212. **Riley, G. A.,** Organic aggregates in sea water and the dynamics of their formation and utilization, *Limnol. Oceanogr.,* 8, 372, 1963.

213. **Riley, G. A.,** Particulate and organic matter in sea water, *Adv. Mar. Biol.,* 8, 1, 1970.

214. **Gordon, D. C.,** Some studies on the distribution and composition of particulate organic carbon in the North Atlantic Ocean, *Deep Sea Res.,* 17, 233, 1970.

215. **Gordon, A. S. and Millero, F. J.,** Adsorption mediated decrease in the biodegradation rate of organic compounds, *Microb. Ecol.,* 11, 289, 1985.

216. **Montagna, P. A.,** *In situ* measurement of meiobenthic grazing rates in sediment bacteria and edaphic diatoms, *Mar. Ecol. Prog. Ser.,* 18, 119, 1984.

217. **Newell, R. C.,** The role of detritus in the nutrition of two deposit feeders, the prosobranch *Hydrobia ulvae* and the bivalve *Macoma balthica, Proc. Zool. Soc. London,* 144, 25, 1965.

218. **McLachlan, A.,** The biomass of macro- and interstitial fauna on clean and wrack-covered beaches in western Australia, *Estuarine Coastal Shelf Sci.,* 21, 587, 1985.

219. **Meyer-Reil, L.-A. and Faubel, A.,** Uptake of organic matter by meiofaunal organisms and interrelationships with bacteria, *Mar. Ecol. Prog. Ser.,* 3, 251, 1980.

220. **Montagna, P. A., Bauer, J. E., Toal, J., Hardin, D., and Spies, R. B.,** Temporal variability and the relationship between benthic meiofaunal and microbial populations of a natural coastal petroleum seep, *J. Mar. Res.,* 45, 761, 1987.

221. **Montagna, P. A., Coull, B. C., Herring, T. L., and Dudley, B. W.,** The relationship between abundances of meiofauna and their suspected microbial food (diatoms and bacteria), *Estuarine Coastal Shelf Sci.,* 17, 381, 1983.

222. **Tietjen, J. H.,** Microbial-meiofaunal interrelationships: a review, *Microbiology,* 1, 335, 1980.

223. **Bjørnsen, P. K., Riemann, B., Horsted, S. J., Nielsen, T. G., and Pock-Sten, J.,** Trophic interactions between heterotrophic nanoflagellates and bacterioplankton in manipulated seawater enclosures, *Limnol. Oceanogr.,* 33, 409, 1988.

224. **Fenchel, T. and Riedl, R. J.,** The sulfide system: a new biotic community underneath the oxidized layer of marine sand bottoms, *Mar. Biol.,* 7, 255, 1971.

225. **Boaden, P. J. S.,** Three new thiobiotic Gastrotricha, *Cah. Biol. Mar.,* 15, 367, 1974.

226. **Jørgensen, B. B.,** The sulfur cycle of a coastal marine sediment (Limfjorden, Denmark), *Limnol. Oceanogr.,* 22, 814, 1977.

227. **Fenchel, T. M.,** The ecology of micro- and meiobenthos, *Annu. Rev. Ecol. Syst.,* 9, 99, 1978.

228. **Wiebe, W. J., Christian, R. R., Hansen, J. A., King, G., Sherr, B., and Skyring, G.,** Anaerobic respiration and fermentation, in *The Ecology of a Salt Marsh,* Pomeroy, L. R. and Wiegert, R. G., Eds., Springer-Verlag, New York, 1981, 137.

229. **Sørensen, J., Jørgensen, B. B., and Revsbech, N. P.,** A comparison of oxygen, nitrate, and sulfate respiration in coastal marine sediments, *Microb. Ecol.,* 5, 105, 1979.

230. **Howarth, R. W. and Hobbie, J. E.,** The regulation of decomposition and heterotrophic microbial activity in salt marsh soils: a review, in *Estuarine Comparisons,* Kennedy, V. S., Ed., Academic Press, New York, 1982, 183.

231. **Jørgensen, B. B.,** Mineralization of organic matter in the sea bed — the role of sulfate reduction, *Nature,* 296, 643, 1982.

232. **Howes, B. L., Dacey, J. W. H., and King, G. M.,** Carbon flow through oxygen and sulfate reduction pathways in salt marsh sediments, *Limnol. Oceanogr.,* 29, 1037, 1984.

233. **Martens, C. S.,** Recycling of organic carbon near the sediment-water interface in coastal environments, *Bull. Mar. Sci.*, 35, 566, 1984.

234. **Dicker, H. J. and Smith, D. W.,** Effects of organic amendments on sulfate reduction activity, H_2 consumption, and H_2 production in salt marsh sediments, *Microb. Ecol.*, 11, 299, 1985.

235. **King, G. M. and Wiebe, W. J.,** Regulation of sulfate concentration and methanogenesis in salt marsh soils, *Estuarine Coastal Mar. Sci.*, 10, 215, 1980.

236. **Senior, E., Lindstrom, E. B., Banat, I. M., and Nedwell, D. B.,** Sulfate reduction and methanogenesis in the sediment of a saltmarsh on the east coast of the United Kingdom, *Appl. Environ. Microbiol.*, 39, 877, 1982.

237. **Jørgensen, B. B. and Fenchel, T.,** The sulphur cycle of a marine sediment model system, *Mar. Biol.*, 24, 189, 1974.

238. **Wiebe, W. J.,** Anaerobic benthic microbial processes: changes from the estuary to the continental shelf, in *Ecological Processes in Coastal and Marine Systems*, Livingston, P. J., Ed., Plenum Press, New York, 1979, 469.

239. **Yelverton, G. F. and Hackney, C. T.,** Flux of dissolved organic carbon and pore water through the substrate of a *Spartina alterniflora* marsh in North Carolina, *Estuarine Coastal Shelf Sci.*, 22, 255, 1986.

240. **Frey, R. W. and Basan, P. B.,** Coastal salt marshes, in *Coastal Sedimentary Environments*, 2nd ed., Davis, R. A., Jr., Ed., Springer-Verlag, New York, 1985, 225.

241. **Koike, I. and Hattori, A.,** Simultaneous determination of nitrification and nitrate reduction in coastal sediments by a ^{15}N dilution technique, *Appl. Environ Microbiol.*, 35, 853, 1978.

242. **Kaspar, H. F., Asher, R. A., and Boyer, I. C.,** Microbial nitrogen transformations in sediments and inorganic nitrogen fluxes across the sediment/water interface on the South Island West Coast, New Zealand, *Estuarine Coastal Shelf Sci.*, 21, 245, 1985.

243. **Seitzinger, S. P., Nixon, S. W., Pilson, M. E., and Burke, S.,** Denitrification and N_2O production in near-shore marine sediments, *Geochim. Cosmochim. Acta*, 44, 1853, 1980.

244. **Seitzinger, S. P., Nixon, S. W., and Pilson, M. E.,** Denitrification and nitrous oxide production in coastal marine ecosystems, *Limnol. Oceanogr.*, 29, 73, 1984.

245. **Seitzinger, S. P. and Nixon, S. W.,** Eutrophication and the rate of denitrification and N_2O production in coastal marine sediments, *Limnol. Oceanogr.*, 30, 1332, 1985.

246. **Jenkins, M. and Kemp, W.,** The coupling of nitrification and dentrification in two estuarine sediments, *Limnol. Oceanogr.*, 29, 609, 1984.

247. **Kemp, W. M., Wetzel, R. L., Boynton, W. R., D'Elia, C. F., and Stevenson, J. C.,** Nitrogen cycling and estuarine interfaces: some current concepts and research directions, in *Estuarine Comparisons*, Kennedy, V. S., Ed., Academic Press, New York, 1982, 209.

248. **Jørgensen, B. B.,** Mineralization and the bacterial cycling of carbon, nitrogen, and sulfur in marine sediments, in *Contemporary Ecology*, Ellwood, D. C., Hedges, J. N., Leatham, M. J., Lynch, J. M., and Slater, J. H., Eds., Academic Press, New York, 1980, 239.

249. **Knox, G. A.,** *Estuarine Ecosystems,* Vol. 1, CRC Press, Boca Raton, FL, 1986.

250. **Winfrey, M. R. and Zeikus, J. B.,** Effect of sulfate on carbon and electron flow during microbial methanogenesis in freshwater sediments, *Appl. Environ. Microbiol.*, 33, 275, 1977.

251. **Rudd, J. W. M. and Taylor, C. D.,** Methane cycling in aquatic environments, *Adv. Aquat. Microbiol.*, 2, 77, 1980.

252. **Kristjansson, J. K., Schonheit, P., and Thauer, R. K.,** Different K_s values for hydrogen of methanogenic bacteria and sulfate reducing bacteria: an explanation for the apparent inhibition of methanogenesis by sulfate, *Arch. Microbiol.*, 131, 278, 1982.

253. **Winfrey, M. R. and Ward, D. M.,** Substrates for sulfate reduction and methane production in intertidal sediments, *Appl. Environ. Microbiol.*, 45, 193, 1983.

254. **Kiene, R. P. and Capone, D. G.,** Effects of organic pollutants on methanogenesis, sulfate reduction and carbon dioxide evolution in salt marsh sediments, *Mar. Environ. Res.*, 13, 141, 1984.

255. **Cappenburg, T. E.,** A study of mixed continuous cultures of sulfate-reducing bacteria and methane-producing bacteria, *Microb. Ecol.*, 2, 60, 1975.

256. **Claypool, G. E. and Kaplan, I. R.,** The origin and distribution of methane in marine sediments, in *Nutrient Gases in Marine Sediments*, Kaplan, I. R., Ed., Plenum Press, New York, 1975, 99.

257. **Lipschultz, F.,** Methane release from a brackish intertidal salt-marsh embayment of Chesapeake Bay, Maryland, *Estuaries*, 4, 143, 1981.

258. **Martens, C. S. and Berner, R. A.,** Interstitial water chemistry of anoxic Long Island Sound sediments. I. Dissolved gases, *Limnol. Oceanogr.*, 22, 10, 1977.

259. **Reeburgh, W. S.,** Observations of gases in Chesapeake Bay sediments, *Limnol. Oceanogr.*, 14, 368, 1969.

260. **Davis, J. B. and Yarborough, H. F.,** Anaerobic oxidation of hydrocarbons by *Desulfovibrio desulfuricans*, *Chem. Geol.*, 1, 137, 1966.

Chapter 2

PHYTOPLANKTON

I. INTRODUCTION

Plankton of estuaries and oceans may be defined as floating or drifting organisms with limited powers of locomotion that are transported primarily by prevailing water movements. Subdivisions of the plankton include bacterioplankton (bacteria), phytoplankton (plants), and zooplankton (animals). Phytoplankton are free-floating, microscopic plants — unicellular, filamentous, or chain-forming species — inhabiting surface waters (photic zone) of open oceanic and coastal environments (Figure 1). Although unicellular forms comprise the bulk of phytoplankton, some green and blue-green algae are filamentous (i.e., develop thread-like cell systems). Colonial diatoms and the blue-green alga, *Trichodesmium thiebautii,* provide examples.[1] In addition, a number of diatoms and dinoflagellates produce chains of loosely associated cells.[2] Not all pelagic photosynthetic organisms are microscopic; for example, macroscopic multicellular brown algae, *Sargassum* spp., yield substantial biomass in the Sargasso Sea.

Despite being composed of single cells or of relatively simply organized, small colonies, phytoplankton encompass a rather wide diversity of algal groups.[3] These diminutive autotrophs, which are largely holoplanktonic, serve a major function in the oceans of the world, being responsible for at least 90% of the photosynthesis, with the remaining 10% mainly attributable to benthic macroalgae and vascular plants (e.g., seagrasses, salt marsh grasses, and mangroves) in intertidal and shallow subtidal environments.[4] Because the ocean covers approximately 72% of the earth's surface, phytoplankton as a group represent the most important primary producers on the planet. They play a vital role in initiating the flow of energy in a useful form through marine ecosystems.

Among classification schemes, phytoplankton are commonly categorized on the basis of size into four classes: (1) ultraplankton (less than 5 µm in diameter), (2) nanoplankton (5 to 70 µm), (3) microphytoplankton (70 to 100 µm), and (4) macrophytoplankton (greater than 100 µm) (Table 1). More than half of all phytoplankton belong to the ultraplankton and nanoplankton. For practical purposes, phycologists often sort the microscopic plants into net plankton and nanoplankton determined by the nominal aperture size of the plankton net deployed in the field. Net plankton samples embody all phytoplankton retained by the finest nets (about 64 µm apertures) that can be conveniently towed, and in coastal waters, tend to be dominated by diatoms and dinoflagellates. The nanoplankton pass through the fine-mesh nets; large numbers of coccolithophores, other flagellates such as those of the Chrysophyceae and Cryptophyceae, and small species of diatoms are the most abundant components.[5] Because nets are now available with mesh dimensions of 20 to 30 µm, some workers recommend limiting the term nanoplankton to all planktonic algae less than 30 µm in size.

In this chapter, the species composition and dynamics of phytoplankton in estuaries are considered, together with standing crop and primary production. Review of the primary production of phytoplankton logically leads to some discussion of the factors which limit it, namely, light, temperature, nutrients, salinity, and zooplankton grazing. As in other floral communities of estuaries, phytoplankton experience seasonal cycles in species composition (succession), biomass, and primary production related to changes in physical, chemical, and biological conditions. This chapter focuses on these cycles and examines the variation of phytoplankton parameters with latitude, climate, season, and time of day. Annual successional patterns are especially well documented for phytoplankton communities in temperate estuaries;[4] the following material draws heavily upon investigations of these communities.

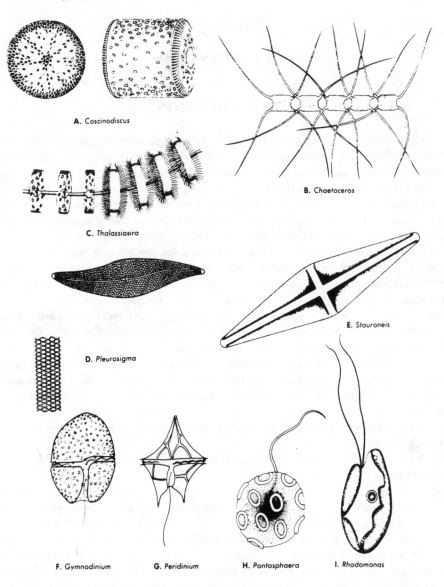

FIGURE 1.   Members of some phytoplankton groups. (A to C) Diatoms, centric types; (D and E) diatoms, pennate types; (F) naked dinoflagellate; (G) thecate, or armored, dinoflagellate; (H) coccolithophore; (I to K) others. (From McConnaughey, B. H. and Zottoli, R., *Introduction to Marine Biology,* C. V. Mosby, St. Louis, 1983. With permission.)

II. TAXONOMY

A. CLASSES

Diatoms (class Bacillariophyceae), dinoflagellates (class Dinophyceae), coccolithophores (class Prymnesiophyceae), silicoflagellates (class Chrysophyceae), and blue-green algae (class Cyanophyceae) constitute the principal taxa of planktonic producers in the ocean.[6] In estuarine or lagoonal systems, other taxonomic assemblages also may be locally important, for example, the green-colored algae (class Chlorophyceae), brown-colored phytoflagellates (class Haptophyceae), and euglenoid flagellates (class Euglenophyceae). The classification of phytoplankton relies on studies of fine cell structure, life-history investigations, and biochemical research.

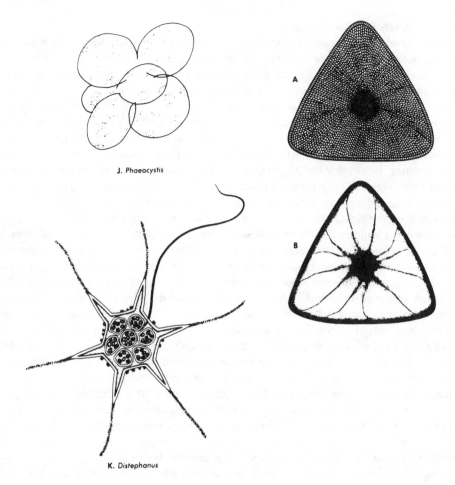

J. *Phaeocystis*

K. *Distephanus*

FIGURE 1 continued.

TABLE 1
Size Classification of Phytoplankton

Floral group	Maximum cell dimension (μm)
Ultraplankton	<5
Nanoplankton	5—70
Microphytoplankton	70—100
Macrophytoplankton	>100

Algalogists utilize certain cell characteristics to identify phytoplankton. Chief among these are the cell shape, cell dimensions, cell wall, mucilage layers, chloroplasts, flagella, reserve substances (e.g., starch, oil, and leucosin), and other cell features (e.g., cell vacuoles and trichocysts).[2,7] Advances in electron microscopy have greatly aided investigators in elucidating the fine structure of cell constituents, which has facilitated taxonomic work. Nevertheless, numerous species, particularly among the nanoplankton, are poorly understood; consequently, revisions to the classification of phytoplankton continually arise.

1. Diatoms (Class Bacillariophyceae)

These microalgae dominate phytoplankton communities in high latitudes of the Arctic and Antarctic, in the neritic zone of boreal and temperate waters, and in areas of upwelling (both coastal and equatorial).[8,9] Many specialists regard diatoms as the most important phytoplankton group, contributing substantially to oceanic productivity, especially in coastal waters. Consisting of a single cell or cell chains, diatoms secrete an external rigid silicate skeleton (pectin impregnated with silica) called a frustule which encases the vegetative protoplast. The frustule or silica cell wall is composed of two valves, the epitheca and hypotheca; the epitheca overlaps the hypotheca in the girdle region, where they link via pectinaceous bands or minute teeth. The valves, therefore, fit together like a petri dish, and they frequently contain highly ornamented sculptured markings. The complex ornamentation may be differentiated into four types, namely, puncta, areolae or alveoli, canaliculi, and costae.[3] Cavities or thin areas of the frustule expedite the exchange of nutrients, gases, and metabolic products across the cell wall.[4]

The shape and symmetry of the frustule assist taxonomists in the classification of diatoms. Based on these features, two orders are recognized, that is, centric diatoms (Centrales or Centricae), having circular or dome-shaped valves and a predominantly planktonic existence, and pennate diatoms, possessing oblong or "boat-shaped" valves and a principally benthic habit. The frustule of centric diatoms has radial symmetry about the pervalar axis, that of pennate diatoms, bilateral symmetry with reference to the apical plane.[8,9] In pennate diatoms, a long slit, the raphe, may extend to both valves and connect to thickenings known as nodules.[10] By protruding their cytoplasm through the raphe and making contact with the substrate, benthic diatoms creep along the seafloor.[4] Most marine diatoms are planktonic, and of the 10,000 species of diatoms that have been identified, about half live in marine environments.

Diatoms range in size from less than 10 μm to approximately 200 μm.[6,10] Bearing no flagella, cilia, or other organs of locomotion, the planktonic species are nonmotile and sink in nonturbulent waters. According to Smayda,[11] sinking rates of diatoms and other phytoplankton depend on cell size and shape, colony dimensions, physiological condition, and age. Live diatom cells descend at a rate of 0 to 30 m/d through the water column, but dead cells fall more quickly, exceeding 60 m/d in some cases.[3] Buoyancy declines with age, resulting in greater sinking rates for senescent populations. As the cell or colony size increases, so does the sinking rate, with the ratio of surface area to volume being a key factor in the sinking process.

Smayda[11] postulated several morphological, physiological, and physical adaptations which foster the suspension of diatoms. Flotation of nonmotile forms may be enhanced by modification of the frustule. For instance, needlelike extensions, spines, and siliceous projections increase the frustule surface area, thereby mitigating sinking rates. Among physiological mechanisms, ion regulation in the cytoplasm and vacuole, gas production via photosynthesis, and secretion of mucilage promote diatom flotation.[1] Water viscosity and vertical circulation in the ocean are two physical factors deemed to be critical to diatom suspension.

Cell constituents also have been implicated in the sinking of phytoplankton populations. Diatoms with large concentrations of lipids sink faster than those with small concentrations. Since older individuals accumulate higher lipid concentrations, they generally sink more rapidly. Higher silica content of the cell walls raises density levels, inducing settlement. The amount of silicon in diatoms varies markedly with species, estimates ranging from 4 to 50% of the dry weight.[3]

The storage products of diatom photosynthesis are chrysolaminarin (a polysaccharide) and lipids. The lipids, which appear as tiny oil droplets dispersed in the cytoplasm, reportedly generate oily patches in some marine waters during final stages of a diatom bloom. The lipid level in the cell is a function of environmental conditions and the species composition. Because of these storage compounds, diatoms furnish zooplankton with high-energy ration.

Most diatoms reproduce by vegetative cell division, leaving two daughter cells of smaller size than the parent cell. The vegetative stage is the only one commonly observed in many species. Vegetative cells may exist independently or develop into distinct colonies.[9] During cell fission, the two valves of the frustule separate slightly, and each retains the products of a mitotic division. Subsequently, each daughter cell secretes a new hypotheca. Due to this type of asexual reproduction executed over several cell generations, the mean size of a population of diatoms gradually diminishes. When a species attains a lower limit in size, restoration of maximum cell size occurs through sexual reproduction and the formation of an auxospore, which then secretes a large-sized frustule. A few diatom species do not undergo a progressive decline in cell size, but seemingly maintain constancy of dimensions. This method of asexual reproduction accounts for the large variability in the size of a population of diatoms of a given species, the smallest being up to 30 times more diminutive than the largest.[4]

Eppley[12] mentioned optimal cell doubling rates in diatoms equal to 0.5 to 6 doublings per day, which allow blooms to develop. Numerous centric diatoms produce resting spores during unfavorable conditions.[13] Resting cells, similar to vegetative cells, also are found in some species, particularly those in temperate waters.[14] As Steidinger and Walker[14] assert, asexual resting spores and resting cells — dormant stages — may be important in species occurrence, succession, and distribution, as well as in population survival.

2. Dinoflagellates (Class Dinophyceae)

Dinoflagellates typically are unicellular, biflagellated, autotrophic forms, and, like diatoms, supply a major fraction of primary production in many regions. Individual dinoflagellate cells range from about 5 to greater than 200 μm in size, but some species (e.g., *Polykrikos* spp.) often grow in larger chains or pseudocolonies.[15] Widely distributed in marine and estuarine environments, dinoflagellates frequently dominate phytoplankton communities in subtropical and tropical waters. In addition, they are usually abundant in temperate and boreal autumnal assemblages.[8] Approximately 1000 to 1500 species of dinoflagellates inhabit marine and freshwater environments;[15] however, most of them (greater than 90%) are marine.[4]

Devoid of an external, siliceous skeleton, many dinoflagellates bear an armor of cellulose thecal plates. This group, the Peridinales, can be readily distinguished with a light microscope from the unarmored or naked forms, the Gymnodiniales, which only have a firm periplast. Representative genera of the Peridinales include *Ceratium, Gonyaulax,* and *Peridinium;* representative genera of the Gymnodiniales are *Amphidinium, Ptychodiscus (Gymnodinium),* and *Gyrodinium.*

Two unequal flagella (longitudinal flagellum and tranverse flagellum) lie in separate grooves on the body surface of most dinoflagellates, propelling and stabilizing the organisms. The two grooves are (1) the girdle or annulus, a transverse groove that surrounds the cell and divides it into two subequal parts (i.e., epicone and hypocone), and (2) the sulcus, a longitudinal furrow passing along the posterior end of the cell. Thecate forms may have a cingulum or perforated plate covering the annulus.

Asexual reproduction via binary cell division is the normal means of reproduction in dinoflagellates, although some individuals reproduce sexually as well (e.g., *Ceratium* and *Glenodinium*). Simple binary fission in dinoflagellates does not translate into a progressive reduction in size of successive generations as in diatoms. The rates of cell division of these plants are comparable to those of diatoms, but can be highly variable contingent upon environmental conditions. Hence, *Peridinium* has a doubling rate of 10 to 50 h; *Exuviaella,* a doubling rate of 15 to 90 h; and *Prorocentrum,* a doubling rate of 12 to 127 h.[3]

The majority of dinoflagellates contain chloroplasts with chlorophyll *a* and *c* and other pigment cells used in photosynthesis. Starch and lipids serve as food reserves. Heterotrophy (e.g., *Kofoidinium* and *Polykrikos*) is not rare, and various species feed on phytoplankton and zooplankton. A number of others consume decaying organic matter (holozoic) or obtain nutrition through parasitic activity (e.g., *Blastodinium* and *Oodinium*). The zooxanthellae enter into a symbiotic relationship with the giant tropical clam, *Tridacna,* and with corals. Various species also are symbiotic with some sea anemones, radiolarians, and fish.

Certain dinoflagellate species generate toxins. When blooms of these microflora reach high densities from 5×10^5 to 2×10^6 cells per liter,[1] the cumulative effect of the toxins liberated periodically induces mass mortality of fish, shellfish, and other organisms. Blooms of dinoflagellates occasionally impart a red or brown color to the water causing the so called "red tide". The genera *Gonyaulax* and *Ptychodiscus* are responsible for occurrences of toxic red tides in estuaries, two noteworthy species promoting this phenomenon being *Gonyaulax polyhedra* and *Ptychodiscus brevis* (=*Gymnodinium breve*). Neurotoxin released by dinoflagellate cells of a red tide can kill fish directly as the cells pass through the gills. In addition, shellfish populations, which accumulate the toxic agent saxitoxin in their siphons and hepatopancreas by filtering the dinoflagellates from the water, cause a neurological disorder, paralytic shellfish poisoning, in humans that consume them.[16] Diarrhetic shellfish poisoning, a transitory gastrointestinal disorder experienced by shellfish eaters, has been traced to planktonic dinoflagellates.[17] Several conditions favor the success of dinoflagellates and the development of red-tide blooms, that is, the sudden input of nutrients by upwelling, washout of nutrients from land, influx of vitamins from shore, and tidal turbulence.[8,18,19]

Some dinoflagellates are bioluminescent, lighting surface waters at night. The genera *Gymnodidium, Noctiluca,* and *Pyrocystis* provide excellent examples. A few, such as the genus *Noctiluca,* while experiencing bioluminesence, are not photosynthetic.[4,20] Spector[21] treats the subject of dinoflagellates comprehensively, and additional information on these microflora can be obtained from his volume.

3. Coccolithophores and Other Brown-Colored Algae (Class Haptophyceae)

The coccolithophores are biflagellate, unicellular algae covered by calcareous plates, called coccoliths, embedded in a gelatinous sheath surrounding the cell.[22] These small phytoplankters (5 to 50 μm in size) reach peak abundance in tropical and subtropical, open-oceanic waters, but sometimes also proliferate in coastal environments. Although most species live in warmer seas, a few (e.g., *Pontosphaera huxleyi* and *Syracosphaera* spp.) attain maximum abundance in colder regions. The major fraction of photosynthesis in certain areas is ascribable to coccolithophores, which periodically comprise a significant portion of calcareous sediment (i.e., *Globigerina* ooze) on the seafloor. Despite the overwhelmingly autotrophic existence of plankters in this group, a few taxa obtain energy heterotrophically below the photic zone.

Other brown-colored algae of this class show a range of body form from unicellular motile species, palmelloid colonial structure, or filamentous microscopic colonies.[3] The brown-colored algae typically possess body scales; however, genera such as *Diacrateria* and *Isochrysis* do not. When present, the scales are unmineralized. Body scales can be used to distinguish members of the Haptophyceae from those of the Chrysophyceae. Most individuals fall in the nanoplanktonic size range and have paired, smooth flagella generally of equal length.[2] They live in inshore as well as offshore waters.[8]

4. Silicoflagellates (Class Chrysophyceae)

These planktonic microflora are single-celled, uniflagellated, or biflagellated organisms, usually less than 30 μm in diameter. They secrete an internal skeleton composed of siliceous spicules. Whereas the bulk of this class consists of photosynthetic plants, some may be heterotrophic. Silicoflagellates can outnumber all other phytoplankton in temperate marine waters, but are most abundant in cold nutrient-rich environments, where they reproduce by simple cell division.

5. Blue-Green Algae (Class Cyanophyceae)

Representatives of this class differ from all other algae in being prokaryotic. The organisms have a relatively simple construction, lacking organized nuclei, nuclear membranes, and chromosomes. They are photosynthetic with chitinous walls. In addition to chlorophyll *a,* blue-green algae contain phycobilins and carotenoids, responsible for the varied color in different species. The pigment phycocyanin causes the blue-green color in many individuals of the group. The coloration of the Red Sea has been attributed, in part, to the presence of *Trichodesmium erythraeum.*

Blue-green algae are frequently encountered in shallow, nearshore tropical seas, but appear in low densities in nearly all regions. Occasionally, they build blooms, especially red tides, in brackish or nearshore habitats. A favored location for blue-green algae is the salt-marsh biotope.

The size of blue-green algae ranges from less than 1 μm for single-celled forms to more than 100 μm for filamentous types. Pelagic Cyanophyceae encompasses species of *Haliarachne, Katagnymene, Oscillatoria,* and *Trichodesmium.*[10] Benthic species often construct mats on mudflats which break free from the substratum and float on the surface of advancing tidal waters. Blue-green algae reproduce by simple cell division, by fragmentation, and by nonmotile spores.[3]

6. Green-Colored Algae (Class Chlorophyceae)

The true green algae exist as unicellular, filamentous, or colonial forms. Both flagellated and nonflagellated cells have been documented. Generally of ultraplanktonic or nanoplanktonic size, these autotrophs can be quite important in estuaries and enclosed seas, producing blooms in both ecosystems from time to time, particularly in late summer and fall.[8]

7. Euglenoid Flagellates (Class Euglenophyceae)

This class characteristically consists of unicellular, biflagellated forms, with a fusiform or cylindrical shape. The flagella emerge from an anterior invagination. Members may be pigmented or colorless, and many species obtain their nutrition saprophytically. Those that photosynthesize normally store paramylum (paramylon), a starch-like carbohydrate.[3] Binary fission is the usual mode of reproduction. Euglenoids, although largely freshwater, also are common in estuaries, especially somewhat polluted systems.[2,3]

III. SAMPLING METHODS

A number of publications deal with methods of phytoplankton sampling.[1,23-26] Plankton nets, plankton pumps, and water bottles represent the three most frequently used types of gear for collecting phytoplankton field samples. Plankton nets, typically with mesh openings ranging from 20 to 64 μm, selectively capture nanoplankton and larger phytoplankters. Fragile forms often suffer cell damage during sampling. Plankton pumps permit a continuous stream of water to be pumped from a desired depth to the surface, where phytoplankton can be concentrated by filtration. Therefore, this method allows phytoplankton samples to be taken at any depth through the water column. Water bottles or closing samplers capture phytoplankton by closing automatically at a selected depth when activated by a solenoid hooked to a pressure-sensing device or by a messenger released from the surface to the sampler.[1] The nansen bottle, niskin sampler, and van dorn sampler are three commonly used closing samplers.

Phytoplankton samples may be stored live for several hours in a refrigerator or ice chest. The samples usually are preserved in a buffered fixative, such as Lugol's solution (10 g of iodine and 20 g of potassium iodide dissolved in 200 ml of distilled water and 20 g of glacial acetic acid), which is added in a ratio of 1 part to 100 parts of seawater sample. Another fixative incorporates a solution of formalin neutralized with hexamine (200 g of hexamine [hexamethylenetetramine] added to 1 l of commercial 40% [37% formaldehyde] formalin). This solution is filtered after about 1 week.[1]

IV. PHYTOPLANKTON BIOMASS AND PRIMARY PRODUCTIVITY

A. BIOMASS

Dawes[1] refers to biomass as the concentration of organic matter present in a unit area per specified time. The standing crop or biomass of phytoplankton can be determined by direct cell counts,[3] by extracting and measuring chlorophyll a,[27,28] and by estimating the content of adenosine triphosphate (ATP)[29] and deoxyribonucleic acid (DNA)[30] in the organisms. Direct counting of cells is tedious and time consuming. Initially, phytoplankton must be concentrated either by settling, which may be accelerated with a Dodson settling tube, or by Millipore filtration and centrifugation, which tend to damage delicate cells. Availing oneself of an inverted microscope and a counting chamber provides an accurate laboratory procedure for elucidating the number of cells of sedimented samples.[25] Alternatively, cell counts can be made on a subsample of a concentrated sample; this technique, although less accurate, is more rapid.[1] Cell volume can be calculated from cell counts, with cell volume serving as an index of standing crop.[3]

A customary method of estimating phytoplankton biomass is the determination of chlorophyll a. The concentration of chlorophyll can be used directly as a measure of biomass, or the standing crop can be computed via a ratio between weights of chlorophyll and biomass, which ranges from a value of 22 to 154.[6] Dawes[1] details the method of converting chlorophyll concentrations to organic matter. This conversion procedure, as emphasized by Mann,[5] is not simple, but the density of chlorophyll a yields a reasonably good index of phytoplankton biomass. Although chlorophyll a determinations are rapid and frequently employed in assessing phytoplankton standing crop, they have limitations, particularly due to fluctuating carbon/chlorophyll ratios[3] that vary with species, light intensity, availability of nutrients, and physiological condition of the cells.[31]

In a summary of the annual mean concentrations of chlorophyll a in 39 estuarine systems, Boynton et al.[32] recorded values of approximately 0.1 to slightly more than 25 mg/m^3. Concentrations peaked during the warmer months of the year (May to October). Elevated chlorophyll a readings occurred concomitantly with phytoplankton blooms in many cases. For example, Palumbo[33] observed a maximum chlorophyll a concentration of 104 mg/m^3 in the Newport River estuary at the time of a phytoplankton bloom (March 14 and 15, 1978).

Phytoplankton biomass in estuaries exhibits marked spatial and temporal variations.[34-44] Changes in biomass along vertical and horizontal planes of an estuary are explainable in terms of fluxes in a few physical-chemical parameters, such as nutrients, salinity, temperature, light, and water column mixing.[5] Phytoplankton typically have a patchy distribution — affecting the uniformity of chlorophyll a from one location to another — owing to their responses to hydrographic, light, and nutrient distributions, to predation and symbiosis, and to mechanical aggregation by physical processes (e.g., currents, eddies, and fronts).[45-47] Constituent species of phytoplankton communities likewise differ in size and growth rates, which influence chlorophyll a concentrations, not only from one geographic area to another but also seasonally in the same geographic area.[48] While net phytoplankton have been treated comprehensively in some investigations of standing crop, ultraplankton and nanoplankton must not be disregarded, because these diminutive forms may predominate in biomass,[34] although the ratio of ultraplankton and nanoplankton to net phytoplankton is highly variable.

Hulburt[49,50] and Hulburt and Rodman[51] compared the density and diversity of phytoplankton in estuaries, coastal waters, and the open ocean. As measurements of chlorophyll concentration indicate, the largest phytoplankton populations (by cell number) inhabit estuarine waters, but the highest diversity corresponds to coastal and open oceanic systems (Figure 2). In addition, an inshore-to-offshore alteration in species composition takes place; diatoms are more abundant in estuarine waters, and populations of coccolithophores, dinoflagellates, and flagellates dominate offshore regions.[52]

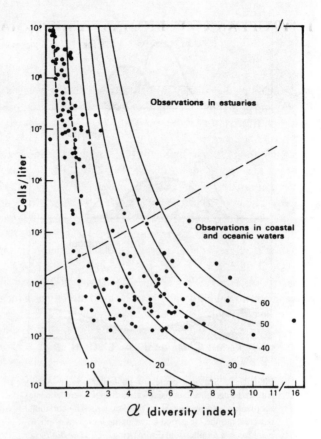

FIGURE 2. The relation of diversity to population size. Curved lines signify 10, 20, 30, 40, 50, and 60 species per liter at various cell number and α values. (From Hulburt, E. M., *J. Mar. Res.*, 21, 81, 1963. With permission.)

A well-defined and predictable seasonal pattern of biomass characterizes temperate-boreal waters. Maximum biomass commonly develops in the spring followed by a decline in the summer and a secondary peak in the fall. This periodicity contrasts with those of Arctic and Antarctic waters, which entail a single peak period of biomass in the summer, and subtropical and tropical waters, which have maximum biomass in the winter. Except in subtropical and tropical regions, zooplankton abundance alternates with that of phytoplankton (Figure 3). Thus, high latitudes have a greater amplitude of biomass change over the year than those of mid and low latitudes. Low latitudes display relatively small temporal fluctuations in standing crop each year.

Accompanying perturbations in standing crop are modifications in the species composition of the phytoplankton which follow seasonal successional patterns. The floristic succession is responsive to temperature, light, and nutrient levels throughout the year. Because greater variations in environmental conditions surface in mid and high latitudes than in low latitudes, the seasonal succession of phytoplankton appears pronounced in temperate, boreal, and Arctic systems, with a less conspicuous succession evident in subtropical and tropical waters. In some regions, the floristic succession has been related to monsoons or periods of upwelling.

Biomass maxima of phytoplankton in estuaries do not necessarily parallel those in nearby coastal waters.[53] However, in most estuaries, phytoplankton blooms arise in the spring or summer months, seasons when biomass peaks also are manifested.[3,32,35,54] As a bloom of phytoplankton diminishes, it is common to delineate a gradual translocation in the position of

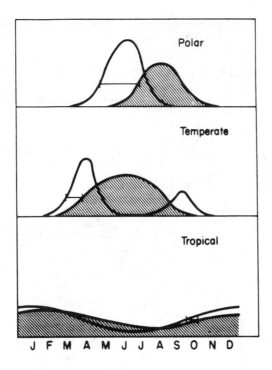

FIGURE 3. Seasonal cycles of phytoplankton in Arctic, temperate-boreal, and tropical waters. Stipled area delineates zooplankton seasonal cycles in these regions. The horizontal bar indicates the lag period between the increases of phytoplankton and zooplankton. (From Dring, M. J., *The Biology of Marine Plants,* Edward Arnold, London, 1982. With permission.)

the chlorophyll maximum, being initially confined to surface waters but progressively sinking to greater depths through time (Figure 4). Hence, the biomass distribution in the water column can change sharply within temporally restricted intervals.

In sum, the biomass or standing crop of phytoplankton relates to the amount of matter in the autotrophic plants at a given time. The standing crop may be presented as biomass (=live weight), as dry plankton (=plankton dried to a constant weight), or as dry organic matter (=dry plankton less ash). The concentration of chlorophyll also yields estimates of phytoplankton standing crop, but the relationship between chlorophyll and total organic matter is variable. Chlorophyll determinations, therefore, must be applied carefully in measurements of phytoplankton standing crop.[55]

B. PRODUCTIVITY

This term pertains to the rate of carbon fixation (organic synthesis) in natural waters and is usually expressed as grams of carbon produced per period of time. Scientists measure primary productivity by assessing rates of photosynthesis and respiration. Oxygen production in photosynthesis serves as a basis for determining productivity, as does carbon dioxide uptake.[56] In addition, the estimation of carbon dioxide released (in respiration) supplies data useful to productivity studies. Other methods of value in investigations of phytoplankton productivity include: (1) the measurement of pH, which is a function of the dissolved carbon dioxide content of estuarine and marine waters, being increased by respiration and decreased by photosynthesis;[37] (2) the rate of appearance of new algal biomass over time to estimate net primary

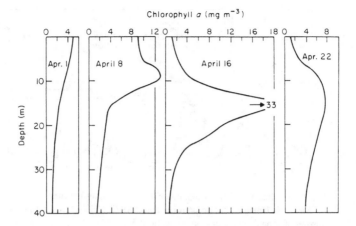

FIGURE 4. Profile of chlorophyll *a* distribution in St. Margaret's Bay, Nova Scotia, reflecting the formation and gradual sinking of the spring diatom maximum. (From Mann, K. H., *Ecology of Coastal Waters: A Systems Approach*, University of California Press, Berkeley, 1982. With permission.)

production;[5] and (3) the use of the radioactive isotope ^{14}C as a tracer for photosynthetic carbon assimilation.[3,6,55,56] Further description of the techniques established for primary productivity research (e.g., light and dark bottle oxygen technique, pH method, ^{14}C method, particle counting method) can be gleaned from several other volumes.[3,6,27,56,57]

Of the various methods directly involved in evaluating phytoplankton productivity in the sea, the most widely adapted one is the ^{14}C technique.[58,59] Initially applied by the Danish *Galathea* expedition around the world from 1950 to 1952,[60,61] this method has since been scrutinized extensively and its reliability and validity questioned repeatedly. It is subject to potential sources of error[62-64] that may manifest themselves at four levels: (1) methodological mistakes, (2) physiological problems, (3) containment problems, and (4) sampling and incubation strategy.[59] Oviatt et al.,[65] Gieskes et al.,[66] and Gieskes and Kraay,[67] discuss the difficulties encountered with this procedure. One major debate has centered on whether the ^{14}C method measures gross photosynthesis, net photosynthesis, or some intermediate value.[68-70] Raymont (p. 392)[3] states that "the method is usually supposed to estimate net rather than gross primary production, or some intermediate value". Most actual measurements, according to Valiela,[6] measure some value closer to net than gross production. Despite the drawbacks associated with the ^{14}C method, it has gained wide acceptance and usage in marine phytoplankton research.

During the past 35 years, thousands of primary production measurements have been taken in estuaries, with estimates of phytoplankton productivity ranging from 6.8 to 530 g C per square meter per year.[71-73] Productivity values near the lower end of this range typically are derived from turbid estuarine waters, and readings at the upper end of the range, from clearer systems. Boynton et al.,[32] in a voluminous review, revealed that the mean annual phytoplankton productivity for 45 estuarine systems equaled 190 g C per square meter per year, which exceeds the value of 100 g C per square meter per year registered for coastal oceanic waters,[74] but is less than the measurement of 300 g C per square meter per year assigned to upwelling regions. At the offings of major estuaries, such as the Hudson, Delaware, and Chesapeake systems, phytoplankton production may surpass that within the estuary. Here, production of phytoplankton can be greater than 350 g C per square meter per year.[75]

Similar to phytoplankton biomass, phytoplankton productivity in estuaries undergoes marked spatial and temporal variations disclosed in numerous publications.[32,34,35,40,53,71,72,76-84] The periodicities of phytoplankton biomass and production in estuaries oftentimes follow the same trend, although maximum biomass and production figures are not necessarily synchro-

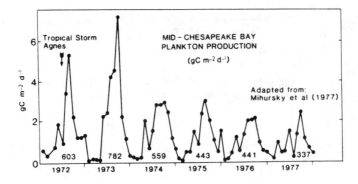

FIGURE 5. Mean monthly phytoplankton production rates in central Chesapeake Bay from January 1972 through December 1977. Values below peaks are estimates of annual phytoplankton production. (From Mihursky, J. A., Heinle, D. R., and Boynton, W. R., Unpubl. Tech. Rep., UMCEES Ref. No. 77-28-CBL, Chesapeake Biological Laboratory, Solomons, MD, 1977. With permission.)

nous.[53] Peak productivity occurs in the warmer months of the year, but high productivity usually persists from spring to fall.[85] This pattern is apparent in Figure 5 — a plot of monthly phytoplankton production at six stations in central Chesapeake Bay between January 1972 and December 1977.[86] In spite of the large differences in the magnitude of annual production in which peak productivity in the bay varies by a factor of 3.6, temporal patterns (i.e., timing, shape, and frequency of events) were strikingly regular over the 6-year study period.

Factors which interact to cause these patterns are nutrients, other chemical parameters, physical conditions (e.g., temperature, light, and mixing conditions), and biological processes (e.g., grazing). Phytoplankton productivity varies substantially among estuaries due to differences in environmental factors. Thus, in high latitude systems, light may be the key factor, whereas in subtropical and tropical estuaries, seasonal nutrient fluctuations may be of paramount significance.[81] However, generalities regarding the relative importance of various biological, chemical, and physical controls on phytoplankton production are difficult to pose when one considers the broad spectrum of estuarine ecosystems. Nevertheless, Cole and Cloern[73] advocate that primary productivity in a nutrient-replete estuary is a function, in simplest terms, of three basic variables (i.e., phytoplankton biomass, biomass-specific carbon assimilation rate, and light availability). According to these authors, productivity should be predictable in respect to these three variables even though they may deviate because of changes in phytoplankton community composition, light adaptation, day length, settling, transport processes, turbulence, temperature, and grazing. Maximum phytoplankton production is rarely achieved in estuaries since high turbidity, commonly existing in these shallow coastal systems, limits light penetration. The shallow waters can arrest bloom development, and the flushing rate can exceed the growth rate of the phytoplankton.[87]

V. SPECIES COMPOSITION AND SUCCESSION

A. GENERAL

In mid- and high-latitude estuaries, where seasonal oscillations in the physical-chemical conditions of the environment are great, the species composition of phytoplankton communities changes progressively during the year. A well-developed seasonal succession of phytoplankton populations typifies many of these communities. Seasonal successions of phytoplankton species are not limited to mid- and high-latitude systems, but occur in low-latitude estuaries as well, despite the smaller amplitude of their environmental variations.[3] Attempts to explicate phytoplankton species succession have delved into several potential causative factors, primarily temperature, nutrient concentrations, light, and biological conditioning of the water by means of organic metabolites released by the organisms.[20,235]

Smayda[35] and Riley[54] deal with the temporal changes in the taxonomic structure of phytoplankton communities for representative estuarine types. The following discussion focuses on the floristic composition and temporal alteration of the taxa in estuaries along the East Coast of the U.S., specifically within the mid-Atlantic Bight. Consideration is given to the phytoplankton community of Chesapeake Bay, Barnegat Bay, Long Island Sound, and Narragansett Bay, drawing heavily on the efforts of Patten et al.,[88] Mulford,[89] and Marshall[90,91] on Chesapeake Bay; Mountford[92] on Barnegat Bay; Riley[54] and Conover[93] on Long Island Sound; and Smayda[35] and Karentz and Smayda[36] on Narragansett Bay. The taxonomic nomenclature by these investigators has been retained in this volume without incorporating recent taxonomic revisions from the literature.

B. TYPE ESTUARINE SYSTEMS
1. Chesapeake Bay

The species composition and seasonal dynamics of phytoplankton have been reported for the upper Chesapeake Bay by Mulford,[89] the middle Chesapeake Bay by Sellner and Kachur,[44] Flemer,[77] Cowles,[94] and Smayda,[95] and the lower Chesapeake Bay by Patten et al.,[88] Marshall,[90,91] Cowles,[94] and Wolfe et al.[96] Sellner[97] has reviewed phytoplankton of the estuary. Several species seasonally dominate the phytoplankton community along the length of the bay. *Skeletonema costatum* and *Asterionella glacialis* are winter dominants; *S. costatum, A. glacialis, Cerataulina pelagica,* and *Rhizosolenia fragilissima,* spring dominants; *Coscinodiscus marginatus, R. calcar avis,* and *Ceratium furca,* summer dominants; and *S. costatum* and *Chaetoceros sociale,* fall dominants.

Marshall[90,91] documented 219 phytoplankton species in lower Chesapeake Bay, noting seasonal occurrence and dominance patterns (Tables 2 and 3). Of the total number of phytoplankton species, 129 were diatoms and 42 dinoflagellates. Other represented groups of phytoplankton consisted of Chrysophyceae (eight species), Cyanophyceae (eight), Cryptophyceae (seven), Euglenophyceae (seven), Prasinophyceae (seven), Haptophyceae (six), Chlorophyceae (three), and Xanthophyceae (two). Phytoplankton dominants were similar in both the lower and upper bay. In general, diatoms (dominated by *S. costatum* and *A. glacialis*) predominated in the winter, spring, and fall, and phytoflagellates in the summer.

An annual pattern of phytoplankton abundance and species composition is discernible in the bay. A successional series of phytoplankton takes place throughout the year, with populations derived from coastal oceanic waters or present as indigenous forms in the estuary and its tributary systems. Chesapeake Bay, while experiencing large year-to-year variations in phytoplankton abundance, exerts a strong influence on the species composition, density, and productivity of phytoplankton in the adjacent coastal waters overlying the continental shelf. Phytoplankton populations are likewise highly concentrated seaward of lower New York and Delaware Bays, reflecting the action of estuarine outflow, which offers nutrients and a "seeding" effect of high phytoplankton numbers to nearby coastal areas.[98]

Seasonal cycles of abundance are a product of the succession of phytoplankters whose growth accounts for pulses at various times of the year. Hence, Marshall[90] conveyed that, in lower Chesapeake Bay, *S. costatum* dominates in the fall and winter, but *A. glacialis* becomes more numerous in the early spring and *Cerataulina pelagica, Cyclotella glomerata,* and *Thalassiosira nordenskioldii,* in the late spring. Phytoflagellates (e.g., *Gymnodinium danicans* and *Gyrodinium estuariale*), in addition to the diatoms *Nitzschia pungens, Coscinodiscus marginatus,* and *Rhizosolenia calcar-avis,* are most abundant in the summer (Table 3).

Marshall[90] ascertained maximum phytoplankton densities of 2.7×10^6 cells per liter during the fall bloom. In middle Chesapeake Bay waters, Cowles[94] attested to a spring peak of 7×10^5 cells per liter. Sellner and Kachur[44] also registered highest phytoplankton densities (greater than

TABLE 2
List of Phytoplankton Taxa from Lower Chesapeake Bay[a,b]

	W	Sp	S	F
Bacillariophyceae				
Achnanthes danica (Floegel) Grun.		X	X	
Actinoptychus senarius Ehrenb.		X	X	X
Amphiprora sp. Ehrenb.			X	
Amphora sp. Ehrenb.		X	X	
A. angusta Greg.			X	
A. ovalis Kutz.			X	
A. proteus Greg.		X		
Asterionella glacialis Castr.	B	A	X	C
Bacillaria paxillifer (Mull.) Hendey	X	X	X	X
Bacteriastrum delicatulum Cleve			X	
Biddulphia alternans (Bail.) Van Heurck		X	X	X
B. longicruris Grev.				X
B. mobiliensis (Bail.) Grun.	X	X	X	X
B. sinensis Grev.	X	X	X	X
Campylodiscus echineis Ehrenb.			X	
Cerataulina pelagica (Cleve) Hendey	X	B	X	
Chaetoceros affine Laud.			X	X
C. breve Schutt				X
C. compressum Laud.	X		X	X
C. debile Cleve	X	X	X	
C. decipiens Cleve	X	X	X	X
C. pendulum Karsten			X	X
C. sociale Laud.			X	C
C. subtile Cleve			X	
Cocconeis sp. Ehrenb.			X	
C. placentula Ehrenb.		X	X	X
C. scutellum ornata Grun.	X	X		X
Coscinodiscus asteromphalus Ehrenb.	X	X	X	X
C. grani Gough		X	X	
C. lineatus Ehrenb.				X
C. marginatus Ehrenb.	X	X	C	X
C. nitidus Greg.	X	X	X	X
C. oculus-iridis Ehrenb.	X			
C. radiatus Ehrenb.	X	X	X	X
Cyclotella caspia Grun.	X	X	X	X
C. glomerata Bach.	C	C	X	
C. meneghiniana Kutz.		X		
C. striata (Kutz.) Grun.		X		
Cylindrotheca closterium (Ehrenb.) Reimann and Lewis		X	X	X
Cymbella excisa Kutz.		X		
C. ventricosa Agardh		X		
Diatoma elongatum (Lyngb.) Agardh		X		
D. elongatum subsalsa Cleve-Euler			X	X
Detonula confervacea (Cleve) Gran		X	X	
Diploneis crabro Ehrenb.			X	X
Ditylum brightwellii (West) Grun.	X	X	X	X
Fragilaria crotonensis Kitt.		X	X	
F. oceanica Cleve		X		
Grammatophora marina (Lyngb.) Kutz.		X		
Guinardia flaccida (Castr.) Perag.		X	X	X
Gyrosigma sp. Hass.		X		
G. balticum (Ehrenb.) Cleve		X	X	X
G. fasciola (Ehrenb.) Cleve	X	X	X	X
G. hippocampus (Ehrenb.) Hass.		X		
G. prolongatum (Wm. Sm.) Cleve			X	

TABLE 2 (continued)
List of Phytoplankton Taxa from Lower Chesapeake Bay[a,b]

	W	Sp	S	F
G. simile Grun.				X
G. spenceri (Wm. Sm.) Cleve		X	X	X
Hemiaulus hauckii Grun.				X
H. membranaceus Cleve				X
Lauderia borealis Gran				X
Leptocylindrus danicus Cleve		X	X	C
L. minimus Gran				X
Licmophora abbreviata Agardh	X	X	X	X
L. paradoxa (Lyngb.) Agardh		X		
Melosira distans (Ehrenb.) Kutz.	X	X	X	
M. moniliformis (O. F. Mull.) Agardh			X	
Navicula sp. #1 Bory		X	X	
Navicula sp. #2 Bory	X	X	X	
N. arvensis Hust.	X	X	X	X
N. delawarensis Grun.		X		
N. gastrum Ehrenb.		X		
N. humerosa Breb.		X		
N. lanceolata (Agardh) Kutz.			X	
N. longa (Greg.) Ralfs		X		
N. maculosa Donk.		X		
N. placenta Ehrenb.		X		
N. pusilla Wm. Sm.		X		
N. salinarum Grun.		X	X	
N. sovereignae Hust.			X	
Nitzschia delicatissima Cleve			X	
N. longissima (Breb.) Ralfs		X	X	
N. obtusa scalpelliformis Grun.			X	X
N. pungens Grun.	X	X	A	X
N. seriata Cleve		X	X	B
N. sigma (Kutz.) Wm. Sm.		X		
N. spathulata Breb.			X	
Paralia sulcata (Ehrenb.) Cleve	X	X	X	X
Pinnularia sp. Ehrenb.			X	
Plagiogramma pulchellum pygmaea (Grev.) Perag.		X	C	
Pleurosigma sp. Wm. Sm.			X	
P. aestuarii (Breb.) Wm. Sm.		X		X
P. angulatum (Quekett) Wm. Sm.	X	X	X	X
P. obscurum Wm. Sm.				X
P. rigidum Wm. Sm.		X		
P. salinarum Grun.			X	
P. strigosum Wm. Sm.	X	X	X	X
Rhabdonema minutum Kutz.	X	X	X	X
Rhaphoneis amphiceros Ehrenb.	X	X	X	X
Rhizosolenia alata Brightw.	X	X	X	X
R. alata indica (Perag.) Gran			X	
R. calcar avis Schultze		X	C	X
R. delicatula Cleve		X		X
R. fragilissima Berg.	X	C		X
R. hebetata semispina (Hensen) Gran		X		
R. imbricata Brightw.	X	X		
R. robusta Norm.				X
R. setigera Brightw.	X	X	X	X
R. stolterfothii H. Perag.			X	X
R. styliformis Brightw.	X	X		X
Schroederella delicatula (Perag.) Pav.	X			X
Skeletonema costatum (Grev.) Cleve	A	C	X	A

TABLE 2 (continued)
List of Phytoplankton Taxa from Lower Chesapeake Bay[a,b]

	W	Sp	S	F
Stauroneis salina Wm. Sm.		X		
Stephanopyxis palmeriana (Grev.) Grun.				X
S. turris (Grev.) Ralfs	X	X		X
Surirella striatula Turp.			X	
Synedra sp. Ehrenb.			X	
S. affinis Kutz.		X		
S. fulgens (Grev.) Wm. Sm.			X	
S. ulna (Nitzsch) Ehrenb.		X		
Tabellaria sp. Ehrenb.			X	X
Thalassionema nitzschioides Hust.	X	C	X	X
Thalassiosira eccentrica (Ehrenb.) Cleve				X
T. gravida Cleve	X	X		
T. nordenskioldii Cleve	C	C		X
T. polychorda (Gran) Jorg.		X		
T. rotula Meun.	X	X		X
Thalassiothrix frauenfeldii Grun.		X		X
T. mediterranea Pav.	X			X
Tropidoneis lepidoptera (Greg.) Cleve			X	
Dinophyceae				
Amphidinium sp. Clap. and Lachm.			X	
A. klebsii Kof. and Swezy		X		
Ceratium furca (Ehrenb.) Clap and Lachm.			X	X
C. fusus (Ehrenb.) Dujard.			X	
C. horridum (Cleve) Grand		X		
C. lineatum (Ehrenb.) Cleve		X		
C. tripos (O. F. Mull.) Nitzsch		X	X	
C. tripos atlanticum Ostenfeld		X		
Dinophysis acuminata Clap. and Lachm.		X		
D. arctica Mereschk.	X			
D. caudata Kent			X	X
D. lachmannii Pauls.			X	
Diplopsalis lenticula Bergh	X			
Dissodium asymmetricum (Mangin) Leoblich III			X	
Gonyaulax sp. Dies.			X	
G. polygramma Stein			X	
G. spinifera (Clap. and Lachm.) Dies.			X	X
Gymnodinium aurantium Campbell		X		X
G. danicans Campbell	X	X	C	
G. galesianum Campbell		X	X	X
G. nelsoni Martin		X	X	X
G. roseostigma Campbell			X	
Gyrodinium aureolum Hulburt		X		
G. dominans Hulburt		X		
G. estuariale Hulburt	X	X	C	X
G. uncatenum Hulburt			X	
Heterocapsa triquetra (Ehrenb.) Stein	X	X	X	
Oblea rotunda (Lebour) Balech	X			
Prorocentrum compressum (Bail.) Abe	X	X	X	X
P. lima (Ehrenb.) Dodge			X	
P. micans Ehrenb.	X	X	X	C
P. minimum (Pav.) Schiller	X	X	X	X
P. redfieldi Bursa			X	
Protoperidinium sp. Bergh			X	
P. aciculiferum Lemm.	X	X	X	
P. brevipes (Pauls.) Balech			X	

TABLE 2 (continued)
List of Phytoplankton Taxa from Lower Chesapeake Bay[a,b]

	W	Sp	S	F
P. depressum (Bail) Balech		X		
P. excavatum (Martin) Balech			X	
P. oblongum (Aurivillius) Parke and Dodge				X
P. pentagonum (Gran) Balech			X	
Scrippsiella trochoidea (Stein) Leoblich III		X	X	X
Zygabikodinium lenticulatum Loeblich Jr. and Loeblich III			X	
Cryptophyceae				
Chroomonas amphioxeia (Conr.) Butch.		X		
C. caroliniana Campbell		X		
Cryptomonas sp. Ehrenb.			X	
C. ovata Ehrenb.		X	X	
C. pseudobaltica Butch.		X		
C. stigmatica Wislouch		X		
Hemiselmis virescens Droop		X		
Haptophyceae				
Chrysochromulina sp. Lackey		X		X
Emiliania huxleyi (Lohm) Hay and Mohler		X	X	X
Hymenomonas carterae (Braarud and Fagerl.) Braarud		X		
H. roseola Stein		X	X	
Pavlova salina (Carter) Green			X	
Pyrmnesium parvum Carter		X		
Chrysophyceae				
Calycomonas ovalis Wulff		X		
C. wulffii Conr. and Kuff.		X		
Dictyocha fibula Ehrenb.		X		
Ebria tripartita (Schumann) Lemm.	X	X	X	X
Ochromonas caroliniana Campbell		X		
O. variabilis H. Meyer		X		
Olisthodiscus carterae Hulburt		X	X	
Synura uvella Ehrenb.		X		
Xanthophyceae				
Monodus guttula Pasch.		X	X	X
Vacuolaria virescens Cienkowski		X		
Euglenophyceae				
Euglena sp. Ehrenb.		X	X	
E. deses Ehrenb.		X		
E. mutabilis Schmitz		X	X	
E. proxima Dangeard		X	X	
Eutreptia lanowii Steuer	X	X	X	
E. viridis Perty		X	X	
Trachelomonas hispida punctata Lemm.			X	
Prasinophyceae				
Pyramimonas sp. Schmarda			X	X
P. amylifera Conr.			X	X
P. grossii Parke	X			
P. micron Conr. and Kuff.			X	X
P. plurioculata Butch.				X
Tetraselmis gracilis (Kylin) Butch.		X		
T. maculata Butch.			X	

TABLE 2 (continued)
List of Phytoplankton Taxa from Lower Chesapeake Bay[a,b]

	W	Sp	S	F
Chlorophyceae				
Chlamydomonas vectensis Butch.		X	X	
Nannochloris atomus Butch.		X	B	X
Scenedesmus quadricauda (Turpin) Breb.	X	X	X	
Cyanophyceae				
Anacystis dimidiata (Kutz.) Dr. and D.			X	
A. marina (Hansg.) Dr. and D.		X	X	X
Gomphosphaeria aponina Kutz.			X	X
Nostoc commune Vauch.			X	X
Oscillatoria erythraea (Ehrenb.) Kutz.			X	X
O. submembranaceae Ard. and Straff.			X	X
Schizothrix calcicola (C. Ag.) Gom.			X	X
Spirulina subsalsa Oerst.			X	

[a]   Seasonal presence noted with X.
[b]   Dominant species are indicated by A, B, C in order of decreasing abundance.

From Marshall, H. G., *Estuaries*, 3, 207, 1980. With permission.

TABLE 3
Seasonally Dominant Species Observed in Lower Chesapeake Bay Waters off Cape Charles (City) and in Old Plantation Creek in 1978 and 1979

Winter (December—February)	Spring (March—May)	Summer (June—August)	Fall (September—November)
Skeletonema costatum	Asterionella glacialis	Nitzschia pungens	Skeletonema costatum
Asterionella glacialis	Cerataulina pelagica	Coscinodiscus marginatus	Leptocylindrus danicus
Cyclotella glomerata	Cyclotella glomerata	Nannochloris atomus	Nitzschia seriata
Thalassiosira nordenskioldii	Thalassiosira nordenskioldii	Gymnodinium danicans	Chaetoceros sociale
	Skeletonema costatum	Gyrodinium estuariale	
	Thalassionema nitzschioides	Cryptomonas sp.	
	Rhizosolenia fragilissima	Chroomonas caroliniana	
		Rhizosolenia calcar avis	
		Ceratium furca	

From Marshall, H. G., *Estuaries*, 3, 207, 1980. With permission.

or equal to 5.3×10^7 cells per liter) in the spring during a bloom in the mesohaline waters along the shallow western shore of the bay. In this region of the estuary, cell numbers consistently increased in the spring and fall over the period 1975 to 1981, concurrently with rising chlorophyll values. Higher cell numbers and chlorophyll readings during these seasons have been ascribed to greater numbers of diatoms. Mid-summer increases in chlorophyll, in turn, have been attributed to a greater contribution by dinoflagellate populations.

Primary production on a daily basis in the estuary averages about 1 to 5 g C per square meter; high production figures correspond to the warmer months of the year (Figure 5). Flemer,[77] evaluating standing crop, listed maximum chlorophyll *a* concentrations of 20 to 60 µg/l.

Nanoplankton — both diatoms and phytoflagellates — are a major component of the total phytoplankton composition and productivity. McCarthy et al.[99] assigned 56.6% to 89.6% of the phytoplankton productivity in Chesapeake Bay to nanoplankton less than 35 μm in size. Nanoplankton less than 10 μm in size generally exceed net plankton in concentration during all seasons in the upper bay.[100] Marshall[90] stressed the presence of diatoms less than 10 μm in size, such as *Navicula* spp., *Cyclotella caspia,* and *Rhabdonema minutum.* Most of the photosynthetic plankton cells in the estuary probably fall in the 2- to 5-μm size range.[101]

Sellner[97] summarized the distributions of phytoplankton taxa, density, biomass, and production, as well as the role of phytoplankton in carbon, oxygen, and nutrient dynamics of Chesapeake Bay. Conclusions of his summation follow. Phytoplankton biomass, as chlorophyll and cell densities, and production exhibit slightly different seasonal distributions. Peak chlorophyll concentrations and cell densities occur in the spring concurrently with the spring bloom. Highest primary production figures also take place concomitantly with the spring bloom, but a secondary maximum develops in the summer. However, phytoplankton blooms form aperiodically throughout the year which can obfuscate these generalized seasonal distribution patterns.

Phytoplankton production in Chesapeake Bay ranges from 74 to 851 g C per square meter per year; production rates in the upper bay (74 to 86 g C per square meter per year) are substantially less than in the mid-bay region (167 to 851 g C per square meter per year).[97] As expounded above, nanoplankton, including microflagellates and small diatoms, are responsible for the major fraction of this production. Aperiodic nutrient limitation (nitrogen, phosphorus, and silicon) constrains phytoplankton production in mesohaline and polyhaline waters. Small coccoid cells about 1 to 2 μm in size may be the most important nanoplankters in terms of total cell numbers, reaching densities of 10^6 to 10^8 cells per liter in the spring and late summer. Much of the primary production by phytoplankton in the estuary supports microheterotrophic metabolism, especially aerobic bacterial populations in the water column and sediments. Low grazing rates on phytoplankton, high sedimentation rates of the cells to the pycnocline, subpycnocline waters, and benthos, and microheterotrophic utilization of the phytoplankton production result in the development and maintenance of hypoxia (less than 2 mg O_2 per liter) and anoxia in the subpycnocline waters in mid Chesapeake Bay.

Cyclotella sp. and *Thalassiosira* sp. dominate the spring bloom in the upper bay, while pennate diatoms and chlorophytes, especially *Scenedesmus* spp., are subdominants at the head of the bay. Dinoflagellates increase in importance as salinity rises toward mesohaline waters, forming 40% of the phytoplankton assemblage in the spring bloom of this area of the estuary. *Prorocentrum minimum (P. mariae-lebouriae)* and *Katodinium rotundatum* are the principal dinoflagellate species constituting the bloom. Marine diatoms become dominant in the mesohaline and polyhaline waters of the bay; some of the most common species are *Cerataulina pelagica, Rhizosolenia fragilissima, Asterionella japonica, Leptocylindrus* spp., *Nitzschia pungens, Skeletonema costatum, Cyclotella glomerata,* and *Thalassiosira nordenskioldii.*

During the spring bloom in 1985 and 1986, the following chlorophyll measurements were registered in surface waters of the estuary. Values at the head of the bay ranged from 6.8 to 12.7 μg/l, increasing to 8.8 to 24.3 μg/l in oligohaline waters in the vicinity of the turbidity maximum and 26.4 to 36.3 μg/l at the head of the deep trough. Lowest chlorophyll readings, approaching 3 μg/l, developed over the deep trough in the mesohaline region, but higher figures of 16.4 to 21 μg/l occurred in shallower polyhaline regions of the lower bay. Subsequent to the spring bloom, chlorophyll concentrations diminished to minima of less than 5 μg/l in November and December.

Cell densities of eukaryotes typically surpass 1×10^7 cells per liter at the time of a spring bloom, but as diatom abundances decline, the densities drop below 1×10^7 cells per liter into the summer. It is not unusual for the cell densities to increase once again in the fall (late September to October), albeit not as dramatically as in the spring bloom. In 1985, for example, diatom densities in the mesohaline region of the bay increased from 1.1×10^6 cells per liter in late September to 6.6×10^6 cells per liter in October. Lowest cell densities, like chlorophyll, are apparent in November and December, amounting to 2×10^6 cells per liter in the upper bay and less than or equal to 0.8×10^6 cells per liter in the lower bay.

Diatom and dinoflagellate blooms appear aperiodically in summer, fall, or winter, with densities exceeding 1×10^7 cells per liter. In winter, for instance, dinoflagellate blooms comprised principally of *Katodinium rotundatum* and *Heterocapsa triquetra* are common. In this season, blooms of the diatoms *Skeletonema costatum* and *Chaetoceros* spp. (especially *C. atlanticum*) also arise. Dinoflagellate blooms consisting of multiple species create impressive red or mahogany tides during the interval of April through October. *Gymnodinium nelsonii, G. stellatum, Gyrodinium uncatenum, G. estuariale, Peridinium* spp., *Cochlidinium catenatum, Prorocentrum* spp., *Oxytoxum*, and *Ceratium* are representative taxa composing these dinoflagellate blooms.

2. Barnegat Bay

Barnegat Bay is a shallow, lagoon-type estuary located along the East Coast of New Jersey between 39°41′ N and 39°56′ N latitude and 74°04′ W and 74°12′ W longitude.[102] It represents a type estuary characteristic of the back bay system of the barrier island coastline of New Jersey. More than 180 species of phytoplankton have been sampled in this estuary, and, as in the case of Chesapeake Bay, the majority of them consist of diatoms and dinoflagellates (Table 4).

Mountford[92] followed the seasonal appearances and disappearances of phytoplankton populations in the bay, interpreting the conspicuous seasonal periodicity of 44 species and groups of species as a response of the organisms to complex interactive effects of temperature, photoperiod, and nutrient supply. Alternating warm-water and cold-water flora appear annually in a successional sequence. Seasonal periodicities in abundance and primary production also typify the phytoplankton community. Phytoplankton abundance is maximum in the summer and minimum from the late fall to mid winter. Phytoplankton densities range from 80,000 to 800,000 cells per milliliter in the summer, with ultraplankton (1 to 3 μm in size) being most abundant. Microflagellates and dinoflagellates are more numerous than diatoms in the summer and fall, attaining densities greater than 10^6 cells per liter in the summer. Common dinoflagellate species at this time include the naked forms *Gymnodinium incoloratum* and *G. punctatum* and the thecate forms *Gonyaulax digitalis, G. spinifera, Prorocentrum micans, P. redfieldi, P. scutellum*, and *P. minimum*. Intense luminescent, dinoflagellate blooms have been denoted during the summer and fall. Only three diatom species remain reasonably abundant in the warmer months — *Skeletonema costatum, Cyclotella* sp., and *Cylindrotheca closterium*. Some summer populations persist into the fall, as diatoms (e.g., *Amphiprora* sp., *Licmophora* sp., and *Thalassionema nitzchioides*) begin to reestablish. Phytoplankton concentrations decrease in the fall with declining insolation and temperature. Diatoms become numerically dominant in early to mid-winter; the standing crop and productivity of phytoplankton drop to minimum levels during periods of ice cover. A diatom bloom manifests itself in mid- to late winter each year subsequent to the breakup of ice cover with rising water temperature and insolation. A succession of diatom populations generate this bloom, with *Thalassiosira nordenskioldii* and *Detonula confervacea* sequentially dominating the phytoplankton in cell numbers and biomass. Zooplankton grazing accompanies the termination of the diatom bloom. *Skeletonema costatum* replaces *T. nordenskioldii* and *D. confervacea* as the principal phytoplankton population in the spring as water temperature approaches 20°C and light intensifies. Over an annual cycle, *S. costatum* is the most common diatom in Barnegat Bay. It seems likely that *T. nordenskioldii, D. confervacea*, and *S. costatum* are inoculated from nearshore oceanic waters into the estuary in the late winter and spring, where nutrient-rich waters enhance their rapid growth.

In terms of standing crop, chlorophyll *a* determinations range from 1 to greater than 35 μg/l. Chlorophyll *a* values as an estimate of biomass are greatest during the winter-spring diatom

TABLE 4

Taxonomic List of Phytoplankton from Barnegat Bay, NJ

Bacillariophyceae
Achnanthes longipes Agardh
Actinoptychus senarius (Ehr.) Ehr.
Amphiprora incompta Hohn and Hellerman
A. surirelloides Hendey
Amphora sp.
Asterionella glacialis Castracane
Bacillaria paxillifer (O. F. Müller) Hendey
Biddulphia spp.
B. arctica (Brightw.)
B. pulchella Gray
B. granulata Roper
B. vesiculosa Agardh
Campylodiscus sp.
C. fastuosus Ehr.
Cerataulina bergoni H. Peragallo
Chaetoceros spp.
C. approximatus Gran and Angst
C. borealis Bailey
C. curvistus Cleve
C. debilis Cleve
C. decipiens Cleve
C. dichaeta Ehr.
C. didymus Ehr.
C. fragilis Meunler
C. secundus Cleve
C. similis Cleve
C. simplex Ostenfeld
C. subtilis Cleve
Cocconeis spp.[a]
Coscinodiscus spp.
C. angstii Gran
C. centralis Ehr.
C. radiatus Ehr.
Cyclotella of meneghiniana Kutzing[a]
Cylindrotheca closterium (Ehr.) Reimann & Lewin
Cymbella spp.
Detonula confervacea Cleve[a]
Diploneis sp.
D. crabro Ehr.
Ditylum brightwelli (West) Grunow
Eucampia groenlandica Cleve
E. zodiacus Ehr.
Fragilaria sp.
F. crotonensis Kitton
F. cylindrus Grunow
Grammatophora spp.
Guinardia flaccida (Castr.) Peragallo
Gyrosigma spp.
Leptocylindrus sp.
L. danicus Cleve
L. minimus Gran
Licmophora sp.
Lithodesmium undulatum Ehr.
Melosira sp.
M. borreri Greville
M. granulata (Ehr.) Ralfs
M. juergensii Agardh

TABLE 4 (continued)
Taxonomic List of Phytoplankton from Barnegat Bay, NJ

M. nummuloides (Dillw.) Agardh
Navicula spp.[a]
N. cruciculoides Brockman
N. distans (W. Smith) Schmidt
N. (Schizonema) gravelei Ag.
N. monilifera Cleve
Nitzschia sp.
N. seriata Cleve
Paralia sulcata (Ehr.) Cleve
Pinnularia sp.
P. ambigua Cleve
Pleurosigma sp.
P. fasciola W. Smith
P. formosum W. Smith
P. marinum Donkin
Porosira glacialis (Grun.) Jörgensen
Rhabdonema adriaticum Kutzing
Rhizosolenia sp.
R. alata Brightwell
R. cylindrus Cleve
R. delicatula Cleve
R. fragilissima Bergon
R. hebetata forma *semispina* (Hensen) Gran
R. setigera Brightw.[a]
R. stolterfothii H. Perag.
Skeletonema costatum (Greville) Cleve
Striatella unipunctata (Lyngbye) Ag.
Surirella sp.
S. smithii Ralfs
Synedra sp.
S. hennedyana Greg.
Tabellaria sp.
Thalassionema nitzschioides Hustedt
Thalassiosira spp.
T. condensata Cleve
T. excentrica (Ehr.) Cleve
T. gravida Cleve
T. hyalina (Grun.) Gran
T. nordenskioldii Cleve[a]
T. rotula Meunier
Thallassiothrix frauenfeldii (West) Grun.
T. longissima Cleve and Grun.
Triceratium favus (Ehr.)

Dinophyceae
Amphidinium spp.
A. carteri Hulburt
A. fusiforme Martin
A. sphenoides Wulff
Ceratium bucephalum? (Cleve) Cleve
C. fusus (Ehr.) Dujardin
C. macroceros (Ehr.) Vanhöffen
C. minutum Jörgensen
C. tripos (O. F. Müller) Nitzsch[a]
Cochlodinium helix (Pouch.) Lemm. ex Lebour
Dinophysis sp.
D. acuminata Clap. and Lach.
D. acuta Ehr.

TABLE 4 (continued)
Taxonomic List of Phytoplankton from Barnegat Bay, NJ

D. ovum Schutt
Dipolopsalis lenticula Bergh
Glenodinium sp.
G. danicum Paulsen
G. foliaceum Stein
Goniodoma sp.
Gonyaulax sp.
G. digitalis (Pouchet) Kofoid[a]
G. polygramma Stein
G. scrippsae Kofoid
G. spinifera (Clap. and Lach.) Diesing[a]
G. tricantha Jörgensen
Gymnodinium spp.
G. incoloratum Conrad & Kufferath[a]
G. nelsoni Martin
G. punctatum Pouchet
G. splendens Lebour[a]
Gyrodinium spp.
G. dominans Hulburt
G. fissum (Levander) Kof. & Swezy
G. pellucidum (Wulff) Schiller
G. pingue (Schutt) Kofoid and Swezy
G. resplendens Hulburt
Hemidinium sp.
Heterocapsa triquetra (Ehr.) Stein
Katodinium sp.
Nematodium sp.
N. armatum (Dogiel) Kofoid & Swezy
Noctiluca scintillans (Macartney) Ehr.
Oblea rotunda (Lebour) Balech[a]
Ostreopsis monotis Schmidt
Peridinium spp.
P. brevipes Paulsen
P. claudicans Paulsen
P. depressum Bailey
P. excavatum Martin
P. granii Ostenfeld
P. leonis Paviilard[a]
P. pallidum Ostenfeld
P. roseum Paulsen
P. trochoideum (Stein) Lemm.[a]
Polykrikos sp.
P. barnegatensis Martin
P. hartmanni Zimmerman
P. kofoidi Chatton
Prorocentrum micans Ehr.[a]
P. minimum (Pavilliard) Schiller[a]
P. redfieldi Bursa[a]
P. scutellum Schroeder

Other flagellate forms
Bipedomonas sp.
Calycomonas gracilis (Lohmann) Wulff
Carteria sp.
Chlamydomonas spp.
Chroomonas sp.
Cryptomonas spp.[a]
Distephanus speculum (Ehr.) Haeckel

TABLE 4 (continued)
Taxonomic List of Phytoplankton from Barnegat Bay, NJ

Ebria tripartita (Schumann) Lemmermann
Euglena spp.[a]
Eutreptia sp.[a]
Ochromonas sp.
Pyramimonas sp.
P. tetrarhynchus Schmarda
P. torta Conrad & Kufferath
Scherefflia dubia Pascher

Other forms
  Aphanothece sp.
  Lyngbya sp.
  Merismopedia sp.
  Nannochloris sp.
  Oscillatoria spp.
  Pediastrum sp.
  Phormidium sp.
  Scenedesmus quadricauda (Turpin) Brebisson
  Spirulina sp.

[a]   Particularly important species, seasonal dominants, or ubiquitous members.

From Mountford, K., *Ecology of Barnegat Bay, New Jersey,* Kennish, M. J. and Lutz, R. A., Eds., Springer-Verlag, New York, 1984, 52. With permission.

bloom or during the period of maximum cell counts in the summer. Phytoplankton productivity also peaks in the summer when maximum gross productivity surpasses 750 mg O_2 per cubic meter per hour. Productivity measurements are minimum in the winter, approaching 0 mg O_2 per cubic meter per hour.

3. Long Island Sound

Phytoplankton have been studied extensively in this large, semienclosed body of water.[3,8,54,93,95,103-105] A taxonomic list of phytoplankton recovered from the sound contains about 200 species, 40 of which are designated as major constituents.[93] Among the chief floral components, 13 species of diatoms and dinoflagellates were deemed to be most important in degree of dominance and frequency of occurrence from year to year during 8 years of observations (Table 5).[54] *Skeletonema costatum* represented the overwhelmingly dominant form, a position commonplace in many New England and mid-Atlantic estuaries and nearshore oceanic waters.

A seasonal periodicity of phytoplankton taxa is conspicuous in Long Island Sound. This predictable alternation of phytoplankters conforms to the definition of succession advanced by Pickett[106] (=changes in species composition and community structure through time). Breitburg[236] suggests that succession may result "from direct interactions between earlier and later colonizers, or from indirect effects by which earlier residents make the physical environment more favorable or less favorable for later animals ... and from non-interactive processes such as chance introduction of propagules or differences in growth rates and longevity among species". Conover[93] advocated the conditioning of the water in Long Island Sound as a principal factor in the phytoplankton species succession.

Species occurrence and abundance of diatoms, dinoflagellates, and silicoflagellates are seasonally cyclical. Diatoms and silicoflagellates attain peak abundance in the winter, whereas dinoflagellates reach numerical dominance in the summer when the rate of production tends to

TABLE 5
The 13 Most Consistently Important Species of
Phytoplankton from Year to Year in Long Island Sound

Species composition	
Diatoms	**Dinoflagellates**
Skeletonema costatum	*Peridinium trochoideum*
Thalassionema nitzschioides	*Prorocentrum triestinum*
Paralia sulcata	*P. scutellum*
Schroederella delicatula	*Exuviaella apora*
Thalassiosira decipiens	
T. gravida	
T. nordenskioldii	
Rhizosolenia setigera	
R. delicatula	

Modified from Riley, G. A., *Estuaries,* Publ. 83, Lauff, G. H., Ed., American Association for the Advancement of Science, Washington, D.C., 1967, 316. With permission.

be high together with zooplankton grazing pressure. At times of maximum phytoplankton numbers, cell counts exceed 10^6 cells per liter (Figure 6). Chlorophyll *a* concentrations average approximately 1 to 10 µg/l, but during the winter-spring diatom bloom, can surpass 20 µg/l (Figure 7). Mandelli et al.[105] obtained highest chlorophyll readings for a region of the sound in the spring and summer (April to August) and lowest readings in the fall and winter. Further seaward of this location, however, these individuals recorded higher quantities of chlorophyll in the fall and winter and lower quantities in the spring and summer.

A well-defined longitudinal gradient in phytoplankton abundance also exists within Long Island Sound. In regard to standing crop, Riley[103] chronicled mean annual chlorophyll values for the entire water column of 9.2, 6.2, and 1.7 µg/l in the western, central, and eastern regions, respectively. In terms of cell numbers, Riley and Conover[104] amassed mean annual surface abundance measurements of 1.7×10^6 cells per liter in the western perimeter, 1.0×10^6 cells per liter in the central sector, and 2.2×10^5 cells per liter in the eastern margin. These regional differences correlate with a gradual diminution of phosphate and inorganic nitrogen concentrations when proceeding toward the Atlantic Ocean. Gross primary production has been estimated at 470 g C per square meter per year.

The winter-spring diatom bloom in this system, flowering between late January and March, is dominated by the diatoms *Thalassiosira nordenskioldii* and *Skeletonema costatum*.[54,95] Early in the season when water temperature and light intensity are reduced, *T. nordenskioldii* predominates and is superseded by *S. costatum* as the primary form at higher temperatures and light intensities. *Asterionella japonica, Chaetoceros* spp., *Lauderia borealis, Leptocylindrus danicus, Peridinium trochoideum,* and other species occur somewhat later. Mandelli et al.,[105] however, ascertained a pulse of diatoms during the late winter and spring comprised of *S. costatum,* either with *Thalassiosira* and *Chaetoceros* spp., or with *Thalassionema* and *Nitzschia* spp. The community composition changes from April to August with the appearance and growth of many populations of dinoflagellates and microflagellates, particularly in the summer. Diatoms may remain dominant until the end of May followed by dinoflagellates in a classic diatom to dinoflagellate succession. *S. costatum* dominates phytoplankton blooms in the early fall,[54] when *Chaetoceros* spp. also increase in numbers.[93]

Similar seasonal patterns have been uncovered among the diverse assemblage of phytoplankton of the New York Bight and adjacent waters.[107] Here, chain-forming diatoms and piconanoplankton dominate the community on a seasonal basis. Marshall and Cohn[107] discovered late

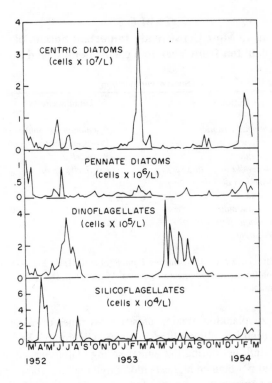

FIGURE 6. Abundance and seasonal occurrence of major taxonomic groups of phytoplankton in Long Island Sound from March 1952 to March 1954. (From Conover, S. M., *Bull. Bingham Oceanogr. Coll.*, 15, 156, 1956. With permission.)

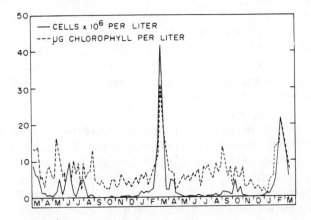

FIGURE 7. Phytoplankton cell numbers and chlorophyll concentrations in the surface waters of Long Island Sound from March 1952 through March 1954. (From Smayda, T. J., *Coastal and Offshore Environmental Inventory, Cape Hatteras to Nantucket Shoals*, Mar. Publ. Ser. No. 2, University of Rhode Island, Kingston, 1973. With permission.)

winter-early spring and fall maxima of phytoplankton in these waters and a general species composition like that of other regions of the northeastern continental shelf of the U.S.

The pattern of seasonal succession of diatoms and dinoflagellates seems to reflect, in part, the dependence of dinoflagellates on diatoms. Levinton (p. 180)[8] clarifies this relationship as follows:

> Conceivably, groups like dinoflagellates are dependent on exudates and nutrients produced in the excretion and decomposition of species earlier in the successional sequence. For example, diatoms early in the successional sequence may be autotrophic, requiring only inorganic nutrients for their survival, whereas species later in the successional sequence might be auxotrophic, requiring nutrients like vitamins that they cannot produce themselves. Later species therefore cannot reach great abundance until the flowering of earlier species. Dinoflagellates are known typically to require more nutrients that they cannot manufacture themselves than diatoms, perhaps explaining the successional sequence in the plankton.

In general, however, the seasonal succession of phytoplankton conforms to the geographical distribution of species; taxa early in succession typify nutrient-rich (eutrophic) coastal environments and those later in succession characterize nutrient-poor (oligotrophic) offshore waters.[8]

Riley[54] postulated that the solar radiation level from mid December to the onset of the winter-spring diatom bloom played an instrumental role in initiating the bloom. Nutrient availability, especially nitrogen, and zooplankton grazing (e.g., by *Acartia hudsonica* and *A. tonsa*) regulate phytoplankton density subsequent to the winter-spring diatom bloom, maintaining lower cell numbers until the fall bloom. Smaller, sporadic pulses of phytoplankton during late spring and early summer probably derive from ammonia formation and utilization by the autotrophs. Two factors may control bloom development in the fall: nutrient supply and light. Subsequent to the breakdown of the summer thermocline, strong vertical mixing of the water column can enrich the surface waters with nutrients, stimulating phytoplankton growth.[103] Nutrient concentrations in the deeper waters of Long Island Sound are minimal, however, spurring Conover[93] to conclude that light, rather than nutrients, is the critical factor regulating fall blooms of phytoplankton.

4. Narragansett Bay

Phytoplankton research on Narragansett Bay has been performed nearly continuously for more than three decades.[34,35,65,95,108-112] Particularly noteworthy is a location in lower Narragansett Bay where a 22-year series of quantitative, weekly phytoplankton samples has been collected to evaluate temporal occurrence and abundance patterns.[36] A shallow, well-mixed estuary, Narragansett Bay represents the site of a major phytoplankton bloom usually between December and April each year, as well as secondary pulses year round.[65,95,110,111,113] Durbin and Durbin[34] point out that small flagellates prevail as the main component of the phytoplankton in the summer and diatoms, in the winter, spring, and fall. Smayda[35] affirms this finding of diatom predominance during most of the year, aside from summer maxima and sporadic occurrences of dinoflagellates such as *Peridinium torchoideum*, *Prorocentrum triangulatum*, and *Massartia rotundata*.

The diatom *Skeletonema costatum* is the most abundant species, accounting for about 80% of all the phytoplankton. Nine diatom species, in addition to four flagellate groups (i.e., *Gymnodinium* spp., a gymnodinoid, *Prorocentrum*, and microflagellates) comprise more than 90% of the total numerical abundance.[3] Of the 138 phytoplankton taxa (excluding microflagellates) identified in the estuary by Karentz and Smayda,[36] 84 diatom and 30 dinoflagellate species were reported. Table 6 lists the frequency of occurrence, total abundance, and maximum recorded values of the 49 most abundant forms during a 22-year sampling period. *S. costatum* clearly was dominate, occurring in 876 of 1000 samples.

The winter-spring diatom bloom commonly commences in December, but may begin as early as November or as late as February. *Detonula confervacea*, an Arctic species, often initiates the bloom and is replaced by *Thalassiosira nordenskioldii* and *S. costatum* as the dominant forms; however, *S. costatum* frequently prevails during most of the bloom period.[95] Subsequently, *Leptocylindrus danicus*, *L. minimus*, *Cerataulina pelagica*, *Asterionella japonica*, and

TABLE 6
List of the 49 Most Abundant Phytoplankton Species from Narragansett Bay, RI

Species	cum	fg	max
Skeletonema costatum (Grev.) Cleve	2,395,720	876	108,749
Detonula confervacea (Cleve) Gran	253,971	264	13,298
Asterionella glacialis Castracane	208,471	355	13,129
Olisthodiscus luteus Carter	138,580	226	31,799
Thalassiosira nordenskioldii Cleve	133,440	323	8,124
Thalassiosira sp. Cleve	119,045	352	6,874
Leptocylindrus minimus Gran	56,635	478	4,951
L. danicus Cleve	49,240	279	9,584
Chaetoceros compressus Lauder	40,859	115	3,302
Rhizosolenia delicatula Cleve	37,638	296	5,448
R. fragilissima Bergon	29,350	201	12,314
Chaetoceros debilis Cleve	20,557	82	3,045
Katodinium rotundatum (Lohmann) Fott	20,254	206	8,749
Porocentrum redfieldii Bursa	19,800	193	3,199
Chaetoceros curvisetum Cleve	19,353	151	2,295
Thalassionema nitzschioides Hustedt	17,896	268	1,512
Cerataulina pelagica (Cleve) Hendy	15,231	120	3,813
Thalassiosira rotula Meunier	15,065	248	1,458
Chroomonas amphioxeia (Conrad) Butcher	12,722	213	629
Gymnodinium sp. Stein	9,644	399	1,899
Chaetoceros gracilis Schütt	9,143	47	1,533
C. diadema (Ehrenb.) Gran	9,004	45	2,495
Pyramimonas torta Conrad et Kuff	7,676	189	3,444
Scrippsiella trochoidea (Stein) Loeblich	7,519	197	1,351
Chaetoceros tortissima Gran	7,456	33	2,089
Prorocentrum triangulatum Martin	7,062	193	579
Chaetoceros didymus Ehrenb.	6,655	131	775
Nitzschia closterium (Ehrenb.) Wm. Smith	6,518	297	533
Chaetoceros affinis Lauder	6,045	58	2,049
C. similis Cleve	5,936	96	557
Nitzschia seriata Cleve	5,908	129	557
Thalassiosira decipiens Cleve	5,423	91	2,037
Rhizosolenia setigera Brightwell	5,166	268	311
Chaetoceros lorenzianus Grunow	4,593	68	524
C. decipiens Cleve	4,489	72	810
Thalassiosira gravida Cleve	4,263	46	878
Chaetoceros socialis Lauder	3,691	19	708
Prorocentrum scutellum Schroder	3,092	150	434
Thalassiosira pseudonana Hasle et Heimdal	2,967	16	1,194
Chaetoceros subtilis Cleve	2,940	60	1,462
Ditylum brightwelli (T. West) Grunow ex. van Heurck	2,536	96	536
Unidentified flagellate	2,550	33	462
Heterocapsa triquetra (Ehrenb.) Stein	2,510	118	258
Lithodesmium undulatum Ehrenb.	2,299	30	1,065
Cyclotella caspia Grunow	1,782	4	993
Peridinium sp. Ehrenb.	1,551	177	216
Phaeocystis pouchetti (Hariot) Lagerh.	1,305	48	499
Eucampia zoodiacus Ehrenb.	1,286	49	155
cf. *Attheya decora* West	1,272	69	72

[a]   Weekly cell/milliliter values.
[b]   Frequency of occurrence in 1000 samples.
[c]   Maximum cells/milliliter observed.

From Karentz, D. and Smayda, T. J., *Mar. Ecol. Prog. Ser.*, 18, 277, 1984. With permission.

Rhizosolenia fragilissima become secondary dominant species.[35] *S. costatum* continues to dominate after the winter-spring diatom bloom, which collapses in May or June, but during the summer the chrysophycean flagellate, *Olisthodiscus luteus,* occasionally replaces *S. costatum* and can form nontoxic red tides. Durbin and Durbin[34] relate successive blooms of *O. luteus, Massartia rotundata,* and *S. costatum* during June, July, and August. In late summer to fall, pulses of *S. costatum, L. danicus, A. japonica,* and *R. fragilissima* usually take place. Figure 8 displays the seasonal abundance and succession of some major phytoplankton species in Narragansett Bay.

Maximum phytoplankton abundance during most years occurs synchronously with the winter-spring diatom bloom; however, in some years (e.g., 1970, 1973—1975), peak cell counts develop in the summer (mid-August). During the winter-spring diatom bloom, maximum phytoplankton numbers range from 4.5×10^6 (1970) to 42×10^6 cells per liter (1963), and during blooms in the late spring through fall, range from 7×10^6 (1967) to 59×10^6 cells per liter (1975).[35] Figure 9 depicts the variability of phytoplankton cell counts over an annual cycle in 1972 and 1973. The magnitude of crop in Narragansett Bay is comparable to or greater than that of other sheltered regions (Table 7).

The standing crop of phytoplankton in terms of chlorophyll concentrations also peaks concurrently with winter-spring and fall blooms. Estimates of chlorophyll *a* concentrations exceed 10 μg/l during the winter-spring diatom bloom and can be greater than 35 μg/l.[114] Phytoplankton blooms generally originate earlier in the shallower areas of Narragansett Bay,[110] where higher chlorophyll values precede those in contiguous deeper waters. When blooms are absent, chlorophyll *a* concentrations normally fall below 10 μg/l.

The mean annual net production for Narragansett Bay has been measured at 269 g C per square meter per year, with highest values in the Providence River (429 g C per square meter per year) and lowest values in the East Passage (218 g C per square meter per year).[65] A gradient in primary production, in accordance with known nutrient gradients, extends from upestuary (higher values) to downestuary (lower values) and from the West Passage (higher values) to the East Passage (lower values). The mean annual net production figure of 269 g C per square meter per year compares favorably with that determined by Smayda[95] for this estuary (220 g C per square meter per year), as well as the production estimates of other temperate estuarine ecosystems.

VI. FACTORS AFFECTING PRIMARY PRODUCTION OF PHYTOPLANKTON

A. GENERAL

Estuaries are highly dynamic coastal environments that fluctuate markedly in physical, chemical, and biological conditions.[115] Diurnal, seasonal, and tidal changes in temperature, salinity, and other physical and chemical factors typify these ecosystems. Light and nutrient availability likewise undergo continual temporal variation. In addition, water circulation and turbidity, often subjected to acute perturbations, influence other parameters critical to the success of planktonic algal populations.

The taxonomic structure of estuarine phytoplankton communities changes through time in response to both abrupt alterations and seasonal oscillations in physical-chemical conditions of the environment.[35,36,116,117] The annual successional patterns established for temperate estuaries, as described above, provide noteworthy examples. In these systems, variations in primary productivity accompany modifications in the species composition and abundance of the communities.

Numerous efforts have been attempted to establish conclusively the factors governing primary production and community dynamics of phytoplankton in estuaries. It has been exceedingly arduous, however, to separate and quantify the effects of factors controlling

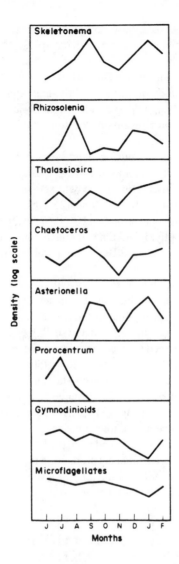

FIGURE 8. Seasonal abundance of
some major phytoplankton species in
Narragansett Bay, depicting general
successional patterns. (From Raymont,
J. E. G., *Plankton and Productivity in the
Oceans*, Vol. 1, 2nd ed., Pergamon
Press, Oxford, 1980. With permission.)

phytoplankton production.[118] Because of fluctuations in environmental conditions, phytoplankton divide at variable rates; consequently, primary productivity differs from one geographic area to another and seasonally within the same geographic region.[53] Until recently, investigations of biological and chemical factors in the development and maintenance of phytoplankton communities have received more attention than those of physical processes.[83]

Both natural and anthropogenic events can modulate environmental factors that affect the abundance and succession of estuarine phytoplankton (Table 8). Rice and Ferguson (p. 2),[31] drawing from the work of McCombie,[119] espouse three types of environmental factors — limiting, controlling, and lethal — that influence the dynamics and physiology of phytoplankton populations.

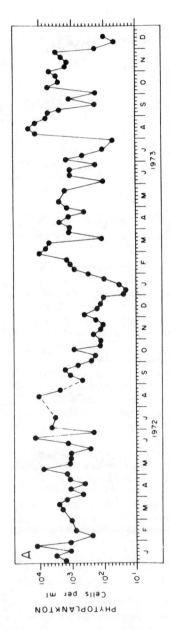

FIGURE 9. Total phytoplankton abundance in Narragansett Bay from January 1972 through December 1973. (From Hulsizer, E. E., *Chesapeake Sci.*, 17, 260, 1976. With permission.)

TABLE 7
Phytoplankton Cell Counts in Narragansett Bay as Compared to Other Temperate Coastal Areas

	Mean crop (cells $\times$ 10^3/1)
Block Island Sound	471
Gulf of Maine	256
Long Island Sound	2500
Lower Narragansett Bay	6700
Vineyard Sound	23
Kiel Bay	627
Loch Striven	1000
California Coast	50

From Smayda, T. J., *Limnol. Oceanogr.*, 2, 342, 1957. With permission.

TABLE 8
Natural and Man-Imposed Conditions Which Determine Levels and Rates of Change of Factors Affecting Abundance and Succession of Estuarine Phytoplankton

Factors	Natural conditions	Man-imposed conditions
Salinity	Precipitation, runoff, evaporation, circulation of water	Water impoundment, channelization, dredge and fill, mosquito ditching
Temperature	Latitude, season, weather, time of day, circulation of water	Heated effluent, dams, canals and waterways, stream channelization
Light intensity		
At surface	Latitude, season, weather, time of day	Air pollution — smog
Below surface	Reflection, absorption, scattering	Dredging, waste dumping, erosion
Nutrients	Drainage, runoff, circulation of water, sediments	Sewage and industrial wastes, urban and agricultural drainage, erosion
Metabolites	Living and dead plants and animals	Sewage, urban and agricultural drainage, erosion
Toxic substances		
Petroleum	Deposits	Leaks and spills during drilling, transport, storage, use or disposal
Radionuclides	Primordial deposits, cosmic ray produced	Fallout, nuclear power reactors, other releases
Heavy metals	Terrestrial deposits, sediments, land drainage	Industrial and domestic wastes, mining, erosion
Synthetic toxicants		Industrial, agricultural, and domestic use

From Rice, T. R. and Ferguson, R. L., *Physiological Ecology of Estuarine Organisms,* Vernberg, F. J., Ed., University of South Carolina Press, Columbia, 1975, 1. With permission.

Environmental factors which affect phytoplankton distribution are referred to as limiting, controlling, or lethal.[119] Limiting factors are light energy and nutrients which may be present at such low levels that no cell division occurs. A controlling factor, such as temperature, affects the rate at which phytoplankton utilize available energy supplies and nutrients. Phytoplankton tolerate a range of concentrations or intensities of all environmental factors. Above or below these ranges or as a result of extreme changes over short periods of time, most factors produce physiological stress, or may become lethal to phytoplankton cells.

This section appraises the major physical (i.e., light, temperature, and water circulation), chemical (i.e., salinity and nutrients), and biological (i.e., grazing) factors affecting the structure of phytoplankton communities.

B. PHYSICAL FACTORS
1. Light

The rate of primary production by phytoplankton is closely coupled to sunlight as a source of radiant energy for photosynthesis.[2,120-122] Four aspects of light have been treated in investigations of phytoplankton primary production: (1) the intensity of incident light, (2) changes in light on passing from air into water, (3) changes in light with increasing water depth, and (4) utilization of radiant energy by phytoplankton cells.[2] Illumination in all aquatic habitats depends on the angle of the sun during the day, the season of the year, latitude, and climatic conditions (e.g., percent of cloud cover). In the atmosphere, some solar radiation is absorbed by water vapor, carbon dioxide, oxygen, and ozone, thereby reducing its intensity. Although the spectral distribution of light incident at the sea surface incorporates a substantial part of the ultraviolet and infrared wavelengths, this radiation is effectively absorbed in the uppermost layers of the water column, leaving only the visible portion of the spectrum to penetrate the deeper layers.[8] For example, 98% of all infrared radiation is absorbed in the upper 2 m of the sea.[3] Smith and Morgan[123] have elucidated the factors responsible for the temporal and spatial variations in light incident upon the sea surface. An additional loss of solar energy occurs by scattering and reflection at the water surface. The amount of solar radiation reflected at the sea surface is a function of the angle of the sun, the degree of cloud cover, and roughness of the water surface.[83]

Light attenuation continues in the water column via absorption and scattering by water molecules, suspended particles, and dissolved matter.[6,8] Absorption, rather than scattering, primarily determines the penetration and distribution of radiant energy underwater.[83,124] Little light attenuation has been ascribed to the water molecules themselves.[125] The diminution of light intensity with increasing water depth is greater in estuaries than oceanic waters due principally to the higher concentrations of suspended particulate matter and dissolved organic substances (e.g., Gelbstoff) which strongly absorb light. In addition, selective absorption of wavelengths of light results in different spectral distributions for coastal and oceanic waters. Gelbstoff, for instance, absorbs more blue light than other wavelengths. The maximum transmittance in turbid coastal waters, therefore, takes place at about 575 nm, and in clearer oceanic systems, at approximately 465 nm.

Glover et al.[126] outlined the relative efficiency of different phytoplankton groups growing under various spectral light fields commonly found in the marine environment, comparing absorption and fluorescence excitation/emission spectra. Experimenting with colored filters and neutral density screens to simulate oceanic and coastal water types, these scientists examined rates of photosynthesis and growth of marine prokaryotic (neritic and oceanic clones of cyanobacteria of the genus *Synechococcus*) and eukaryotic phytoplankton under variable conditions of light intensity and quality. The resultant rates of photosynthesis and growth in these phytoplankton groups were a function, in part, of the two variables.

Jerlov[124] has formulated an optical classification of seawater based on variations in spectral transmittance (Figure 10). Three oceanic water types and nine coastal water types have been designated. As seen in Figure 10, the higher the numerical value in both the oceanic and coastal water types, the poorer the light transmittance and the longer the wavelengths of maximum penetration of the water column. The effects of turbidity in coastal waters are clearly reflected in these transmittance curves which project poorer light penetration and a progressive shift in maximum transmittance from the blue to the green and orange segments of the spectrum, with increasing turbidity for coastal water types 1 through 9. The curves for the oceanic water types indicate greater light penetration than those for the coastal water types and a maximum transmittance in the blue-wavelength portion of the spectrum. Because the spectrum of underwater irradiance in clear oceanic waters is compressed among the blue wavelengths and

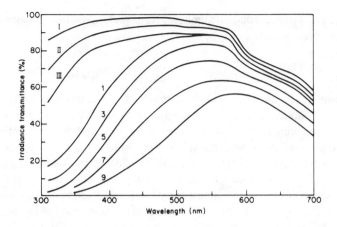

FIGURE 10.   Spectral transmission per meter of downward irradiance in oceanic (I to III) and coastal (1 to 9) water types. (From Dring, M. J., *The Biology of Marine Plants,* Edward Arnold, London, 1982. With permission.)

that of coastal waters is shifted to the green wavelengths, oceanic waters have been broadly categorized as "blue" and coastal waters as "green".[68]

Beer's law can be used to calculate the total amount of light penetrating to any depth of water.[6] This law is expressed as

$$I_z = I_0 e^{-kz} \tag{1}$$

where I_z is the intensity of light at depth z; I_0, the intensity of light at the surface; and k, the extinction coefficient of water that varies from place to place, being wavelength specific. While light becomes increasingly monochromatic as it penetrates the water column, changes in the intensity of light probably have more significance for phytoplankton growth.[68] The peak transmittance of oceanic water type I (Figure 10) equals 98%/m at 465 nm, but at a depth of 254 m, radiation of this wavelength declines to 1% of its surface irradiance. In contrast, the transmittance of coastal water type 9 reaches a maximum value of 56%/m at 575 nm; the radiation decreases to 1% of its surface irradiance at a depth of only 7.9 m. The attenuation of solar radiation proceeds exponentially with increasing water depth.[8] The intensity of light at a particular depth can be measured as irradiance, that is, energy per unit area of surface (units: mol/m²/s, where mol = 1 mol or Avogadro's number of photons).[68]

Phycologists are mainly concerned with photosynthetically active radiation (PAR), approximately 400 to 700 nm, which embodies that portion of the spectrum where photosynthesis in ordinary photoautotrophic plants occurs.[127] Within this range of usable wavelengths, phytoplankton mainly utilize chlorophyll pigments in their chloroplasts to absorb light of greater than 600 nm and accessory pigments (e.g., fucoxanthin and peridinin) to absorb light of less than 600 nm. These pigments enable phytoplankton to use all light in the visible spectrum. Although chlorophyll *a* is the primary pigment in photosynthesis, phytoplankton also possess chlorophylls *b* and *c* which differ in molecular structure and absorption spectra. Accessory pigments absorb energy and transfer it to chlorophyll *a*. Table 9 specifies the primary and accessory pigments found among the algal classes.

Drew[128] invokes seven irradiance factors that affect photosynthesis of marine plants:

1.   The absorption spectrum of the plant photosynthetic pigments
2.   The action spectrum of photosynthesis resulting therefrom
3.   The compensation irradiance at which the photosynthetic rate just equals the respiratory rate, resulting in no net gain of carbon and hence no plant growth

4. The saturation irradiance above which no extra photosynthesis occurs because other factors such as enzyme activity or inorganic carbon supply have become limiting rather than irradiance
5. Possible photoinhibition due to damage to the photosynthetic pigments by very high or particularly energetic irradiance, resulting in a reduction of photosynthesis
6. Possible photomorphogenic responses controlled by irradiance at specific wavelengths
7. Possible intrinsic factors such as diurnal rhythms in photosynthetic performance

Much has been written concerning the relationship between light intensity and photosynthetic rate. Laboratory research discloses that photosynthesis by phytoplankton increases logarithmically with increasing light intensity until a maximum value, P_{max} (the light saturation value), is reached. In nature, photosynthesis tends to be inhibited near the sea surface, and because light intensity decreases exponentially with increasing depth, the rate of photosynthesis gradually rises to a maximum value and then gradually diminishes until a point where photosynthesis just balances the metabolic losses due to respiration. This point is termed the compensation depth (Figure 11), and it lies approximately at the depth at which the light intensity drops to 1% of the surface radiation.[5] The photic or photosynthetic zone encompasses the area of the sea above the compensation point; photosynthesis exceeds respiration in this zone. The compensation depth and thickness of the photic zone vary greatly, but in most estuaries, high concentrations of suspended particles and dissolved substances usually limit both substantially. Thus, in turbid estuarine waters, the compensation depth can be 1 m or less, whereas in turbid nearshore oceanic waters it approaches 10 m, and in clear oceanic waters, 120 m.[2,35]

Phytoplankton have different tolerances to light intensity. Diatoms, for example, are light saturated at a less intense light level than dinoflagellates. Therefore, dinoflagellates might be expected to be more numerous nearer the sea surface.

These microscopic plants alter their reproductive rate, assimilation number, and chemical composition in response to different light intensities. Adaptations of phytoplankton to variable light intensities include a change in the amount of pigments or photosynthetic enzymes in the cells.[119] Physiological adjustment involves several morphological and biochemical alterations as compiled by Levinton (p. 186):[8]

1. Change in total photosynthetic pigment content
2. Change in pigment proportions
3. Change in the morphology of the chloroplast
4. Change in chloroplast arrangement
5. Change in availability of dark reaction enzymes

Photosynthesis-irradiance curves are most helpful in studies of the physiological adjustments and primary production of phytoplankton to changing light intensity and quality.[121,129-131] Work by Harris[132] and Lewis et al.[133] demonstrates conclusively that phytoplankton respond on a number of different time scales to light regime variations. Falkowski,[134] exploring the time-scale relationship between irradiance, vertical motion, and primary production, clarified phytoplankton responses on short time scales (e.g., the "flicker" effect and state transitions), on medium time scales (e.g., diel changes and light-shade adaptation), and on long time scales (e.g., species succession). Phycologists have a poor understanding of the mechanisms controlling these responses. Absolute time scales of change in light fields relevant to phytoplankton cover about 10^{-1} s for the "flicker" effect (generated by variations in light intensity because of focusing and defocusing of light by waves at the air-seawater interface), 10^0 to 10^2 s for clouds passing across the sun, 10^4 s for diel effects, and 10^5 to 10^6 s for storm associated phenomena. Variations in light regimes over longer time scales evoke those operating on a seasonal basis. Turbulent mixing superimposes additional flux on atmospheric variations in light regimes.

TABLE 9
Distribution of Pigments in Algal Classes[a]

Pigments	Algal classes									
	Chlorophyceae	Prasinophyceae	Euglenophyceae	Chrysophyceae	Haptophyceae	Xanthophyceae[b]	Bacillariophyceae	Dinophyceae	Cyanophyceae	Cryptophyceae
Chlorophyll *a*	+	+	+	+	+	+	+	+	+	+
Chlorophyll *b*	+	+	+	−	−	−	−	−	−	−
Chlorophyll *c*	−	−	−	+	+	−	+	+	−	−
Chlorophyll *e*	−	−	−	−	−	+	−	−	−	−
α-Carotene	(+)	(+)	−	−	+	−	(+)	(+)	−	+
β-Carotene	+	+	+	+	+	+	+	+	+	−
γ-Carotene	+	(+)	−	−	−	−	−	−	−	−
ε-Carotene	−	−	−	−	−	−	(+)	−	−	−
Lutein	+	+	+	+	−	+	(+)	−	(+)	−
Zeaxanthin	+	+	?	−	−	−	−	−	+	+
Violaxanthin	+	+	−	−	−	+?	−	−	−	−
Neoxanthin	+	+	+	−	−	+?	−	−	−	−
Fucoxanthin	−	−	−	+	+	+	+	−	−	−
Diatoxanthin	−	−	−	−	−	−	+	−	−	+
Diadinoxanthin	−	−	−	−	−	−	+	+	−	−
Neodiadinoxanthin	−	−	−	−	−	−	−	+	−	−
Dinoxanthin	−	−	−	−	−	−	−	+	−	−
Neodinoxanthin	−	−	−	−	−	−	−	+	−	−
Peridinin	−	−	−	−	−	−	−	+	−	−
Neoperidinin	−	−	−	−	−	−	−	−	−	−
Myxoxanthin	−	−	−	−	−	−	−	−	+	−
Oscilloxanthin	−	−	−	−	−	−	−	−	+	−

C-Phycoerythrin		—	—	—	—	—	—	—	—	—	+	Phycoerythrin } found in
C-Phycocyanin		—	—	—	—	—	—	—	—	—	+	Phycocyanin } some genera

[a] Classes with planktonic representatives

[b] Representative genera planktonic, but not extensively reported on in plankton studies.

From Boney, A. D., *Phytoplankton*, Edward Arnold, London, 1975. With permission.

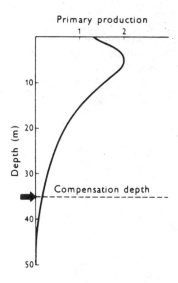

FIGURE 11. Generalized diagram depicting changes in phytoplankton production with depth. (From Boney, A. D., *Phytoplankton,* Edward Arnold, London, 1975. With permission.)

The physiological responses of phytoplankton to light changes on medium time scales have elicited the greatest interest of aquatic ecologists. Physiological adjustments of phytoplankton to low-frequency variations in light intensity (light-shade adaptation) or water quality (chromatic adaptation) are manifested in variations in the amounts and ratios of photosynthetic pigments, changes in photosynthetic responses, gross chemical composition (e.g., C/N ratios), cell volume, dark respiration rates, and levels of enzyme activity associated with carbon fixation.[135] More data have been collected on light-shade adaptation than on chromatic adaptation in phytoplankton.[134]

The relationship between irradiance and phytoplankton growth is species dependent. Five hypotheses have been postulated to explain interspecific variations in the relationship between irradiance and growth: (1) differences in the light-harvesting properties among the species; (2) differences in photosynthetic machinery which result in changes in the quantum requirement of photosynthesis; (3) differences in photosynthesis: respiration ratios; (4) varying proportions of photosynthetically fixed carbon which is excreted or secreted; and (5) the chemical composition of different species influencing photosynthetic quotients.[136] Falkowski et al.[136] evaluated the relative importance of these five factors on growth-irradiance relationships.

Light limitation of phytoplankton growth and production is common in estuaries[78,85] as conveyed above for several temperate systems on the East Coast of the U.S. The triggering of spring outbursts of phytoplankton populations in temperate zones, coupled to an increase in light availability, has been well documented.[1,81,115] Over an annual cycle, the limitation of phytoplankton growth may be caused by factors other than light, such as temperature, mixing processes, nutrients, and grazing, but increasing light acts as a controlling factor in the initiation of spring blooms in mid and high latitudes. In deeper temperate systems, spring growth proceeds only after a thermocline forms, isolating the mixed layer in the photic zone.[1]

At mid latitudes, of course, incident light varies considerably during the year, but in a more or less regular pattern to yield a characteristic annual cycle that regulates phytoplankton primary production. The intensity of daily illumination in temperate waters of the Northern Hemisphere peaks in May and June and drops to a minimum level in December and January.[2] Parsons and

Takahashi[23] computed minimum and maximum incident radiation values of 131 and 769 ly/d in June and July, respectively, at 50 N latitude. Harvey,[137] however, cited somewhat lower figures of incident radiation at this latitude — approximately 216 ly/d during the May-to-June period and an average of 24 ly/d during mid-winter. The mean daily estimates of incident radiation depend largely on weather conditions, with cloud cover mitigating light intensity as much as one third or less than that of clear skies.[3] The gradational increase in light intensity and penetration from December through March ultimately produces conditions suitable for a phytoplankton bloom.

A somewhat different light regime typifies high-latitude regions. Here, the magnitude of change of irradiance is most extreme between January and December, and large seasonal variations in irradiance (both long periods of low incident radiation) limit phytoplankton production. Ryther[74] substantiates low annual production for the true Arctic Ocean (1 g C per square meter per year) and the Antarctic (100 g C per square meter per year). Production in mid- and high latitude waters is generally negligible in winter.

In the tropics, excluding the rainy season, incident radiation has greater constancy. An estimated 216 ly/d of visible light energy impinges on tropical waters each day.[137] Raymont[3] quotes total incident radiation measurements for the Gulf of Panama approaching a maximum of 600 ly/d during the dry season and a minimum of 350 to 400 ly/d during the rainy season. Thus, the overall incident radiation is much larger in tropical latitudes than in temperate and polar latitudes, even for inshore waters where irradiance values may change markedly. Nevertheless, the presence of a permanent thermocline in many tropical regions results in low nutrient concentrations that limit phytoplankton production despite favorable light intensities.[1]

2. Temperature

Together with light intensity, temperature influences the photosynthesis and growth of phytoplankton.[132,138-142] Investigations of the effects of temperature on phytoplankton have involved assessments of systems subjected to both natural and anthropogenic perturbations of the thermal regime.[143-147] Research clearly has shown that division rates of phytoplankton increase with temperature within defined limits.[92,148] In laboratory cultures, the division rates of marine phytoplankton generally increase by two to four times with a 10°C rise in temperature, as long as the temperatures lie within a range favorable to growth.[31,149] In addition, as temperature increases, the carbon per cell and carbon per unit chlorophyll *a* in marine phytoplankton typically rise.[31,150,151] Phytoplankton development is thought to be most significantly impacted by water temperatures below 10°C, which leads to diminishing growth in many species.[83,149]

Eppley[148] formulated an empirical equation to predict the maximum expected specific growth rate of marine phytoplankton as a function of temperature. It places upper limits on expected rates of marine phytoplankton photosynthesis. The relationship between temperature and phytoplankton division rate is given by

$$\log_{10} u = 0.0275T - 0.070 \tag{2}$$

where u is the rate of cell division in doublings per day, and T is the temperature in degrees Celsius. The Q_{10} of this equation — defined as the ratio of the cell division rate with a 10°C rise in temperature — equals 1.88.[31]

Conflicting conclusions have been proffered on the criticalness of temperature to phytoplankton growth. Although some findings suggest that phytoplankton development is most greatly affected by water temperatures below 10°C, which can cause declining growth and the elimination of certain species in mid-latitude estuaries,[83,149] other studies downplay the importance of temperature in governing the abundance and growth of phytoplankton in marine waters.[6] Yet, many phytoplankton species exhibit highest growth rates within specific temperature ranges.[6,144,152] In general, diatoms are better adapted than dinoflagellates to growth at lower temperatures.[92,153] Whereas maximum growth rates of many phytoplankton species in temperate, subtropical, and tropical waters take place at temperatures above 10°C, phytoplankton in polar seas, such as Antarctic habitats, display peak growth rates at much lower temperatures.[154] However, exceedingly low water temperatures in the Antarctic Ocean can restrict growth and photosynthesis of the microflora. Because spring blooms flower in most temperate estuaries at the time of minimum temperatures, low temperatures are not believed to be a chief limiting factor in phytoplankton growth and photosynthesis in these systems.[6]

Temperature exerts indirect effects on phytoplankton aside from the direct influence it has on their growth, enzymatic activities, and other metabolic processes. The development of a thermocline, for example, has significant consequences on the phytoplankton community.[2] The formation of a thermocline accompanies stratification of the water column, which mitigates the downward passage of algal cells from the photic zone.[3] In mid and high latitudes, rising temperature in the spring and summer fosters thermal stratification and greater stability of the water column, with water circulation above the thermocline capable of transporting phytoplankton into well-lit areas.[2] The thermocline tends to form at the depth of the halocline,[155] existing as a discontinuity layer of rapidly changing temperature that separates an upper, warm mixed layer from a cooler, lower layer.[5] Venrick[156] reiterated that nearly all subsurface phytoplankton maxima are associated with the presence of a thermocline or pycnocline. High wind stress and strong continuous turbulence usually preclude the development of the thermocline or severely limit its persistence to perhaps a few days or even less. In some mid- and high-latitude estuaries, such as Long Island Sound and Narragansett Bay,[8,157] the thermocline is a seasonal phenomenon, but in low-latitude systems with sufficient depth, the thermal stratification may be present year-round.

The occurrence of a thermocline has been related to improved growth of phytoplankton and seasonal primary production cycles,[1] although a clear relationship may be obscured by concomitant changes in other factors, including zooplankton grazing and nutrient concentrations.[3] Indeed, thermal stratification of the water column suppresses nutrient influx to the photic zone from deeper waters which inhibits phytoplankton growth if the replenishment of essential nutrients from terrestrial drainage does not maintain sufficient nutrient levels in the surface layers.

3. Water Circulation

Phytoplankton in the photic zone are vulnerable to displacement or washout to nearshore oceanic waters by the continual seaward flow of surface waters in estuaries. The seaward flowing, upper layer not only subjects entrained populations to washout but also to osmotic stress.[35] The lower layer of estuarine flow acts to concentrate phytoplankton in a manner similar to suspended sediments of a turbidity maximum;[158] as net seaward flowing river water converges with net landward flowing seawater, it concentrates the algal cells, some of which have settled through surface waters.[83] The sinking of phytoplankton cells is critical to their survival in nature.[53,159]

The relation between water residence and phytoplankton turnover time regulates, in part, phytoplankton development in estuaries.[160] Peterson and Festa[83] define *in situ* phytoplankton turnover as the ratio of the standing stock to its integral photosynthetic production. They signify that rates of *in situ* production can be repressed when rates of physical removal are great. Hence, not enough time may be available for phytoplankton to build large densities by *in situ* processes if the water residence time is small. Ketchum[161] analyzes the ability of an endemic phytoplankton population to maintain itself in an estuary or a given segment of an estuarine salinity gradient. This ability is deemed to be a function of the growth rate of the population relative to the proportion of the population removed via flushing from the estuary or a given segment during each tidal cycle.[35] Rates of reproduction and recruitment relative to rates of attrition by estuarine circulation play a vital role in regulating the population level.

When the flushing rate of an estuary exceeds the growth rate of the phytoplankton, maximum production is likely to be unattainable, especially if high turbidity limits light penetration and shallowness reduces the occurrence of blooms.[87] Estuarine waters having a short residence time generally harbor a mixture of allochthonous phytoplankton populations from nearshore oceanic and riverine systems. Autochthonous forms develop most frequently in those systems where the residence time of the estuarine water is greater than a few weeks.[85]

C. CHEMICAL FACTORS
1. Salinity

Like temperature, salinity indirectly affects phytoplankton through its modulation of the density of estuarine waters and the stability of the water column.[68] It also directly influences the rate of cell division of these plants as well as their occurrence, distribution, and productivity. Salinity changes temper photosynthesis in one of two ways: (1) by changes of the carbon dioxide system, and (2) by changes of osmotic pressure.[127,162] Since phytoplankton inhabit waters of highly variable salinities in the estuarine environment, they commonly experience widely fluctuating osmotic stress.[35] These microalgal cells retain a somewhat elevated internal salt concentration and osmotic pressure than that of the surrounding medium,[163] causing water flow into the cell and sustaining turgor.[31] Marine phytoplankton carried into estuaries require osmotic mechanisms to minimize water uptake that can increase turgor and possibly burst cells. Conversely, freshwater forms translocated downstream into estuarine zones necessitate osmotic mechanisms to prevent excess loss of water that can result in plasmolysis and death.[35] As osmotic pressure and ionic composition change in the cell, other cellular processes (e.g., synthesis of chlorophyll and rates of photosynthesis) may also be altered.[164-166]

Research on the effects of salinity on the rate of cell division and growth of phytoplankton has unclosed a general euryhaline tolerance among the organisms ascribable to their ability to osmoregulate.[85,167,168] Despite this ability, phytoplankton still remain susceptible to problems pertaining to osmotic shock and stress.[35] The optimum salinity for growth and reproduction of a given species typically is the salinity at which the species normally lives in the natural environment. For instance, the optimum salinity for cell growth of *Skeletonema costatum* ranges from about 10 to 40‰ and that of *S. subsalsum,* from approximately 2 to 20‰. These values embody the salinities at which the two populations are encountered in estuaries and coastal waters. For *Cryptomonas* sp., a cryptomonad flagellate, *Syracosphaera carterae,* a coccolithophorid, and *Amphidinium* sp. and *Exuviaella baltica,* two dinoflagellates, cell division proceeds over the range from 5 to 10‰ to greater than 35‰ (Figure 12).[169] Salinities between 15 and 40‰ do not affect the cell division rate of *Isochrysis galbana.*[170] Cells of *Prorocentrum micans* divide at a maximum rate at 20‰; cell division declines by 25% when the salinity is increased to 40‰.[171] Below 15‰, cell division of *Asterionella japonica* ceases, and it peaks between 30 and 35‰.[170] Guillard,[163] experimenting with estuarine, neritic, and oceanic forms, commented that the lowest salinities at which they divided at over half maximum rates ranged from 0.5 to 8, 0.5 to 20, and 18 to 24‰ for estuarine, neritic, and oceanic species, respectively.

In a laboratory study of 46 marine isolates, Brand[172] attested to wide limits of salinity between which reproduction of the populations occurred. Estuarine phytoplankton species, on the average, were more euryhaline than coastal and oceanic forms (Figure 13). While most coastal populations tolerated much lower salinities than those from which they were isolated, estuarine and oceanic species tolerated salinities similar to those observed in their natural habitats.

Among planktonic microalgal groups, differences in general distributional trends in space reflect, in part, tolerances to salinity. Thus, euglenoids and planktonic blue-green algae are more important in freshwater habitats (less than 0.5‰) than in mixohaline and marine habitats. In contrast, diatoms, coccolithophores, and dinoflagellates attain greater abundance as salinity increases, with diatoms and coccolithophores being better represented at intermediate salinities (0.5 to 18‰) than dinoflagellates. Chlorophytes have nearly equal representation in limnetic, estuarine, and marine waters.[35]

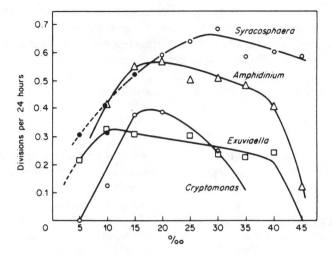

FIGURE 12.   Effect of salinity on cell division rates of *Syracosphaera carterae, Amphidinium* sp., *Exuviaella baltica,* and *Cryptomonas* sp. (From Perkins, E. J., *The Biology of Estuaries and Coastal Waters,* Academic Press, London, 1974. With permission.)

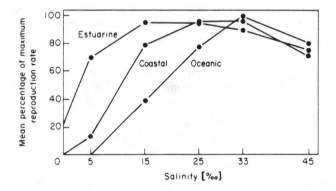

FIGURE 13.   Mean percentage of the maximum reproduction rates of estuarine, coastal, and oceanic groups of phytoplankton species. (From Brand, L. E., *Estuarine Coastal Shelf Sci.,* 18, 543, 1984. With permission.)

2. Nutrients

Phytoplankton require a number of substances for growth and reproduction, the most important of which are the macronutrient elements nitrogen, phosphorus, and silicon. Major elements essential for phytoplankton growth and, hence, production (e.g., carbon, oxygen, magnesium, potassium, and calcium) are abundant relative to what is needed by the organism.[3] Other nutrients, including the trace elements iron, manganese, zinc, copper, cobalt, and molybdenum, may limit phytoplankton growth if present in insufficient quantities. Certain organic compounds, particularly the three vitamins cobalamine, thiamin, and biotin, are necessary for adequate growth of some phytoplankton species. Phytoplankton can utilize many essential elements in both particulate and dissolved forms.[8] Fogg[173] mentions at least 18 minerals and various organic growth factors required for the growth of planktonic microalgae; any one of these substances, when deficient, may compromise the growth process.[5]

The Michaelis-Menten equation has been employed to describe the uptake of nutrient elements by phytoplankton. The uptake rate, most often measured in chemostats, traces a

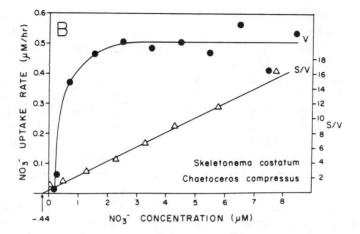

FIGURE 14.   Nitrate uptake rate of phytoplankton (in chemostats) dominated by *Skeletonema costatum* and *Chaetoceros compressus* concentrated from the surface waters of Narragansett Bay. Nitrate uptake rate (V) of this phytoplankton vs. nitrate concentration (S) of the medium employing the Michaelis-Menten equation. The half-saturation constant (0.44) is given as the negative S-intercept of the linear regression of S/V vs. S. (From Furnas, M. J., Hitchcock, G. L., and Smayda, T. J., *Estuarine Processes*, Vol. 1, Wiley, M. L., Ed., Academic Press, New York, 1976, 118. With permission.)

hyperbolic curve when plotted (Figure 14). The rate of uptake of a limiting nutrient increases with increasing external concentration, but it ultimately reaches an asymptote, at which point the nutrient is no longer limiting. This equation is represented as

$$V = \frac{V_{max} \cdot S}{K_s + S} \tag{3}$$

where V is the rate of nutrient uptake; V_{max}, the maximum rate of uptake; S, the concentration of the limiting nutrient; and K_s, the substrate concentration at which $V = V_{max}/2$.[6]

An empirically derived expression, the Monad equation can define the growth rate of phytoplankton populations in chemostats. Analogous to the theoretical Michaelis-Menten equation for nutrient uptake, the Monad equation circumscribes a hyperbolic relationship of the growth rate of a phytoplankton species to the concentration of a limiting nutrient. The Monad equation is written as

$$u = \frac{u_{max} \cdot S}{K_s + S} \tag{4}$$

where u_{max} is the maximum growth rate, and all other terms are the same as in Equation 3. Although the concentration of the limiting nutrient in the environment most strongly modulates the uptake rate, other factors (e.g., cell size and water temperature) also can be influential.[6]

a. Nitrogen

Terry et al. (p. 323)[174] note, "It is widely accepted that algal growth rate and yield are determined by the abundance of the single nutrient which is in shortest supply relative to the cells requirements, and are independent of all other nutrients, as long as the concentrations of excess nutrients do not reach toxic levels". In estuaries and coastal oceanic waters, nitrogen rather than phosphorus is regarded as the limiting element to algal growth.[6,175,176] The supply of nitrogen likewise limits phytoplankton productivity in the oligotrophic open ocean.[177] In contrast to nitrogen limitation of phytoplankton productivity in estuarine and marine waters, phosphorus supply controls the primary productivity of most lacustrine systems.[176,178-180]

Howarth and Cole[176] recently proposed that nitrogen limitation of phytoplankton growth in coastal marine waters results from low molybdenum availability. According to their findings, sulfate in seawater inhibits molybdate assimilation by phytoplankton, and the molybdenum deficiency which ensues leads to low rates of nitrogen fixation. Coastal marine ecosystems may be nitrogen limited, therefore, because nitrogen-fixing organisms respond much more slowly than those in most lacustrine environments and are less effective in alleviating nitrogen limitation.

Nitrogen occurs in a number of different forms in seawater. It is present as molecular nitrogen (N_2) and nitrous oxide (N_2O). Ammonia (lumped with ammonium — NH_4^+ — for discussions in this chapter), nitrite, and nitrate constitute the primary dissolved inorganic forms. Dissolved organic forms (e.g., urea, amino acids, and peptides) also have been detected. In addition, nitrogen exists in particulate forms, largely organic.

Due to the wide diversity of inorganic and organic forms of nitrogen, its behavior in estuarine waters is much more complex than that of phosphorus and silicon, the two other major nutrient elements. The number of oxidation states of nitrogen (–5 to +3), which are biologically active, obfuscates studies of the nitrogen cycle in estuaries.[181] Phytoplankton assimilate all three major dissolved inorganic forms of nitrogen for growth. While most phytoplankton grow well on nitrate, some do not.[3] Various euglenids, cryptomonads, and green algae, for example, require ammonia and amino acids (reduced nitrogen) for growth.

Ammonia and organic nitrogenous substances (e.g., urea) are the preferred nitrogen forms of phytoplankton.[182] The uptake of ammonia confers a significant advantage on these organisms since the nutrient can be used directly in the synthesis of amino acids. Wheeler[183] addresses the aspects of phytoplankton nitrogen metabolism that partially control nutrient uptake and growth.

Nixon and Pilson[184] survey the concentrations and distributions of different nitrogen forms in estuaries, explain nitrogen cycling processes operating in sediments and the water column, and assess the role of nitrogen in estuarine productivity. Biogeochemical cycling of nitrogen entails several vital processes in the water column and bottom sediments. In the water column, these processes include uptake, remineralization, and oxidation, whereas in estuarine sediments they involve burial, remineralization, biological uptake, oxidation, reduction, nitrous oxide production, and denitrification.[184] A strong coupling has been established between the benthic system and phytoplankton nutrient dynamics.[8] Water column production correlates well with benthic mineralization.[184] An overview of the nitrogen cycling processes in estuaries is given elsewhere.[115]

A distinction has been drawn between new and regenerated forms of nitrogen for supporting phytoplankton production. Dugdale and Goering[185] differentiated new and regenerated nitrogen in coastal and oceanic waters, the new nitrogen referring to nitrate allochthonously entering the photic zone from coastal or upwelling sources and the regenerated nitrogen, ammonia, and urea autochthonously regenerated by zooplankton in the upper part of the water column. Kemp et al.[186] have extended this concept to estuarine waters, defining new nitrogen as that entering allochthonously from the watershed or from offshore waters and regenerated nitrogen as that recycled either within the water column or across the sediment-water interface. Raymont (p. 338)[3] dictates that "This distinction is fundamentally important in relation to the transfer of material to higher trophic levels ... only the fraction produced from new nutrients represents an increase in phytoplankton population and is potentially available to support increased production at secondary and higher levels." [15]N labeling of different sources of nitrogen has been applied successfully to distinguish new and regenerated nitrogen fractions.[185,187-191]

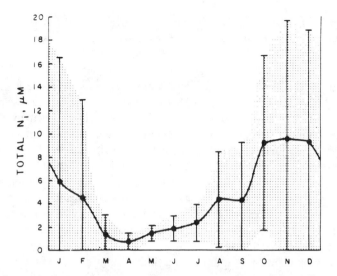

FIGURE 15. Annual cycle of dissolved inorganic nitrogen in the lower West Passage of Narragansett Bay, RI during the years 1976—1981. The points reflect average values of the monthly means for each year calculated from weekly measurements. The bars represent ±2 standard deviations. Note the greater consistency of values during the spring and early summer compared to those during the fall and winter. The mean concentrations peak in late fall or early winter and are minimum in spring or early summer. The timing and magnitude of the higher values are most variable. (From Pilson, M. E. Q., EOS, Trans. Am. Geophys. Union, 63, 347, 1982. With permission.)

Only a few estuaries have been reasonably well described with respect to spatial and temporal variations in the concentration of dissolved inorganic nitrogen — nitrite, nitrate, and ammonia — over at least one annual cycle.[184] Dissolved inorganic nitrogen concentrations in estuaries generally range from near 0 to greater than 100 μM. Considerable variation of inorganic nitrogen is evident over the year, and, although regular seasonal cycles may occur, the annual pattern differs somewhat from year to year, too (Figure 15). A seasonal variation of certain inorganic nitrogen forms takes place in the upper 10 to 20 cm of estuarine seafloor sediments, and spatial changes associated with sediment type are apparent as well.[184] A much smaller data base exists on seasonal concentrations of organic forms of this element in the estuarine environment. Whereas controversy surrounds nitrogen budgets of estuaries, the importance of nitrogenous nutrients to phytoplankton production is indisputable.[189,192-194]

In spite of the high seasonal variability in the amount of dissolved inorganic nitrogen in estuaries, a simplified cycle can be devised for many of the systems. In temperate estuaries, for example, nitrogenous nutrients accumulate in the winter when phytoplankton are light limited. An increase in light intensity in the late winter and early spring drives the critical depth in vertically stratified systems below the depth of mixing of surface waters. Consequently, a phytoplankton bloom ensues. From February to September, nitrogenous nutrients are depleted by the rapid growth of algae and vascular plants; grazing pressure by zooplankton on phytoplankton accelerates during these months. The rise in zooplankton abundance may lag only slightly behind phytoplankton blooms. Recycled nitrogen seems to be a major source of nitrogen for phytoplankton at this time. Light duration and intensity diminish in the fall, creating conditions of reduced phytoplankton production. Nitrogenous nutrients begin to accumulate once again with the curtailment of phytoplankton growth. While seasonal cycles of nitrogen may be borne by seasonally varying river flow rather than by seasonal mixing processes in many estuaries,[6] tidally induced vertical advection and turbulence in some systems make significant quantities of nitrogenous nutrients available for phytoplankton production.

b. Phosphorus

This essential macronutrient may ultimately control organic production in certain marine environments.[175,195] Recent research on phosphorus dynamics in marine waters using radioisotopic tracers (^{32}P and ^{33}P)[196-200] has demonstrated rapid uptake of this element by phytoplankton and its swift recycling between various inorganic and organic compartments.[195] Dissolved inorganic phosphorus is quickly assimilated by marine plankton. Total dissolved phosphorus can be partitioned into a dissolved organic phosphorus pool (i.e., nonreactive phosphorus) and a dissolved inorganic phosphorus pool (i.e., soluble reactive phosphorus or phosphate, PO_4^{3-}). In surface seawater, the steady state concentration of dissolved organic phosphorus commonly exceeds that of dissolved inorganic phosphorus, and at times may be as much as an order of magnitude greater.[195,201] Dissolved organic phosphorus probably serves as a valuable source of phosphorus for microbial metabolism.[202,203]

In estuarine environments, phosphorus is partitioned into dissolved inorganic, dissolved organic, and particulate pools. The principal form of dissolved inorganic phosphorus occurs as orthophosphate (phosphoric acid, H_3PO_4, and its dissociation products $H_2PO_4^-$, HPO_4^{2-}, and PO_4^{3-}). The main ion in seawater is HPO_4^{2-}, which exceeds PO_4^{3-} in concentration by about one order of magnitude.[6] Orthophosphate appears to be the preferred form of phytoplankton,[8] but these autotrophs no doubt utilize dissolved organic phosphorus during periods of deficiency.[2] Dissolved organic phosphorus can turn over quite rapidly.[204]

Bottom sediments represent a sink for phosphate, which adsorbs readily to amorphous oxyhydroxides, calcium carbonate, and clay minerals under aerobic conditions.[6,205,206] Estimates of phosphate release from marine sediments range from -15 to 50 $\mu m/m^2/h$.[6,207,208] Concentrations may be very low in surface waters (often less than 1 μg at P per liter), mainly due to uptake by bacteria and algae. Indeed, the concentration of phosphate in sediments and the water column has been attributed largely to biological activity.[6,115] The availability of dissolved inorganic phosphorus follows well defined temporal cycles in some estuaries, but not in others.[204,209]

The ratio of nitrogen to phosphorus in phytoplankton is about 16:1, approximating the ratio of N/P in marine waters. N/P ratios can vary widely; over a 12-month period, for instance, Harris and Riley[210] calculated ratios of 12.6:1 to 19.8:1 in phytoplankton of Long Island Sound. Smith (p. 1149)[175] advises that "phosphorus vs. nitrogen limitation is a function of the relative rates of water exchange and internal biochemical processes acting to adjust the ratio of ecosystem N:P availability". Nitrogen shortages rather than phosphorus limitation are generally responsible for arresting the growth of phytoplankton populations in marine systems.[8]

c. Silicon

Silicon, although not an essential requirement of all estuarine organisms, is critical for the skeletal growth of diatoms, and in at least one species, *Cylindrotheca fusiformis,* is needed for DNA synthesis. This nutrient element enters estuaries in three principal forms, that is, as detrital quartz, alumino-silicates (clays), and dissolved silicon. The concentration of dissolved silicon fluctuates over an annual cycle, being subject to substantial depletion by biological uptake. The rapid growth of phytoplankton in the spring and summer months corresponds in many cases to episodes of low dissolved silicon concentrations in the water column. In some estuaries (e.g., San Francisco Bay), dissolved silicon levels also have been related to variations in rates of riverine influx.[211,212] D'Elia et al.[213] recount the seasonal cycle of dissolved silicon in the mesohaline region of Chesapeake Bay, where high concentrations of this element have been monitored in the spring and from late summer to fall, and low concentrations in the late spring when levels are potentially limiting to diatom growth.

Dissolved silicon in estuarine waters is present as silicic acid (H_4SiO_4). When depleted, phytoplankton cell division can be inhibited and metabolic activity of the cell suppressed.[8,214] Silicon availability frequently impacts the abundance and productivity of diatoms[213] and may limit other phytoplankton populations.[109]

In the silicon cycle of estuaries, riverine input and biological uptake are major factors regulating silicic acid concentrations. Chemical dissolution, especially at the sediment-water interface, also constitutes a key link in the cycle. However, few dissolved silicon budgets have been formulated for estuaries because of the paucity of quantitative data. The recycling of biogenic silicon probably is a chief pathway of silicic acid genesis for use by diatoms, although the amount of dissolved silicon regenerated by the decomposition of siliceous remains has not received sufficient attention in the literature. Because of recycling, the depletion of dissolved silicon in estuarine and oceanic waters appears to be less critical than in lacustrine environments.[2]

d. Trace Metals

Viarengo[215] divides trace metals into two subclasses: (1) iron, magnesium, manganese, cobalt, zinc, and copper which are essential for the correct functioning of biochemical processes; and (2) cadmium, mercury, chromium, lead, and other trace elements without any established biological function but which are potentially important contaminants in the aquatic environment. Iron, manganese, cobalt, copper, and zinc are universally essential for phytoplankton growth albeit in small concentrations.[3] Some trace metals (e.g., copper and zinc), although essential for growth of phytoplankton, can be toxic at low concentrations and may inhibit productivity.[3,216-218] Toxicity of trace metals has been coupled to the free metal ion activity regardless of the total metal concentration.[217] Most data on trace element requirements of phytoplankton have been acquired from laboratory culture experiments.[219] Phytoplankton need more iron than any other trace metal,[220] with the biologically available species being dissolved ionic species[221] and dissolved organic complexes.[222] A shortage of iron lowers photosynthetic potential, but phytoplankton utilize chelating substances to alter the availability of iron as well as other trace metals.[8] Iron, manganese, and zinc play a significant role in oxidase systems (i.e., the oxygen evolution step of photosynthesis containing iron as the cofactor) and as cofactors of enzymes necessary for plant growth.[8]

e. Organic Trace Substances

The bulk of phytoplankton in estuaries is auxotrophic, assimilating organic nutrients in association with photosynthesis.[8] Among organic trace substances, three vitamins — vitamin B_{12} (cobalamine), thiamin, and biotin — appear to be necessary for phytoplankton growth. While some phytoplankton populations are capable of synthesizing these vitamins themselves, the majority relies on bacteria as the major producers. Cobalamine has received the greatest attention relative to thiamin and biotin principally because phytoplankton have displayed a distinct need for this substance in culture experiments. The quantity of cobalamine in coastal waters amounts to 5 to 10 ng/l, far in excess of oceanic regions (i.e., 0.1 to 2 ng/l).[2,3] Similar to cobalamine, thiamin also has greater concentrations in coastal waters than in the open ocean. Thiamin concentrations may undergo seasonal cycles, as exemplified by the North Sea where the quantities of this vitamin attain peak values in the winter and minimum values in the summer. The primary effect of vitamin limitation is at times manifested most clearly in changes in species composition rather than production alterations of the phytoplankton community.[3]

D. BIOLOGICAL FACTORS
1. Grazing

Estimates of the biomass or standing crop of a phytoplankton community may not reflect actual production as a result of grazing by herbivorous organisms, especially zooplankton.[1] Some phytoplankton losses arise via sinking below the photic zone; however, the bulk of the cells disappear by zooplankton ingestion. Intensity of grazing is highly variable in space and time. In some waters, a distinctive alternation of abundance of phytoplankton and zooplankton has been recognized, with the decrease in phytoplankton often ascribed to zooplankton grazing. In other waters, however, grazing by zooplankton has been discounted as a major factor in the decline of phytoplankton populations. Raymont (p. 352)[3] insists that "Only in somewhat peculiar marine environments, often in very shallow waters, where phytoplankton crop can be excessively rich, and where the zooplankton population, due frequently to unusually unfavorable hydrographic factors, is low, does grazing appear to exert little influence on algal density". A lack of control by zooplankton grazing on phytoplankton is most often evident for inshore waters of greater environmental variability than for offshore waters.[8]

The potential effectiveness of zooplankton grazing becomes apparent when examining the daily demands of zooplankton populations for maintenance and growth. Daily requirements of zooplankton normally approach 30 to 50% of their body weight each day. In exceptional cases (e.g., *Calanus*), the daily demand may surpass 300% of a species' body weight every day.[3] Tintinnids in Long Island Sound, the dominant component of the microzooplankton community, consume approximately 27% of the annual primary production, and copepods remove about 44%.[223] Herbivorous macrozooplankton reportedly are the primary consumers of phytoplankton crop, but microzooplankton remain vital as trophic intermediaries in energy flow through the estuarine ecosystem.[224]

Selective feeding by zooplankton populations can regulate the composition of the phytoplankton community.[8] Copepods, for example, prefer to crop larger-size phytoplankton cells. They also seemingly have an ability to graze the most abundant size classes, thereby favoring the proliferation of the less common size classes and less abundant phytoplankton species. When grazing is intense and phytoplankton abundance drops below a critical level, the zooplankton abundance likewise diminishes after a lag time. The short generation time of phytoplankton relative to that of zooplankton enhances rapid recovery of the microflora.

In seasonal plankton cycles of estuaries, phytoplankton blooms are commonly superseded by a peak in zooplankton abundance, although strong oscillations in both phytoplankton and zooplankton abundance tend to be suppressed.[8] The time lag between a phytoplankton bloom and subsequent zooplankton expansion in these systems can be variable, leading to differences in phytoplankton concentrations. The time lag in open oceanic environments may be slight, particularly in the tropics. Zooplankton grazing in open oceanic, tropical waters acts as a stabilizing force on phytoplankton standing crop throughout the year.[225] Where utilization of phytoplankton is less effective in the water column, benthic grazing can limit phytoplankton biomass, as has been advanced for San Francisco Bay[226,227] and elsewhere.[228-230]

Both phytoplankton and zooplankton exhibit a patchy distribution,[231,232] the patches varying in dimensions from a few meters to many kilometers.[3] Phytoplankton concentrate more often in large patches than zooplankton.[233] The patchiness of phytoplankton density is especially noteworthy in estuaries and coastal waters characterized by abrupt changes in environmental conditions (e.g., areas of wind-induced turbulence and river runoff). Parsons and Takahashi[23] view six processes that promote patchiness: (1) physical-chemical boundary conditions, (2) advective effects of water movement, (3) grazing, (4) reproductive rates within a population, (5) social behavior in populations of the same species, and (6) intraspecific interactions causing attraction or repulsion between species. Of these six processes, Levinton[8] considers hydrography to be the principal mechanism generating patchy distributions of plankton. However, grazing largely contributes to patchiness under certain conditions; the tendency for denser patches of phytoplankton to spatially separate or alternate with zooplankton, for example, has been attributed to zooplankton grazing.[5]

Patchiness of zooplankton has been scrutinized in terms of vertical migration behavior. Stimulated by changes in the level of ambient illumination, zooplankton groups descend to greater depths of the water column during the day and ascend toward the surface during the night.

This diel vertical migration confers advantages to zooplankton populations, enabling them to avoid predators in well-lit surface waters and to use cooler, deeper waters to lower metabolic rates. Zooplankton grazing of phytoplankton peaks at night. This nocturnal feeding and subsequent migration to cooler waters is advantageous to the organisms in the maximization of food intake and utilization, fecundity, and various strategies associated with horizontal dispersion and transport.[5,234] Diel vertical migration fosters demographic zooplankton populations in estuaries.

VII. SUMMARY AND CONCLUSIONS

Phytoplankton are microscopic, free-floating plants ubiquitous in the oceans and estuaries of the world. Encompassing a variety of algal groups, the predominantly autotrophic organisms consist of single cells or relatively simply organized filamentous or chain-forming cell systems. They commonly have been subdivided on the basis of size into ultraplankton (less than 5 μm in diameter), nanoplankton (5 to 70 μm), microphytoplankton (70 to 100 μm), and macrophytoplankton (greater than 100 μm). However, two broad fractions, the net plankton and nanoplankton, often are differentiated by means of the nominal aperture size of the plankton net deployed during field sampling. All phytoplankton retained by the net (approximately 64 μm apertures) comprise the net plankton, and those passing through the net constitute the nanoplankton.

Diatoms (class Bacillariophyceae), dinoflagellates (class Dinophyceae), coccolithophores (class Prymnesiophyceae), silicoflagellates (class Chrysophyceae), and blue-green algae (class Cyanophyceae) represent the principal phytoplankton taxa of estuaries, although other taxonomic groups, including the green-colored algae (class Chlorophyceae), brown-colored phytoflagellates (class Haptophyceae), and euglenoid flagellates (class Euglenophyceae), also may be locally important in these coastal ecosystems. Algalogists classify phytoplankton by using various criteria, notably biochemistry, life history, and fine cellular structure. Characteristics of the cell have greatly aided scientists in the identification of phytoplankton. Such features as the cell shape, cell dimensions, cell wall, mucilage layers, chloroplasts, flagella, reserve substances (e.g., starch, oil, and leucosin), cell vacuoles, and trichocysts have facilitated taxonomic work.

Diatoms, forming a frustule or external skeleton composed primarily of silica, frequently dominate phytoplankton communities in high latitude regions, temperate inshore waters, and upwelling systems. Most marine diatoms are planktonic and nonmotile, ranging in size from 10 to 200 μm. They tend to sink in nonturbulent waters, although a number of morphological, physiological, and physical adaptations promote their suspension. Storage products of photosynthesis include chrysolaminarin (a polysaccharide) and lipids, making these microflora a high energy ration for zooplankton. Most diatoms reproduce by vegetative cell division, and under optimal conditions, cells double at a rate of 0.5 to 6 doublings per day, which enables blooms to develop.

Dinoflagellates are also widely distributed in estuarine and oceanic environments, being dominant in many subtropical and tropical regions and abundant in temperate and boreal autumnal assemblages. Typically unicellular, biflagellated, and autotrophic, these plankters usually compose a major fraction of the phytoplankton communities of estuaries. Two groups can be distinguished by the presence or absence of thecal plates: (1) the Peridinales (thecate or armored types) and (2) the Gymnodinales (nonthecate or naked forms). While some dinoflagellates reproduce sexually, most undergo asexual reproduction by binary cell division. Rates of reproduction are variable depending on environmental conditions, but approach those of the diatoms. The vast majority of dinoflagellates consists of autotrophic forms; however, heterotrophic (e.g., *Kofoidinium* and *Polykrikos*) and parasitic (e.g., *Blastodinium* and *Oodinium*) types also exist. A number of species produce toxins which, when liberated, occasionally cause mass mortality of fish, shellfish, and other organisms. These effects become most pronounced during toxic, "red tide" blooms in estuaries associated with the proliferaton of such genera as

Gonyaulax and *Ptychodiscus* (=*Gymnodinium*). Paralytic shellfish poisoning, a neurological disorder in humans, is ascribable to the toxic agent saxitoxin (produced by dinoflagellates) that accumulates in shellfish tissues.

Coccolithophores are small, biflagellate, unicellular algae characteristically covered by coccoliths or small calcareous plates embedded in a gelatinous sheath surrounding the cell. They sometimes attain large numbers in estuaries, especially in tropical and subtropical regions. Another phytoplankton group, the silicoflagellates, occasionally is abundant in estuaries, too. These small uniflagellate or biflagellate forms secrete an internal skeleton of siliceous spicules.

Plankton nets, plankton pumps, and water bottles are three kinds of gear typically employed for collecting phytoplankton field samples. Once samples are obtained, the standing crop or biomass of phytoplankton can be determined by direct cell counts, chlorophyll *a* determinations, or ATP content. The standing crop of phytoplankton is usually estimated by chlorophyll *a* measurements which in estuaries commonly show marked spatial and temporal variations generated, in part, by the patchy distribution of the microflora. The patchiness of phytoplankton populations develops from organismal responses to hydrographic, light, and nutrient distributions, to predation and symbiosis, and to mechanical aggregation by physical processes.

A seasonal cycle of phytoplankton biomass occurs in many estuaries, with maximum values observed in the late winter, spring, or summer concomitantly with phytoplankton blooms. This temporal pattern is more evident in temperate, boreal, and polar latitudes than in subtropical and tropical regions. A floristic succession of phytoplankton species ordinarily accompanies temporal fluctuations in the standing crop.

Phytoplankton productivity refers to the rate of carbon fixation (organic synthesis), and is determined by measuring rates of photosynthesis or respiration. Methods used to assess primary productivity of phytoplankton are the measurements of: (1) oxygen released during photosynthesis, (2) carbon dioxide uptake, (3) pH, (4) the rate of appearance of new algal biomass over time, and (5) uptake of the radioactive isotope, ^{14}C. The ^{14}C method represents the most widely adapted technique of estimating primary planktonic productivity in the sea.

Phytoplankton productivity in estuaries ranges from about 6.8 to 530 g C per square meter per year. Lower values are encountered in turbid estuaries, and higher values in clearer systems. The productivity of phytoplankton varies widely both in space and time; highest productivity readings generally take place during the warmer months of the year. In many mid-latitude estuaries, productivity levels remain rather high from spring to fall.

The species composition and seasonal dynamics of phytoplankton communities of several temperate estuaries on the East Coast of the U.S. have been the focus of comprehensive investigations. Four of these systems — Chesapeake Bay, Barnegat Bay, Long Island Sound, and Narragansett Bay — have been investigated intensely. The phytoplankton communities of these mid-latitude estuaries display similar patterns in species composition, production, and seasonal cycles of abundance and occurrence.

The primary production of phytoplankton is a function of the interaction of a number of physical, chemical, and biological factors, the most important of which are light, temperature, water circulation, salinity, nutrients, and grazing. Light energy has been deemed to be a limiting factor controlling the distribution of phytoplankton. Seasonal variations in illumination with latitude result in different seasonal production patterns in tropical, temperate, boreal, and polar regions.

The amount of illumination in an aquatic habitat depends on the angle of the sun during the day, the season of the year, latitude, and local climatic conditions (e.g., percent of cloud cover). A portion of the solar radiation incident at the sea surface is lost by scattering and reflection. In the water column, absorption and scattering of light by water molecules, suspended particles, and dissolved matter further attenuate light. Light attenuation is greater in estuaries than coastal or open oceanic environments where turbidity tends to be lower. In addition, different spectral distributions characterize estuarine, coastal, and oceanic waters. In turbid coastal systems, the maximum transmittance occurs at approximately 575 nm compared to 465 nm in clearer oceanic systems.

Within the range of PAR, about 400 to 700 nm, phytoplankton utilize chlorophyll pigments as well as accessory pigments (e.g., fucoxanthin and peridinin) to absorb all wavelengths of light. Photosynthesis increases logarithmically with increasing light intensity until a maximum value is reached. Near the sea surface where light saturation exists, however, photosynthesis remains inhibited. With increasing depth below the sea surface, photosynthesis rises to a maximum amount. At greater depths, photosynthesis diminishes; at the compensation depth, it just balances the metabolic losses due to respiration. Because of high turbidity in estuaries, the compensation depth may be very shallow, in some cases 1 m or less. Photosynthesis-irradiance curves have proven to be useful in studies of phytoplankton production in relationship to changing light intensity and quality.

Light has been implicated as a major factor triggering winter-spring phytoplankton blooms in mid- and high-latitude systems. Nutrients that accumulate during the winter in these estuaries are assimilated by phytoplankton when light increases, thereby accelerating growth and initiating the algal blooms. Phytoplankton pulses commonly appear in estuaries at various times during the spring and summer months.

Temperature affects enzymatic activities (respiration and photosynthesis) and growth rate processes of phytoplankton. Laboratory cultures indicate that the cell division rates of phytoplankton generally increase by two to four times with a 10°C increase in temperature, as long as the temperatures lie within a range favorable to growth. The development of a thermocline in some deeper estuaries due to seasonal temperature changes has been related to improved growth of phytoplankton and seasonal production cycles.

Salinity variations present problems of osmotic shock and stress for estuarine organisms. Phytoplankton employ several osmotic mechanisms to ameliorate the effects of highly variable salinity levels. Because of their ability to osmoregulate, phytoplankton typically have euryhaline tolerances. Experimental work demonstrates that estuarine phytoplankton are more euryhaline than coastal and oceanic forms. The rate of cell division of these microflora, as well as their occurrence, distribution, and productivity, is influenced by salinity.

Nutrients are necessary for adequate growth and production of phytoplankton. The major nutrient elements include nitrogen, phosphorus, and silicon. Trace elements, such as iron, manganese, zinc, copper, cobalt, and molybdenum, may limit phytoplankton growth if present in insufficient concentrations. However, some trace metals (e.g., copper and zinc) can be toxic to phytoplankton even at low concentrations and, consequently, may hinder their productivity. Some species require the vitamins cobalamine, thiamin, and biotin, as well as other organic compounds.

In estuaries, phytoplankton productivity normally is limited by the supply of nitrogen. Ammonia (lumped here with ammonium), nitrite, and nitrate comprise the principal dissolved inorganic forms of nitrogen. Urea, amino acids, and peptides are important dissolved organic forms. Phytoplankton incorporate both dissolved inorganic and organic nitrogen for growth.

Nitrogen cycling in estuaries involves a number of dynamic processes operating in the water column and bottom sediments. Uptake, remineralization, and oxidation in the water column, and burial, remineralization, biological uptake, oxidation, reduction, nitrous oxide production, and denitrification in seafloor sediments contribute to the complex biogeochemical cycling of nitrogen in these coastal systems. In the recycling of nitrogenous material, it is necessary to distinguish between new and regenerated nitrogen.

The concentration of dissolved inorganic nitrogen usually ranges from 0 to 100 μM in estuaries. Seasonal cycles in the concentration of dissolved inorganic nitrogen are evident in many estuaries, particularly those in middle to high latitudes. A seasonal pattern in the amount of certain forms of inorganic nitrogen also may occur within the upper 10 to 20 cm of the bottom sediments. Nitrogen tends to accumulate during the winter months in estuaries when phytoplankton growth is light limited. In contrast, the rapid growth of plants during the spring and summer months depletes nitrogen in estuarine waters. Seasonal variations of river flow account for temporal cycles of nitrogen in many estuaries.

Orthophosphate is the primary form of dissolved inorganic phosphorus in estuaries and the preferred form of phytoplankton. While the concentration of orthophosphate ions may be reduced in surface waters because of uptake by bacteria and algae, bottom sediments accumulate phosphorus compounds and serve as a source for overlying waters. Nitrogen to phosphorus ratios in marine waters and phytoplankton average about 16:1. In some estuaries during certain seasons, phosphorus rather than nitrogen may be the principal limiting nutrient to phytoplankton growth.

Diatoms require silicon for skeletal growth as do other organisms (e.g., radiolarians and siliceous sponges). Dissolved silicon is present in estuaries as silicic acid (H_4SiO_4), which can be depleted by phytoplankton blooms dominated by diatoms. The concentration of dissolved silicon undergoes seasonal cycles associated with seasonal fluxes of riverine input and silicon removal by phytoplankton. Low silicon availability can depress the abundance and productivity of diatoms.

Zooplankton grazing on phytoplankton populations periodically places a check on their standing crop. An acute alternation in phytoplankton and zooplankton abundance characterizes many coastal and estuarine waters. Variations in the time lag between large increases in phytoplankton and zooplankton can foster substantial differences in phytoplankton abundance. Phytoplankton, as well as zooplankton, have a patchy distribution due to physical-chemical boundary conditions, advective effects of water movement, social behavior in populations of the same species, intraspecific interactions causing attraction or repulsion between species, grazing, and reproduction. Although some of these effects may be most pronounced in oceanic environments, they also have been discerned in estuarine systems.

REFERENCES

1. **Dawes, C. J.,** *Marine Botany,* John Wiley & Sons, New York, 1981.
2. **Boney, A. D.,** *Phytoplankton,* Edward Arnold, London, 1975.
3. **Raymont, J. E. G.,** *Plankton and Productivity in the Oceans,* Vol. 1, 2nd ed., Pergamon Press, Oxford, 1980.
4. **Thurman, H. V. and Webber, H. H.,** *Marine Biology,* Charles E. Merrill Publishing, Columbus, OH, 1984.
5. **Mann, K. H.,** *Ecology of Coastal Waters: A Systems Approach,* University of California Press, Berkeley, 1982.
6. **Valiela, I.,** *Marine Ecological Processes,* Springer-Verlag, New York, 1984.
7. **Dodge, J. D.,** *The Fine Structure of Algal Cells,* Academic Press, New York, 1973.
8. **Levinton, J. S.,** *Marine Ecology,* Prentice-Hall, Englewood Cliffs, NJ, 1982.
9. **Garrison, D. L.,** Planktonic diatoms, in *Marine Planktonic Life Cycle Strategies,* Steidinger, K. A. and Walker, L. M., Eds., CRC Press, Boca Raton, FL, 1984, 1.
10. **McConnaughey, B. H. and Zottoli, R.,** *Introduction to Marine Biology,* C. V. Mosby, St. Louis, 1983.
11. **Smayda, T. J.,** The suspension and sinking of phytoplankton in the sea, *Oceanogr. Mar. Biol. Annu. Rev.,* 8, 853, 1970.
12. **Eppley, R. W.,** The growth and culture of diatoms, in *The Biology of Diatoms,* Werner, D., Ed., Blackwell Scientific, Oxford, 1977, 24.
13. **Hargraves, P. E.,** Studies on marine planktonic diatoms. II. Resting spore morphology, *J. Phycol.,* 12, 118, 1976.
14. **Steidinger, K. A. and Walker, L. M.,** Introduction, in *Marine Plankton Life Cycle Strategies,* Steidinger, K. A. and Walker, L. M., Eds., CRC Press, Boca Raton, FL, 1984.
15. **Walker, L. M.,** Life histories, dispersal, and survival in marine, planktonic dinoflagellates, in *Marine Plankton Life Cycle Strategies,* Steidinger, K. A. and Walker, L. M., Eds., CRC Press, Boca Raton, Fla., 1984, 19.
16. **Steidinger, K. A.,** A re-evaluation of toxic dinoflagellate biology and ecology, in *Progress in Phycological Research,* Vol. 2, Round, F. and Chapman, D., Eds., Elsevier North Holland, New York, 1983, 147.
17. **Maranda, L. and Shimizu, Y.,** Diarrhetic shellfish poisoning in Narragansett Bay, *Estuaries,* 10, 298, 1987.

18. **Hutner, S. H. and McLaughlin, J. J. A.,** Poisonous tides, *Sci. Am.,* 199, 92, 1958.
19. **Steidinger, K. A. and Ingle, R. M.,** Observations of the 1971 summer red tide in Tampa Bay, Florida, *Environ. Lett.,* 3, 271, 1972.
20. **Nybakken, J. W.,** *Marine Biology: An Ecological Approach,* Harper & Row, New York, 1982.
21. **Spector, D. L.,** Ed., *Dinoflagellates,* Academic Press, Orlando, FL, 1984.
22. **Gross, M. G.,** *Oceanography: A View of the Earth,* 3rd ed., Prentice-Hall, Englewood Cliffs, NJ, 1982.
23. **Parsons, T. R. and Takahashi, M.,** *Biological Oceanographic Processes,* Pergamon Press, Elmsford, NY, 1973.
24. **Stein, J. R.,** Ed., *Handbook of Phycological Methods: Culture Methods and Growth Measurements,* Cambridge University Press, London, 1973.
25. **UNESCO,** A Review of Methods Used for Quantitative Phytoplankton Studies, UNESCO Tech. Pap. Mar. Sci., Paris, 1974, 18.
26. **Hellebust, J. A. and Craigie, J. S.,** *Handbook on Phycological Methods: Physiological and Biochemical Methods,* Cambridge University Press, London, 1978.
27. **Strickland, J. D. H. and Parsons, T. R.,** *A Practical Handbook of Seawater Analysis,* Fisheries Research Board of Canada, Ottawa, 1968, 167.
28. **Holm-Hansen, O. and Rieman, B.,** Chlorophyll *a* determination: improvements in methodology, *Oikos,* 30, 438, 1978.
29. **Holm-Hansen, O. and Booth, C. R.,** The measurement of adenosine triphosphate in the ocean and its ecological significance, *Limnol. Oceanogr.,* 11, 510, 1966.
30. **Hobbie, J. E., Holm-Hansen, O., Packard, T. T., Pomeroy, L. R., Sheldon, R. W., Thomas, J. P., and Wiebe, W. J.,** A study of the distribution and activity of microorganisms in ocean waters, *Limnol. Oceanogr.,* 17, 544, 1972.
31. **Rice, T. R. and Ferguson, R. L.,** Response of estuarine phytoplankton to environmental conditions, in *Physiological Ecology of Estuarine Organisms,* Vernberg, F. J., Ed., University of South Carolina Press, Columbia, 1975, 1.
32. **Boynton, W. R., Kemp, W. M., and Keefe, C. W.,** A comparative analysis of nutrients and other factors influencing estuarine phytoplankton production, in *Estuarine Comparisons,* Kennedy, V. S., Ed., Academic Press, New York, 1982, 69.
33. **Palumbo, A. V.,** Dynamics of Bacterioplankton in the Newport River Estuary, Ph.D. thesis, North Carolina State University, Raleigh, 1980.
34. **Durbin, A. G. and Durbin, E. G.,** Standing stock and estimated production rates of phytoplankton and zooplankton in Narragansett Bay, Rhode Island, *Estuaries,* 4, 24, 1981.
35. **Smayda, T. J.,** The phytoplankton of estuaries, in *Estuaries and Enclosed Seas,* Ketchum, B. H., Ed., Elsevier, Amsterdam, 1983, 65.
36. **Karentz, D. and Smayda, T. J.,** Temperature and seasonal occurrence patterns of 30 dominant phytoplankton species in Narragansett Bay over a 22-year period (1959—1980), *Mar. Ecol. Prog. Ser.,* 18, 277, 1984.
37. **Alpine, A. E. and Cloern, J. E.,** Spatial variability of phytoplankton growth rate in San Francisco Bay: responses to light availability, *Estuaries,* 8, 76A, 1985.
38. **Guerrero, H. G. D.,** Short term patterns of chlorophyll *a* and copepod abundance in Long Island Sound, *Estuaries,* 8, 77, 1985.
39. **Horne, A. J. and Roth, J. C.,** Phytoplankton spatial distribution in San Francisco Bay measured by a helicopter synoptic technique, *Estuaries,* 8, 77A, 1985.
40. **Oviatt, C. A.,** Physical factors controlling the ecology of northeast coastal waters, *Estuaries,* 8, 86A, 1985.
41. **Roth, J. C. and Horne, A. J.,** Diurnal plankton fluctuations in deep and shallow waters of San Francisco Bay, *Estuaries,* 8, 77A, 1985.
42. **Williams, J., Foerster, J., and Skove, F.,** Diurnal variation of surface plankton in the Patuxent River, *Estuaries,* 8, 38A, 1985.
43. **Harding, L. W., Meeson, B. W., Jr., and Fisher, T. R., Jr.,** Phytoplankton production in two east coast estuaries: photosynthesis-light functions and patterns of carbon assimilation in Chesapeake and Delaware Bays, *Estuarine Coastal Shelf Sci.,* 23, 773, 1986.
44. **Sellner, K. G. and Kachur, M. E.,** Phytoplankton: relationships between phytoplankton, nutrients, oxygen flux, and secondary producers, in *Ecological Studies in the Middle Reach of Chesapeake Bay,* Heck, K. L., Jr., Ed., Springer-Verlag, Berlin, 1987, 12.
45. **Bowman, M. J., Esaias, W. E., and Schnitzer, M. B.,** Tidal stirring and the distribution of phytoplankton in Long Island Sound and Block Island Sound, *J. Mar. Res.,* 39, 587, 1981.
46. **McCarthy, J. J. and Altabet, M. A.,** Patchiness in nutrient supply: implications for phytoplankton ecology, in *Trophic Interactions within Aquatic Ecosystems,* Selected Symp. 85, Meyers, D. G. and Strickler, J. R., Eds., American Association for the Advancement of Science, Washington, D. C., 1984, 29.
47. **Bennett, A. F. and Denman, K. L.,** Phytoplankton patchiness: inferences from particle statistics, *J. Mar. Res.,* 43, 307, 1985.

48. **Sephton, D. H. and Harris, G. P.,** Physical variability and phytoplankton communities. VI. Day to day changes in primary productivity and species abundance, *Arch. Hydrobiol.,* 102, 155, 1984.

49. **Hulburt, E. M.,** The diversity of phytoplankton populations in oceanic, coastal, and estuarine regions, *J. Mar. Res.,* 21, 81, 1963.

50. **Hulburt, E. M.,** Competition for nutrients by marine phytoplankton in oceanic, coastal, and estuarine regions, *Ecology,* 51, 475, 1970.

51. **Hulburt, E. M. and Rodman, J.,** Distribution of phytoplankton species with respect to salinity between the coast of southern New England and Bermuda, *Limnol. Oceanogr.,* 8, 263, 1963.

52. **Yentsch, C. S.,** Plankton Production, Mesa New York Bight Atlas Monograph 12, New York Sea Grant Institute, Albany, 1977.

53. **Ferguson, R. L., Thayer, G. W., and Rice, T. R.,** Marine primary producers, in *Functional Adaptations of Marine Organisms,* Vernberg, F. J. and Vernberg, W. B., Eds., Academic Press, New York, 1981, 9.

54. **Riley, G. A.,** The plankton of estuaries, in *Estuaries,* Publ. 83, Lauff, G. H., Ed., American Association for the Advancement of Science, Washington, D.C., 1967, 316.

55. **Steemann Nielsen, E.,** Productivity, definition, and measurement, in *The Sea,* Vol. 2, Hill, M. N., Ed., Wiley-Interscience, New York, 1963, 129.

56. **Vollenweider, R. A.,** Ed., *A Manual on Methods for Measuring Primary Production in Aquatic Environments,* 2nd ed., IBP Handbook No. 12, Blackwell Scientific, Oxford, 1974.

57. **Odum, E. P.,** *Fundamentals of Ecology,* 3rd ed., W. B. Saunders, Philadelphia, 1971.

58. **Colijn, F. and de Jonge, V. N.,** Primary production of microphytobenthos in the Ems-Dollard estuary, *Mar. Ecol. Prog. Ser.,* 14, 185, 1984.

59. **Davies, J. M. and Williams, P. J. le B.,** Verification of ^{14}C and O_2 derived primary organic production measurements using an enclosed ecosystem, *J. Plankton Res.,* 6, 457, 1984.

60. **Steemann Nielsen, E.,** The use of radioactive carbon (C^{14}) for measuring organic production in the sea, *J. Cons. Perm. Int. Explor. Mer,* 18, 117, 1952.

61. **Steemann Nielsen, E. and Aabye Jensen, E.,** Primary oceanic production, the autotrophic production of organic matter in the oceans, *Galathea Rep.,* 1, 49, 1957.

62. **Sieburth, J. McN.,** International Helgoland Symposium: convener's report on informal session on biomass and productivity of microorganisms in planktonic ecosystems, *Helgol. Wiss. Meeresunters.,* 30, 697, 1977.

63. **Peterson, B. J.,** Aquatic primary productivity and the ^{14}C-CO_2 method: a history of the productivity problem, *Annu. Rev. Ecol. Syst.,* 11, 359, 1980.

64. **Colijn, F., Gieskes, W. W. C., and Zevenboom, W.,** The measurement of primary production: problems and recommendations, *Hydrobiol. Bull.,* 17, 29, 1983.

65. **Oviatt, C., Buckley, B., and Nixon, S.,** Annual phytoplankton metabolism in Narragansett Bay calculated from survey field measurements and microcosm observations, *Estuaries,* 4, 167, 1981.

66. **Gieskes, W. W. C., Kraay, G. W., and Baars, M. A.,** Current ^{14}C methods for measuring primary production: gross underestimates in oceanic waters, *Neth. J. Sea Res.,* 13, 58, 1979.

67. **Gieskes, W. W. C. and Kraay, G. W.,** State-of-the-art in the measurement of primary production, in *Flows of Energy and Materials in Marine Ecosystems: Theory and Practice,* Fasham, M. J. R., Ed., Plenum Press, New York, 1984, 171.

68. **Dring, M. J.,** *The Biology of Marine Plants,* Edward Arnold, London, 1982.

69. **Dring, M. J. and Jewson, D. H.,** What does ^{14}C uptake by phytoplankton really measure? A theoretical modelling approach, *Proc. R. Soc. London Ser. B.,* 214, 351, 1982.

70. **Holligan, P. M., Williams, P. J. le B., Purdie, D., and Harris, R. P.,** Photosynthesis, respiration and nitrogen supply of plankton populations in stratified, frontal and tidally mixed shelf waters, *Mar. Ecol. Prog. Ser.,* 17, 201, 1984.

71. **Stockner, J. G. and Cliff, D. D.,** Phytoplankton ecology of Vancouver Harbor, *J. Fish. Res. Board Can.,* 36, 1, 1979.

72. **Joint, I. R. and Pomroy, A. J.,** Primary production in a turbid estuary, *Estuarine Coastal Shelf Sci.,* 13, 303, 1981.

73. **Cole, B. E. and Cloern, J. E.,** Significance of biomass and light availability to phytoplankton productivity in San Francisco Bay, *Mar. Ecol. Prog. Ser.,* 17, 15, 1984.

74. **Ryther, J. H.,** Geographic variations in productivity, in *The Sea,* Vol. 2, Hill, M. N., Ed., Wiley-Interscience, New York, 1963, 347.

75. **O'Reilly, J. E. and Busch, D. A.,** Phytoplankton primary production for the northeastern Atlantic shelf, Rapp. *P. Réun. Cons. Int. Explor. Mer,* 183, 255, 1983.

76. **Williams, R. B.,** Annual phytoplanktonic production in a system of shallow temperate estuaries, in *Some Contemporary Studies in Marine Science,* Barnes, H., Ed., Hafner, New York, 1966, 699.

77. **Flemer, D.,** Primary production in Chesapeake Bay, *Chesapeake Sci.,* 11, 117, 1970.

78. **Cadée, G. C. and Hegeman, J.,** Primary production of phytoplankton in the Dutch Wadden Sea, *Neth. J. Sea Res.,* 8, 240, 1974.

79. **Malone, T. C.,** Environmental regulation of phytoplankton productivity in the lower Hudson estuary, *Estuarine Coastal Mar. Sci.,* 5, 157, 1977.
80. **Cadée, G. C.,** Primary production and chlorophyll in the Zaire River, estuary and plume, *Neth. J. Sea Res.,* 12, 368, 1978.
81. **Grindley, J. R.,** Estuarine plankton, in *Estuarine Ecology: With Particular Reference to Southern Africa,* Day, J. H., Ed., A. A. Balkema, Rotterdam, 1981, 117.
82. **Flint, R. W.,** Phytoplankton production in the Corpus Christi Bay estuary, *Cont. Mar. Sci.,* 27, 65, 1984.
83. **Peterson, D. H. and Festa, J. F.,** Numerical simulation of phytoplankton productivity in partially mixed estuaries, *Estuarine Coastal Shelf Sci.,* 19, 563, 1984.
84. **Sellner, K. G.,** Plankton productivity and biomass in a tributary of the upper Chesapeake Bay. I. Importance of size-fractionated phytoplankton productivity, biomass and species composition in carbon export, *Estuarine Coastal Shelf Sci.,* 17, 197, 1983.
85. **Wolff, W. J.,** Biotic aspects of the chemistry of estuaries, in *Chemistry and Biogeochemistry of Estuaries,* Olausson, E. and Cato, I., Eds., John Wiley & Sons, Chichester, 1980, 263.
86. **Mihursky, J. A., Heinle, D. R., and Boynton, W. R.,** Ecological Effects of Nuclear Steam Electric Station Operations on Estuarine Systems, Unpubl. Tech. Rep., UMCEES Ref. No. 77-28-CBL, Chesapeake Biological Laboratory, Solomons, MD, 1977.
87. **McLusky, D. S.,** *The Estuarine Ecosystem,* Halsted Press, New York, 1981.
88. **Patten, R., Mulford, R., and Warinner, J.,** An annual phytoplankton cycle in the lower Chesapeake Bay, *Chesapeake Sci.,* 4, 1, 1963.
89. **Mulford, R. A.,** An annual plankton cycle on the Chesapeake Bay in the vicinity of Calvert Cliffs, Maryland: June 1969—May 1970, *Proc. Acad. Nat. Sci. (Philadelphia),* 124, 17, 1972.
90. **Marshall, H. G.,** Seasonal phytoplankton composition in the lower Chesapeake Bay and Old Plantation Creek, Cape Charles, Virginia, *Estuaries,* 3, 207, 1980.
91. **Marshall, H. G.,** The composition of phytoplankton within the Chesapeake Bay plume and adjacent waters off the Virginia coast, U.S.A., *Estuarine Coastal Shelf Sci.,* 15, 29, 1982.
92. **Mountford, K.,** Phytoplankton, in *Ecology of Barnegat Bay, New Jersey,* Kennish, M. J. and Lutz, R. A., Eds., Springer-Verlag, New York, 1984, 52.
93. **Conover, S. M.,** Oceanography of Long Island Sound, 1952—1954. IV. Phytoplankton, *Bull. Bingham Oceanogr. Coll.,* 15, 62, 1956.
94. **Cowles, R.,** A biological study of the offshore waters of the Chesapeake Bay, *Fish. Bull. (U.S.),* 46, 277, 1930.
95. **Smayda, T. J.,** A survey of phytoplankton dynamics in the coastal waters from Cape Hatteras to Nantucket, in *Coastal and Offshore Environmental Inventory, Cape Hatteras to Nantucket Shoals,* Mar. Publ. Ser. No. 2, University of Rhode Island, Kingston, 1973.
96. **Wolfe, J. J., Cunningham, B., Wilkerson, N., and Barnes, J.,** An investigation of the microplankton of Chesapeake Bay, *J. Elisha Mitchell Sci. Soc.,* 42, 25, 1926.
97. **Sellner, K. G.,** Phytoplankton in Chesapeake Bay: role in carbon, oxygen and nutrient dynamics, in *Contaminant Problems and Management of Living Chesapeake Bay Resources,* Majumdar, S. K., Hall, L. W., Jr., and Austin, H. M., Eds., Pennsylvania Academy of Science, Easton, 1987, 134.
98. **Marshall, H. G.,** Phytoplankton distribution along the eastern coast of the USA. V. Seasonal density and all volume patterns for the northeastern continental shelf, *J. Plankton Res.,* 6, 169, 1984.
99. **McCarthy, J. J., Taylor, W. R., and Loftus, J.,** Significance of nanoplankton in the Chesapeake Bay estuary and problems associated with the measurement of nanoplankton productivity, *Mar. Biol.,* 24, 7, 1974.
100. **Van Valkenburg, S. D. and Flemer, D. A.,** The distribution and productivity of nanoplankton in a temperate estuarine area, *Estuarine Coastal Mar. Sci.,* 2, 311, 1974.
101. **Van Valkenburg, S. D., Jones, J. K., and Heinle, D. R.,** A comparison by size class and volume of detritus versus phytoplankton in Chesapeake Bay, *Estuarine Coastal Mar. Sci.,* 6, 569, 1978.
102. **Chizmadia, P. A., Kennish, M. J., and Ohori, V. L.,** Physical description of Barnegat Bay, in *Ecology of Barnegat Bay, New Jersey,* Kennish, M. J. and Lutz, R. A., Eds., Springer-Verlag, New York, 1984, 1.
103. **Riley, G. A.,** Environmental control of autumn and winter diatom flowerings in Long Island Sound, in *International Oceanography Congress,* Preprints, Sears, M. E., Ed., American Association for the Advancement of Science, Washington, D. C., 1959, 850.
104. **Riley, G. A. and Conover, S. M.,** Phytoplankton of Long Island Sound, 1954—1955, *Bull. Bingham Oceanogr. Coll.,* 19, 5, 1967.
105. **Mandelli, E. F., Burkholder, T. E., Doheny, T. E., and Brody, J.,** Studies of primary productivity in coastal waters of southern Long Island, New York, *Mar. Biol.,* 7, 153, 1970.
106. **Pickett, S. T. A.,** Succession: an evolutionary interpretation, *Am. Nat.,* 110, 197, 1976.
107. **Marshall, H. G. and Cohn, M. S.,** Phytoplankton composition of the New York Bight and adjacent waters, *J. Plankton Res.,* 9, 267, 1987.
108. **Smayda, T. J.,** Phytoplankton studies in lower Narragansett Bay, *Limnol. Oceanogr.,* 2, 342, 1957.
109. **Smayda, T. J.,** The growth of *Skeletonema costatum* during a winter-spring bloom in Narragansett Bay, Rhode Island, *J. Bot.,* 20, 219, 1973.

110. **Pratt, D. M.,** The phytoplankton of Narragansett Bay, *Limnol. Oceanogr.,* 4, 425, 1959.

111. **Pratt, D. M.,** The winter-spring diatom flowering in Narragansett Bay, *Limnol. Oceanogr.,* 10, 173, 1965.

112. **Durbin, E. G., Durbin, A. G., Smayda, T. J., and Verity, P. G.,** Food limitation of production by adult *Acartia tonsa* in Narragansett Bay, Rhode Island, *Limnol. Oceanogr.,* 28, 1199, 1983.

113. **Hulsizer, E. E.,** Zooplankton of lower Narragansett Bay, 1972—1973, *Chesapeake Sci.,* 17, 260, 1976.

114. **Krout, J. E.,** Pigment and Pigment Ratio (430 NM/665 NM) Distributions in Several Marine Environments, M.S. thesis, University of Rhode Island, Kingston, 1971.

115. **Kennish, M. J.,** *Ecology of Estuaries,* Vol. 1, CRC Press, Boca Raton, FL, 1986.

116. **Smayda, T. J.,** Phytoplankton species succession, in *The Physiological Ecology of Phytoplankton,* Morris, I., Ed., University of California Press, Berkeley, 1980, 493.

117. **Watling, L., Bottom, D., Pembroke, A., and Maurer, D.,** Seasonal variations in Delaware Bay phytoplankton community structure, *Mar. Biol.,* 52, 207, 1979.

118. **Côté, B. and Platt, T.,** Utility of the light-saturation curve as an operational model for quantifying the effects of environmental conditions on phytoplankton photosynthesis, *Mar. Ecol. Prog. Ser.,* 18, 57, 1984.

119. **McCombie, A. M.,** Actions and interactions of temperature, light intensity and nutrient concentration on the growth of the green algae, *Chlamydomonas reinhardi* Dangeard, *J. Fish. Res. Board Can.,* 17, 871, 1960.

120. **Lewis, M. R., Ulloa, O., and Platt, T.,** Photosynthetic action, absorption, and quantum yield spectra for a natural population of *Oscillatoria* in the North Atlantic, *Limnol. Oceanogr.,* 33, 92, 1988.

121. **Marra, J., Heinemann, K., and Landriau, G., Jr.,** Observed and predicted measurements of photosynthesis in a phytoplankton culture exposed to natural irradiance, *Mar. Ecol. Prog. Ser.,* 24, 43, 1985.

122. **Platt, T.,** Primary production of the ocean water column as a function of surface light intensity: algorithms for remote sensing, *Deep Sea Res.,* 33, 149, 1986.

123. **Smith, H. and Morgan, D. C.,** The spectral characteristics of the visible radiation incident upon the surface of the earth, in *Plants and the Daylight Spectrum,* Smith, H., Ed., Academic Press, London, 1981, 1.

124. **Jerlov, N. G.,** *Marine Optics, Elsevier Oceanography Series,* Vol. 14, Elsevier, Amsterdam, 1976.

125. **Smith, R. C. and Baker, K. S.,** Optical properties of the clearest natural waters (200—800 nm), *Appl. Opt.,* 20, 177, 1981.

126. **Glover, H. E., Keller, M. D., and Spinrad, R. W.,** The effects of light quality and intensity on photosynthesis and growth of marine eukaryotic and prokaryotic phytoplankton clones, *J. Exp. Mar. Biol. Ecol.,* 105, 137, 1987.

127. **Steemann Nielsen, E.,** *Marine Photosynthesis: With Special Emphasis on the Ecological Aspects, Elsevier Oceanography Series,* Vol. 13, Elsevier Scientific, Amsterdam, 1975.

128. **Drew, E. A.,** Light, in *Sublittoral Ecology: The Ecology of the Shallow Sublittoral Benthos,* Earll, R. and Erwin, D. G., Eds., Clarendon Press, Oxford, 1983, 10.

129. **Lewis, M. R., Warnock, R. E., and Platt, T.,** Absorption and photosynthetic action spectra for natural phytoplankton populations: implications for production in the open ocean, *Limnol. Oceanogr.,* 30, 794, 1985.

130. **Lewis, M. R., Warnock, R. E., Irwin, B., and Platt, T.,** Measuring photosynthetic action spectra of natural phytoplankton populations, *J. Phycol.,* 21, 310, 1985.

131. **Sathyendranath, S. and Platt, T.,** Computation of aquatic primary production: extended formalism to include effect of angular and spectral distribution of light, *Limnol. Oceanogr.,* 34, 188, 1989.

132. **Harris, G. P.,** Photosynthesis, productivity and growth: the physiological ecology of phytoplankton, *Arch. Hydrobiol.,* 10, 1, 1978.

133. **Lewis, M. R., Cullen, J. J., and Platt, T.,** Relationships between vertical mixing and photoadaptation of phytoplankton: similarity criteria, *Mar. Ecol. Prog. Ser.,* 15, 141, 1984.

134. **Falkowski, P. G.,** Physiological responses of phytoplankton to natural light regimes, *J. Plankton Res.,* 6, 295, 1984.

135. **Falkowski, P. G.,** Light-shade adaptation in marine phytoplankton, in *Primary Productivity in the Sea,* Falkowski, P. G., Ed., Plenum Press, New York, 1980, 99.

136. **Falkowski, P. G., Dubinsky, Z., and Wyman, K.,** Growth-irradiance relationships in phytoplankton, *Limnol. Oceanogr.,* 30, 311, 1985.

137. **Harvey, H. W.,** *The Chemistry and Fertility of Sea Waters,* Cambridge University Press, London, 1955.

138. **Jones, R. I.,** Factors controlling phytoplankton production and succession in a highly eutrophic lake (Kinnego Bay, Lough Neagh). III. Interspecific competition in relation to irradiance and temperature, *J. Ecol.,* 65, 579, 1977.

139. **Li, W. K. W.,** Temperature adaptation in phytoplankton: cellular and photosynthetic characteristics, in *Primary Productivity in the Sea,* Falkowski, P. G., Ed., Plenum Press, New York, 1980, 259.

140. **Morris, I.,** Ed., *The Physiological Ecology of Phytoplankton,* University of California Press, Berkeley, 1980.

141. **Foy, R. H.,** Interaction of temperature and light on the growth rates of two planktonic *Oscillatoria* species under a short photoperiod regime, *Br. Phycol. J.,* 18, 267, 1983.

142. **Post, A. F., de Wit, R., and Mur, L. R.,** Interactions between temperature and light intensity on growth and photosynthesis of the cyanobacterium *Oscillatoria agardhii, J. Plankton Res.,* 7, 487, 1985.

143. **Haertel, L., Osterberg, C., Curl, H. C., and Park, P. K.,** Nutrient and plankton ecology of the Columbia River estuary, *Ecology,* 50, 962, 1969.

144. **Goldman, J. C. and Carpenter, E. J.,** A kinetic approach to the effect of temperature on algal growth, *Limnol. Oceanogr.,* 19, 756, 1974.

145. **Goldman, J. C. and Ryther, J. H.,** Temperature-influenced species competition in mass cultures of marine phytoplankton, *Biotechnol. Bioeng.,* 18, 1125, 1976.

146. **Goldman, J. C.,** Temperature effects on phytoplankton growth in continuous culture, *Limnol. Oceanogr.,* 22, 932, 1977.

147. **Wilde, E. W., Olmstead, L. L., and Gnilka, A.,** Some observations concerning the effects of a power station's thermal effluent on phytoplankton dynamics, *J. Tenn. Acad. Sci.,* 52, 10, 1977.

148. **Eppley, R. W.,** Temperature and phytoplankton growth in the sea, *Fish. Bull. (U.S.),* 70, 1063, 1972.

149. **Fogg, G. E.,** *Algal Cultures and Phytoplankton Ecology,* University of Wisconsin Press, Madison, 1965.

150. **Eppley, R. W. and Sloan, P. K.,** Growth rates of marine phytoplankton: correlation with light adsorption by cell chlorophyll *a, Physiol. Plant.,* 19, 47, 1966.

151. **Jørgensen, E. G.,** The adaptation of plankton algae. II. Aspects of the temperature adaptation of *Skeletonema costatum, Physiol. Plant.,* 21, 423, 1968.

152. **Hulburt, E. M. and Guillard, R. R. L.,** The relationship of the distribution of the diatom *Skeletonema tropicum* to temperature, *Ecology,* 49, 337, 1968.

153. **Hand, W. G., Collard, P. A., and Davenport, D.,** The effects of temperature and salinity change on swimming rate in the dinoflagellates, *Gonyaulax* and *Gyrodinium, Biol. Bull.,* 128, 90, 1965.

154. **El-Sayed, S.,** Productivity of the Antarctic waters — a reappraisal, in *Marine Phytoplankton and Productivity,* Holm-Hansen, O., Bolis, L., and Gilles, R., Springer-Verlag, New York, 1984, 19.

155. **Bowden, K. F.,** Physical factors: salinity, temperature, circulation, and mixing processes, in *Chemistry and Biogeochemistry of Estuaries,* Olausson, E. and Cato, I., Eds., John Wiley & Sons, Chichester, 1980, 37.

156. **Venrick, E. L.,** Winter mixing and the vertical stratification of phytoplankton — another look, *Limnol. Oceanogr.,* 29, 636, 1984.

157. **Kremer, J. N. and Nixon, S. W.,** *A Coastal Marine Ecosystem: Simulation and Analysis, Ecological Studies 24,* Springer-Verlag, New York, 1978.

158. **Cloern, J. E., Arthur, J. F., Ball, M. D., Cole, B. E., Wong, R. L., and Alpine, A. E.,** River discharge controls phytoplankton dynamics in the northern San Francisco Bay estuary, *Estuarine Coastal Shelf Sci.,* 16, 415, 1983.

159. **Smetacek, V. S.,** Role of sinking in diatom life-history cycles: ecological, evolutionary and geological significance, *Mar. Biol.,* 84, 239, 1985.

160. **Callaway, R. J. and Specht, D. T.,** Dissolved silicon in the Yaquina estuary, Oregon, *Estuarine Coastal Shelf Sci.,* 15, 561, 1982.

161. **Ketchum, B. H.,** Relation between circulation and planktonic populations in estuaries, *Ecology,* 35, 191, 1954.

162. **Hammer, L.,** Salzgehalt und photosynthese bei marinen pflanzen, *Mar. Biol.,* 1, 185, 1968.

163. **Guillard, R. R. L.,** Salt and osmotic balance, in *Physiology and Biochemistry of Algae,* Lewin, R. A., Ed., Academic Press, New York, 1962, 529.

164. **McLachlan, J.,** The effect of salinity on growth and chlorophyll content in representative classes of unicellular marine algae, *Can. J. Microbiol.,* 7, 399, 1961.

165. **Nakanishi, M. and Monsi, M.,** Effect of variation in salinity on photosynthesis of phytoplankton growing in estuaries, *J. Fac. Sci. Univ. Tokyo Sect. 3,* 9, 19, 1965.

166. **Qasim, S. Z., Blattalhiri, P. M. A., and Devassy, V. P.,** The influence of salinity on the rate of photosynthesis and abundance of some tropical phytoplankton, *Mar. Biol.,* 12, 200, 1972.

167. **Nordli, E.,** Experimental studies on the ecology of *Ceratia, Oikos,* 8, 200, 1957.

168. **Vosjan, J. H. and Siezen, R. J.,** Relation between primary production and salinity of algal cultures, *Neth. J. Sea Res.,* 4, 11, 1968.

169. **Braarud, T.,** Salinity as an ecological factor in marine phytoplankton, *Physiol. Plant.,* 4, 28, 1951.

170. **Kain, J. M. and Fogg, G. W.,** Studies on the growth of marine phytoplankton. I. *Asterionella japonica* Gran, *J. Mar. Biol. Assoc. U.K.,* 37, 397, 1958.

171. **Kain, J. M. and Fogg, G. W.,** Studies on the growth of marine phytoplankton. III. *Prorocentrum micans* Ehrenberg, *J. Mar. Biol. Assoc. U.K.,* 39, 33, 1960.

172. **Brand, L. E.,** The salinity tolerance of forty-six marine phytoplankton isolates, *Estuarine Coastal Shelf Sci.,* 18, 543, 1984.

173. **Fogg, G. E.,** Primary productivity, in *Chemical Oceanography,* Vol. 2, 2nd ed., Riley, J. P. and Skirrow, G., Eds., Academic Press, London, 1975, 385.

174. **Terry, K. L., Laws, E. A., and Burns, D. J.,** Growth rate variation in the N-P requirement ratio of phytoplankton, *J. Phycol.,* 21, 323, 1985.

175. **Smith, S. V.,** Phosphorus versus nitrogen limitation in the marine environment, *Limnol. Oceanogr.,* 29, 1149, 1984.

176. **Howarth, R. W. and Cole, J. J.,** Molybdenum availability, nitrogen limitation, and phytoplankton growth in natural waters, *Science,* 229, 653, 1985.

177. **Kanda, J., Saino, T., and Hattori, A.,** Nitrogen uptake by natural populations of phytoplankton and primary production in the Pacific Ocean: regional variability of uptake capacity, *Limnol. Oceanogr.,* 30, 987, 1985.

178. **Caraco, N., Tamse, A., Boutros, O., and Valiela, I.,** Nutrient limitation of phytoplankton growth in brackish coastal ponds, *Can. J. Fish. Aquat. Sci.,* 44, 473, 1987.

179. **Wetzel, R. G.,** *Limnology,* 2nd ed., Saunders College Publishing, Philadelphia, 1983.

180. **Elser, J. J., Elser, M. M., MacKay, N. A., and Carpenter, S. R.,** Zooplankton-mediated transitions between N- and P-limited algal growth, *Limnol. Oceanogr.,* 33, 1, 1988.

181. **Webb, K. L.,** Conceptual models and processes of nutrient cycling in estuaries, in *Estuaries and Nutrients,* Neilson, B. J. and Cronin, L. E., Eds., Humana Press, Clifton, NJ, 1981, 25.

182. **Lippson, A. J., Haire, M. S., Holland, A. F., Jacobs, F., Jensen, J., Moran-Johnson, R. L., Polgar, T. T., and Richkus, W. A.,** *Environmental Atlas of the Potomac Estuary,* Johns Hopkins University Press, Baltimore, 1981.

183. **Wheeler, P. A.,** Phytoplankton nitrogen metabolism, in *Nitrogen in the Marine Environment,* Carpenter, E. J. and Capone, D. G., Eds., Academic Press, New York, 1983, 309.

184. **Nixon, S. W. and Pilson, M. E. Q.,** Nitrogen in estuarine and coastal marine ecosystems, in *Nitrogen in the Marine Environment,* Carpenter, E. J. and Capone, D. G., Eds., Academic Press, New York, 1983, 565.

185. **Dugdale, R. C. and Goering, J. J.,** Uptake of new and regenerated forms of nitrogen in primary productivity, *Limnol. Oceanogr.,* 12, 196, 1967.

186. **Kemp, W. M., Wetzel, R. L., Boynton, W. R., D´Elia, C. F., and Stevenson, J. C.,** Nitrogen cycling and estuarine interfaces: some current concepts and research directions, in *Estuarine Comparisons,* Kennedy, V. S., Ed., Academic Press, New York, 1982, 209.

187. **MacIsaac, J. J. and Dugdale, R. C.,** The kinetics of nitrate and ammonia uptake by natural populations of marine phytoplankton, *Deep Sea Res.,* 16, 45, 1969.

188. **McCarthy, J. J.,** Nitrogen, in *The Physiological Ecology of Phytoplankton,* Morris, I., Ed., Blackwell Scientific, Oxford, 1980, 191.

189. **Paasche, E. and Kristiansen, S.,** Nitrogen nutrition of the phytoplankton in the Oslofjord, *Estuarine Coastal Shelf Sci.,* 14, 237, 1982.

190. **Dugdale, R. C. and Wilkerson, F. P.,** The use of ^{15}N to measure nitrogen uptake in eutrophic oceans; experimental consideration, *Limnol. Oceanogr.,* 31, 673, 1986.

191. **Kokkinakis, S. A. and Wheeler, P. A.,** Nitrogen uptake and phytoplankton growth in coastal upwelling regions, *Limnol. Oceanogr.,* 32, 1112, 1987.

192. **Carpenter, E. J. and Capone, D. G.,** Eds., *Nitrogen in the Marine Environment,* Academic Press, New York, 1983.

193. **Kaufman, Z. G., Lively, J. S., and Carpenter, E. J.,** Uptake of nitrogenous nutrients by phytoplankton in a barrier island estuary: Great South Bay, New York, *Estuarine Coastal Shelf Sci.,* 17, 483, 1983.

194. **Carpenter, E. J. and Dunham, S.,** Nitrogenous nutrient uptake, primary production, and species composition of phytoplankton in the Carmans River estuary, Long Island, New York, *Limnol. Oceanogr.,* 30, 513, 1985.

195. **Orrett, K. and Karl, D. M.,** Dissolved organic phosphorus production in surface seawaters, *Limnol. Oceanogr.,* 32, 383, 1987.

196. **Perry, M. J. and Eppley, R. W.,** Phosphate uptake by phytoplankton in the central North Pacific Ocean, *Deep Sea Res.,* 28, 39, 1981.

197. **Harrison, W. G.,** Uptake and recycling of soluble reactive phosphorus by marine microplankton, *Mar. Ecol. Prog. Ser.,* 10, 127, 1983.

198. **Herbland, A.,** Phosphate uptake in the euphotic layer of the equatorial Atlantic Ocean: methodological observations and ecological significance, *Oceanogr. Trop.,* 19, 25, 1984.

199. **Sorokin, Y. I.,** Phosphorus metabolism in planktonic communities of the eastern tropical Pacific Ocean, *Mar. Ecol. Prog. Ser.,* 21, 87, 1985.

200. **Harrison, W. G. and Harris, L. R.,** Isotope-dilution and its effects on measurements of nitrogen and phosphorus uptake by oceanic microplankton, *Mar. Ecol. Prog. Ser.,* 27, 253, 1986.

201. **Smith, R. E. H., Harrison, W. G., and Harris, L.,** Phosphorus exchange in marine microplankton communities near Hawaii, *Mar. Biol.,* 86, 75, 1985.

202. **Jackson, G. A. and Williams, P. M.,** Importance of dissolved organic nitrogen and phosphorus to biological nutrient cycling, *Deep Sea Res.,* 32, 223, 1985.

203. **Smith, V. S., Kimmerer, W. J., and Walsh, T. W.,** Vertical flux and biogeochemical turnover regulate nutrient limitation of net organic production in the North Pacific Gyre, *Limnol. Oceanogr.,* 31, 161, 1986.

204. **McCarthy, J. J.,** Uptake of major nutrients by estuarine plants, in *Estuaries and Nutrients,* Neilson, B. J. and Cronin, L. E., Eds., Humana Press, Clifton, NJ, 1981, 139.

205. **Krom, M. D. and Berner, R. A.,** Adsorption of phosphate in anoxic marine sediments, *Limnol. Oceanogr.,* 25, 797, 1980.

206. **Stumm, W. and Morgan, J. J.,** *Aquatic Chemistry,* 2nd ed., John Wiley & Sons, New York, 1981.
207. **Nixon, S. W.,** Remineralization and nutrient cycling in coastal marine ecosystems, in *Estuaries and Nutrients,* Neilson, B. J. and Cronin, L. E., Eds., Humana Press, Clifton, NJ, 1981, 111.
208. **Fisher, T. R., Carlson, P. R., and Barker, R. T.,** Sediment nutrient regeneration in three North Carolina estuaries, *Estuarine Coastal Shelf Sci.,* 14, 101, 1982.
209. **Taft, J. L. and Taylor, W. R.,** Phosphorus dynamics in some coastal plain estuaries, in *Estuarine Processes,* Vol. 1, Wiley, M. L., Ed., Academic Press, New York, 1976, 79.
210. **Harris, E. and Riley, G. A.,** Oceanography of Long Island Sound, 1952—1954. VIII. Chemical composition of the plankton, *Bull. Bingham Oceanogr. Coll.,* 15, 315, 1956.
211. **Peterson, D. H., Conomos, T. J., Broenkow, W. W., and Scrivani, E. P.,** Processes controlling the dissolved silica distribution in San Francisco Bay, in *Estuarine Research,* Vol. 1, Cronin, L. E., Ed., Academic Press, New York, 1975, 153.
212. **Aston, S. R.,** Nutrients, dissolved gases, and general biogeochemistry in estuaries, in *Chemistry and Biogeochemistry of Estuaries,* Olausson, E. and Cato, I., Eds., John Wiley & Sons, Chichester, 1980, 233.
213. **D'Elia, C. F., Nelson, D. M., and Boynton, W. R.,** Chesapeake Bay nutrient and plankton dynamics. III. The annual cycle of dissolved silicon, *Geochim. Cosmochim. Acta,* 47, 1945, 1983.
214. **Werner, D.,** Silicate metabolism, in *The Biology of Diatoms,* Werner, D., Ed., Blackwell Scientific, Oxford, 1977, 110.
215. **Viarengo, A.,** Biochemical effects of trace metals, *Mar. Pollut. Bull.,* 16, 153, 1985.
216. **Thomas, W. H., Holm-Hansen, O., Seibert, D. L. R., Azam, F., Hodson, R., and Takahashi, M.,** Effects of copper on phytoplankton standing crop and productivity: controlled ecosystem pollution experiment, *Bull. Mar. Sci.,* 27, 34, 1977.
217. **Gavis, J.,** Toxic binding of cupric ion by marine phytoplankton, *J. Mar. Res.,* 41, 53, 1983.
218. **Anderson, D. M., Lively, J. S., and Vaccaro, R. F.,** Copper complexation during spring phytoplankton blooms in coastal waters, *J. Mar. Res.,* 42, 677, 1984.
219. **Romeo, M. and Gnassia-Barelli, M.,** Metal uptake by different species of phytoplankton in culture, *Hydrobiologia,* 123, 205, 1985.
220. **Wells, M. L., Zorkin, N. G., and Lewis, A. G.,** The role of colloid chemistry in providing a source of iron to phytoplankton, *J. Mar. Res.,* 41, 731, 1983.
221. **Anderson, M. A. and Morel, F. M. N.,** The influence of aqueous iron chemistry on the uptake of iron by the coastal diatom *Thalassiosira weissflogii, Limnol. Oceanogr.,* 27, 789, 1982.
222. **Trick, C. G., Andersen, R. J., Gillam, A., and Harrison, P. J.,** Prorocentrin: an extracellular siderophore produced by the marine dinoflagellate *Prorocentrum minimum, Science,* 219, 306, 1983.
223. **Capriulo, G. M. and Carpenter, E. J.,** Grazing by 35 to 202 μm microzooplankton in Long Island Sound, *Mar. Biol.,* 56, 319, 1980.
224. **Paranjape, M. A., Conover, R. J., Harding, G. C., and Prouse, N. J.,** Micro- and macrozooplankton on the Nova Scotian shelf in the prespring bloom period: a comparison of their potential resource utilization, *Can. J. Fish. Aquat. Sci.,* 42, 1484, 1985.
225. **Steemann Nielsen, E.,** The balance between phytoplankton and zooplankton in the sea, *J. Cons. Perm. Int. Explor. Mer,* 23, 178, 1958.
226. **Cloern, J. E.,** Phytoplankton ecology of the San Francisco Bay system: the status of our current understanding, in *San Francisco Bay: The Urbanized Estuary,* Conomos, T. J., Ed., Pacific Division, American Association for the Advancement of Science, San Francisco, 1979, 247.
227. **Nichols, F. H.,** Increased benthic grazing: an alternative exploration for low phytoplankton biomass in northern San Francisco Bay during the 1976—1977 drought, *Estuarine Coastal Shelf Sci.,* 21, 379, 1985.
228. **Dame, R., Zingmark, R., Stevenson, H., and Nelson, D.,** Filter feeder coupling between the estuarine water column and benthic subsystems, in *Estuarine Perspectives,* Kennedy, V. S., Ed., Academic Press, New York, 1980, 521.
229. **Officer, C. B., Smayda, T. J., and Mann, R.,** Benthic filter feeding: a natural eutrophication control, *Mar. Ecol. Prog. Ser.,* 9, 203, 1982.
230. **Cohen, R. R. H., Dressler, P. V., Phillips, E. J. P., and Cory, R. L.,** The effect of the Asiatic clam, *Corbicula fluminea,* on phytoplankton of the Potomac River, Maryland, *Limnol. Oceanogr.,* 29, 170, 1984.
231. **Steele, J. H.,** Ed., *Spatial Pattern in Plankton Communities,* Plenum Press, New York, 1978.
232. **Bennett, A. F. and Denman, K. L.,** Phytoplankton patchiness: inferences from particle statistics, *J. Mar. Res.,* 43, 307, 1985.
233. **Mackas, D. L. and Boyd, C. M.,** Spectral analysis of zooplankton spatial heterogeneity, *Science,* 204, 62, 1979.
234. **Longhurst, A. R.,** Vertical migration, in *The Ecology of the Seas,* Cushing, D. H. and Walsh, J. J., Eds., Blackwell Scientific, Oxford, 1976, 116.
235. **Sommer, U.,** *Plankton Ecology: Succession in Plankton Communities,* Springer-Verlag, Berlin, 1989.
236. **Breitburg, D. L.,** Development of a subtidal epibenthic community: factors affecting species composition and the mechanisms of succession, *Oecologia,* 65, 173, 1985.

Chapter 3

ZOOPLANKTON

I. INTRODUCTION

Among the planktonic faunal assemblages of estuaries are zooplankton — volumetrically abundant animals typically several microns to 2 cm in size — that drift passively in currents due to limited capabilities of locomotion. These lower-trophic-level consumers constitute the principal herbivorous component of estuarine ecosystems. Whereas most zooplankton consume phytoplankton or detritus and serve as an essential link in aquatic food chains by converting plant to animal matter, others are primary carnivores. Some species obtain nutrition by the direct uptake of dissolved organic nutrients. Zooplankton basically gather food via filter feeding or raptorial feeding.[1] Raptorial feeders seize and eat individual cells, removing a few selected prey. Grazing pressure by herbivorous zooplankton commonly regulates the standing crop of phytoplankton populations (See Chapter 2, Section VI.D.1).

Both biological and physical-chemical conditions in estuaries control the species composition, abundance, and distribution of zooplankton. These microfauna must adapt to varying stresses associated with biological (e.g., scarcity of food, competition, and predation) and physical-chemical (e.g., temperature, salinity, mass movements of water, and dissolved oxygen levels) factors.[2,3] General accounts of zooplankton ecology are contained in a number of recent publications.[1-11]

This chapter deals with various aspects of estuarine zooplankton, including their classification, taxonomy, and ecology. The role of zooplankton in estuarine food webs is investigated. In addition, the factors influencing the distribution and abundance of these organisms are assessed.

II. CLASSIFICATION

Marine scientists usually classify zooplankton according to their size or length of planktonic life. Three major size categories of zooplankton are recognized, namely, microzooplankton, mesozooplankton, and macrozooplankton. Microzooplankton comprise those forms which pass through plankton nets with a mesh size of 202 μm, and mesozooplankton, those forms retained by these nets. Larger zooplankton captured by plankton nets with a mesh size of 505 μm embody the macrozooplankton.[12,13]

In regard to the duration of planktonic life, zooplankton may be grouped into three classes: (1) holoplankton, (2) meroplankton, and (3) tychoplankton. Holoplankton are those organisms which spend their entire life in the plankton, in contrast to meroplankton which remain planktonic for only a portion of their life cycle. Tychoplankton refer to small animals, primarily benthic organisms, temporarily translocated into the water column by current action, behavioral activity (e.g., diurnal vertical migration), or other mechanisms.

A. CLASSIFICATION BY SIZE
1. Microzooplankton

Microzooplankton smaller than approximately 60 μm consist primarily of protozoans, such as foraminiferans, radiolarians, and tintinnids. Tintinnids and other diminutive, ciliated protozoans numerically dominate the zooplankton community of some estuaries during certain seasons of the year. High abundances of tintinnids and rotifers have been documented in a number of systems.[14-18] Trochophore larvae of polychaetes, veliger larvae of mollusks, copepod nauplii, and arthropod larvae (e.g., barnacle nauplii and crab zoeae) seasonally compose a substantial portion of the total microzooplankton. Many protozoans graze heavily on bacteria while serving as a significant food source for larger microzooplankton (see Chapter 1, Section IV.B.3);[13] both free-living bacteria and bacteria attached to detrital particles are ingested by the zooplankters.[19] Some microzooplankton — for example, planktonic ciliates — compete with macrozooplankton for microzooplanktonic foods,[20] and provide a trophic link between nano-phytoplankton and macrozooplankton.[18,21-23] Larger planktonic ciliates, although consuming much phytoplankton,[24,25] also reportedly subsist on ciliates.[26] Omnivory and carnivory among these organisms can govern the trophic dynamics and structure of microplanktonic communities.[27] In sum, therefore, protozoans as well as small metazoans, particularly larval stages, encompass the microzooplankton of estuaries.[28]

2. Mesozooplankton

Cladocerans, copepods, rotifers, and meroplankton represent the predominant mesozooplankton in estuarine waters.[29] Together with the macrozooplankton, the mesozooplankton are considered to be the principal components of the zooplankton.[9] Microzooplankton less than about 200 μm in size have been difficult to sample quantitatively due to the clogging of plankton nets, which hinders accurate assessment of this fraction.

Copepods clearly dominate the mesozooplankton, being the most abundant group of zooplankton in estuaries. They serve as a vital coupling between phytoplankton and heterotrophic consumers. Free-living planktonic species of the orders Calanoida, Cyclopoida, and Harpacticoida frequently appear in zooplankton samples. Herbivorous copepods, especially calanoids, tend to have the greatest absolute abundance and biomass,[4,30] supplying ration for numerous estuarine animals. However, omnivorous and carnivorous copepods likewise play a critical role in the flow of energy to higher trophic levels. Indeed, planktonic copepods may also be detritivores.

In general, herbivorous copepods filter water to collect phytoplankton during feeding. Carnivorous forms (e.g., some cyclopoids and calanoids) actively prey on other zooplankton by seizing the individual with their appendages.[31] Davis[30] and Odhner[32] present lucid descriptions of the feeding mechanisms of various life-history stages of copepods.

Copepods experience pronounced spatial and temporal changes in distribution in response to variations in biotic and abiotic factors.[8] Perturbations in physical factors, such as fluctuations in temperature or photoperiod and seasonal variations in riverine discharges and coastal hydrography, have been implicated in the temporal alterations of zooplankton communities.[29] Other physical factors (e.g., tidal currents, fronts, and stratification) can modulate organismal position. Chief among the biotic factors affecting zooplankton distribution are the environmental preferences of the species, food availability, predator-prey interactions, larval behavior, larval abundance, and vertical migration patterns. Hence, food restrictions may limit copepod production,[33] and feeding rhythms can regulate vertical distribution.[34] Seasonal cycles of phytoplankton biomass, for instance, foster seasonal cycles of zooplankton biomass, as is evident in the Kiel Bight.[35] In Narragansett Bay, RI, both seasonal and annual variations occur in phytoplankton/zooplankton interactions.[35] Predation by the ctenophore, *Mnemiopsis leidyi,* seems to control the abundance of copepod populations in this Rhode Island estuary.[36-38] Throughout the inshore waters of the East Coast of the U.S., in fact, lobate ctenophores of the genus *Mnemiopsis* (i.e., *M. leidyi* and *M. mccradyi*) are voracious predators of mesozooplankton.[39] Planktivorous fishes also add to increased mortality rates of *Acartia* spp.[40,41] Because zooplankton predators like *M. leidyi* proliferate seasonally, predation contributes to the conspicuous seasonal cycles of abundance of copepods observed in many estuaries.[42]

In respect to spatial variations, vertical migrations and heterogeneous vertical distributions of copepod populations are well established. For example, the calanoid copepod, *A. tonsa,* undergoes an acute vertical migration, ascending the water column at night and descending to greater depths during the day.[43,44] By migrating, the copepods avoid predation pressure. Spatial variation in the density of copepods is often most notable along the longitudinal axis of an estuary. Estuarine circulation causes a landward transport of copepods that have migrated into the near-bottom layer, thereby concentrating the zooplankton in specific areas.[29] Zooplankton characteristically exhibit a patchy distribution from the head to the mouth of an estuary, attributable to hydrography, grazing, and other factors.[4,5,45] Ueda[46] attested to the spatial disparity in the distribution of the two closely related *Acartia* species, *A. omorii* and *A. hudsonica,* in Maizuru Bay, Japan. Spatial segregation of these two copepod species, with *A. hudsonica* confined to the inner perimeter of the bay, not only reflects the species preferences for different hydrographic conditions, but also interspecific competition between the two species. Spatial variations in copepod numbers can be exacerbated by increased variability of riverine discharges which is coupled to advective residence time and the salinity distribution of the system.[29]

3. Macrozooplankton

Sampling with a 505-μm mesh plankton net generally collects two groups of zooplankton. The first group — jellyfish — mainly consists of hydromedusae, comb jellies, and true jellyfishes.[47] This group may dominate the zooplankton in terms of total volume.[48] The second group — crustaceans — incorporates a variety of different forms (e.g., amphipods, isopods, mysid shrimp, and true shrimp). Polychaete worms and insect larvae also have representative members among the macrozooplankton (Figure 1).

B. CLASSIFICATION BY LENGTH OF PLANKTONIC LIFE
1. Meroplankton

These organisms, as recounted above, spend only a portion of their life in the plankton. They are primarily planktonic larvae of the benthic invertebrates, benthic chordates, and nekton (i.e., ichthyoplankton). Sexual jellyfish stages of the hydrozoan and scyphozoan coelenterates (e.g., *Sarsia* spp. and *Cyanea* spp.) form an additional meroplanktonic component.[49] Nearly all animal phyla contribute individuals to the meroplankton.

In estuaries and coastal oceanic waters, the meroplankton of benthic invertebrates are numerically significant. More than 1.25×10^5 species of benthic fauna have been identified, the bulk of them typified by a free-swimming larval stage persisting for a few weeks.[2] Many species pass through a succession of larval stages prior to becoming an adult; therefore, multiple stages of the same species may be present concurrently in the plankton. Most taxa of the Crustacea, for instance, usually have more than one larval stage, and some decapods exhibit as many as 18 stages.[6]

Benthic ecologists discriminate between two major types of planktonic larvae of marine benthic invertebrates, that is, planktotrophic and lecithotrophic forms. Planktotrophic larvae actively ingest particulate food while in the plankton. Lecithotrophic larvae, however, utilize stored food reserves from the egg for their development.[7] A third type of planktonic larval development, termed facultative planktotrophy, has been enlisted to define a larva capable of feeding while in the plankton, but harboring sufficient yolk reserves to develop and metamorphose.[50,51] Emlet[52] alluded to several advantages of facultative planktotrophy, which may be a transition between planktotrophy and lecithotrophy, explaining that the augmentation of nutritive stores by larval feeding could allow an extended competent period or could enhance growth and survivorship subsequent to metamorphosis. He espoused the following advantages of planktotrophy and lecithotrophy (p. 183): "An advantage attributed to planktotrophy is less parental investment per offspring, thus resulting in a greater number of offspring. Advantages of lecithotrophy are a shorter planktonic period which reduces exposure to dangers such as predation or dispersal away from appropriate settling areas".

For planktotrophic larvae, development to metamorphosis generally takes place in 2 to 6 weeks. Lecithotrophic larvae usually require less time to reach metamorphosis, typically several days to about 2 weeks.[53] When a larva of a benthic marine invertebrate attains metamorphic

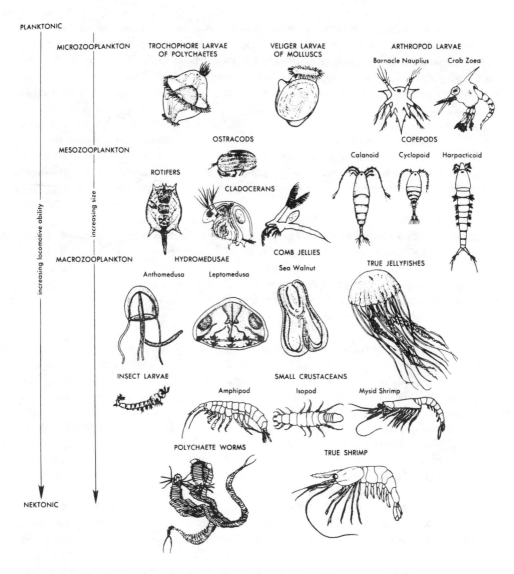

FIGURE 1. Micro-, meso-, and macrozooplankton commonly found in estuaries. (From Lippson, A. J., Haire, M. S., Holland, A. F., Jacobs, F., Jensen, J., Moran-Johnson, R. L., Polgar, T. T., and Richkus, W. A., *Environmental Atlas of the Potomac Estuary,* Johns Hopkins University Press, Baltimore, 1981. With permission.)

competency, it is developmentally capable of settlement to the seafloor and metamorphosis. However, metamorphosis may be delayed during the competency period because of the lack of an appropriate stimulus.[53] A competent larva can delay metamorphosis for weeks while searching the seafloor for a suitable habitat. During this search of the substrate, the larva is subject to predation from a host of sources, such as protozoans, larger omnivorous and carnivorous zooplankton, benthic invertebrates, and fish.[54]

In addition to planktotrophy and lecithotrophy, three alternate developmental patterns of benthic marine invertebrate larvae have been delineated.[55] These are (1) demersal development, (2) direct development, and (3) viviparity. Chia,[51] Mileikovsky,[56,57] and Crisp[58] provide a description of them. An estimated 70% of the species of benthic marine invertebrates experience pelagic planktotrophic development.[53,59]

Occasionally overlooked in meroplanktonic investigations are the ichthyoplankton. Although adult fishes may be benthic, benthopelagic, mesopelagic, or pelagic, most produce planktonic eggs and/or larvae.[60,61] Leiby[60] details the methods by which the eggs and larvae of estuarine and marine fishes enter the planktonic community. Recruitment to the adult population is dependent on the survivorship of these early life-history stages.[61]

The meroplankton of benthic marine invertebrates and the ichthyoplankton are susceptible to the vagaries of environmental conditions that often severely deplete their numbers. Temperature, salinity, turbidity, circulation, and seafloor conditions, as well as other physical and chemical factors, influence larval development, distribution, and survivorship. Biological factors, including predation, availability of food, and seasonal abundance of adults and larvae, also affect meroplankton success.[62] Predation alone causes larval mortality to exceed 90%.[2] In spite of immense numbers of eggs and larvae generated by benthic marine invertebrates (e.g., barnacles, bivalves, cyphonautes, gastropods, and polychaetes) in coastal waters, the majority does not survive due to the stresses imposed by the physical and biological conditions of the estuary. The fecundity of the oviparous American oyster, *Crassostrea virginica* Gmelin, for example, can exceed 100×10^6 eggs in one spawn.[63] The hard clam, *Mercenaria mercenaria* Linné, has somewhat lower fecundity values. Davis and Chanley[64] recorded fecundities of 8 to 39.5×10^6 eggs ($\bar{X} = 24 \times 10^6$ eggs) per female per spawning season, and Bricelj and Malouf[65] registered a maximum fecundity figure of 16.8×10^6 eggs by a single female per spawning season. Carriker[66] projected that larval survival of *M. mercenaria* in Little Egg Harbor, NJ, equaled 2.6%. Larval and early juvenile mortality of the American oyster and hard clam can surpass 99%, approaching 100% under certain conditions in nature.

The abundance of meroplankton in estuaries fluctuates greatly because of the different reproductive strategies of benthic invertebrate populations. In temperate systems, it is not uncommon for pulses of meroplankton to arise from spring through summer when many of the benthic populations spawn. However, spawning usually does not occur synchronously and the duration of the planktonic stage of life differs among these populations. Consequently, the time of peak numbers of barnacle, bivalve, cyphonaute, gastropod, and polychaete larvae deviates during the year as demonstrated by Sandine[67] in Barnegat Bay, NJ.

2. Holoplankton

The predominant holoplanktonic groups in estuaries are copepods, cladocerans, and rotifers.[13] The "oar-footed crustaceans" or copepods are the most important constituents, not only numerically, but also in terms of their influence on nutrient recycling and energy transfer in these coastal ecosystems.[68] Several genera of the Copepoda inhabit estuarine waters, with species of *Acartia, Eurytemora, Pseudodiaptomus,* and *Tortanus* being commonplace.[69] Smaller species of copepods frequently dominate the holoplanktonic fauna. The upper, middle, and lower reaches of estuaries are unalike in copepod species composition, although some overlap in seasonal distributions is apparent. *Eurytemora* spp. typically occur in the upper reaches; *Acartia* spp. (*A. bifilosa, A. discaudata, A. hudsonica, A. tonsa*) usually characterize the middle reaches.[49] The lower reaches consist of a greater diversity of forms, and near the mouth, euryhaline marine species are noteworthy. Species of *Centropages, Oithona, Paracalanus,* and *Pseudcalanus* remain prominent here. Marine species enter estuaries and proliferate in the lower reaches when conditions become favorable. In contrast, estuarine species generally dominate in the upper reaches.[69]

Calanoid copepods, nearly entirely planktonic, outnumber cyclopoids and harpacticoids in estuaries. The cyclopoids are principally littoral and benthic, but include several abundant planktonic species. Repeatedly associated with detritus, algae, or other substrates, harpacticoids frequently attain high numbers in shallow, detritus-based systems.[70]

The spatial distribution of copepod populations is salinity dependent. On the basis of salinity tolerance, Grindley (p. 123)[69] differentiated four components of the holoplankton:

1. A stenohaline marine component penetrating only into the mouth (e.g., *Corycaeus* spp.)

TABLE 1
Salinity Ranges of Various Copepod Species Observed in Nature

Copepod species	0	10	20	30	40	50	60	70	80
Acartia (Paracartia) africana				——	—				
A. (Paracartia) longipatella		———	——	——	—				
A. (Arcartiella) natalensis		———	——	——	——	——	——	——	——
Calanoides carinatus				—					
Centropages brachiatus				—					
C. chierchiae				—					
C. furcatus				—					
Clausidium sp.				—					
Clausocalanus furcatus				—					
Corycaeus spp.				—					
Ctenocalanus vanus				—					
Euterpina acutifrons				—					
Halicyclops spp.		——	——	——	——	——	——	——	——
Harpacticus ? gracilis			——	——	——	—			
Hemicyclops sp.			——	——	——	——	——		
Nannocalanus minor				—					
Oithona brevicornis/nana	——	——	——	——	——	——	——	——	——
O. plumifera				—					
O. similis				——					
Paracalanus aculeatus				——					
P. crassirostris		——	——	——	——	—			
P. parvus			——	——	—				
Porcellidium sp.				—					
Pseudodiaptomus stuhlmanni	——	——	——	——	——	——	——		
P. hessei	——	——	——	——	——	——	——	——	
P. nudus				—					
Rhincalanus nasutus				—					
Saphirella stages				——	——				
Temora turbinata				—					
Tortanus capensis				——	——				
Tegastes sp.				——					

From Grindley, J. R., *Estuarine Ecology: With Particular Reference to Southern Africa*, Day, J. H., Ed., A. A. Balkema, Rotterdam, 1981, 117. With permission.

2. A euryhaline marine component penetrating farther up the estuary (e.g., *Paracalanus* spp.)
3. A true estuarine component incorporating species confined to estuaries (e.g., *Pseudodiaptomus hessei*)
4. A freshwater component encompassing species normally found in freshwater (e.g., *Diaptomus* spp.)

Along a longitudinal salinity gradient of an estuary, copepods from all four groups form a chain of overlapping species populations.[68] Table 1 gives salinity ranges of various copepod species. The distribution of cladocerans, like that of copepods, is salinity dependent.[49]

Aside from salinity, temperature acts as a major factor limiting the distribution and production of holoplankton. Much information on the effects of temperature on estuarine copepods has been culled from research on zooplankton communities along the East Coast of the U.S.[3,8,14,33,37,68,71] In the mid-Atlantic Bight region, the copepod community experiences a seasonal succession of populations, most conspicuously the calanoid copepods *A. hudsonica* and *A. tonsa. A. hudsonica* attains peak abundance during winter-spring. Its temperate-tropical

TABLE 2
Distribution and Occurrence of Major Holoplanktonic
Species within the Mid-Atlantic Bight Area

Species	Distribution	Occurrence
Dominant		
Acartia hudsonica	Estuarine and marine[a]	Winter-spring
A. tonsa	Estuarine and marine[a]	Summer-fall
Calanus finmarchicus	Stenohaline marine[b]	Year-round
Centropages hamatus	Euryhaline marine[c]	Year-round
C. typicus	Stenohaline marine[b]	Summer-fall
Pseudocalanus minutus	Euryhaline marine[c]	Winter-spring
Temora longicornis	Euryhaline marine[c]	Winter-spring
Common		
Labidocera aestiva	Euryhaline marine[c]	Summer-fall
Tortanus discaudatus	Estuarine and marine[a]	Spring-summer
Sagitta elegans	Stenohaline marine[b]	Winter-spring
Frequent		
Oithona brevicornis	Estuarine and marine[a]	Summer-fall
O. similis	Euryhaline marine[c]	Year-round

[a]   Estuarine and marine — widely distributed in an estuary and reproducing
    to a limited extent in open coastal waters.
[b]   Stenohaline marine — reproducing in open neritic waters.
[c]   Euryhaline marine — reaching maximum population densities near the
    coast, close to the mouths of estuaries.

From Jeffries, H. P. and Johnson, W. C., *Coastal and Offshore Environmental Inventory: Cape Hatteras to Nantucket Shoals,* Marine Publ. Ser. No. 2, University of Rhode Island, Kingston, 1973, 4—1. With permission.

congenere, *A. tonsa,* reaches maximum numbers during summer-fall, replacing *A. hudsonica* as the predominant form. The seasonal alternation of these two *Acartia* species has been ascribed to: (1) temperature shifts in the relative competitive coefficients of the two species, although salinity also affects the species interaction; and (2) the responses of the population increase rates to suitable conditions within the seasonal cycle.[8] The succession of other congeneric associates, such as *Centropages hamatus* (spring-summer) and *C. typicus* (summer-fall), may also be evident.[68] Table 2 chronicles the distributions of major holoplanktonic species in the mid-Atlantic Bight and the seasons of their occurrence.

3. Tychoplankton

Some demersal zooplankton are periodically inoculated into the plankton by bottom currents, wave action, and bioturbation. These organisms do not normally constitute a quantitatively significant fraction of the zooplankton community. However, they supply ration for planktivorous fishes and carnivorous zooplankton. Tychopelagic species continually recovered from the stomachs of planktivorous fishes are amphipods, cumaceans, isopods, and mysids.[72-74] The recruitment process of planktonic populations, as well as the aperiodic entry of tychoplankton into the water column of shallow regions, can be enhanced by the bioturbation of benthic fauna, for example, the conveyor-belt-feeding polychaetes *Cistenides (=Pectinaria) gouldii* and *Clymenella torquata*.[75,76] Storms and currents, which roil bottom sediments, likewise promote the upward translocation and lateral distribution of the tychoplankton.

III. TAXONOMY

Omori and Ikeda[9] present a classification of marine zooplankton, categorizing representative taxa. This classification scheme is reproduced below. It treats taxa sampled from estuarine as well as marine ecosystems. As expressed by Omori and Ikeda (p. 4),[9] the word "Part" signifies that some species of a class, order, or genus are holoplanktonic.

A. Phylum Protozoa
    Class Mastigophora (Flagellata)*
        Subclass Zoomastigophorea
            Order Choanoflagellida
                Part; *Diaphanoeca, Monosiga, Stephanoeca*
    Class Sarcodinea
        Subclass Rhizopoda
            Order Foraminifera
                Important as fossils; Part; *Globigerina, Globorotalia*
        Subclass Actinopoda
            Order Radiolaria
                Part; *Acanthometron, Aulosphaera*
    Class Ciliata**
        Order Holotricha
            Some species cause red tides; *Mesodinium*
        Order Spirotricha
            Part; suborder Tintinnina is an important group of microzooplankton;
                Condonella, Favella, Parafavella, Tintinnopsis, Tintinnus

B. Phylum Cnidaria (Coelenterata)
    Class Hydrozoa
        Order Hydroida
            Suborder Athecata (Anthomedusae)
                Many are meroplanktonic; *Leuckartiara, Sarsia*
            Suborder Thecata (Leptomedusae)
                Aequorea, Obelia
        Order Limnomedusae
            Brackish water; *Craspedacusta*
        Order Trachylina (Trachymedusae)
            Aglantha, Geryonia, Rhopalonema
        Order Siphonophora
            Suborder Calycophorae
                Abyla, Muggiaea
            Suborder Physophorae
                Agalma
            Suborder Rhizophysaliae (Cystonectae)
                Physalia
            Suborder Chondrophorae
                Porpita, Velella

*   Subclass Phytomastigophorea not included.
** According to Corliss (p. 189),[77] the systematic position of Tintinnina is as follows:
    Phylum Ciliophora
      Class Polyhymenophora
        Order Oligotrichida
          Suborder Tintinnida

Class Scyphozoa
  Order Stauromedusae
    Haliclystus
  Order Cubomedusae
    Carybdea, Tamoya
  Order Coronatae
    Deep water; *Atolla, Atrella*
  Order Semaeostomeae
    Aurelia, Dactylometra
  Order Lobata
    Bolinopsis, Leucothea, Mnemiopsis
  Order Cestida
    Cestum
Class Atentaculata (Nuda)
  Order Beroida
    Beroe

C. Phylum Nemertinea
  Class Enopla
    Order Hoplonemertinea
      Deep water; *Nectonemertes, Pelagonemertes*

D. Phylum Aschelminthes
  Class Rotatoria
    Order Monogononta
      Brackish water; *Brachionus, Keratella, Notholca*

E. Phylum Mollusca
  Class Gastropoda
    Subclass Prosobranchia
      Order Mesogastropoda
        All of suborder Heteropoda: *Atlanta, Carinaria, Hydrobia, Janthina, Pterotrachea*
    Subclass Opisthobranchia
      Order Thecosomata (Pteropoda)
        Cavolina, Clio, Creseis, Spiratella (Limacina)
      Order Gymnosomata (Pteropoda)
        Clione, Pneumoderma
      Order Nudibranchia
        Part; warm water; *Glaucus*

F. Phylum Annelida
  Class Polychaeta
    Order Errantia
      Part; *Aliciopa, Lepidametria, Poeobius, Sagitella, Tomopteris, Vanadis*

G. Phylum Arthropoda
  Class Crustacea
    Subclass Branchiopoda
      Order Cladocera
        Part; coastal and brackish water; *Evadne, Penilia, Podon*

Subclass Ostracoda
    Order Myodocopida
        Part; *Archiconchoecia, Conchoecia, Gigantocypris*
Subclass Copepoda
    Order Calanoida
        One of the most important marine zooplankton taxa; *Acartia,*
        Calanus, Candacia, Centropages, Eucalanus, Euchaeta, Eurytemora,
        Haloptilus, Metridia, Paracalanus, Pseudocalanus, Scolecithrix,
        Sinocalanus, Temora, Undinula
    Order Cyclopoida
        Part; *Corycaeus, Oithona, Oncaea, Sapphirina*
    Order Harpacticoida
        Part; majority are coastal or epibenthic; *Diosuccus, Euterpina,*
        Harpacticus, Microsetella, Tigriopus, Tisbe
    Order Monstrilloida
        Part; *Monstrilla*
Subclass Malacostraca
    Order Mysidacea
        Part; many are coastal or brackish and epibenthic; some are meso- and
        bathypelagic; *Archiomysis, Holmesiella, Lophogaster, Mysis,*
        Neomysis, Siriella
    Order Cumacea
        Part; many are epibenthic; *Dimorphostylis*
    Order Amphipoda
        Part of suborder Gammaridea and all Hyperiidea are planktonic or
        parasitic; some scientists consider that hyperiids are not free-living
        amphipods but are all parasitoids, which develop obligatorily on
        gelatinous hosts (see Laval, 1980);
        Cyphocaris, Hyperia, Parathemisto, Phronima, Themisto, Vibilia
    Order Euphausiacea
        Important taxa in epi- and mesopelagic plankton; *Euphausia,*
        Meganyctiphanes, Nematoscelis, Thysanoëssa, Thysanopoda
    Order Decapoda
        Suborder Dendrobrachiata
            Bentheogennema, Gennadas, Acetes, Lucifer, Sergestes, Sergia
        Suborder Pleocyemata
            Acanthephyra, Hymenodora

H. Phylum Chaetognatha
  Class Sagittoidea
    Eukrohnia, Krohnitta, Pterosagitta, Sagitta

I. Phylum Echinodermata
  Class Holothuroidea
        Part; deep water; *Enypniastes, Pelagothuria*

J. Phylum Chordata
  Class Appendiculata (Larvacea)
    Order Appendicularia (Copelata)
        Fritillaria, Oikopleura
  Class Thaliacea

Order Pyrosomata
  Pyrosoma
Order Cyclomyaria (Doliolida)
  Doliolum
Order Desmomyaria (Salpida)
  Salpa, Thalia, Thetys

Among invertebrate larvae that appear as plankton, some have specific names. The following are representative groups of organisms and the phyla to which they belong:

A. Phylum Porifera
    1. Parenchymula, amphiblastula, olynthus
B. Phylum Coelenterata
    1. Planula, actinula, ..., larvae of Hydrozoa
    2. Scyphistoma, strobila, ephyra, ..., larvae of Scyphozoa
    3. Diconula, conaria, rataria, ..., larvae of Chondrophorae
C. Phylum Platyhelminthes
    1. Müller's larva, Götte's larva, ..., larvae of Polycladida
D. Phylum Nemertinea
    1. Desor's larva, pilidium, ..., larvae of Heteronemertea
E. Phylum Mollusca
    1. Trochophore, veliger, ..., larvae of Gastropoda and Bivalvia
    2. Rhynchoteuthion, ..., larvae of Cephalopoda Ommastrephidae
F. Phylum Annelida
    1. Loven's larva, trochophore
G. Phylum Arthropoda
    1. Nauplius, metanauplius, ..., larvae of Crustacea
    2. Cypris, pupa, ..., larvae of Cirripedia
    3. Protozoea, zoea, metazoea, mysis, ..., larvae of Malacostraca
    4. Manca, ..., larvae of Isopoda and Cumacea
    5. Calyptopis, furcilia, ..., larvae of Euphausiacea
    6. Elaphocaris, acanthosoma, ..., larvae of Sergestidae
    7. Phyllosoma, puerulus, ..., larvae of Palinuridae
    8. Glaucothoë, ..., larvae of Paguroidea
    9. Megalopa, ..., larvae of Brachyura
    10. Erichthoidina, erichthus, alima, pseudozoea, ..., larvae of Stomatopoda
H. Phylum Tentaculata
    1. Actinotrocha, ..., larvae of Phoronidea
    2. Cyphonautes, ..., larvae of Bryozoa
I. Phylum Echinodermata
    1. Doliolaria, pentacrinoid, ..., larvae of Crinoidea
    2. Bipinnaria, brachiolaria, ..., larvae of Asteroidea
    3. Ophiopluteus, pluteus, ..., larvae of Ophiuroidea
    4. Pluteus, ..., larvae of Echinoidea
    5. Auricularia, doliolaria, ..., larvae of Holothuroidea
J. Phylum Hemichordata
    1. Tornaria, ..., larvae of Enteropneusta

Some of the aforementioned taxonomic groups are particularly prevalent in estuaries. Members of the phyla Protozoa, Cnidaria, Ctenophora, and Arthropoda (class Crustacea) can far exceed other taxa of holoplankton in numerical importance in these systems. Larvae of taxa among the Annelida, Arthropoda, Bivalvia, and Tentaculata constitute seasonally abundant meroplanktonic forms, as do larvae of the Chordata (class Larvacea; appendicularians). Various orders of the subclass Malacostraca, including Amphipoda, Isopoda, Cumacea, and Mysidacea, have representatives with an epibenthic habit that occasionally enter the water column as part of the tychoplankton.

IV. ZOOPLANKTON COMMUNITIES

Investigations on the zooplankton community of estuaries have lagged behind those on the benthos and nekton.[48] The most detailed studies of zooplankton have been performed on the East Coast of the U.S., where zooplankton species composition, spatial and temporal abundance, and feeding relationships have been determined in many estuaries. This section focuses on the zooplankton of a few of these systems, namely, Narragansett Bay (RI), Long Island Sound (NY), Barnegat Bay (NJ), Chesapeake Bay (MD), and North Inlet (SC). Additionally, the zooplankton community of San Francisco Bay on the West Coast of the U.S. is reviewed.

A. NARRAGANSETT BAY

The zooplankton community of Narragansett Bay — a shallow, glacially formed embayment extending roughly north-south into the coastal waters of Rhode Island — is dominated by the copepods *Acartia hudsonica* (Bradford) (formerly *A. clausi* Giesbrecht) and *A. tonsa* Dana.[8,78] As in other East Coast estuaries of the U.S., *A. hudsonica* predominates during the winter and spring and *A. tonsa*, during the summer and fall.[37] Copepods, in general, account for an average of 80 to 95% of the total number of zooplankton over an annual cycle.[8,68,78,79] Other copepod species, while abundant, may be one order of magnitude less numerous than the *Acartia* spp.; these populations consist primarily of *Oithona* spp., *Parvocalanus crassirostris*, *Pseudocalanus minutus*, and *Saphirella* sp.[8,37] In addition to copepods, cladocerans (e.g., *Evadne* sp. and *Podon* sp.) attain high numbers in early summer, and rotifers become abundant in late winter.[8] Meroplankton of benthic invertebrates are seasonally abundant, with peak densities during spring, summer, or fall. Figure 2 depicts abundance cycles of *A. hudsonica* and *A. tonsa* in Narragansett Bay, and Figure 3 displays abundance cycles of the copepods *Oithona* spp., *Paracalanus parvus*, and *Pseudocalanus minutus*, as well as cladocerans and rotifers.

A distinctive feature of the zooplankton community in the bay is the seasonal alternation of the copepod dominants *Acartia hudsonica* and *A. tonsa*. From January through June, Hulsizer[14] recorded older copepodite and adult stages of *A. hudsonica* in densities of about $5 \times 10^3/m^3$, which precipitously declined to near zero in July. Copepodite and adult stages of *A. tonsa* appeared in June or early July and quickly replaced *A. hudsonica*, reaching densities of approximately 1×10^2 to $1 \times 10^3/m^3$ from July to December, but disappearing in January or early February. Meanwhile, *A. hudsonica* reappeared in October and increased in abundance through the fall.

Durbin and Durbin[37] illustrated that the concentrations of *Acartia* nauplii were much greater than the numbers of copepodites and adults. The maximum density of nauplii equaled 1.97 $\times 10^5/m^3$ and often exceeded $1 \times 10^5/m^3$. The density of *Acartia* nauplii usually surpassed that of *Acartia* adults by more than one order of magnitude. Copepodites, in turn, had concentrations intermediate between the nauplii and adult forms.

Hulsizer[14] and Kremer[36] uncovered a regular pattern of zooplankton abundance in Narragansett Bay; large-standing stocks occurred in early summer, but dropped to low levels in late summer (Figure 4). *Acartia* spp. constituted about 50% of the total biomass at most times.[37] Kremer[36] calculated a decline in the zooplankton standing crop of 10% per day from a biomass peak, usually in mid-July, through a biomass low, typically in mid-August (Figure 4). Both the diminution in total zooplankton numbers and total biomass to minimum values in August have been attributed to predation by the ctenophore, *Mnemiopsis leidyi*, which is most abundant from

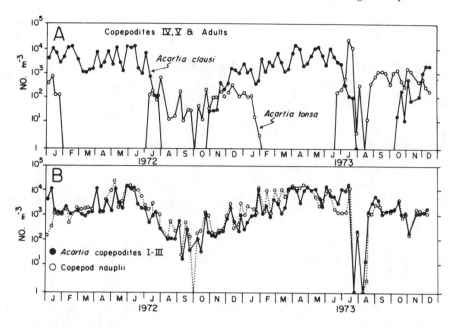

FIGURE 2.   Abundance cycles of *Acartia hudsonica* (= *A. clausi*) and *A. tonsa* in Narragansett Bay. (A) Densities of older copepodites (summed) and adults; (B) densities of nauplii and younger copepodites. Scales are linear between each logarithmic interval. (From Miller, C. B., *Estuaries and Enclosed Seas,* Ketchum, B. H., Ed., Elsevier, Amsterdam, 1983, 103. With permission.)

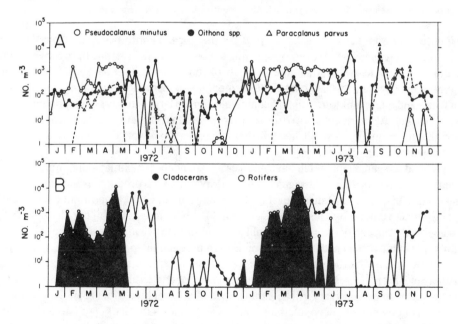

FIGURE 3.   (A) Abundance cycles of *Pseudocalanus minutus, Oithona* spp., and *Paracalanus parvus* in Narragansett Bay; (B) abundance cycles of cladocerans and rotifers. Scales are linear between each logarithmic interval. (From Miller, C. B., *Estuaries and Enclosed Seas,* Ketchum, B. H., Ed., Elsevier, Amsterdam, 1983, 103. With permission.)

July through October.[36,80] This insatiable predator increases markedly in abundance during the summer, approaching densities of 50/m³.[14] *M. leidyi,* together with larval fish, significantly crop the zooplankton populations of temperate estuaries during certain seasons. Cropping by *M.*

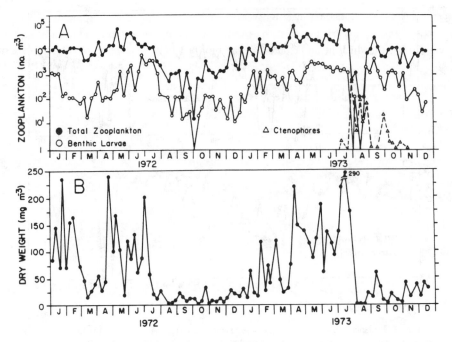

FIGURE 4. (A) Abundance cycles of holoplankton in Narragansett Bay with comparisons to abundance of benthic larvae and ctenophores. Scale is linear between each logarithmic interval. (B) Seasonal cycles of dry weight biomass of zooplankton in Narragansett Bay. Note linear scale. (From Miller, C. B., *Estuaries and Enclosed Seas,* Ketchum, B. H., Ed., Elsevier, Amsterdam, 1983, 103. With permission.)

leidyi during the summer averages 5 to 10% per day for the entire bay, indicating that this carnivore has the capability of severely depleting the zooplankton.[36] Despite the overwhelming predatory role of *M. leidyi* in the summer decrease of zooplankton in the bay, other carnivorous and omnivorous organisms also contribute to zooplankton depletion, including juvenile and adult fishes and meroplankton. Valiela[11] commented on the large numbers of ctenophores, fish larvae, and adult Atlantic menhaden (*Brevoortia tyrannus*) that primarily consume zooplankton, adding that it has not been established whether this predation lowers the concentration of the grazed populations.

Phytoplankton and zooplankton have mid-season abundance maxima, more pronounced at the head than at the mouth of Narragansett Bay.[11] Phytoplankton production and zooplankton biomass may be greater upestuary during the summer months, at which time zooplankton cropping reduces chlorophyll levels.[11,81] Durbin and Durbin[37] monitored higher phytoplankton biomasses in the upper bay ($\overline{X}$ = 16.95 mg chl *a* per cubic meter) than in the lower bay ($\overline{X}$ = 6.37 mg chl *a* per cubic meter) and higher phytoplankton production up the bay ($\overline{X}$ = 3.15 g C per square meter per day) than down the bay ($\overline{X}$ = 1.79 g C per square meter per day). These different measurements probably result, in part, from a gradient of decreasing nutrient concentrations from upestuary to downestuary.[78] The significance of zooplankton grazing in the control or attenuation of phytoplankton populations remains a complex subject of continuing research.[8] Grazing by zooplankton should be considered along with other factors (e.g., light, temperature, and nutrients) in the analysis of phytoplankton dynamics of estuaries (see Chapter 2).

B. LONG ISLAND SOUND

The genera *Acartia* and *Temora* dominate the zooplankton community of Long Island Sound.[82] *A. hudsonica* is the numerically dominant species in the Sound, but *T. longicornis* replaces *A. hudsonica* as the most abundant form in the spring and early summer. Peterson[82] noted that *T. longicornis* ranked first on a biomass basis, attaining a maximum value of 500 µg dry weight per liter in June and exceeding the biomass of *A. hudsonica* by a factor of 2 to 5. Despite the significance of *T. longicornis* in Long Island Sound, it is a minor constituent of the zooplankton community of Narragansett Bay during most months of the year.[79]

According to Tiselius and Peterson,[83] Long Island Sound harbors two copepod assemblages: (1) a boreal assemblage characterized by the copepods *A. hudsonica, T. longicornis,* and *Pseudocalanus* sp.; and (2) a subtropical assemblage dominated by *A. tonsa, Paracalanus crassirostris, Oithona* sp., and *Labidocera aestiva.* Peak abundance of the boreal copepods takes place in April or May; their numbers diminish in mid-June, and they disappear by the end of July.[83,84] The subtropical assemblage begins to increase in abundance in the summer, usually mid-July. While the boreal copepod populations vanish during the summer and fall, the subtropical forms persist in the plankton year-round.[82] The boreal and subtropical assemblages coexist in Long Island Sound during the winter and spring. Pulsing of the boreal forms in the spring contributes to their ecological dominance early in the year. The quantity of chaetognaths (*Sagitta elegans*) and cladocerans rapidly rises during June and July, subsequent to the diminution of boreal copepods but prior to the peak development of subtropical species. In late spring and early summer, meroplankton compose a significant fraction of the plankton.[5]

Aside from the well-defined seasonal variation in copepod species composition in Long Island Sound, exemplified by the succession of *A. hudsonica* and *A. tonsa,*[85] spatial changes in the (vertical) distribution of zooplankton are perceptible. For instance, Peterson[82] observed successive life-cycle stages of *T. longicornis* in progressively deeper sections of the water column. *Temora* eggs concentrated within the upper 5 m of the water column, and nauplii, at depths of 3 to 15 m. Copepodites 1 to 5 generally reached peak numbers between 10 and 20 m. Adults were most abundant within the lower 10 m of the water column or on the estuarine seafloor.

C. BARNEGAT BAY

Calanoid copepods, most notably *Acartia hudsonica, A. tonsa,* and *Oithona colcarva,* dominate the microzooplankton of Barnegat Bay, a shallow, lagoon-type estuary along the East Coast of New Jersey.[67,83] During the winter months, *A. hudsonica* predominates, and during the summer months, *A. tonsa* or *O. colcarva.* Sandine[67] identified ten additional copepod species in Barnegat Bay samples: *Paracalanus crassirostris, P. parvus, Oithona similis, Centropages hamatus, C. typicus, Temora longicornis, Pseudocalanus minutus, Pseudodiaptomus coronatus, Tortanus discautatus,* and *Labidocera aestiva.* Of these forms, *P. crassirostris, O. similis,* and *P. coronatus* are most abundant.

Rotifers comprised 11% of the mean annual density of microzooplankton collected from September 1975 to August 1977, yielding a peak density of $3 \times 10^5/m^3$ in February 1976. The maximum density of tintinnids was $1.6 \times 10^5/m^3$ during the 2-year period. Meroplanktonic larvae accounted for 2 to 49% of the total mean monthly microzooplanktonic density. The quantity of meroplankton approached maximum figures in the spring, especially during April when a maximum mean monthly density of $6.7 \times 10^4/m^3$ was registered. In spite of the continued reproduction of various benthic invertebrate species, an overall decline in meroplanktonic density occurred through the summer.

The number of microzooplankton is greatest in the spring and summer (Figure 5). The maximum mean monthly densities surpass $1 \times 10^5/m^3$ at these times. Copepods are responsible for a substantial portion of the total microzooplankton (Figure 6). Seasonal maxima in the spring and summer, however, often involve pulses of meroplankton (Figure 7).

Macrozooplankton in the estuary are dominated by *Rathkea octopunctata, Neomysis americana, Crangon septemspinosa, Neopanope texana, Jassa falcata, Sagitta* spp., and *Sarsia* spp.

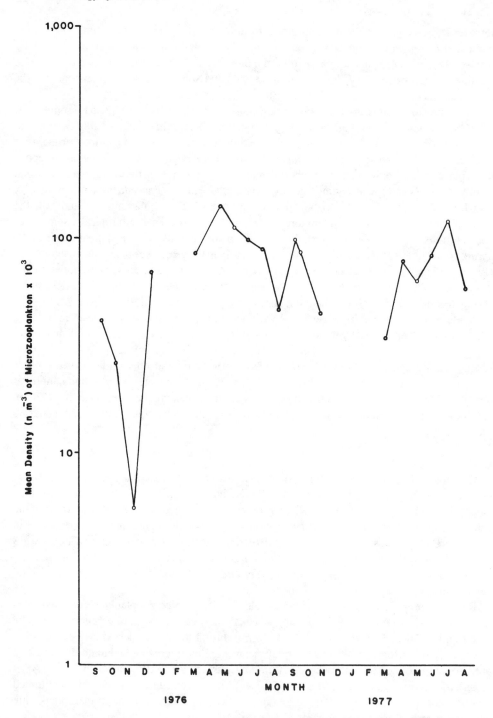

FIGURE 5. Mean monthly density of microzooplankton in Barnegat Bay, NJ from September 1975 to August 1977. No samples collected in January, February, and December 1976 and in January and February 1977. (From Sandine, P. H., *Ecology of Barnegat Bay, New Jersey,* Kennish, M. J. and Lutz, R. A., Eds., Springer-Verlag, New York, 1984, 95. With permission.)

The abundance of macrozooplankton varies markedly from year to year.[67] For example, the hydromedusa, *R. octopunctata,* had a maximum mean monthly density of less than 1/m³ during 1975/1976 but greater than 200/m³ during 1976/1977. Macrozooplankton abundance, similar to

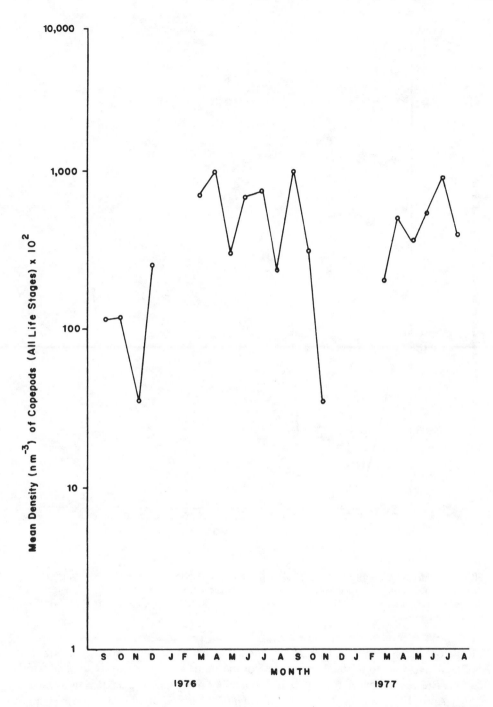

FIGURE 6.   Mean monthly density of copepods in Barnegat Bay, NJ from September 1975 to August 1977. No samples collected in January, February, and December 1976, and in January and February 1977. (From Sandine, P. H., *Ecology of Barnegat Bay, New Jersey,* Kennish, M. J. and Lutz, R. A., Eds., Springer-Verlag, New York, 1984, 95. With permission.)

that of microzooplankton abundance, peaks during the spring and summer. Among the macrozooplankton, the ctenophore, *Mnemiopsis leidyi,* and the arrow worms, *Sagitta* spp., prey heavily on microzooplankton, particularly copepods. The mean annual density of macrozo-oplankton from September 1975 to September 1977 ranged from 51 to 115/m³.

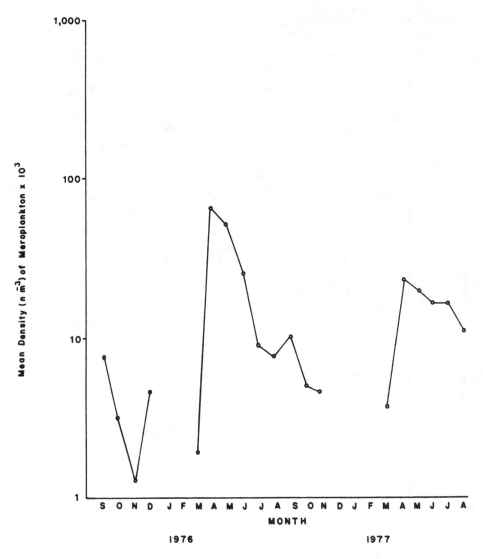

FIGURE 7. Mean monthly density of meroplankton in Barnegat Bay, NJ from September 1975 to August 1977. No samples collected in January, February, and December 1976, and in January and February 1977. (From Sandine, P. H., *Ecology of Barnegat Bay, New Jersey,* Kennish, M. J. and Lutz, R. A., Eds., Springer-Verlag, New York, 1984, 95. With permission.)

Ichthyoplankton abundance tends to be bimodal, with highest values from January through April and from June to September.[67] Larvae of the sand lance, *Ammodytes* sp., and winter flounder, *Pseudopleuronectes americanus,* outnumber other ichthyoplankters from January through April. Elvers of the American eel, *Anguilla rostrata,* are the only other common ichthyoplankton in the bay during winter-spring.[67] The most abundant ichthyoplankton forms between June and September include eggs and larvae of the bay anchovy, *Anchoa mitchilli,* and larvae of the gobies, *Gobiosoma* spp. Other ichthyoplankton retrieved in samples during the warmer months of the year, albeit in lesser numbers, are eggs, larvae, or juveniles of the American eel (*A. rostrata*), the Atlantic menhaden (*Brevoortia tyrannus*), atherinids, blennids, the cunner (*Tautogolabrus adspersus*), the hogchoker (*Trinectes maculatus*), the northern pipefish (*Syngnathus fuscus*), and the northern puffer (*Sphoeroides maculatus*).[86] By far the most numerous fish egg is that of the bay anchovy, responsible for more than 90% of all fish eggs recovered in the bay in previous sampling periods.[67] The majority of ichthyoplankton in the estuary exhibits greater densities at night, corroborating the findings of investigators working on other systems.[87-89]

D. CHESAPEAKE BAY

Brownlee and Jacobs[90] studied the micro- and mesozooplankton of Chesapeake Bay. The principal components of the microzooplankton are the protozoan phyla Sarcodina and Ciliophora, the Rotifera, and the nauplii of the crustacean order Copepoda. The main taxa of the mesozooplankton, in turn, consist of juveniles and adults of the crustacean orders Copepoda and Cladocera, the Chaetognatha, the gelatinous zooplankton, notably, the Ctenophora and Schyphozoa, and the meroplanktonic stages of benthic invertebrates (e.g., barnacles, oysters, and polychaetes). Abundance and biomass of microzooplankton are salinity dependent; both decrease with increasing salinity. This relationship not only holds for the main body of the Chesapeake, but also for the tributary systems of the Potomac, Patuxent, and Choptank Rivers. Maximum abundance and biomass of the total microzooplankton arise in late summer and fall. Peaks of microzooplankton in the spring and fall either coincide with maximum counts of diatom-rich phytoplankton or follow pulses of these microscopic plants by about 1 month. Summer and winter microzooplankton maxima, meanwhile, seem to be associated with blooms of small flagellates and 1- to 2-μm-diameter coccoid cells, perhaps blue-green algae. Summertime peaks also correspond with high primary production. Among the mesozooplankton, the calanoid copepods, *Acartia tonsa, A. hudsonica,* and *Eurytemora affinis,* predominate. Cladocerans, barnacle nauplii, and polychaete larvae become important at specific times of the year and at certain salinity regimes. Ctenophores exert predation pressure on copepod populations in the summer.

The spatial distribution of microzooplankton in the bay can be divided into freshwater, oligohaline, and mesohaline-polyhaline zonal patterns.[90] Rotifers, a major microzooplankton group, are important in all salinity regimes. The highest abundance of tintinnid ciliates occurs in the transition zones. The Sarcodina primarily inhabit freshwater habitats. The most abundant microzooplankton taxa recovered during sampling in 1984 and 1985 were *Synchaeta* spp. (mean density = 117.1/l), *Tintinnopsis fimbriata* (mean density = 106.0/l), and copepod nauplii (mean density = 39.2/l) (Table 3).

Peaks in mesozooplankton abundance during 1984 and 1985 appeared 1 to 2 months subsequent to maximum counts of both microzooplankton and phytoplankton.[90] Based on field as well as laboratory observations, the mesozooplankton may be important predators of microzooplankton, particularly in late winter in the oligohaline zone and in late summer in the mesohaline zone. Sampling of the mesozooplankton from August 1984 to September 1985 revealed that *A. tonsa* was the most abundant form (mean density = 5504/m³), followed by *E. affinis* (mean density = 2032/m³) and *Bosmina longirostris* (mean density = 715/m³) (Table 4).

Jacobs et al.,[91] reviewing the distributional patterns of zooplankton taxa in upper Chesapeake Bay, disclosed that the cladocerans *B. longirostris, Diaphanosoma leuchtenbergianum, Moina micrura,* the calanoid copepod, *Eurytemora affinis,* and the cyclopoid copepod, *Cyclops vernalis,* dominate the community in freshwater regions. This community has a higher diversity than the estuarine community in mesohaline environments dominated by the calanoid copepod, *A. tonsa. A. tonsa* is present year-round at all salinities, but greatest abundances take place from April to October when densities oftentimes equal 1 to 2×10^5/m³ and may surpass 1×10^6/m³.[92] Olson[93] attested to the significance of *A. tonsa* in the total zooplankton community at Calvert Cliffs over the interval 1975 to 1980. While this species is subject to severe depletion in numbers due to predation, its generation time, measured in days and weeks, enables the copepod to recover quickly from short-term perturbations attributable to vagaries in environmental condi-

TABLE 3
The 30 Most Abundant Microzooplankton Taxa Collected in
Chesapeake Bay from August 1984 to December 1985

Taxon	Taxonomic group	Density (#/l)	Maximum density (#/l)	Percent of total
Synchaeta spp.	Rotifera	117.1	2513	29.38
Tintinnopsis fimbriata	Tintinnina	106.0	4334	20.59
Copepod nauplii	Copepoda	39.2	1006	19.45
Synchaeta sp. A	Rotifera	34.2	609	2.89
Tintinnopsis dadayi	Tintinnina	17.6	486	2.89
Keratella cochlearis cochlearis	Rotifera	27.2	481	2.80
Polyarthra sp.	Rotifera	36.8	378	2.38
Difflugiidae	Sarcodina	14.3	270	2.06
Synchaeta sp. B	Rotifera	65.2	296	1.80
Tintinnopsis subactua	Tintinnina	11.7	442	1.76
T. radix	Tintinnina	22.1	439	1.37
Brachionus angularis	Rotifera	24.9	205	1.21
Trichocerca sp.	Rotifera	14.5	264	1.11
B. calyciflorus	Rotifera	16.2	170	1.05
Pelecypoda larvae	Pelecypoda	7.4	63	1.02
Keratella cochlearis tecta	Rotifera	16.0	232	0.89
Filinia sp.	Rotifera	21.0	223	0.66
Polyarthra vulgaris	Rotifera	17.4	101	0.62
Notholca acuminata	Rotifera	12.3	93	0.48
B. plicatilus	Rotifera	11.4	66	0.33
Hexarthra mira	Rotifera	16.3	132	0.30
Tintinnopsis subacuta	Tintinnina	4.5	28	0.24
Filinia longiseta	Rotifera	10.9	47	0.24
Acineta sp.	Ciliophora	15.6	149	0.22
B. caudatus	Rotifera	10.0	63	0.22
Keratella sp.	Rotifera	15.2	132	0.22
Cyphoderiidae	Sarcodina	1.8	20	0.17
Centrophyxidae	Sarcodina	2.7	30	0.17
Conochilus unicornis	Rotifera	15.6	68	0.16
Arcella sp.	Sarcodina	2.9	46	0.15

From Brownlee, D. C. and Jacobs, F., *Contaminant Problems and Management of Living Chesapeake Bay Resources,* Majumdar, S. K., Hall, L. W., Jr., and Austin, H. M., Eds., Pennsylvania Academy of Science, Easton, 1987, 217. With permission.

tions (e.g., storms) as well as biotic interactions. *A. hudsonica* and *E. affinis* replace *A. tonsa* as the dominant copepod at higher and lower salinities, respectively, in the winter and early spring. These two copepods commonly have densities above $2 \times 10^5/m^3$ and, in exceptional periods, *E. affinis* densities may exceed $3 \times 10^6/m^3$.[92]

The dominant estuarine copepods display annual abundance cycles that alternate.[8] In the Patuxent River estuary, maximum zooplankton numbers take place in the spring (March to April). Rotifer counts during spring are high (1×10^4 to $1 \times 10^6/m^3$) in the estuary, and copepod densities generally range from 1×10^4 to $1 \times 10^5/m^3$.[13] Cladocerans likewise inhabit freshwater zones in densities between 1×10^4 and $1 \times 10^5/m^3$ during the spring.

Variable meroplankton pulses from spring to fall signal episodic spawning events of benthic invertebrates. For instance, Butt et al.,[94] surveying commercially important meroplankton of lower Chesapeake Bay, expounded on mid- to late summer maxima of blue crab (*Callinectes sapidus*) larvae in the deeper channels of the Chesapeake Bay mouth. American oyster (*Crassostrea virginica*) larvae were most conspicuous in July and August and occasionally September, mostly in the vicinity of the confluence of the James and Elizabeth Rivers.

TABLE 4
Mean Abundance, Percent Composition, and Ubiquity of Mesozooplankton
Collected in Chesapeake Bay from August 1984 to December 1985[a]

Taxon	Density (5 s/m²)	Percent of total	Ubiquity (% of total samples in which taxon occurred)
Acartia tonsa	5503.82	54.13	91.25
Eurytemora affinis	2032.20	19.99	53.61
Bosmina longirostris	715.45	7.04	32.89
Polychaete larvae	506.94	4.99	59.51
Barnacle nauplii	336.12	3.31	73.57
A. clausi	254.32	2.50	9.70
Diaphanosoma leuchtenbergianum	153.73	1.51	14.07
Moina micrura	86.48	0.85	13.88
Podon polyphemoides	82.21	0.81	24.71
Acartia sp.	81.54	0.80	7.22
Mesocyclops edax	80.57	0.79	14.26
Copepod nauplii	76.93	0.76	61.60
Cyclops vernalis	43.18	0.42	25.10
Ostracoda	27.71	0.27	57.03
Oithona colcarva	26.03	0.26	28.71
Daphnia retrocurva	23.06	0.23	18.63
Scapholeberis kingi	17.80	0.18	6.46
Gastropod larvae	16.41	0.16	12.93
Centropages hamatus	16.13	0.16	9.32
Pseudodiaptomus coronatus	13.41	0.13	22.62
Harpacticoida	10.99	0.11	30.80
Halicyclops magnaceps	9.82	0.10	4.75
Mollusca	8.49	0.08	13.12
Alonella sp.	4.68	0.05	8.17
Alona spp.	4.33	0.04	1.33
Cyclops bicuspidatus	4.29	0.04	10.08
Diaptomus sp.	3.97	0.04	15.59
Mysid	3.17	0.03	46.20
Ergasilus sp.	2.98	0.03	13.31
Polychaeta	2.96	0.03	3.04
Unid. fish eggs	2.69	0.03	11.79
Ilyocryptus spinifer	1.76	0.02	10.65
Sida crystallina	1.37	0.01	6.84
Brachyurian zoea	1.33	0.01	12.17
Paracyclops fimbriatus poppei	1.09	0.01	5.32
Sagitta sp.	0.97	0.01	14.83
Chydorius sp.	0.84	0.01	9.89
Camptocercus rectirostris	0.82	0.01	3.61
Hydracarina	0.79	0.01	4.75
Leptodora kindtii	0.75	0.01	2.66
Ilyocryptus sp.	0.74	0.01	2.28
Eucyclops agilis	0.70	0.01	9.89
Palaemonetes sp.	0.62	0.01	12.36
Unid. crab zoea	0.49	0.00	1.71
Unid. fish larvae	0.46	0.00	18.25
Alona affinis	0.43	0.00	8.56
Temora turbinata	0.41	0.00	1.33
Gammarus fasciatus	0.40	0.00	17.30
Centropages furcatus	0.38	0.00	0.57
Paracalanus crassirostris	0.32	0.00	1.14
Cyclopoida	0.16	0.00	0.76
Oligochaeta	0.13	0.00	0.19
Morone americana	0.13	0.00	1.14

TABLE 4 (continued)
Mean Abundance, Percent Composition, and Ubiquity of Mesozooplankton
Collected in Chesapeake Bay from August 1984 to December 1985[a]

Taxon	Density (5 s/m²)	Percent of total	Ubiquity (% of total samples in which taxon occurred)
Corophium lacustre	0.07	0.00	8.17
Alona costata	0.06	0.00	3.42
Argulus sp.	0.06	0.00	13.50
Morone sp.	0.04	0.00	0.57
Alona sp.	0.04	0.00	0.57
Sapherella sp.	0.04	0.00	0.76
Bosmina sp.	0.03	0.00	0.38
Clupeidae	0.02	0.00	0.95
M. saxatilis	0.02	0.00	0.76
Monoculodes edwardsi	0.02	0.00	4.94
Eubosmina coregoni	0.01	0.00	0.95
Leptocheirus plumulosus	0.01	0.00	1.33
Chaoborus sp.	0.01	0.00	3.23
Chironomid larvae	0.01	0.00	6.46
Isopoda	0.01	0.00	7.22
Parathemisto compressa	0.01	0.00	1.90
Micropogon undulatus	0.00	0.00	2.28
Anchoa mitchilli	0.00	0.00	1.33
Brachyurian megalops	0.00	0.00	0.95
Pseudopleuronectes americanus	0.00	0.00	0.38
Menidia	0.00	0.00	0.76
Gobiosoma bosci	0.00	0.00	0.76
Piscicolidae	0.00	0.00	0.38
Lucifer faxoni	0.00	0.00	0.57
Dipteran larvae	0.00	0.00	0.38
Euceramus praelongus	0.00	0.00	0.19

[a]   Mean, maximum, and percent of total abundance over all stations and dates.

From Brownlee, D. C. and Jacobs, F., *Contaminant Problems and Management of Living Chesapeake Bay Resources,* Majumdar, S. K., Hall, L. W., Jr., and Austin, H. M., Eds., Pennsylvania Academy of Science, Easton, 1987, 217. With permission.

E. NORTH INLET

Lonsdale and Coull[95] described the zooplankton assemblages of the North Inlet estuary near Georgetown, SC. In this system, as in many other estuarine systems, copepods predominate, constituting 64 to 69% of the total zooplankton numbers and biomass. During a 20-month sampling interval from January 1974 to August 1975, zooplankton density varied between 3.77 $\times 10^2$ and 8.44×10^4/m³, with a mean value of 9.76×10^3/m³, and zooplankton biomass, between 6.4×10^2 and 1.40×10^5 µg dry weight per cubic meter, with a mean value of 1.62×10^4 µg dry weight per cubic meter. Zooplankton abundance was highest from June through mid-July which does not conform to the spring and fall maxima apparent in many temperate systems. Figure 8 illustrates the mean density and biomass of zooplankton in the North Inlet estuary from January 1974 to August 1975.

Paravocalanus crassirostris (Dahl) was the most numerous copepod sampled, accounting for 16% of all zooplankton and 13% of the total biomass. *Acartia tonsa, Oithona colcarva,* and *Euterpina acutifrons* followed *P. crassirostris* in numerical importance, being common members of the community. Several copepod populations were seasonal inhabitants, including

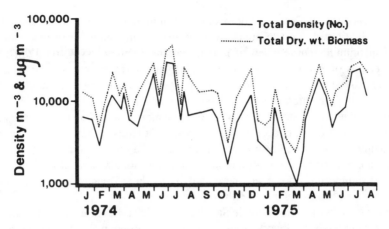

FIGURE 8. Total zooplankton numbers and dry weight biomass in North Inlet, SC during 1974 and 1975. (From Lonsdale, D. J. and Coull, B. C., *Chesapeake Sci.,* 18, 272, 1977. With permission.)

Corycaeus sp. and *Temora turbinata* (July to December), *Centropages typicus* and *Paracyclopina* sp. (summer species), and *C. hamatus* (winter species).

Meroplankton comprised a larger fraction of the total zooplankton here than in other estuaries in New Jersey,[96] Maryland,[97] and North Carolina.[98] Cirripedia nauplii were most plentiful, making up 13% of the total zooplankton. Bivalve, gastropod, and polychaete larvae composed an additional 12%. Lonsdale and Coull[95] contended that the zooplankton community of North Inlet is most closely allied to the zooplankton communities of Florida waters rather than those of temperate estuarine systems.

F. SAN FRANCISCO BAY

Ambler et al.[29] discussed results of monthly and semimonthly zooplankton sampling in San Francisco Bay from 1978 through the spring of 1981. This project was initiated to ascertain: (1) the numerical and biomass dominants of the zooplankton community, (2) seasonal changes in abundance and distribution of copepods and net microzooplankton, and (3) interannual variations in the community. It also dealt with the mechanisms of temporal variability in the zooplankton.

San Francisco Bay embodies two estuarine systems — San Pablo Bay and South Bay — which communicate with the Pacific Ocean via the inlet of Golden Gate. San Pablo Bay and South Bay harbor different zooplankton assemblages. In the northern reach, copepod populations are distributed according to their salinity tolerances; hence, *Sinocalanus doerrii* inhabits the Sacramento-San Joaquin Rivers, *Eurytemora affinis* Suisun Bay, *Acartia* spp. San Pablo Bay, and *Paracalanus parvus* Central Bay. The magnitude of river inflow modulates the distribution of the zooplankton such that *S. doerrii* is found at the riverine boundary, *E. affinis* in the oligohaline mixing zone, *Acartia* spp. in polyhaline waters, and *P. parvus* at the seaward boundary. *A. clausi* s. l. concentrates in San Pablo Bay nearly year-round, but *A. californiensis* appears there only from August to October. South Bay contains larger densities of these two *Acartia* species. *A. clausi* s. l. attains maximum densities during the wet season and *A. californiensis,* during the dry season. A seasonal succession is evident among these two populations in South Bay, the warm-water form (*A. californiensis*) replacing the cold-water form (*A. clausi* s. l.). The copepod *Oithona davisae* is most conspicuous in the fall.

San Pablo Bay and South Bay not only differ in the succession of *Acartia* species, but also in overall microzooplankton structure. Rotifers (*Synchaeta* sp.) and tintinnids (*Tintinnopsis* sp. A, *Tintinnopsis* sp. B, and *Eutintinnus neriticus*), together with *Acartia* spp. nauplii, are the most frequently occurring and abundant microzooplankton in San Francisco Bay. The rotifers and tintinnids are distributed baywide, but in lower numbers in oligohaline areas (e.g., Suisun Bay). The dominant tintinnid, *Tintinnopsis* sp. B, reaches densities in excess of $10^5/m^3$. In South Bay, maximum counts of *Tintinnopsis* sp. B are more variable from year to year than those of *Tintinnopsis* sp. A which peak with the timing of phytoplankton blooms. Tintinnids in San Pablo Bay persist nearly year-round, attaining highest numbers in shoal waters. Seasonal abundance distributions of the microzooplankton, in general, are more patchy than those of the copepods.

Seasonal occurrences of meroplankton contribute to the temporal patchiness of the zooplankton community. In the winter-spring, meroplankton of barnacles, bivalves, gastropods, and polychaetes are responsible for a substantial portion of the total zooplankton biomass in South Bay. During summer-fall, however, meroplankton of these organisms account for a large fraction of the zooplankton biomass in the northern reach.

Seasonal dynamics of zooplankton in San Francisco Bay reflect various physical, chemical, and biological mechanisms at work. Some of the key factors controlling zooplankton community dynamics are river discharge, salinity, temperature, coastal hydrography, and seasonal phytoplankton cycles. The zooplankton respond to these major regulating factors in a complex fashion to yield the temporal and spatial patterns of the taxa observed in the estuary.

V. ZOOPLANKTON DYNAMICS

A. BIOTIC AND ABIOTIC FACTORS
1. Effects of Light

Forward[96] reviews zooplankton photoresponses and the involvement of endogenous rhythms in migration. Three general vertical migration patterns characterize most zooplankton species, specifically nocturnal, twilight, and reverse migration. Zooplankton undergoing a nocturnal migration commence their ascent of the water column near sunset, reach a minimum depth between sunset and sunrise, and subsequently descend to a maximum depth during the day. Their descent begins near sunrise. The single daily ascent of a nocturnal migration contrasts with an opposite daily pattern of motion typified by a reverse migration in which a zooplankton species ascends to a minimum depth of the water column during the day and descends to a maximum depth at night. Reverse migration is the least common type of daily vertical movement displayed by zooplankton. In a twilight migration, in turn, a zooplankter initiates its ascent of the water column about sunrise; however, upon attaining a minimum depth, it later descends at night in a migration termed the midnight sink or nocturnal sinking. At sunrise, the organism rises once again, but subsequently descends to the daytime depth.

Light is a major environmental factor regulating diel vertical migration of zooplankton.[96] Acting as an environmental cue which triggers faunal migration, changes in illumination at sunrise and sunset elicit vertical movements of the organisms.[43,97,98] Vertical migration seems to be responsive to the following light cues: (1) a change in depth of a particular light intensity, (2) a change in underwater spectra, (3) a change in the polarized light pattern, (4) an absolute amount of change in light intensity, and (5) a relative rate of intensity change.[44,99,100] As summarized by Forward et al. (p. 146),[101] "The cue for initiating vertical movements is the rate and direction of change in light intensity from the ambient level (adaptation intensity) which itself can change over a day". Factors other than light (i.e., temperature, salinity, organism's age, physiological condition, and reproductive stage) may alter the diel vertical migration pattern, thereby adding complexity to this area of study.[102] Thus, Levinton,[5] dealing with problematical aspects of vertical migration in the copepod genus *Calanus*, cited differences associated with season, sex, and molt stage. Jacobs[103] and Stickney and Knowles[104] related vertical migrations of *Acartia tonsa* to the stage of the tide.

Hypotheses have been formulated to explain the adaptive value of vertical migration,[105] as covered briefly in Chapter 2, Section VI. Possible benefits are an energetic advantage to migrants,[3,106] predator avoidance,[3,5,30] greater food intake and utilization,[31] maximization of fecundity,[31] and greater transport and dispersal within the estuary.[5,31] Increased lateral dispersal can provide vulnerable larval stages with an adequate food supply.[107]

2. Effects of Temperature

Temperature influences zooplankton physiology and ecology.[3,108] Metabolic rates are a function of temperature, with fecundity, duration of life, adult size, and other parameters being modulated by temperature levels. Lower temperatures, for example, favor zooplankton with higher fecundity.[109] The vertical migration of zooplankton populations into deeper, colder areas of a thermally stratified system, therefore, confers an advantage to the organism in regard to fecundity.

McLaren[110] illustrated the importance of temperature in the development times of several copepod species, determining that the development times were accurately modeled by the Belehrádek equation given as

$$V = a(T + \alpha)^b \tag{1}$$

where V is the velocity of a process; a is a constant providing a scale correction related to the units of V and describing the Y-intercept on a log-log plot; b is the slope of a log-log plot and delineates the curvature of an arithmetic plot; and α defines a biological zero, the temperature at which V becomes zero.[3] The Belehrádek function has not only been used successfully in studies of the embryonic development of copepods, but also in investigations of the size of adults. According to the assessment by Deevey,[111,112] various taxa of marine copepods grow to smaller sizes at high rather than low temperatures within their range due, as proffered by Heinle,[3] to changes in the molting rate as temperatures rise or fall. At higher temperatures, molting is faster, and less growth takes place in the intermolt period, resulting in smaller adult sizes. Hence, zooplankton living in warmer seas usually grow to a smaller size than individuals occupying cooler waters.[3] Smaller individuals tend to lay fewer eggs; consequently, water temperatures will also affect fecundity. The development times, adult size, and fecundity of *Pseudocalanus minutus*, as an example, increase at lower temperatures.[109] Landry[113] detected a similar effect of temperature on fecundity of *Acartia clausi* (=*A. hudsonica*) from Jackles lagoon (San Juan Island, WA).

A number of studies support the findings of Deevey.[111,112] Data from Landry[113] on *A. clausi* demonstrate a change in mean prosome length from 0.580 to 0.830 mm from summer to winter. Unpublished data of J. Flynn on *A. clausi* s. l. from Yaquina Bay, OR, quoted by Miller,[8] indicate the existence of populations of different sizes in the lower and upper bay ascribable to a longitudinal temperature gradient. *A. clausi* s. l. in the lower bay, where temperatures approximate 12 C year-round, have prosome lengths of about 0.79 mm. In the upper bay, however, summer temperatures approach 23 C, and the mean length of the copepods is about 0.55 mm. Changes in the size of *A. clausi* s. l. have also been discerned in Narragansett Bay. Durbin and Durbin[114] discovered sudden population shifts from large to small adults in the spring caused by rapidly rising temperatures in the bay.

The hatching of copepod eggs depends on temperature as well. The length of dormancy and the termination of diapause in copepods are temperature regulated.[30] Johnson,[115] experimenting with *A. californiensis*, deemed temperature to be a controlling factor in the hatching rates of the species.

In addition to the physiological influence of temperature on zooplankton perhaps manifested most clearly in changing metabolic rates, ecological effects linked to temperature changes include seasonal variations in the species composition and abundance of zooplankton in shallow-water systems, particularly well documented for mid-latitude estuaries. Seasonal changes in community structure arise from successions of constituent populations, leading to fluctuating dominance patterns during the year. The community of zooplankton in mid-Atlantic Bight estuaries, for example, follows a succession of species populations, with winter-spring dominants (e.g., *A. hudsonica*) being replaced by summer-fall dominants (e.g., *A. tonsa*). This successional pattern is succinctly appraised by Jeffries and Johnson (p. 4-2):[68] "As in the ocean, seasonal temperature change in estuaries causes a succession of species populations, which fall into two major groups, the winter-spring and summer-fall components, representing the boreal and temperate-tropical provinces. As one group wanes, the other waxes, so production is maintained throughout the year, despite annual temperature ranges far exceeding the reproductive tolerances of either group. In the open ocean, the same zoogeographic groupings apply to a different, more complex species assemblage."

Temperature triggers spawning of many benthic invertebrates resulting in pulses of meroplankton during the warm months. These pulses bolster the high standing crops of zooplankton delineated in many estuaries and coastal waters during the summer.[48] A number of resident and migratory fish populations spawn seasonally in estuaries, adding icthyoplankton to the total concentration of zooplankters.[116]

Whereas the holoplankton of the open ocean generally peak in abundance in spring and fall, the community in estuaries often experiences a different seasonal pattern of events.[49] Holoplanktonic species subject to variable seasonal densities may survive through the off seasons by having a few individuals persist to proliferate as environmental conditions improve. Alternatively, the propagation of populations takes place via individuals developing from resting eggs generated the previous year.[68] The seasonal success of copepods in terms of growth and development hinges largely on temperature and food supply.[117]

3. Effects of Salinity

The salinity tolerance of zooplankton limits their distribution within an estuary. These diminutive fauna are responsive to salinity levels encountered along the longitudinal axis of an estuarine system. They respond to salinity changes in the vertical plane as well.[48] The salinity tolerances of holoplankton and meroplankton differ among species and may vary among ontogenetic stages of a single species.[13] The patterns of species succession and dominance from the head to the mouth of an estuary, therefore, are contingent upon salinity concentrations. Salinity extremes often arrest growth and increase mortality of susceptible forms. However, species experiencing widely fluctuating salinities can survive these conditions by active osmoregulation or by toleration of low internal osmolality.[8]

The salinity tolerances of holoplankton and meroplankton have been investigated both in the field and laboratory. *Eurytemora affinis*, a fresh-to-brackish water copepod, will not reproduce in salinities above 22.5‰, but will survive salinities up to 35‰.[13,117] Also in fresh-to-brackish water areas, cladocerans and rotifers may be well represented, although these two groups encompass species that have adapted to a broad range of salinities downestuary. Two zooplankters having a widespread salinity tolerance are *Pseudodiaptomus hessei* and *P. stuhlmanni*; *P. hessei* is found in salinities of 1 to 74‰, and *P. stuhlmanni*, in salinities of 1 to 75‰. These species dominate the hypo- and hypersaline regions of estuaries.[69] Lance,[118,119] working experimentally with *Acartia bifilosa*, *A. discaudata*, and *A. tonsa*, detected reduced survival for all three forms below 9‰. *A. tonsa* exhibits the greatest tolerance to reduced salinities, *A. bifilosa* an intermediate tolerance, and *A. discaudata* a low tolerance. Reduction in salinity by dilution to 5‰ caused higher mortality of *A. bifilosa* and *A. discaudata* than of *A. tonsa*. Males and copepodite stages of the three *Acartia* species have less tolerance to lower salinities than females.[120]

Effects of salinity on zooplankton populations are evident in mid-Atlantic estuaries as elsewhere. Differences in the tolerances of *A. tonsa* and *A. hudsonica* to reduced salinity, for example, play a role in the seasonal dominance patterns of the copepods in the mid-Atlantic.[68] Although temperature is the main factor causing the seasonal exchanges of dominance in the

TABLE 5
Growth Rates of *Acartia tonsa* at Different Salinities[3]

Salinity (‰)	Experimental temperature (°C)	Days from egg to adult	Corrected to 20°C	Ref. (sources from Ref. 3)
5	20.0	12.12	12.12	D. R. Heinle and M. A. Ross, unpublished
10	20.0	8.19	8.19	D. R. Heinle and M. A. Ross, unpublished
12	20.0	8.58	8.58	Heinle (1969b)
12	20.0	8.92	8.92	Heinle (1969b)
20	20.0	9.36	9.36	D. R. Heinle and M. A. Ross, unpublished
25	21.5	11.5	13.34	Johnson (1974)
25	21.5	11.6	13.45	Johnson (1974)
31	17.0	25.0	22.0	Zillioux and Wilson (1966)

From Heinle, D. R., *Functional Adaptations of Marine Organisms*, Vernberg, F. J. and Vernberg, W. B., Eds., Academic Press, New York, 1981, 85. With permission.

lower reaches of these systems, the sensitivity of the copepods to salinities upestuary promotes succession of the species. *A. hudsonica*, being more sensitive to brackish water conditions than *A. tonsa*, is replaced by *A. tonsa* far upestuary early in the spring, with replacement gradually spreading seaward and taking one to several months to progress to the estuarine mouth.

Heinle[3] examined the growth rates of *A. tonsa* over a range of salinities, recording its more rapid growth at lower levels (Table 5). Maximum growth rates occur at approximately 10 to 12‰.[3,108,121-123] This species does not seem to osmoregulate to any appreciable extent. It may osmoregulate at a cellular level or simply tolerate internal osmolality.[8]

The development of meroplankton and ichthyoplankton in estuaries is, like that of holoplankton, dependent on salinity.[49,124] The larvae of the barnacles *Balanus balanoides, Chthamalus stellatus*, and *Elminius modestus* are euryhaline, but immobilize at salinities less than 12‰ and greater than 50‰. The nauplii of *B. amphitrite* and *B. eburneus* can tolerate salinities in the range of 10 to 20‰. The nauplii of *Chelonobia patula* have a slightly narrower tolerance range (15 to 20‰ and possibly up to 50‰). Larvae of the calico crab, *Hepatus epheliticus*, require salinities less than 50‰ to complete their development. *Rhithropanopeus harrisii* is another crab whose larval development is a function of salinity. Similarly, the development of the eggs and larvae of *Crassostrea virginica* and *Mercenaria mercenaria* is salinity dependent.[63,86]

Salinity gradients govern the distribution of eggs and larvae of fishes in some estuaries through physiological effects on development.[125] Such is the case in the Delaware River estuary.[124] The effect has also been demonstrated in other regions.[126]

Perkins[49] gives additional examples of salinity controls on the spatial distribution of zooplankton taxa in estuaries. In the Brisbane River estuary, Australia, *Pseudodiaptomus* spp. predominate in the lower reaches, but are replaced successively upestuary by *Acartia* sp. and *Isias uncipes* and near the estuarine head by the euryhaline forms *Gladioferens pectinatus* and *Sulcanus conflictus*. The upestuary penetration of *Evadne tergestina, Penilia avirostris,* and *Podon polyphemoides* in Chesapeake Bay is constrained by salinity concentrations; *E. tergestina, P. avirostris,* and *P. polyphemoides* do not extend into waters where salinities fall below 15.75, 18.12, and 3.00‰, respectively.

Salinity restrictions likewise limit the distribution of zooplankton in the vertical plane of an estuary. Reduced salinity in the surface layer of a stratified system, for instance, can inhibit or preclude the vertical migration of susceptible forms. This is true for a number of species of the genus *Acartia* (e.g., *A. bifilosa, A. discaudata,* and *A. hudsonica*) in addition to other copepods (e.g., *Centropages hamatus* and *Temora longicornis*).[49]

4. Water Circulation

Hydrographic conditions in an estuary — riverine discharges, tidal currents, and waves — strongly affect zooplankton position. Tidal exchange, as perceived by Grindley,[69] is the single most important control on zooplankton distribution. The currents in estuaries displace zooplankton populations, especially in small systems or those characterized by strong river inflow which may translocate larvae to coastal oceanic waters, thereby reducing standing crops in the main body of the estuary.[127] The vertical migration behavior of many zooplankton species interacts with the two-way estuarine flow to enable the populations to maintain their location in the estuary. However, circulation patterns typically exacerbate spatial heterogeneity in zooplankton communities.

In marine ecosystems, the spatial heterogeneity of zooplankton communities may be manifested as shifts in the absolute abundance of a single species or as alterations in species composition.[30] At least six scales of spatial patterns are observable in zooplankton communities: (1) micro (1 cm to 1 m), (2) fine (1 m to 1 km), (3) coarse (1 km to 100 km), (4) meso (100 km to 1000 km), (5) macro (1000 km to 3000 km), and (6) mega (greater than 3000 km).[30,128] The processes generating these spatial patterns have been treated in modeling studies.[128] Most models of single species populations have keynoted smaller-scale processes less than 50 km.[129,130]

The vertical distribution of meroplankton and ichthyoplankton ameliorates their losses due to advective processes. The biological adaptations of barnacle larvae[131] and winter flounder larvae[132] to concentrate in bottom waters, for instance, mitigate the threat of lateral displacement through the mouth of the estuary via physical dispersion because of the landward flow of water in deeper zones.[133] Most studies on estuarine tidal transport or retention of organisms have centered on invertebrate taxa, yielding an unbalanced data base heavily skewed toward certain groups, such as the crustaceans.[134] Despite the emphasis on observations of specific invertebrate groups, work on larval-fish transport in estuarine systems has expanded in recent years. In regard to the migration of larval fishes in estuaries, the principal movement is in the vertical rather than the horizontal plane; the migratory behavior maximizes the benefits of flushing, transport, or retention of larvae.[62] The behavioral responses of larvae of the Atlantic croaker (*Micropogonias undulatus*), flounder (*Paralichthys* sp.), and spot (*Leiostomus xanthurus*) in the Cape Fear River, a tidal estuary in North Carolina, show how active larval movements provide an aid to the transport and retention of the larvae within an estuary.[62,135]

While many organisms have adapted strategies in early life-history stages to increase the likelihood of their retention in an estuary, others have developed behavioral patterns which enhance their transport away from retention sites to the nearshore ocean. Hence, recent research on the blue crab, *Callinectes sapidus* Rathbun, in Chesapeake Bay has revealed a large spawning population in the lower bay that releases larvae just prior to maximum ebb tide.[136] This spawning activity is followed by the upward migration of the first stage larvae of the species into the surface waters of the estuary where they get transported out of the bay into inshore coastal areas. Larval development continues offshore in the upper 1 m of the water column for the next 30 to 40 d until the postlarval (megalopal) stage is reached, at which time the megalopae move into the estuary to enter the shellfishery 12 to 15 months later.[137]

The larvae of some marine fishes do not complete their development in oceanic waters, but enter estuaries for shelter and food. They accomplish this entry either by active swimming or passive drift in tidal currents.[138] Multiple factors may impact the recruitment of these early life stages into the estuarine environment.[139-141] Beckley[138] looks at early juveniles of marine finfish species which use the Swartkops estuary (South Africa) as a nursery area by entering the system on the flood tide. She details flood-tide immigration of larval and juvenile *Heteromycteris capensis, Liza richardsoni,* and *Rhabdosargus holubi.*

Hiscock[142] evaluates the effects of water movement on organisms in subtidal systems. His work focuses on the action of residual currents in the spread of species through their planktonic larval stages. Water circulation is clearly paramount to the distribution of estuarine zooplankton populations.

5. Zooplankton Feeding
a. Copepods

The success of zooplankton populations naturally depends on their ability to obtain food. Despite the rather extensive data base on the ecology of these organisms, the feeding habits of even the most common species remained enigmatic until the recent past. Indeed, the exact mechanisms by which they capture and select different food items have been unclear.[143] Whereas most zooplankton appear to be omnivorous,[143] the literature is replete with studies of zooplankton grazing of phytoplankton. Copepods, in particular, serve as a principal avenue of energy flow in the grazing food web of estuarine ecosystems, consuming phytoplankton and other zooplankton.[144] For example, *Centropages hamatus*, a coastal marine copepod, not only ingests phytoplankton, but also copepod nauplii. *Labidocera aestiva*, primarily a carnivorous zooplankter, grazes on phytoplankton as well.[144] Some copepods (e.g., *Anomalocera ornata* and *C. typicus*) effectively eat fish larvae. In contrast, copepods may incur heavy losses due to predation by ichthyoplankton. Hence, in a survey of the diets of 76 species of fish larvae, Turner[143] identified six genera of calanoid copepods (i.e., *Acartia, Calanus, Centropages, Paracalanus, Pseudocalanus,* and *Temora*), three genera of cyclopoid copepods (i.e., *Corycaeus, Oithona,* and *Oncaea*), and harpacticoid copepods that were habitually important in their nutrition. The diet of some fish larvae changes during ontogeny; with increasing age, more animal matter is consumed. Larvae of the sand lance, *Ammodytes americanus*, provide an example. The smaller larvae of this species graze on phytoplankton while the larger larvae ingest copepods of increasingly older developmental stages.[145] Because some copepods consume fish larvae and vice versa, grazing food webs in estuarine environments can become, not surprisingly, quite complicated and highly problematical.

Observations on copepods indicate that they feed by means of passive filtration of small particles or by active capture of large particles.[8,146-149] In the active capture of prey, a copepod initially detects a food particle by chemo- or mechanoreception, and subsequently seizes it. The prey can be detected up to about 1 mm away from the animal via mechanoreception.[149] Within this distance, the copepod has the capability of sensing the shape and dimensions of its prey. Once captured, the particle is oriented by the copepod to determine its suitability for consumption. Some particles are ingested and others rejected. While the mechanical properties of the feeding apparatus may be partially responsible for some of the particle selectivity, decision making and chemosensation probably play a critical role.[150] Indeed, distance and contact chemoreception, along with mechanoreception, possibly are all involved in the detection and selection process.[149,151] Particle size has been invoked as a primary factor influencing the selectivity of copepod feeding.[152] However, other factors might also affect the selectivity process, including particle shape or smell.[147] Nevertheless, selective feeding by copepods has significant ramifications for plankton communities since it can result in one population of prey being more heavily grazed than the others in a community. Thus, a potential exists for the species composition of a phytoplankton community to be altered by selective copepod feeding which may, in turn, affect the success of zooplankton populations.

An early perspective of copepod feeding did not consider behaviorally mediated particle selection, but advocated indiscriminate filter feeding or a simple sieving process whereby particle retention depends on the efficiency of action of the second maxillae.[153-155] Feeding currents generated by cephalic appendages transport the food particles to the mouth. Setae filter the particles, and they are subsequently consumed.[156] This filtration feeding method is certainly an oversimplification as evidenced by various types of copepods, especially certain populations of cyclopoids and harpacticoids, which ingest particles much differently. Even among the calanoid copepods, where filter feeding is widespread,[30] a complex of appendage movements accompanies the feeding process. The application of high-speed microcinematography has greatly improved resolution of this problem. Koehl (p. 135)[157] states, "High-speed movies of water movement near feeding calanoid copepods reveal that the copepods propel water past themselves by flapping their second antennae, mandibular palps, first maxillae, and maxillipeds, and that the animals actively capture with their second maxillae parcels of that water that contain food particles. Copepods are not simple on-off filtering machines, but rather have a repertoire of behaviors associated with feeding: they can create scanning currents, fling and close their second maxillae to capture individual large particles, repeatedly flap their second maxillae to feed on small particles, reorient and comb particles from the second maxillae into the mouth, and groom their feeding appendages."

Diel vertical migration of zooplankton with greater nocturnal feeding rates has been assessed with respect to grazing impacts on phytoplankton and bacteria.[4,34,158-160] With increased densities of zooplankton in surface waters at night, a decline in the concentration of food particles reportedly occurs that is inversely proportional to the length of time the zooplankton remain in the upper layers.[161,162] Conclusions from other investigations discredit this view, suggesting that diel vertical migration behavior and feeding activity may not be strongly coupled, as illustrated by the copepods *Calanus* and *Pseudocalanus*.[163] Still, laboratory research[164,165] and field work[166,167] have been performed, elucidating diel feeding patterns of zooplankton.[168] On theoretical grounds, an energetic advantage has been postulated in which nocturnal grazing imparts a greater energy gain for the organism than continuous feeding.[73]

Enright[169] hypothesized that nocturnal grazing by copepods is due to the organism's migration into layers of high food concentration at night. The timing of nocturnal grazing in copepods may be controlled by light cues, relegating migration and feeding to two independent behaviors.[98] Daro[34] conveyed that zooplankton adjust their diel rhythm according to the environmental conditions present in order to obtain maximum energy at minimal cost. Consequently, a diversity of vertical distribution patterns of zooplankton species exists, each species having a preference for a certain depth in the water column.[30,34] The migration of zooplankton, however, may also lead to intensification of patchiness.[170]

b. Ingestion Rates and Assimilation

The rate of ingestion of zooplankton increases with increasing (phytoplankton) prey density up to a limit above which the ingestion rate is nearly constant. Two models can be utilized to describe the relationship of food intake vs. food concentration. The first is the Ivlev equation expressed as

$$I = I_m\left[1 - e^{-\delta(p - p')}\right] \qquad (2)$$

where I is the rate of ingestion, I_m is the maximum rate of ingestion, is a constant, p is the phytoplankton concentration, and p′ is the phytoplankton concentration at which no ingestion results. The second is the Michaelis-Menten equation given as

$$I = \frac{I_m(p - p')}{K + (p - p')} \qquad (3)$$

where K is a constant, and the other parameters are the same as above.

That no feeding of zooplankton occurs below a threshold concentration of food particles has been emphasized by Parsons et al.[171,172] and McAllister.[173] Such feeding thresholds bear major ecological significance because they preclude total exploitation and extermination of phytoplankton populations by the grazers, thereby yielding a more stable system.[174] The phytoplankton, not being completely depleted in numbers, regenerate as the zooplankton grazing pressure

TABLE 6
Percentage of Primary Production Consumed by Herbivores in Marine and Terrestrial Environments[a]

Coastal environments	Percentage of production eaten by herbivores	Number of trophic steps	Ref. (sources from Ref. 11)
Vascular plants			
Eelgrass, North Sea	4	3	Nienhuis and van Ierland (1978)
Salt marsh, Georgia	4.6	3—4	Teal (1962)
Salt marsh, North Carolina	58		Smith and Odum (1981)
Mangrove swamp, Florida	9—27[b]		Onuf et al. (1977)
Phytoplankton			
Long Island Sound	73[c]	4	Riley (1956)
Narragansett Bay	0—30[d]	4	Martin (1970)
Cochin Backwater, India	10—40		Qasim and Odum (1981)
Beaufort Sound	1.9—8.9		Williams et al. (1968)
Offshore California	7—52 (average 23)		Beers and Stewart (1971)
Peruvian upwelling	92, 54—61	3	Walsh (1975), Whiteledge (1978)
Open seas (all phytoplankton)			
Georges Bank	50—54	4	Riley (1963), Cohen et al. (1981)
North Sea	75—80	4—6	Crisp (1975)
Sargasso Sea	100	5	Menzel and Ryther (1971)
Eastern Tropical Pacific	39—140 (average 70)[e]	5	Beers and Stewart (1971)

[a] Annual consumption except where indicated otherwise. These values are rough, but best possible estimates based on many assumptions and extrapolations.

[b] Leaves and buds only.

[c] This is an estimate of consumption of organic matter in the water column. Larger zooplankton consume about 20%, microzooplankton and bacteria an additional 43%. In the bottom, benthic animals use an estimated 31% of net primary production.

[d] Of standing stock of algae.

[e] Includes only microzooplankton that passed through a 202-μm mesh. The biomass of these small species was about 24% of that of the larger zooplankton. Total consumption could easily be larger than reported if any of the larger species are herbivorous.

From Valiela, I., *Marine Ecological Processes,* Springer-Verlag, New York, 1984. With permission.

diminishes. Feeding thresholds incorporated into the aforementioned models consequently furnish a means for recovery of the phytoplankton that has been cropped.

In the open ocean, zooplankton grazing typically removes 50% or more of the net production of phytoplankton. The intensity of grazing pressure of zooplankton in many shallow coastal and estuarine areas in contrast is much lower (Table 6). In these systems, consumption by zooplankton frequently amounts to less than 30% of the net production of phytoplankton throughout the year.

Zooplankton assimilate only a fraction of the food ingested, and the remainder is egested. A number of techniques have been adapted to measure the assimilation efficiency of zooplankton. A review of the procedures is available in Heinle[3] which serves as the basis of this discussion. Food requirements of an animal can be computed from the following equation:[175]

$$R = G + T + E \tag{4}$$

where G equals the amount of food used for growth and reproduction, T is the amount of food utilized for metabolism, and E connotes the amount of food excreted and egested. The efficiency of assimilation can be calculated from

TABLE 7
Assimilation of Phytoplankton by Zooplankton

Species	Food	Method	Assimilation (%)	Reference (sources from Ref. 3)
Calanus finmarchicus	Various diatoms and flagellates	Tracer-isotope [^{32}P]	15—99	Marshall and Orr (1961)
	Skeletonema costatum	Tracer-isotope [^{14}C]	60—78	Marshall and Orr (1955b)
Temora longicornis	*S. costatum*	Tracer-isotope [^{32}P]	50—98	Berner (1962)
Euphausia pacifica	*Dunaniella primolecta*	Tracer-isotope [^{14}C]	85—99	Lasker (1960)
Ostrea edulis (larvae)	*Isochrysis galbana*	Tracer-isotope [^{32}P]	13—50	Walne (1965)
Calanus helgolandicus	Natural Particulate Matter	Chemical analyses	74—91	Corner (1961)
Metridia lucens	*T. nordenskioldii*	"Ratio method"	50—84	
	Ditylum spp.		35	Haq (1967)
	Artemia nauplii		59	
C. hyperboreus	*Exuviella* spp.	"Ratio method"	39.0—85.6	
		Chemical analyses	54.6—84.6	
Natural zooplankton	Natural Particulate Material	"Ratio method"	32.5—92.1	Conover (1966a)
C. finmarchicus	*Skeletonema costatum*	"Ratio method"	53.8—64.4	Corner et al. (1967)
	S. costatum	Chemical analyses	57.5—67.5	

From Heinle, D. R., *Functional Adaptations of Marine Organisms,* Vernberg, F. J. and Vernberg, W. B., Eds., Academic Press, New York, 1981, 85. With permission.

$$A = (R - E)/R \qquad (5)$$

By multiplying Equation 5 by 100, the assimilation efficiency is expressed as a percent.

Conover[176] estimated assimilation efficiency indirectly by determining the percent ash content of the food and the percent ash content of the feces and applying the relationship

$$A = \frac{\left(F^1 - E^1\right)}{\left(1 - E^1\right)\left(F^1\right)} \times 100 \qquad (6)$$

where A represents the percent assimilation efficiency, F^1 represents the ratio of ash-free dry weight to dry weight of the food, and E^1 represents the ratio of ash-free dry weight to dry weight of the feces. Values of carbon assimilation derived by this method probably yield more accurate estimates than those of nitrogen and phosphorus.[177-179]

Other attempts at estimating assimilation efficiency incorporate: (1) calculations of the ratios of nitrogen and phosphorus in new animal tissue, food excretory products, and fecal pellets; and (2) measurements of the uptake of radioactively labeled foods by zooplankton and their eggs and fecal pellets. Both of these techniques require an accurate accounting of either the nitrogen and phosphorus or tracer-isotope content of the organisms and their excretion products. Unfortunately, radioactive labeling experiments at times yield inaccurate results.[180]

Most assimilation efficiencies of zooplankton computed by the aforementioned methods exceed 50%; values as high as 99% have been registered (Table 7). Petipa[180] recommended an

TABLE 8
Ingestion, Assimilation, and Metabolic Expenditures of Calanoid Copepods

Species	Temperature (°C)	Amount ingested (μg C/d at μg C/l)	Amount assimilated (μg C/d)	Amount respired (μg C/d)
Calanus helgolandicus female, (Corner et al., 1971)	12.5	24.4 at 37 3.88 at 11	8.54 1.36	4.2 (Mullin and Brooks, 1970)
Eucalanus pileatus female, June 2—4, 1980	20			
h 24—36		2.26 at 109	2.03	4.0 (Ivleva, 1980)
h 36—48		1.01 at 103	0.91	2.3 (Mullin and Brooks, 1970)
Paracalanus sp. female, May 16—18, 1980	20			
h 24—36		1.68 at 91	1.51	1.03 (Ivleva, 1980)
h 36—48		107 at 66	0.96	

From Paffenhöfer, G.-A., *Trophic Interactions within Aquatic Ecosystems,* Meyers, D. G. and Strickler, J. R., Eds., American Association for the Advancement of Science, Selected Symp. 85, Westview Press, Boulder, CO, 1984, 75. With permission.

assimilation efficiency of 80% for *Acartia tonsa,* but there has been scant treatment of the assimilation efficiencies of small estuarine and coastal forms.[3] The daily food requirements of these zooplankton can be very high. It is not unusual for the daily ration to be greater than 100% of their body weight, although many species consume food in amounts less than 10 to 100% of their body weight each day.

Both secondary production and nutrient cycling rates are affected by rates of ingestion and assimilation.[150] Table 8 compiles ingestion, assimilation, and metabolic expenditure values for three calanoid copepods. Based on these data, the zooplankters generally assimilate between 1 and 2 and up to 8.54 μg C per day. Egested material from the zooplankton yields a source of nutrients for primary producers, either released directly as dissolved components to the inorganic pool or, subsequently, via microbial decomposition of the particulate matter.[150] Zooplankton release nitrogen and phosphorus mainly as NH_3 and PO_4^{3-} — two nutrients rapidly assimilated by phytoplankton.[181] The effect of nutrient release from zooplankton on nutrient pools cannot be discounted; algal demands for nitrogen and phosphorus may be met by nutrient turnover of zooplankton alone.[150] Butter et al.[178] denoted that approximately 60% of the phosphorus ingested by *Calanus* was excreted as dissolved phosphorus in the water column. In Narragansett Bay, net zooplankton account for about 35% of the total nitrogen regenerated by four major faunal groups (i.e., Atlantic menhaden, ctenophores, net zooplankton, and benthos) (Table 9). The regeneration rates of nutrients by zooplankton, however, are not constant in estuarine systems.[11]

6. Secondary Production

Production of a population of animals may be defined as the formation of new biomass within a specified period of time.[37] Two basic approaches have been applied in production studies, namely, the population dynamics and energy budget methods. The work of Valiela,[11] Mann,[31] Edmondson and Winberg,[182] Winberg,[183] and Greze[184] covers the procedures used to estimate production. The population dynamics approach concentrates on growth in estimating production, whereas the energy budget approach measures key components — consumption, respira-

TABLE 9
Regeneration of Nitrogen in
Narragansett Bay, RI

Source	Annual inputs (10⁶ g-atoms/year) nitrogen
Atlantic menhaden	0.8
Ctenophores	8.1
Net zooplankton	98.5
Benthos	264.0
Total	371.4

From Nixon, S. W., *Estuaries and Nutrients,*
Neilson, B. J. and Cronin, L. E., Eds., Humana
Press, Clifton, NJ, 1981, 111. With permission.

tion, egestion, and excretion — in the calculation of production.[11,37] Cohort, cumulative growth, and life-table methods of estimating production are effective population dynamics approaches.[11] They seem to be simpler and subject to fewer errors than the energy budget approach.

Durbin and Durbin[37] employed the population dynamics method to investigate copepod production in Narragansett Bay. Adapting the mathematical formulations of Ricker,[185,186] they calculated instantaneous growth rates from

$$\frac{W_t}{W_0} = e^{Gt} \tag{7}$$

where W_t is the weight of the organism at time t, and W_0 is its weight at time 0. Instantaneous mortality rates were derived from

$$\frac{N_t}{N_0} = e^{Zt} \tag{8}$$

where N_t equals the number of organisms present at time t, and N_0 represents the number present at time 0. If G and Z do not change, the production (P) of a population during a time interval can be determined from the equation

$$P = G\bar{B} \tag{9}$$

where G gives the instantaneous rate of growth and $\bar{B}$ specifies the mean biomass during the interval of time, and where

$$\bar{B} = \int_{t=0}^{t=1} B_0 e^{(G-Z)t} dt \tag{10}$$

$$= \frac{B_0\left(e^{(G-Z)} - 1\right)}{(G-Z)} \tag{11}$$

where B_0 is the biomass of the population at time 0. By neglecting mortality (i.e., $Z = 0$) in this equation, Durbin and Durbin[37] were able to estimate the potential production (P^1) of two dominant copepods (i.e., *Acartia hudsonica* and *A. tonsa*) in Narragansett Bay. A second estimate of production was made using a vertical life table that enabled the calculation of an apparent mortality rate Z^1. Z^1 yielded an estimate of the significance of mortality.

Applying Equations 7 to 11, Durbin and Durbin[37] calculated potential production rates of *A. hudsonica* in the lower and upper bay amounting to a mean of 7.25 and 10.77 mg C per cubic meter per day, respectively. Higher potential production rates were realized for *A. tonsa* in both regions of the estuary ($\overline{X} = 19.0$ mg C per cubic meter per day in the lower bay; $\overline{X} = 22.9$ mg C per cubic meter per day in the upper bay). The authors ascribed the higher potential production rates of *A. tonsa* to elevated temperatures in summer when the species was dominant, in contrast to the lower temperatures of the spring when *A. hudsonica* attained peak abundance. They showed that these potential production measurements compared favorably to the production values for *A. clausi* in Jackles Lagoon during the period of maximum growth from April to July in 1973 (21 mg C per cubic meter per day) and 1974 (16.7 mg C per cubic meter per day).[40]

Heinle[187] ascertained copepod production in several subsections of the Patuxent River estuary from the mouth to its head for the period of July 1965 to July 1966. Annual production was based on totals for the holoplankton *A. tonsa*, *Eurytemora affinis*, and *Scottolana canadensis*. Secondary production estimates of these three forms in three zones along the longitudinal axis of the estuary were as follows:

Zone I (mouth)	= 1.89 g C per cubic meter per year
Zone II	= 2.77 g C per cubic meter per year
Zone III (upestuary)	= 4.33 g C per cubic meter per year

Miller[8] notes that best estimates of secondary production by area in temperate estuarine systems range from about 5 to 10 g C per square meter per year.

VI. SUMMARY AND CONCLUSIONS

The plankton communities of estuaries consist of bacteria (bacterioplankton), microscopic plants (phytoplankton), and diminutive animals (zooplankton). These organisms drift passively in estuarine waters, having at best only limited mobility. It has become axiomatic in estuarine and marine ecology that zooplankton are a vital part of the food webs of coastal and open oceanic waters; they convert plant to animal matter and serve as the principal herbivorous component by grazing on a large crop of phytoplankton.

Organismal size and length of planktonic existence constitute the two primary bases for classifying zooplankton populations. Three size categories — micro-, meso-, and macrozooplankton — are typically utilized in classification schemes of zooplankton. Those forms passing through a plankton net with a mesh size of 202 μm comprise the microzooplankton, and individuals retained by this net embody the mesozooplankton. Still larger members of the zooplankton community collected with plankton nets constructed of 505-μm mesh incorporate the macrozooplankton. Among the microzooplankton less than about 60 μm in size, protozoans and tintinnids, together with meroplanktonic larvae of benthic invertebrates and copepod nauplii, are most numerous. Important groups of the mesozooplankton include cladocerans, copepods, rotifers, and larger meroplankton. Copepods, especially, may completely overshadow other mesozooplankton of estuaries in terms of absolute abundance and biomass. The major fraction of the macrozooplankton commonly derives from four faunal groups, that is, the jellyfish group (hydromedusae, comb jellies, true jellyfishes), crustaceans (amphipods, isopods, mysid shrimp, true shrimp), polychaete worms, and insect larvae.

The classification of zooplankton by means of the duration of planktonic life creates two broad subdivisions — meroplankton and holoplankton. Meroplankton are those animals which live only a part of their life cycle in the plankton and largely consist of larvae of the benthic invertebrates, benthic chordates, and fishes. In contrast, holoplankton refer to those organisms which remain in the plankton their entire lives. Copepods, cladocerans, and rotifers exemplify a holoplanktonic existence, with copepods generally regarded as the dominant group of holoplankton in many estuarine systems. A third group of zooplankton, the tychoplankton, is frequently overlooked, but periodically contributes substantial numbers to the plankton. These demersal zooplankton occasionally enter the water column due to wave action, bottom currents, or bioturbation by the benthos. Some migrate into bottom waters to feed. Species of amphipods, isopods, cumaceans, and mysids are common members of the tychoplankton, and all represent potential food sources for omnivorous and carnivorous zooplankton as well as planktivorous fish.

The taxonomic list of organisms comprising the zooplankton of estuaries is extensive. Certain phyla (e.g., Protozoa, Cnidaria, Ctenophora, and Arthropoda), however, often are more conspicuous in the holoplankton. Characteristic meroplankton taxa of benthic invertebrates include those of the Annelida, Arthropoda, Bivalvia, and Tentaculata.

Some of the most detailed investigations of zooplankton communities have been conducted on estuaries along the East Coast of the U.S. Scientists have examined the species composition, spatial and temporal density patterns, and trophic relationships of zooplankton in many of these systems. A persistent successional pattern evident from this research involves the seasonal alternation of dominant zooplankton populations, notably copepods. Hence, in Narragansett Bay, RI, where copepods account for 80 to 95% of all zooplankton in some years, *Acartia hudsonica* predominates during the winter and spring and *A. tonsa*, during the summer and fall. Standing stocks peak in the early summer. Other numerically significant copepod species in this estuary are *Oithona* spp., *Parvocalanus crassirostris*, *Pseudocalanus minutus*, and *Saphirella* sp. Maximum densities of meroplankton develop in spring, summer, or fall. Grazing by the ctenophore, *Mnemiopsis leidyi*, is responsible for the acute decline in abundance and biomass of zooplankton in mid-summer, usually August.

A similar alternation of copepod dominants occurs in Long Island Sound; here, *A. hudsonica* predominates both in total numbers and biomass. However, the copepod *Temora longicornis* also becomes a dominant zooplankter in the spring and early summer. Two copepod assemblages inhabit the estuary: (1) a boreal assemblage typified by *A. hudsonica, T. longicornis,* and *Pseudocalanus* sp.; and (2) a subtropical assemblage characterized by the dominate populations *A. tonsa, P. crassirostris, Oithona* sp., and *Labidocera aestiva*. The boreal assemblage reaches maximum concentrations in April and May, but the number of boreal copepods decreases in June and disappears by the end of July. The subtropical assemblage, in turn, increases in abundance in summer, generally about mid-July. This assemblage maintains year-round occupancy of the estuary.

In Barnegat Bay, NJ, *A. hudsonica* is the most abundant holoplankter during the winter, with *A. tonsa* and *O. colcarva* predominating in the summer. Microzooplankton generally display peak numbers in the spring and fall. Pulses of meroplanktonic larvae, in part, cause high densities of microzooplankton in the spring which gradually diminish through the summer season. Macrozooplankton likewise attain maximum concentrations during the spring and summer when *Rathkea octopunctata, Neomysis americana, Crangon septemspinosa, Neopanope texana, Jassa falcata, Sagitta* spp., and *Sarsia* spp. proliferate. *Mnemiopsis leidyi* and *Sagitta* spp. consume large quantities of microzooplankton, possibly regulating the population size of these smaller forms. Larval fishes have two annual abundance peaks, from January through April and from June to September. From January through April, larvae of the sand lance (*Ammodytes* sp.) and winter flounder (*Pseudopleuronectes americanus*) outnumber other ichthyoplankton. Eggs and larvae of the bay anchovy (*Anchoa mitchilli*) and larvae of gobies (*Gobiosoma* spp.) dominate from June to September.

The abundances of zooplankton in Chesapeake Bay are strongly salinity dependent. In the freshwater zone of the estuarine system, cladocerans (*Bosmina longirostris, Diaphanosoma leuchtenbergianum, Moina micrura*), the calanoid copepod *Eurytemora affinis,* and the cyclopoid copepod *Cyclops vernalis* dominate the zooplankton community. Mesohaline regions of the estuary have a lower diversity of forms than freshwater habitats; the most numerous zooplankter is the calanoid copepod *A. tonsa* which peaks in density between April and October. In higher salinity areas, *A. hudsonica* replaces *A. tonsa* as the principal copepod species in the winter and early spring, while *E. affinis* concurrently replaces *A. tonsa* in lower salinity regions. Peak zooplankton abundance in the Patuxent River estuary, a tributary system of Chesapeake Bay, arises in March or April. Once again reproduction by benthic populations in the spring add meroplankton larvae to the total zooplankton count. Abundance patterns of meroplankton circumscribe bimodal curves, with maximum values taking place in the spring and secondary maxima in the fall. Phytoplankton blooms sequentially followed by microzooplankton and mesozooplankton peaks delineate the chain of predator-prey relationships prevalent in the bay.

As in other East Coast estuaries of the U.S., copepods in North Inlet, SC, form the largest part of the zooplankton community, accounting for 64 to 69% of all the animals. The most common copepods in decreasing numerical importance are *P. crassirostris, A. tonsa, O. colcarva,* and *Euterpina acutifrons.* Maximum zooplankton numbers extend from June through mid-July.

The zooplankton community of San Francisco Bay differs markedly from that of East Coast estuaries. This bay is actually structured by two estuarine systems (i.e., San Pablo Bay and South Bay) that harbor distinctive zooplankton assemblages. *Acartia* spp. inhabit San Pablo Bay, with *A. clausi* s. l. found there nearly year-round. Copepods undergo a succession of forms in South Bay. Highest densities of *A. clausi* s. l. (cold-water form) are recorded during the wet season, and maximum densities of *A. californiensis* (warm-water form) during the dry season. *Oithona davisae* replaces *A. californiensis* as the dominant zooplankton in the fall. San Pablo Bay and South Bay also are unalike in respect to abundances of rotifers, tintinnids, and meroplankton.

A number of biotic and abiotic factors affect zooplankton dynamics and community structure. Light is a major environmental factor regulating diel vertical migration of these organisms. Acting as an environmental cue which triggers the migration process, illumination changes at sunrise and sunset elicit vertical movements of zooplankton populations. Three general vertical migration patterns are recognized, specifically nocturnal, twilight, and reverse migration. The following light cues may precipitate a vertical migration: (1) a change in depth of a specific light intensity, (2) a change in underwater spectra, (3) a change in the polarized light pattern, (4) an absolute amount of change in light intensity, and (5) a relative rate of intensity change.

Metabolic rates of zooplankton are a function of temperature, which influences the growth, fecundity, longevity, as well as other life processes of zooplankton. Because higher temperatures accelerate molting with less growth occurring during intermolt periods, susceptible zooplankton inhabiting warmer waters usually grow to a smaller size than individuals living in cooler regions. Fecundity is lower in adults of smaller size; therefore, lower temperature favors zooplankton with greater fecundity. In regard to ecological effects attributable to temperature changes, seasonal variations in the species composition and abundance of zooplankton in estuaries are well documented, particularly in temperate waters. The community of zooplankton in mid-Atlantic Bight estuaries, for example, experiences a succession of species populations in which winter-spring dominants (e.g., *A. hudsonica*) are replaced by summer-fall dominants (e.g., *A. tonsa*). From spring to fall, increased spawning of benthic invertebrates triggered by temperature changes occasionally results in pulses of meroplankton that often contribute to peak concentrations of zooplankton during these seasons.

Both holoplankton and meroplankton are responsive to salinity levels encountered along the longitudinal and vertical axes of estuarine systems. The salinity tolerance of zooplankton varies among species and may change among ontogenetic stages of a single species. As a result, the patterns of species succession and dominance along the longtiudinal axis of an estuary are contingent upon salinity concentrations. In the vertical plane, salinity limitations may restrain the vertical migration of zooplankton populations. Salinity extremes often arrest growth and increase mortality of susceptible forms.

Because zooplankton have only limited capabilities of locomotion, their position in an estuary greatly depends on water circulation patterns. As the magnitude of river inflow and tidal currents rises, zooplankton are easily displaced in the estuary and can be advected out of the estuarine basin to the nearshore ocean. Holoplankton or meroplankton transported out of the lower end of an estuary may be permanently lost from the system. Within the estuary, alterations of current patterns lead to a heterogeneous distribution of zooplankton populations. Some zooplankters have adaptive strategies that either increase the probability of retention in the estuary, or enhance the likelihood of dispersal to the nearshore ocean.

Zooplankton have an essential role in estuarine food chains as a intermediate link between primary producers (i.e., phytoplankton) and secondary consumers. The grazing of zooplankton or phytoplankton provides energy for higher-trophic-level organisms; however, many zooplankters are omnivorous which complicates the structure of estuarine food webs. Observations of copepods indicate that they feed by means of a passive filtration process of small particles or by active capture of large particles. In the active capture of prey, copepods initially detect a food particle by chemo- or mechanoreception, and, subsequently, they seize it. Feeding appears to be much more sophisticated among copepods than once believed; until recently, consumption was considered to entail indiscriminate filter feeding or simple sieving of phytoplankton. Some copepods clearly display selective feeding behavior. Some workers have advocated a coupling between diel vertical migration behavior and feeding activity of these organisms.

The rate of ingestion of zooplankton increases with increasing prey density up to a limit above which the ingestion rate is nearly constant. This relationship can be described by using a number of models, such as the Ivlev equation and Michaelis-Menten equation. A feeding threshold of zooplankton also exists below which phytoplankton are no longer exploited. Feeding thresholds preclude the elimination of phytoplankton populations via grazing pressure. While zooplankton ingest large amounts of food, only a portion is assimilated, with the remainder being egested. Assimilation efficiencies for zooplankton measured by several different methods often exceed 50% and may equal 99% in some cases.

Secondary production of zooplankton — the formation of new biomass within a specified period of time — has been calculated for a number of estuarine systems. Methods used to determine secondary production include the population dynamics approach and the energy budget approach. In temperate estuaries, secondary production generally ranges from 5 to 10 g C per square meter per year.

REFERENCES

1. **Kennish, M. J.,** Ed., *Practical Handbook of Marine Science,* CRC Press, Boca Raton, FL, 1989, Sect. 8.
2. **Gross, M. G.,** *Oceanography: A View of the Earth,* 3rd ed., Prentice-Hall, Englewood Cliffs, NJ, 1982.
3. **Heinle, D. R.,** Zooplankton, in *Functional Adaptations of Marine Organisms,* Vernberg, F. J. and Vernberg, W. B., Eds., Academic Press, New York, 1981, 85.
4. **Raymont, J. E. G.,** *Plankton and Productivity in the Oceans,* Vol. 1, 2nd ed., Pergamon Press, Oxford, 1980.
5. **Levinton, J. S.,** *Marine Ecology,* Prentice-Hall, Englewood Cliffs, NJ, 1982.
6. **Nybakken, J. W.,** *Marine Biology: An Ecological Approach,* Harper & Row, New York, 1982.
7. **McConnaughey, B. H. and Zottoli, R.,** *Introduction to Marine Biology,* C. V. Mosby, St. Louis, 1983.

8. **Miller, C. B.,** The zooplankton of estuaries, in *Estuaries and Enclosed Seas,* Ketchum, B. H., Ed., Elsevier, Amsterdam, 1983, 103.
9. **Omori, M. and Ikeda, T.,** *Methods in Marine Zooplankton Ecology,* John Wiley & Sons, New York, 1984.
10. **Steidinger, K. A. and Walker, L. M.,** Eds., *Marine Plankton Life Cycle Strategies,* CRC Press, Boca Raton, FL, 1984.
11. **Valiela, I.,** *Marine Ecological Processes,* Springer-Verlag, New York, 1984.
12. **Biological Methods Panel Committee on Oceanography,** *Recommended Procedures for Measuring the Productivity of Plankton Standing Stock and Related Oceanic Properties,* National Academy of Sciences, Washington, D. C., 1969.
13. **Lippson, A. J., Haire, M. S., Holland, A. F., Jacobs, F., Jensen, J., Moran-Johnson, R. L., Polgar, T. T., and Richkus, W. A.,** *Environmental Atlas of the Potomac Estuary,* Johns Hopkins University Press, Baltimore, 1981.
14. **Hulsizer, E. E.,** Zooplankton of lower Narragansett Bay, 1972—1973, *Chesapeake Sci.,* 17, 260, 1976.
15. **Hargraves, P. E.,** Seasonal variations of tintinnids (Ciliophore: Oligotrichidae) in Narragansett Bay, Rhode Island, U.S.A., *J. Plankton Res.,* 3, 81, 1981.
16. **Henroth, L.,** Marine pelagic rotifers and tintinnids — important trophic links in the spring plankton community of the Gullmar Fjord, Sweden, *J. Plankton Res.,* 5, 835, 1983.
17. **Verity, P. G.,** Grazing of phototrophic nanoplankton by microzooplankton in Narragansett Bay, *Mar. Ecol. Prog. Ser.,* 29, 105, 1986.
18. **Verity, P. G.,** Growth rates of natural tintinnid populations in Narragansett Bay, *Mar. Ecol. Prog. Ser.,* 29, 117, 1986.
19. **Roman, M. R.,** Ingestion of detritus and microheterotrophs by pelagic marine zooplankton, *Bull. Mar. Sci.,* 35, 477, 1984.
20. **Smetacek, V.,** The annual cycle of protozooplankton in the Kiel Bight, *Mar. Biol.,* 63, 1, 1981.
21. **Berk, S. G., Brownlee, D. C., Heinle, D. R., Kling, H. J., and Colwell, R. R.,** Ciliates as a food source for marine planktonic copepods, *Microb. Ecol.,* 4, 27, 1977.
22. **Heinbokel, J. F. and Beers, J. R.,** Studies on the functional role of tintinnids in the southern California Bight. III. Grazing impact of natural assemblages, *Mar. Biol.,* 52, 23, 1979.
23. **Stoecker, D. K. and Sanders, N. K.,** Differential grazing by *Acartia tonsa* on a dinoflagellate and a tintinnid, *J. Plankton Res.,* 7, 85, 1985.
24. **Capriulo, G. M. and Carpenter, E. J.,** Abundance, species composition, and feeding impact of tintinnid microzooplankton in central Long Island Sound, *Mar. Ecol. Prog. Ser.,* 10, 277, 1983.
25. **Verity, P. G.,** The Physiology and Ecology of Tintinnids in Narragansett Bay, Rhode Island, Ph.D. thesis, University of Rhode Island, Kingston, 1984.
26. **Robertson, J. R.,** Predation by estuarine zooplankton on tintinnid ciliates, *Estuarine Coastal Shelf Sci.,* 16, 27, 1983.
27. **Stoecker, D. K. and Evans, G. T.,** Effects of protozoan herbivory and carnivory in a microplankton food web, *Mar. Ecol. Prog. Ser.,* 25, 159, 1985.
28. **Beers, J. R., Stewart, G. L., and Hoskins, K. D.,** Dynamics of micro-zooplankton populations treated with copper: controlled ecosystem pollution experiment, *Bull. Mar. Sci.,* 27, 66, 1977.
29. **Ambler, J. W., Cloern, J. E., and Hutchinson, A.,** Seasonal cycles of zooplankton from San Francisco Bay, *Hydrobiologia,* 129, 177, 1985.
30. **Davis, C. C.,** Planktonic Copepoda (including Monstrilloida), in *Marine Plankton Life Cycle Strategies,* Steidinger, K. A. and Walker, L. M., Eds., CRC Press, Boca Raton, FL, 1984, 67.
31. **Mann, K. H.,** *Ecology of Coastal Waters: A Systems Approach,* University of California Press, Berkeley, 1982.
32. **Odhner, M. X.,** Life Histories and Nutrition of Copepods at Calvert Cliffs, Technical Report for Baltimore Gas and Electric Company, Academy of Natural Sciences, Philadelphia, 1977.
33. **Durbin, E. G., Durbin, A. G., Smayda, T. J., and Verity, P. G.,** Food limitation of production by adult *Acartia tonsa* in Narragansett Bay, Rhode Island, *Limnol. Oceanogr.,* 28, 1199, 1983.
34. **Daro, M. H.,** Feeding rhythms and vertical distribution of marine copepods, *Bull. Mar. Sci.,* 37, 487, 1985.
35. **Smayda, T. J.,** The phytoplankton of estuaries, in *Estuaries and Enclosed Seas,* Ketchum, B. H., Ed., Elsevier, Amsterdam, 1983, 65.
36. **Kremer, P.,** Predation by the ctenophore *Mnemiopsis leidyi* in Narragansett Bay, Rhode Island, *Estuaries,* 2, 97, 1979.
37. **Durbin, A. G. and Durbin, E. G.,** Standing stock and estimated production rates of phytoplankton and zooplankton in Narragansett Bay, R. I., *Estuaries,* 4, 24, 1981.
38. **Deason, E. E. and Smayda, T. J.,** Ctenophore-zooplankton interactions in Narragansett Bay, Rhode Island, USA, during 1972—1977, *J. Plankton Res.,* 4, 203, 1982.
39. **Larson, R. J.,** In situ feeding rates of the ctenophore *Mnemiopsis mccradyi, Estuaries,* 10, 87, 1987.
40. **Landry, M. R.,** Population dynamics and production of a planktonic marine copepod, *Acartia clausi,* in a small temperate lagoon on San Juan Island, Washington, *Int. Rev. Gesamten Hydrobiol.,* 63, 77, 1978.

41. **Johnson, J. K.,** Population Dynamics and Cohort Persistence of *Acartia californiensis* (Copepoda: Calanoida) in Yaquina Bay, Oregon, Ph.D. thesis, Oregon State University, Corvallis, 1980.

42. **Deason, E. E.,** *Mnemiopsis leidyi* (Ctenophora) in Narragansett Bay, 1975—79: abundance, size composition and estimation of grazing, *Estuarine Coastal Shelf Sci.,* 15, 121, 1982.

43. **Stearns, D. E. and Forward, R. B., Jr.,** Photosensitivity of the calanoid copepod *Acartia tonsa, Mar. Biol.,* 82, 85, 1984.

44. **Stearns, D. E. and Forward, R. B., Jr.,** Copepod photobehavior in a simulated natural light environment and its relation to nocturnal vertical migration, *Mar. Biol.,* 82, 91, 1984.

45. **Omori, M. and Hamner, W. M.,** Patchy distribution of zooplankton: behavior, population assessment and sampling problems, *Mar. Biol.,* 72, 193, 1982.

46. **Ueda, H.,** Temporal and spatial distribution of the two closely related *Acartia* species *A. omorii* and *A. hudsonica* (Copepoda, Calanoida) in a small inlet water of Japan, *Estuarine Coastal Shelf Sci.,* 24, 691, 1987.

47. **Lippson, A. J. and Lippson, R. L.,** *Life in the Chesapeake Bay,* Johns Hopkins University Press, Baltimore, MD, 1984.

48. **Stickney, R. R.,** *Estuarine Ecology of the Southeastern United States and Gulf of Mexico,* Texas A & M University Press, College Station, 1984.

49. **Perkins, E. J.,** *Biology of Estuaries and Coastal Waters,* Academic Press, London, 1974.

50. **Vance, R. R.,** On reproductive strategies in marine benthic invertebrates, *Am. Nat.,* 107, 339, 1973.

51. **Chia, F. S.,** Classification and adaptive significance of developmental patterns in marine invertebrates, *Thalassia Jugosl.,* 10, 121, 1974.

52. **Emlet, R. B.,** Facultative planktotrophy in the tropical echinoid *Clypeaster rosaceus* (Linnaeus) and a comparison with obligate planktotrophy in *Clypeaster subdepressus* (Gray) (Clypeasteroida: Echinoidea), *J. Exp. Mar. Biol. Ecol.,* 95, 183, 1986.

53. **Day, R. and McEdward, L.,** Aspects of the physiology and ecology of pelagic larvae of marine benthic invertebrates, in *Marine Plankton Life Cycle Strategies,* Steidinger, K. A. and Walker, L. M., Eds., CRC Press, Boca Raton, FL, 1984, 93.

54. **Rumrill, S. S., Pennington, J. T., and Chia, F. S.,** Differential susceptibility of marine invertebrate larvae: laboratory predation of sand dollar, *Dendraster excentricus* (Eschscholtz), embryos and larvae by zoeae of the red crab, *Cancer productus* Randall, *J. Exp. Mar. Biol. Ecol.,* 90, 193, 1985.

55. **Jablonski, D. and Lutz, R. A.,** Larval ecology of marine benthic invertebrates: paleobiological implications, *Biol. Rev.,* 58, 21, 1983.

56. **Mileikovsky, S. A.,** Types of larval development in marine bottom invertebrates, their distribution and ecological significance: a reevaluation, *Mar. Biol.,* 10, 193, 1971.

57. **Mileikovsky, S. A.,** Types of larval development in marine bottom invertebrates: an integrated ecological scheme, *Thalassia Jugosl.,* 10, 171, 1974.

58. **Crisp, D. J.,** The role of the pelagic larva, in *Perspectives in Experimental Zoology,* Spencer-Davies, T., Ed., Pergamon Press, Oxford, 1976, 145.

59. **Thorson, G.,** Reproduction and larval ecology of marine bottom invertebrates, *Biol. Bull.,* 25, 1, 1950.

60. **Leiby, M. M.,** Life history and ecology of pelagic fish eggs and larvae, in *Marine Plankton Life Cycle Strategies,* Steidinger, K. A. and Walker, L. M., Eds., CRC Press, Boca Raton, FL, 1984, 121.

61. **Fortier, L. and Leggett, W. C.,** A drift study of larval fish survival, *Mar. Ecol. Prog. Ser.,* 25, 245, 1985.

62. **Norcross, B. L. and Shaw, R. F.,** Oceanic and estuarine transport of fish eggs and larvae: a review, *Trans. Am. Fish. Soc.,* 113, 153, 1984.

63. **Galtsoff, P. S.,** The American oyster *Crassostrea virginica* Gmelin, *Fish. Bull. (U.S.),* Vol. 64, 1964.

64. **Davis, H. C. and Chanley, P. E.,** Spawning and egg production of oysters and clams, *Biol. Bull.,* 110, 117, 1956.

65. **Bricelj, V. M. and Malouf, R. E.,** Aspects of reproduction of hard clams (*Mercenaria mercenaria*) in Great South Bay, New York, *Proc. Natl. Shellfish. Assoc.,* 70, 216, 1980.

66. **Carriker, M. R.,** Interrelation of functional morphology, behavior and autecology in early stages of the bivalve *Mercenaria mercenaria, J. Elisha Mitchell Sci. Soc.,* 77, 168, 1961.

67. **Sandine, P. H.,** Zooplankton, in *Ecology of Barnegat Bay, New Jersey,* Kennish, M. J. and Lutz, R. A., Eds., Springer-Verlag, New York, 1984, 95.

68. **Jeffries, H. P. and Johnson, W. C.,** Distribution and abundance of zooplankton, in *Coastal and Offshore Environmental Inventory: Cape Hatteras to Nantucket Shoals,* Marine Publ Ser. No. 2, University of Rhode Island, Kingston, 1973, 4-1.

69. **Grindley, J. R.,** Estuarine plankton, in *Estuarine Ecology: With Particular Reference to Southern Africa,* Day, J. H., Ed., A. A. Balkema, Rotterdam, 1981, 117.

70. **Gilbert, J. J. and Williamson, C. E.,** Sexual dimorphism in zooplankton (Copepoda, Cladocera, and Rotifera), *Annu. Rev. Ecol. Syst.,* 14, 1, 1983.

71. **Fulton, R. S., III,** Predation, production and the organization of an estuarine copepod community, *J. Plankton Res.,* 6, 399, 1984.

72. **Hobson, E. S.,** Diurnal-nocturnal activity of some inshore fishes in the Gulf of California, *Copeia,* 1965, 291, 1965.

73. **Robertson, A. I. and Howard, R. K.,** Diel trophic interactions between vertically-migrating zooplankton and their fish predators in an eelgrass community, *Mar. Biol.,* 48, 207, 1978.

74. **Alldredge, A. L. and King, J. M.,** The distance demersal zooplankton migrate above the benthos: implications for predation, *Mar. Biol.,* 84, 253, 1985.

75. **Marcus, N. H.,** Recruitment of copepod nauplii into the plankton: importance of diapause eggs and benthic processes, *Mar. Ecol. Prog. Ser.,* 15, 47, 1984.

76. **Marcus, N. H. and Schmidt-Gengenbach, J.,** Recruitment of individuals into the plankton: the importance of bioturbation, *Limnol. Oceanogr.,* 31, 206, 1986.

77. **Corliss, J. O.,** *The Ciliated Protozoa: Characterization, Classification and Guide to the Literature,* 2nd ed., Pergamon Press, Elmsford, NY, 1979.

78. **Kremer, P. and Nixon, S. W.,** *A Coastal Marine Ecosystem: Simulation and Analysis,* Ecological Studies 24, Springer-Verlag, Heidelberg, 1978.

79. **Martin, J. H.,** Phytoplankton-zooplankton relationships in Narragansett Bay, *Limnol. Oceanogr.,* 10, 185, 1965.

80. **Kremer, P. and Nixon, S. W.,** Distribution and abundance of the ctenophore, *Mnemiopsis leidyi,* in Narragansett Bay, *Estuarine Coastal Mar. Sci.,* 4, 627, 1976.

81. **Martin, J. H.,** Phytoplankton-zooplankton relationships in Narragansett Bay. III. Seasonal changes in zooplankton excretion rates in relation to phytoplankton abundance, *Limnol. Oceanogr.,* 13, 63, 1968.

82. **Peterson, W. T.,** Abundance, age structure and in situ egg production rates of the copepod *Temora longicornis* in Long Island Sound, New York, *Bull. Mar. Sci.,* 37, 726, 1985.

83. **Tiselius, P. T. and Peterson, W. T.,** Life history and population dynamics of the chaetognath *Sagitta elegans* in central Long Island Sound, *J. Plankton Res.,* 8, 183, 1986.

84. **Deevey, G. B.,** Oceanography of Long Island Sound, 1952—1954. V. Zooplankton, *Bull. Bingham Oceanogr. Coll.,* 15, 113, 1956.

85. **Conover, S. A.,** Oceanography of Long Island Sound, 1952—1954. IV. Biology of *Acartia clausi* and *A. tonsa,* *Bull. Bingham Oceanogr. Coll.,* 15, 156, 1956.

86. **Kennish, M. J.,** Summary and conclusions, in *Ecology of Barnegat Bay, New Jersey,* Kennish, M. J. and Lutz, R. A., Eds., Springer-Verlag, New York, 1984, 339.

87. **Bridges, J. P.,** On day and night catches of fish larvae, *J. Cons. Perm. Int. Explor. Mer.,* 22, 42, 1956.

88. **Daiber, F. C.,** Tidal creeks and fish eggs, *Estuarine Bull.,* 7, 6, 1963.

89. **Isaacs, J. D.,** Night-caught and day-caught larvae of the California Sardine, *Science,* 144, 1132, 1964.

90. **Brownlee, D. C. and Jacobs, F.,** Mesozooplankton and microzooplankton in the Chesapeake Bay, in *Contaminant Problems and Management of Living Chesapeake Bay Resources,* Majumdar, S. K., Hall, L. W., Jr., and Austin, H. M., Eds., Pennsylvania Academy of Science, Easton, 1987, 217.

91. **Jacobs, F., Burton, W., and Moss, I.,** Patterns of zooplankton abundance, species composition, and biomass in upper Chesapeake Bay, *Estuaries,* 8, 78A, 1985.

92. **Heinle, D. R.,** Free-living Copepoda of the Chesapeake Bay, *Chesapeake Sci.,* 13 (Suppl.), S117, 1972.

93. **Olson, M. M.,** Zooplankton, in *Ecological Studies in the Middle Reach of Chesapeake Bay: Calvert Cliffs,* Heck, K. L., Jr., Ed., Springer-Verlag, Berlin, 1987, 38.

94. **Butt, A. J., Alden, R. W., III, and Young, R. J.,** Commercially Important Meroplankton of the Lower Chesapeake Bay and Proposed Norfolk Disposal Site. 1. Blue Crabs, Rock Crabs, and Oysters, Unpubl. Tech. Rep., Old Dominion University, Norfolk, 1985.

95. **Lonsdale, D. J. and Coull, B. C.,** Composition and seasonality of zooplankton of North Inlet, South Carolina, *Chesapeake Sci.,* 18, 272, 1977.

96. **Forward, R. B., Jr.,** Diel vertical migration: zooplankton photobiology and behaviour, *Oceanogr. Mar. Biol. Annu. Rev.,* 26, 361, 1988.

97. **Forward, R. B., Jr.,** A reconsideration of the shadow response of a larval crustacean, *Mar. Behav. Physiol.,* 12, 99, 1986.

98. **Stearns, D. E.,** Copepod grazing behavior in simulated natural light and its relation to nocturnal feeding, *Mar. Ecol. Prog. Ser.,* 30, 65, 1986.

99. **Swift, M. C. and Forward, R. B., Jr.,** Absolute light intensity vs. rate of relative change in light intensity: the role of light in the vertical migration of *Chaoborus punctipennis* larvae, *Bull. Mar. Sci.,* 43, 604, 1988.

100. **Forward, R. B., Jr.,** Behavioral responses of larvae of the crab *Rhithropanopeus harrisii* (Brachyura: Xanthidae) during diel vertical migration, *Mar. Biol.,* 90, 9, 1985.

101. **Forward, R. B., Jr., Cronin, T. W., and Stearns, D. E.,** Control of diel vertical migration: photoresponses of a larval crustacean, *Limnol. Oceanogr.,* 29, 146, 1984.

102. **Davis, C. S.,** Interaction of a copepod population with the mean circulation on Georges Bank, *J. Mar. Res.,* 42, 573, 1984.

103. **Jacobs, J.,** Animal behaviour and water movement as co-determinants of plankton distribution in a tidal system, *Sarsia*, 34, 355, 1968.

104. **Stickney, R. R. and Knowles, S. C.,** Summer zooplankton distribution in a Georgia estuary, *Mar. Biol.*, 33, 147, 1975.

105. **Longhurst, A. R.,** Vertical migration, in *The Ecology of the Seas*, Cushing, D. H. and Walsh, J. J., Eds., Blackwell Scientific, Oxford, 1976, 116.

106. **McLaren, I. A.,** Effects of temperature on growth of zooplankton, and the adaptive value of vertical migration, *J. Fish. Res. Board Can.*, 26, 199, 1963.

107. **Fortier, L. and Leggett, W. C.,** Vertical migrations and transport of larval fish in a partially mixed estuary, *Can. J. Fish. Aquat. Sci.*, 40, 1543, 1983.

108. **Heinle, D. R.,** Temperature and zooplankton, *Chesapeake Sci.*, 10, 186, 1969.

109. **McLaren, I. A.,** Demographic strategy of vertical migration by a marine copepod, *Am. Nat.*, 108, 91, 1974.

110. **McLaren, I. A.,** Predicting development rate of copepod eggs, *Biol. Bull.*, 131, 457, 1966.

111. **Deevey, G. B.,** Relative effects of temperature and food on seasonal variations in length of marine copepods in some eastern American and western European waters, *Bull. Bingham Oceanogr. Coll.*, 17, 55, 1960.

112. **Deevey, G. B.,** Seasonal variations in length of copepods in South Pacific New Zealand waters, *Aust. J. Mar. Freshwater Res.*, 17, 155, 1966.

113. **Landry, M. R.,** Population Dynamics of the Planktonic Marine Copepod *Acartia clausi* Giesbrecht in a Small Temperate Lagoon, Ph.D. thesis, University of Washington, Seattle, 1976.

114. **Durbin, E. G. and Durbin, A. G.,** Length and weight relationships of *Acartia clausi* from Narragansett Bay, R. I., *Limnol. Oceanogr.*, 23, 958, 1978.

115. **Johnson, J. K.,** Effects of temperature and salinity on the production and hatching of dormant eggs of *Acartia californiensis* (Copepoda) in an Oregon estuary, *Fish. Bull. (U.S.)*, 77, 567, 1979.

116. **Haedrich, R. L.,** Estuarine fishes, in *Estuaries and Enclosed Seas*, Ketchum, B. H., Ed., Elsevier, Amsterdam, 1983, 183.

117. **Van Vaupel-Klein, J. C. and Weber, R. E.,** Distribution of *Eurytemora affinis* (Copepoda: Calanoida) in relation to salinity: field and laboratory observations, *Neth. J. Sea Res.*, 9, 297, 1975.

118. **Lance, J.,** The salinity tolerance of some estuarine planktonic copepods, *Limnol. Oceanogr.*, 8, 440, 1963.

119. **Lance, J.,** Respiration and osmotic behavior of the copepod *Acartia tonsa* in diluted sea water, *Comp. Biochem. Physiol.*, 14, 15, 1965.

120. **Lance, J.,** The salinity tolerance of some estuarine planktonic Crustacea, *Biol. Bull.*, 127, 108, 1964.

121. **Heinle, D. R.,** Production of a calanoid copepod, *Acartia tonsa*, in the Patuxent River estuary, *Chesapeake Sci.*, 7, 59, 1966.

122. **Zillioux, E. J. and Wilson, D. F.,** Culture of a planktonic calanoid copepod through multiple generations, *Science*, 151, 996, 1966.

123. **Johnson, J. K.,** The Dynamics of an Isolated Population of *Acartia tonsa* Dava (Copepoda) in Yaquina Bay, Oregon, M.S. thesis, Oregon State University, Corvallis, 1974.

124. **Wang, J. C. S. and Kernehan, R. J.,** *Fishes of the Delaware Estuaries. A Guide to the Early Life History Stages*, E. A. Communications, Towson, MD, 1979.

125. **Little, M. C., Reay, P. J., and Grove, S. J.,** Distribution gradients of ichthyoplankton in an East African mangrove creek, *Estuarine Coastal Shelf Sci.*, 26, 669, 1988.

126. **Tabb, D. C.,** The estuary as a habitat for spotted seatrout, *Cynoscion nebulosus*, *Am. Fish. Soc. Spec. Publ.*, 3, 37, 1966.

127. **McLusky, D. S.,** *The Estuarine Ecosystem*, Halsted Press, New York, 1981.

128. **Haury, L. R., McGowan, J. A., and Wiebe, P. H.,** Patterns and processes in the time-space scales of plankton distributions, in *Spatial Pattern in Plankton Communities*, Steele, J. H., Ed., Plenum Press, New York, 1978, 277.

129. **Wroblewski, J. S.,** A simulation of the distribution of *Acartia clausi* during the Oregon USA upwelling August, 1973, *J. Plankton Res.*, 2, 46, 1980.

130. **Wroblewski, J. S.,** Interaction of currents and vertical migration in maintaining *Calanus marshallae* in the Oregon upwelling zone — a simulation, *Deep Sea Res.*, 29, 665, 1982.

131. **Bousfield, E. L.,** Ecological control of the occurrence of barnacles in the Miramichi estuary, *Bull. Natl. Museum Can.*, 131, 1, 1955.

132. **Pearcy, W. G.,** Ecology of an estuarine population of winter flounder, *Pseudopleuronectes americanus* (Walbaum). II. Distribution and dynamics of larvae, *Bull. Bingham Oceanogr. Coll.*, 18, 16, 1962.

133. **Riley, G. A.,** The plankton of estuaries, in *Estuaries*, Lauff, G. H., Ed., Publ. 83, American Association for the Advancement of Science, Washington, D. C., 1967, 316.

134. **Kennedy, V. S.,** Ed., *Estuarine Comparisons*, Academic Press, New York, 1982.

135. **Weinstein, M. P., Weiss, S. L., Hodson, R. G., and Gerry, L. R.,** Retention of three taxa of post larval fishes in an intensively flushed tidal estuary, Cape Fear River, North Carolina, *Fish. Bull. (U.S.)*, 78, 419, 1980.

136. **Provenzano, A. J., McConaugha, J. R., Phillips, K. B., Johnson, D. R., and Clark, J.,** Diurnal vertical distribution of first stage larvae of the blue crab, *Callinectes sapidus,* at the mouth of the Chesapeake Bay, *Estuarine Coastal Shelf Sci.,* 16, 489, 1983.

137. **Johnson, D. R., Hester, B. S., and McConaugha, J. R.,** Studies of a wind mechanism influencing the recruitment of blue crabs in Middle Atlantic Bight, *Cont. Shelf Res.,* 3, 425, 1984.

138. **Beckley, L. E.,** Tidal exchange of ichthyoplankton in the Swartkops estuary mouth, South Africa, *S. Afr. Tydskr. Dierk,* 20, 15, 1985.

139. **Blaber, S. J. M. and Blaber, T. G.,** Factors affecting the distribution of juvenile estuarine and inshore fish, *J. Fish. Biol.,* 17, 143, 1980.

140. **Whitfield, A. K.,** Factors influencing the recruitment of juvenile fishes into the Mhlanga estuary, *S. Afr. J. Zool.,* 15, 166, 1980.

141. **Whitfield, A. K.,** Factors influencing the utilization of southern African estuaries by fishes, *S. Afr. J. Sci.,* 79, 362, 1983.

142. **Hiscock, K.,** Water movement, in *Sublittoral Ecology: The Ecology of the Shallow Sublittoral Benthos,* Earll, R. and Erwin, D. G., Eds., Clarendon Press, Oxford, 1983, 58.

143. **Turner, J. T.,** The Feeding Ecology of Some Zooplankters That Are Important Prey Items of Larval Fish, NOAA Tech. Rep. NMFS 7, U.S. Department of Commerce, Washington, D. C., 1984, 28 pp.

144. **Conley, W. J. and Turner, J. T.,** Omnivory by the coastal marine copepods *Centropages hamatus* and *Labidocera aestiva, Mar. Ecol. Prog. Ser.,* 21, 113, 1985.

145. **Monteleone, D. M. and Peterson, W. T.,** Feeding ecology of American sand lance *Ammodytes americanus* larvae from Long Island Sound, *Mar. Ecol. Prog. Ser.,* 30, 133, 1986.

146. **Price, H. J., Paffenhöfer, G.-A., and Strickler, J. R.,** Modes of cell capture in calanoid copepods, *Limnol. Oceanogr.,* 28, 116, 1983.

147. **Price, H. J. and Paffenhöfer, G.-A.,** Effects of feeding experience in the copepod *Eucalanus pileatus:* a cinematographic study, *Mar. Biol.,* 84, 35, 1984.

148. **Price, H. J.,** Feeding mechanisms in marine and freshwater zooplankton, *Bull. Mar. Sci.,* 43, 327, 1988.

149. **Legier-Visser, M. F., Mitchell, J. G., Okubo, A., and Fuhrman, J. A.,** Mechanoreception in calanoid copepods: a mechanism for prey detection, *Mar. Biol.,* 90, 529, 1986.

150. **Lehman, J. T.,** Grazing, nutrient release, and their impacts on the structure of phytoplankton communities, in *Trophic Interactions within Aquatic Ecosystems,* Meyers, D. G. and Strickler, J. R., Eds., American Association for the Advancement of Science, Selected Symp. 85, Westview Press, Boulder, CO, 1984, 49.

151. **Strickler, J. R.,** Sticky water: a selective force in copepod evolution, in *Trophic Interactions within Aquatic Ecosystems,* Meyers, D. G. and Strickler, J. R., Eds., American Association for the Advancement of Science, Selected Symp. 85, Westview Press, Boulder, CO, 1984, 187.

152. **Harris, R. P.,** Comparison of the feeding behavior of *Calanus* and *Pseudocalanus* in two experimentally manipulated enclosed ecosystems, *J. Mar. Biol. Assoc. U.K.,* 62, 71, 1982.

153. **Boyd, C. M.,** Selection of particle sizes by filter-feeding copepods: a plea for reason, *Limnol. Oceanogr.,* 21, 175, 1976.

154. **Nival, P. and Nival, S.,** Particle retention efficiencies of an herbivorous copepod, *Acartia clausi* (adult and copepodite stages): effects on grazing, *Limnol. Oceanogr.,* 21, 24, 1976.

155. **Frost, B. W.,** Feeding behavior of *Calanus pacificus* in mixtures of food particles, *Limnol. Oceanogr.,* 22, 125, 1977.

156. **Gauld, D. T.,** The swimming and feeding of planktonic copepods, in *Some Contemporary Studies in Marine Science,* Barnes, H., Ed., Hafner, Darien, CT, 1966, 313.

157. **Koehl, M. A. R.,** Mechanisms of particle capture by copepods at low Reynolds numbers: possible modes of selective feeding, in *Trophic Interactions within Aquatic Ecosystems,* Meyers, D. G. and Strickler, J. R., American Association for the Advancement of Science, Selected Symp. 85, Westview Press, Boulder, CO, 1984, 135.

158. **Gauld, D. T.,** Diurnal variations in the grazing of planktonic copepods, *J. Mar. Biol. Assoc. U.K.,* 31, 461, 1951.

159. **Lampert, W. and Taylor, B. E.,** Zooplankton grazing in a eutrophic lake: implications of diel vertical migration, *Ecology,* 66, 68, 1985.

160. **Daro, M. H.,** Migratory and grazing behavior of copepods and vertical distribution of phytoplankton, *Bull. Mar. Sci.,* 43, 710, 1988.

161. **Mackas, D. L. and Bohrer, R.,** Fluorescence analysis of zooplankton gut contents and an investigation of diel feeding patterns, *J. Exp. Mar. Biol. Ecol.,* 25, 77, 1976.

162. **Arashkevich, Y. C.,** Relationship between the feeding rhythm and the vertical migrations of *Cypridina sinuosa* (Ostracoda, Crustacea) in the western part of the equatorial Pacific, *Oceanology,* 17, 466, 1977.

163. **Bohrer, R. N.,** Experimental studies on diel vertical migration in evolution and ecology of zooplankton communities, in *Evolution and Ecology of Zooplankton Communities,* Kerfoot, W. C., Ed., University Press, London, 1980, 111.

164. **Starkweather, P. L.,** Diel variation in feeding behavior in *Daphnia pulex:* influences of food density and nutritional history on mandibular activity, *Limnol. Oceanogr.,* 23, 307, 1978.

165. **Starkweather, P. L.,** Daily patterns of feeding behavior in *Daphnia* and related microcrustacea: implications for cladoceran autecology and the zooplankton community, *Hydrobiologia,* 100, 203, 1983.

166. **Duval, W. S. and Green, G. H.,** Diel feeding and respiration rhythms in zooplankton, *Limnol. Oceanogr.,* 21, 823, 1976.

167. **Haney, J. F.,** Diel patterns of zooplankton behavior, *Bull. Mar. Sci.,* 43, 583, 1988.

168. **Haney, J. F.,** Regulation of cladoceran filtering rates in nature by body size, food concentration, and diel feeding patterns, *Limnol. Oceanogr.,* 30, 397, 1985.

169. **Enright, J. T.,** Diurnal vertical migration: adaptive significance and timing. I. Selective advantage: a metabolic model, *Limnol. Oceanogr.,* 22, 856, 1977.

170. **Genin, A., Haury, L., and Greenblatt, P.,** Interactions of migrating zooplankton with shallow topography: predation by rockfishes and intensification of patchiness, *Deep Sea Res. Part A,* 35, 151, 1988.

171. **Parsons, T. R., LeBrasseur, R. T., and Fulton, J. D.,** Some observations on the dependence of zooplankton grazing on the cell size and concentration of phytoplankton blooms, *J. Oceanogr. Soc. Jpn.,* 23, 10, 1967.

172. **Parsons, T. R., LeBrasseur, R. T., Fulton, J. D., and Kennedy, O. D.,** Production studies in the Strait of Georgia. II. Secondary production under the Fraser River plume, February to May, 1967, *J. Exp. Mar. Biol. Ecol.,* 3, 39, 1969.

173. **McAllister, C. D.,** Zooplankton rations, phytoplankton mortality, and the estimation of marine production, in *Marine Food Chains,* Steele, J. H., Ed., University of California Press, Berkeley, 1970, 419.

174. **Steele, J. H.,** *The Structure of Marine Ecosystems,* Harvard University Press, Cambridge, 1974.

175. **Richman, S.,** The transformation of energy by *Daphnia pulex, Ecol. Monogr.,* 28, 273, 1958.

176. **Conover, R. J.,** Assimilation of organic matter by zooplankton, *Limnol. Oceanogr.,* 11, 338, 1966.

177. **Butter, E. I., Corner, E. D. S., and Marshall, S. M.,** On the nutrition and metabolism of zooplankton. VI. Feeding efficiency of *Calanus* in terms of nitrogen and phosphorus, *J. Mar. Biol. Assoc. U.K.,* 49, 997, 1969.

178. **Butter, E. I., Corner, E. D. S., and Marshall, S. M.,** On the nutrition and metabolism of zooplankton. VII. Seasonal survey of nitrogen and phosphorus excretion by *Calanus* in the Clyde Sea area, *J. Mar. Biol. Assoc. U.K.,* 50, 525, 1970.

179. **Parsons, T. R. and Takahashi, M.,** *Biological Oceanographic Processes,* Pergamon Press, Oxford, 1973.

180. **Petipa, T. S.,** Relationship between growth energy metabolism, and ration in *Acartia clausi* Giesbrecht, in *Physiology of Marine Animals,* Akad. Nauk SSR, Moscow, 1966, 82.

181. **Neilson, B. J. and Cronin, L. E.,** Eds., *Estuaries and Nutrients,* Humana Press, Clifton, NJ, 1981.

182. **Edmondson, W. T. and Winberg, G. G.,** Eds., *A Manual on Methods for the Assessment of Secondary Productivity in Fresh Waters,* Blackwell Scientific, Oxford, 1971.

183. **Winberg, G. G.,** *Methods for the Estimates of Production of Aquatic Animals,* Academic Press, New York, 1971.

184. **Greze, V. N.,** Production in animal populations, in *Marine Ecology,* Vol. 4, Kinne, O., Ed., John Wiley & Sons, New York, 1978.

185. **Ricker, W. E.,** Production and utilization of fish populations, *Ecol. Monogr.,* 16, 373, 1946.

186. **Ricker, W. E.,** Computation and interpretation of biological statistics of fish populations, *Fish. Res. Board Can.,* Vol. 91, 1975.

187. **Heinle, D. R.,** Effects of Temperature on the Population Dynamics of Estuarine Copepods, Ph.D. thesis, University of Maryland, College Park, 1969.

188. **Heinle, D. R. and Ross, M. A.,** unpublished.

189. **Paffenhöfer, G. -A.,** Calanoid copepod feeding: grazing on small and large particles, in *Trophic Interactions within Aquatic Ecosystems,* Meyers, D. G. and Strickler, J. R., Eds., American Association for the Advancement of Science, Selected Symp. 85, Westview Press, Boulder, CO, 1984, 75.

Chapter 4

BENTHOS

I. INTRODUCTION

The benthos refers to those organisms, both plant and animal, attached to, living on, in, or near the sea bed.[1] While phytoplankton are critical to the primary production of the open ocean, attached macrophytes (e.g., seagrasses, salt marsh grasses, and mangroves) dominate the autotrophic biomass in estuaries and shallow coastal waters and frequently the total production of these ecosystems as well.[2] In addition, they serve as habitat formers, supporting a high diversity of aquatic fauna. Among the benthic macrophytes of estuaries, algae generally contribute a smaller fraction of organic production than submergent and emergent vascular plants. However, algae may be key components of some systems, being present as single-cell, motile species commonly observed on the surfaces of tidal flats, multicellular forms attached to soft or hard substrates, and passively drifting populations unaffixed to the estuarine seafloor.[3] Benthic macroalgae usually prefer attachment on rocky shores, stones, shells, or man-made structures than on soft bottom sediments.[4] They are distributed along both horizontal and vertical gradients in response to light, temperature, salinity, and nutrient conditions in the environment.[2]

Benthic fauna have been broadly differentiated into populations residing on the seafloor or on a firm substrate (epifauna) and populations living in the sediment (infauna). Studies of bottom-dwelling invertebrates have progressed on two basic levels: (1) the species level, whereby data are gathered on the life history and population dynamics of the fauna; and (2) the community level, whereby an assemblage of organisms is assessed in terms of multiple variables, such as species composition, abundance, diversity, and productivity.[5] According to the community concept as recounted by Erwin,[6] the patterns of faunal and floral assemblages can be investigated using two approaches: (1) the level bottom approach in which assemblages are largely constant and related to the physical environment, primarily the substrate; and (2) the zonation approach applicable, in particular, to intertidal and shallow subtidal zones where assemblages are related to a physical gradient or a combination of physical gradients. As a result of the ebb and flow of the tides, intertidal habitats present a unique set of problems — for example, thermal and dessication stress, wave action and exposure, sediment instability, and variations in salinity and carbon dioxide[7] — that affect organismal distribution.

Physical and chemical factors in the estuarine environment clearly influence the functional morphology and behavior of the benthos.[8] Salinity changes along the longitudinal axis of an estuary modulate abundance and diversity of the benthos, although salinity profiles tend to be more stable in interstitial than overlying waters, and, consequently, the benthic infauna may be less impacted than the epifauna by salinity perturbations in the water column. The species composition of benthic communities depends greatly on the sediment type, which often varies appreciably within short distances. Fluctuations in other physical and chemical factors (e.g., dissolved oxygen, temperature, turbidity, wave action, and turbulence) can also alter the structure of a benthic community. The availability of organic matter and oxygen below the sediment-water interface has profound effects on the vertical distribution of macrobenthic species.[9] Day[10] notes that it is not the average but the extreme conditions of environmental factors which limit the abundance and distribution of the benthic macrofauna of estuaries.

Biotic factors, such as predation and species competition, cannot be disregarded in investigations of the occurrence and distribution of benthic fauna. They also can act as limiting factors;[10] hence, the mere tolerance of a species to physical and chemical conditions oftentimes does not provide sufficient explanation of an observed distribution pattern. The occurrence of a species depends on biological adaptation as well. Similarities in reproductive seasons, modes of feeding, size, and other biotic factors, for instance, may enable cohabitation of several species in the same general environment.[11,12]

The objective of this chapter is to examine the benthic organisms of estuaries, including their composition, abundance, distribution, and biotic interactions. Assessment will be made of the physical and chemical factors to which the benthos are adapted. Because of the importance of substrate type as a major control of the distribution of bottom-dwelling flora and fauna,[8] the physical and chemical characteristics of estuarine bottom sediments, as well as organism-sediment relations, will be covered. Interactions between benthic populations and seafloor sediments are critical to the physical and chemical properties of granular substrates in estuaries. Many benthic flora and fauna adapt to life on hard substrates; these biota will also be discussed.

II. GENERAL FEATURES OF THE BENTHOS

A. BENTHIC FLORA

The benthic flora of estuaries consist of micro- and macrophytic populations. Microfloral populations are generally most conspicuous in intertidal zones where prolific growth of minute, single-celled algae belonging to the Bacillariophyceae (diatoms), Cyanophyceae (blue-green algae), and Dinophyceae (dinoflagellates) commonly discolor the sediment. On tidal flats and salt marshes, the development of large mats of green and blue-green algae add significantly to the total algal biomass of some systems. The assemblage of microflora can be extensive in certain biotopes; the salt marshes of Georgia, for example, contain several hundred species of pennate diatoms alone.[13,14] Blue-green algal mats typically occur in warm-water regions (e.g., Gulf of Mexico), most frequently in shallow waters less than about 10 cm.[15] However, blue-green algal mats also grow along the East Coast of the U.S. as far north as New Jersey. In these mats, other microflora (i.e., pink and purple bacteria, diatoms, unicellular green algae, and flagellates) usually proliferate.[12] European systems — especially tidal flat estuaries — likewise have well-developed algal mats in intertidal zones.[16]

The microflora live on substrates as individual cells or filamentous colonies and adhere to the surfaces of other plants, animals, sediments, rocks, and man-made structures. Benthic diatoms, potentially major primary producers in inner estuarine areas, show strong motility, vertically migrating within sediments in response to variable light intensities. Some benthic diatoms attach to sedimentary grains and impart a brownish hue to the sediments.

Benthic macrophytes in estuaries concentrate in three vascular plant communities: (1) salt marshes (mainly temperate to subpolar latitudes), (2) seagrasses (tropical to temperate latitudes), and (3) mangroves (tropical and subtropical latitudes). Salt marsh plants underlie most of the 2×10^6 ha of estuarine wetlands estimated in the lower 48 states. Many of the species are broadly distributed, belonging to a few cosmopolitan genera (e.g., *Spartina* (cordgrasses), *Juncus* (rushes), and *Salicornia* (glassworts).[17] Seagrasses, marine monocots forming widespread and recurrent biotopes in shallow subtidal estuarine waters worldwide, are highly productive plants, especially in tropical regions. Whereas mangrove swamps occupy intertidal zones along coastlines between 25° N and 25° S latitudes in the U.S., only southern Florida has expansive mangroves, amounting to approximately 6.5×10^5 ha.[18] Macrobenthic algae, in general, are poorly represented in estuaries, as alluded to above, and most primary production among the macrophytes is attributable to rooted vegetation in the three aforementioned communities of vascular plants.

Some of the most capacious salt marsh vegetation borders the protected coastlines of Canada and the U.S. The southern shore of Hudson Bay, the East Coast of Canada and the U.S., and the Gulf Coast of the U.S. have particularly well-developed salt marshes.[19-21] Chapman[19-21] subdivided salt marshes into nine geographical units, although only three apply to North America according to Mitsch and Gosselink.[18] These three encompass: (1) Arctic marshes, including those in northern Canada, Greenland, Iceland, northern Scandinavia, and Russia; (2) eastern

North America marshes inhabiting the East Coast of Canada, the U.S., and the Gulf Coast of the U.S.; and (3) western North America marshes. The eastern North America marshes are further subdivided into a Bay of Fundy group, New England group (Maine to New Jersey), and coastal plain group (New Jersey to Louisiana). Only a few species compose the Arctic marshes due to the severe climate. Here, the most abundant plants are species of the sedge, *Carex,* and the grass, *Puccinellia phryganodes.* In eastern North America marshes, species of *Spartina* and *Juncus* predominate. Much less prominant than the salt marshes of eastern North America are the restricted stretches of marshes found on the West Coast of Canada and the U.S. Vast tracts of *Salicornia* and *Suaeda* border narrow zones of *Spartina foliosa* in this geographical area.

Tidal salt marshes grade inland into tidal freshwater marshes which, in turn, grade into nontidal freshwater wetlands.[22] Plant and animal communities of tidal freshwater marshes are dominated by freshwater species.[23] These marshes experience tidal fluctuations, but typically have salinities of 0.5‰ or less. As in the case of tidal salt marshes, tidal freshwater marshes are larger on the East and Gulf Coasts than on the West Coast of the U.S., covering an area of about 650,000 ha.[18] The highest biomasses correspond to marshes in the mid-Atlantic states, South Carolina, and Georgia.[23]

Plant communities in tidal freshwater marshes are more diverse than in tidal salt marshes, with a variety of grasses, as well as annual and perennial broad-leaved aquatic species, predominating. These wetland plant communities undergo noteworthy seasonal changes in species composition.[18] In addition, the communities exhibit a horizontal zonation of constituent populations. For example, on the East Coast of the U.S., the low marsh area represents an ideal environment for the growth of broad-leaved, fleshy plants, such as *Nuphur luteum* (spatterdock), *Zizania aquatica* (wild rice), and *Zizaniopsis miliaceae* (giant cutgrass). The diversity of plants increases in middle and upper sections of the tidal freshwater marsh; *Typha* spp. (cattail), *Polygonums* spp. (smartweeds), *Hibiscus moscheutos* (rosemallow), *Acorus calanus* (sweet flag), and *Biden* spp. (burmarigold) are notable members of the community at higher elevations.[18]

Progressing seaward of many tidal salt marshes, dense stands of submerged seagrass meadows create habitat not only for epiphytic flora and fauna, but also for infaunal organisms. Moreover, seagrasses stabilize seafloor sediments via a complex network of roots and rhizomes. Waterfowl and aquatic fauna also use seagrasses or associated algae as a source of food.[24-26]

Worldwide, nearly 50 species of seagrasses have been classified among 12 genera (i.e., *Amphibolis, Cymodocea, Enhalus, Halodule, Halophila, Heterozostera, Posidonia, Phyllosopadix, Syringodium, Thalassia, Thalassodendron,* and *Zostera).* On the Atlantic Coast of Europe, only *Zostera* is observed.[27]

Eelgrass, *Z. marina* L., has a wide distribution in temperate estuaries. Benthic organisms in *Z. marina* beds have been studied in detail in different geographic regions. Kikuchi[26] tabulated faunal communities associated with *Z. marina* beds, identifying subunits as a function of microhabitat structure and mode of life of the fauna. He cites the following biota which live on the leaves: (1) epiphytic felt flora and micro- and meiofauna, (2) sessile fauna attached to the leaves, (3) mobile epifauna moving on the leaf surfaces, and (4) swimming fauna resting on the leaves. Stems and rhizomes harbor a second category of fauna (e.g., nest-building polychaetes and amphipods). Fishes, cephalopods, and other highly mobile animals (e.g., decapod crustaceans), which swim under the leafage, embody a third group. The fauna inhabiting the sediment — a fourth category — can be further divided into epibenthic and infaunal populations. Many infaunal species collected in seagrass meadows are not endemic to the beds, but occur in the adjacent unvegetated areas from which they pervade the beds. The total density, biomass, and diversity of the infauna of eelgrass beds commonly surpass those of surrounding estuarine areas.[26]

Mangroves replace tidal salt marshes as the primary wetland community in subtropical and tropical regions.[18,28] Also known as mangals, mangroves consist of assemblages of halophytic trees and shrubs (monocots and dicots), as well as other plants, which grow as forested vegetation in intertidal and shallow subtidal zones. Research on mangroves has expanded markedly in the last two decades, especially in regard to the functional aspects of the communities.[29] Mitsch and Gosselink[18] and Lugo and Snedaker[30] classify mangrove ecosystems on the basis of their hydrodynamics and topography. They differentiate six types of mangrove wetlands, namely: (1) overwash mangrove islands, (2) fringe mangrove wetlands, (3) riverine mangrove wetlands, (4) basin mangrove wetlands, (5) hammock mangrove wetlands, and (6) dwarf mangrove wetlands. Rützler and Feller (p. 17)[29] and Tomlinson[31] invoke the following features of mangrove plants: (1) they are ecologically restricted to tidal swamps; (2) the major element of the community frequently forms pure stands; (3) the plants are morphologically adapted with aerial roots and viviparity (producing new plants instead of seeds); (4) they are physiologically adapted for salt exclusion or salt excretion; and (5) they are taxonomically isolated from terrestrial relatives, at least at the generic level.

About 80 species of mangroves belonging to 16 genera have been documented in the literature,[27] but only about 34 species in 9 genera are considered to be true mangroves.[31] The Old World mangrove swamps, specifically the Indo-Pacific region, contain the greatest diversity (i.e., 30 to 40 species). Far fewer species (about 10 species) exist in the New World of the Americas. Although mangroves may be observed as far as 32°N in the Northern Hemisphere in the U.S., they principally live on the Atlantic and Gulf Coasts of Florida, south of 28 to 29° N latitude.[18] Craighead[32] estimates that mangrove vegetation covers approximately 1750 km² of the Florida Coast. Three species dominate the mangroves of south Florida, that is, *Avicennia germinans* L. (black mangrove), *Laguncularia racemosa* L. Gaerntn. (white mangrove), and *Rhizophora mangle* L. (red mangrove).

Mangroves in south Florida, as elsewhere, characteristically display a zonal pattern. *R. mangle* is best adapted to conditions of submergence and, therefore, forms a fringe below the mean low tide level. Inland of the red mangroves are black mangroves known as basin mangrove wetlands; they cover an area from the lower to the middle or upper intertidal zone. White mangroves range from the middle to upper intertidal zone landward of the black mangroves. Upland ecosystems are frequently separated from the white mangroves by a zone of buttonwoods, *Conocarpus erecta*. The buttonwoods also proliferate in the intertidal zone in some regions.

B. BENTHIC FAUNA

The benthic fauna of estuaries may be grouped according to size into three categories: (1) macrofauna, (2) meiofauna, and (3) microfauna. The macrofauna include larger metazoans retained by sieves of 0.5- to 2-mm mesh. The meiofauna comprise smaller forms captured on sieves of 0.04- to 0.1-mm mesh. The microfauna consist of those bottom dwelling animals which pass through sieves of 0.04- to 0.1-mm mesh.

The term meiofauna, coined by Mare,[33] has also been used to define benthic metazoans that weigh less than 10^{-4} g (wet weight).[34,35] These organisms can be further subdivided into permanent and temporary members. While the permanent meiofauna incorporate adults of sufficiently small size to be classified in this group, the temporary meiofauna constitute juvenile stages of the macrofauna.[35] Among the permanent meiofauna are nearly all gastrotrichs, kinorhynchs, nematodes, rotifers, archiannelids, halacarines, harpacticoid copepods, ostracods, mystacocarids, and tardigrades as well as representatives of the bryozoans, gastropods, holothurians, hydrozoans, oligochaetes, polychaetes, turbellarians, nemertines, and tunicates.[3] The microfauna are composed essentially of protozoans,[35] although some workers also enlist bacteria.[3]

Other classifications of the benthic fauna are based on their life habits and adaptations. Nonparasitic species, for example, have been separated into epibenthic, infaunal, interstitial, boring, swimming, and commensal-mutualistic types.[8] Epibenthic animals may attach to a substrate by basally cemented structures (e.g., oysters and serpulid polychaetes), holdfasts (e.g., stalked barnacles), or roots (e.g., stalked crinoids). Whereas some epifauna have a sessile habit, living permanently attached to a substrate, others are vagile with considerable mobility over a surface. The infauna live in burrows and tubes, or they freely move through unconsolidated sediment.[5] *Arenicola marina,* the lugworm, and *Corophium volutator,* an amphipod, build U-shaped burrows. The fiddler crabs, *Uca pugilator* and *U. pugnax,* excavate burrows, which in the case of *U. pugilator* have single openings that are almost always plugged at high tide on tidal flats. In contrast, *U. pugnax* rarely plugs its burrow.[36] The polychaete, *Cistenides (= Pectinaria) gouldii,* constructs a tube of fine sand grains. *Nephtys incisa,* another polychaete, is motile and traverses through meters of sediment in search of food. Interstitial animals typically range from 0.2 to 3 mm in size.[35] Many interstitial faunal species have an oblong shape which greatly facilitates their movement through the grain interstices below the sediment-water interface. Meiobenthic organisms — nematodes, gastrotrichs, and harpacticoid copepods — are important interstitial forms. Animals that bore into hard substrates do so via chemical or mechanical processes. Hence, the boring sponge, *Cliona celata,* and the boring turbellarian, *Stylochus ellipticus,* produce bore holes by means of chemical attack, and the rock borer, *Zirphaea crispata,* and woodborer, *Teredo carcinata,* generate burrows mechanically by rasping with their valves to excavate substrates. *Urosalpinx cinerea,* the common oyster drill, utilizes both chemical attack and mechanical abrasion to bore through the valves of its prey, frequently juvenile American oysters *(Crassostrea virginica).* Examples of swimming benthic inverte-brates are species of polychaetes *(Nereis* and *Nephtys)* and bay scallops *(Argopecten irradians).* Occasionally, benthic populations enter into symbiotic relationships in which one species benefits while its host does not, with the host being unaffected by the relationship (commensal-ism). In some cases, both species benefit (mutualism). These biotic relationships are more evident in low than in middle or high latitude regions.[8]

Benthic fauna can be broadly classified according to their mode of obtaining food.[8] Five categories are recognizable, specifically deposit feeders, suspension feeders, parasites, herbi-vores, and carnivores-scavengers. Table 1 lists the feeding types of the ten numerically dominant macrobenthic species at the mouth of Stouts Creek in Barnegat Bay, NJ. The most abundant deposit feeder at this location is *Cistenides (=Pectinaria) gouldii* (burrowing deposit feeder); only one surface deposit feeder, *Polycirrus eximius,* is listed. *Cyathura polita* represents the only abundant scavenger in the bay. The suspension-feeding coot clam *(Mulinia lateralis),* an opportunistic species, consistently ranks high, remaining one of the three numerically dominant species through time. *Turbonilla interrupta* is the predominant benthic parasitic form. Of the carnivorous species, *Acteocina canaliculata* and *Mitrella lunata* have the highest ranking. Herbivores are not tabulated, but the grazing periwinkle, *Littorina saxatilis,* in addition to several other grazing species, occurs in the estuary.[37]

III. BENTHIC FLORA

A. BENTHIC ALGAE
1. Benthic Microalgae

The benthic microalgae of estuaries have received less attention than vascular plants, which are generally much more productive. However, these diminutive plants contribute significantly to the total primary production of some habitats, such as tidal flats (Table 2). Diatoms and blue-green algae are principal constituents of the estuarine benthic microalgal communities. In the Potomac River estuary, for instance, unicellular and colonial blue-green algae and diatoms coat substrate surfaces (e.g., submerged rocks, bulkheads, driftwood, and metals). Although less abundant, microalgae of other groups (e.g., golden-brown algae) also grow on submerged materials in estuaries.[38]

Benthic microalgae may live attached to submerged plants as epiphytes, live in and on the

TABLE 1
Rank, Mean Density,[a] and Feeding Type of the Ten Numerically Dominant Macrobenthic Species at the Mouth of Stouts Creek, Barnegat Bay, NJ

Species	Feeding type	Rank and mean density				
		1969	1970	1971	1972	1973
Mulinia lateralis	S[b] (Maurer et al., 1974)	1 (5943)	3 (386)	3 (65)	3 (100)	2 (51)
Pectinaria gouldii	BD[c] (Young and Young, 1977)	2 (1604)	2 (667)	1 (441)	2 (342)	1 (55)
Acteocina canaliculata	C[d] (Young and Young, 1977)	3 (601)	5 (42)	(4)	4 (33)	3 (27)
Glycera americana	BD (Gosner, 1971)	4 (425)	(18)	(6)	(8)	(1)
Turbonilla interrupta	P[e] (Gosner, 1971)	5 (213)	(16)	(2)	6 (20)	9 (7)
Golfingia improvisum	BD (Barnes, 1968)	6 (102)	4 (63)	(0)	(0)	(0)
Mitrella lunata	C (Young and Young, 1977)	7 (98)	9 (22)	(7)	(12)	(1)
Cyathura polita	BS[f] (Gosner, 1971)	8 (78)	10 (21)	6 (21)	(11)	(2)
Ampelisca spp.	D[g] (Mills, 1967)	9 (46)	1 (933)	2 (132)	1 (843)	4 (13)
Maldanopsis elongata	BD (Gosner, 1971)	10 (33)	(12)	(7)	(4)	7 (8)
Goniadidae	C (Gosner, 1971)	(32)	8 (23)	4 (34)	5 (29)	(1)
Scoloplos fragilis	BD (Gosner, 1971)	(32)	6 (28)	5 (24)	7 (16)	7 (8)
Haminoea solitaria	C (Young and Young, 1977)	(20)	(17)	7 (20)	(4)	(0)
Diopatra cuprea	C (Gosner, 1971)	(26)	(13)	8 (19)	(9)	5 (9)
S. robustus	BD (Gosner, 1971)	(0)	(0)	9 (14)	10 (14)	(5)
Leptosynapta tenuis	D (Barnes, 1968)	(1)	(7)	10 (13)	(2)	(1)
Molgula manhattensis[h]	S (Barnes, 1968)	(0)	7 (26)	(1)	(1)	(2)
Polycirrus eximius	SD[i] (Gosner, 1971)	(0)	(0)	(7)	8 (15)	(3)
Cerebratulus lacteus	C (Barnes, 1968)	(1)	(0)	(1)	9 (14)	(1)
Maldane sarsi	BD (Gosner, 1971)	(12)	(3)	(2)	(6)	5 (9)
Glycera dibranchiata	BD (Gosner, 1971)	(1)	(7)	(12)	(11)	9 (7)
Tellina agilis	BD and S (Maurer et al., 1974)	(23)	(13)	(13)	(4)	(5)

[a]   Number/m^2.
[b]   Suspension feeder.
[c]   Burrowing deposit feeder.
[d]   Carnivore.
[e]   Parasite.
[f]   Burrowing scavenger.
[g]   Detritus feeder.
[h]   Urochordate.
[i]   Surface deposit feeder.

From Loveland, R. E. and Vouglitois, J. J., *Ecology of Barnegat Bay, New Jersey,* Kennish, M. J. and Lutz, R. A., Eds., Springer-Verlag, New York, 1984, 135. With permission.

upper few millimeters of sediment, live on stones and rocks as lithophytes, and live on man-made structures. On bottom sediments, microalgae can be temporarily benthic. Planktonic species that periodically get deposited on the estuarine seafloor by currents and populations that sink to the bottom to reproduce or form resting stages are examples.

The microflora form patches or mats, especially in intertidal zones. Extensive mats of filamentous green and blue-green algae have been reported in salt marshes of Delaware,[39] New York,[40] and Massachusetts.[41] However, in other regions (e.g., Georgia salt marshes), the epibenthic algae appear largely as small patches on mud substrates.[14] While a diverse assemblage of species comprises the benthic microalgal communities of many estuaries, only a few taxa dominate the standing crop. For example, Williams[13] conveyed that four genera (i.e., *Cylindrotheca, Gyrosigma, Navicula,* and *Nitzschia)* accounted for 90% of the cells of the diatom assemblage in the salt marshes of Sapelo Island, GA.

TABLE 2
Trophic Types of Tidal Flats Indicating Benthic Microalgal Production

Trophic type	Primary production	Pool of detritus; decomposition	Consumption biomass
Exposed sandy shore	Inhibited by unstable sediment	Small, irregular import; oxic	Interstitial fauna, some large scavengers
Beds of suspension feeders	High in phytoplankton, variable in benthic micro- and macroalgae	Moderate to large, accumulation of fecal material, variable import; oxic and anoxic	Large suspension feeders, particularly bivalves; few grazers and deposit feeders
Mudflat	High in benthic microalgae[a], occasionally drifting green algae	Accumulation of allochthonous detritus; primarily anoxic	Deposit feeders and algal grazers dominate; often small (nematodes)
Sandflat	High in benthic microalgae[a]	Variable, small import; oxic and anoxic	Grazers dominate; large deposit feeders
Seagrass bed	High in aquatic phanerogams, benthic and epiphytic microalgae[a]	Large, variable import; oxic and anoxic	Trophically very diverse, many grazers on epiphytes
Saltmarsh and mangroves	High in halophytes, variable in benthic microalgae[a]	Very large, high import; primarily anoxic	Trophically diverse; very often small

[a] Includes photosynthetic bacteria.

From Reise, K., *Tidal Flat Ecology: An Experimental Approach to Species Interactions, Ecological Studies.* Vol. 54, Springer-Verlag, New York, 1985. With permission.

Of the motile benthic microalgae, diatoms — particularly naviculoid genera — and dinoflagellates experience acute vertical migration.[10] At high tide and nocturnal low tide, they migrate downward in sediments, perhaps to escape grazing organisms and scour by tidal currents, whereas during daytime low tides, they migrate upward and emerge at the surface. The total distance of migration only extends through the top few millimeters of seafloor sediments. *Hantzschia virgata,* a diatom, moves upward in sediment during daytime low tides as a consequence of the combination of an endogenic lunar-day rhythm with a solar-day rhythm.[42] The intensity of light and water content of the sediment are additional factors responsible for the migration patterns.[3,10] Brown or green patches of diatoms, dinoflagellates, and euglenoids spread over muddy intertidal flats and sand banks. Extracellular mucous films secreted by these microflora bind grains and inhibit their erosion and transport, thereby stabilizing bottom sediments.[10,43] Cammen and Walker[44] attributed the release of these extracellular substances to the presence of sediment bacteria. Admiraal et al.[45] dealt with the characteristics of epipelic diatom growth on intertidal sediments. Baillie[46] elaborated on the organization of an intertidal epipelic diatom community in terms of the vertical cell-size distributions of its constituent populations, ascertaining that small cells amassed at the sediment-water interface, while large cells concentrated deeper in the microhabitat far below light and oxygen penetration. Explanations of the general biology of sediment-inhabiting diatoms are given by Round[47] and Werner.[48]

Periodic inundation and exposure in the intertidal zone subjects epibenthic algae to harsh environmental conditions. Alterations of temperature, light intensity, salinity, pH, nutrient levels, grazing, and sediment stability forge a highly unstable environment for many of these organisms which oftentimes limits their standing crop biomasses and productivity.[14] Microgradients of sediment grain sizes and organic particle sizes, percentage of organics, porosity, light attenuation, and oxygen govern vertical microalgal distribution patterns.[46] Gradients in parameters along the length of an estuary regulate the lateral distribution of species as well. Thus,

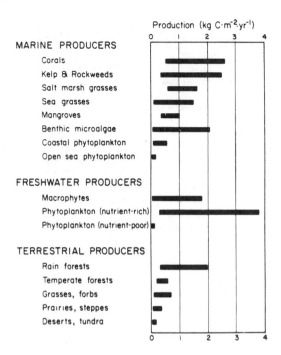

FIGURE 1. Annual net production rates of marine, freshwater, and terrestrial producers. (From Valiela, I., *Marine Ecological Processes,* Springer-Verlag, New York, 1984. With permission.)

Lippson et al.[38] correlated differences in the composition of the benthic microalgal communities along the Potomac River estuary with progressive salinity changes. In the tidal fresh portion of the estuary, the colonial diatom, *Melosira* sp., the filamentous blue-green algae, *Oscillatoria* spp. and *Schizothrix* sp., and the gelatinous blue-green algae, *Anabaena* spp. and *Anacystis* spp., dominate during the summer. The golden-brown alga, *Chromulina pasheri,* is common, and phytoplankton (e.g., *Cosmarium* spp. and *Closterium* spp.) also attach to the sediment surface during this season. Diatom genera (i.e., *Asterionella, Fragilaria, and Navicula)* abound among attached benthic macroalgae in the early spring and fall. The dinoflagellates, *Gyrodinium* sp. and *Gonium* sp., and the euglenoid, *Euglena* sp., are found in association with the epibenthic floral assemblages at various times.

As salinity levels rise in the mid-estuary, diatoms gradually attain greater abundances in the benthic microfloral community, with blue-green algae being present in much lower numbers. Diatoms increase in dominance in the Potomac River estuary during the winter months as other algal groups diminish. Dinoflagellates and euglenoids likewise inhabit the central segment of the estuary. The benthic microalgae in the lower estuary have not been adequately surveyed to describe their assemblages.

Annual production by benthic microalgae, although slight in many estuaries, approaches or exceeds that of salt marshes, seagrasses, or mangroves in some systems. Values range from less than 25 g C per square meter per year to somewhat greater than 2000 g C per square meter per year (Figure 1). Production tends to be greatest in the upper 1 cm of mudflat sediments. Large populations of microalgae stretch across the lower intertidal zone, where they may be extremely productive.[49] Table 3 contains estimates of benthic microalgal production from different environments.

The abundance, distribution, and species composition of attached microalgae are controlled, in part, by grazing.[50,51] While both light and nutrients limit microalgal production, grazing serves

TABLE 3
Annual Production Rates of Benthic Microalgae in
Selected Estuarine Ecosystems

Location	Production	Ref. (sources from Ref. 130)
Danish fjords	116	Grøntved (1960)
Wadden Sea, Netherlands	115—178	Cadee and Hegeman (1974)
	101 ± 39	McLusky (1981)
Grevelingen, Netherlands	25—37	Wolff (1977)
Ythan estuary, U.K.	31	Sibert and Naiman (1980)
Lynher, U.K.	143	Joint (1978)
Alewife Cove	45	Welsh et al. (1982)
Barataria Bay	240	Welsh et al. (1982)
Bissel Cove	52	Welsh et al. (1982)
Charlestown River	41	Welsh et al. (1982)
Delaware	160	Gallagher and Daiber (1974)
False Bay	143—226	Pamatmat (1968)
Flax Pond	52	Welsh et al. (1982)
Hempstead Bay	62	Welsh et al. (1982)
Jordan Cove	41	Welsh et al. (1982)
Niantic River	32	Welsh et al. (1982)
Sapelo Island	180	Pomeroy (1959)

a   Production values in g C per square meter per year.

From Kennish, M. J., *Ecology of Estuaries,* Vol. 1, CRC Press, Boca Raton, FL, 1986. With permission.

as an avenue of considerable loss of plant biomass.[14] Mud snails (e.g., *Ilyanassa obsoleta*),[52,53] fiddler crabs (*Uca* spp.),[54,55] and herbivorous fish (e.g., mullet)[56] purportedly remove significant quantities of epibenthic algae. *Littorina scutulata,* a grazing gastropod, reduces the percentage cover of diatom carpets in the intertidal zone of southern Oregon during the warmer months of the year. Snails in densities of approximately three/dm^2 are capable of eliminating algal patches.[51]

Together with filamentous algae (e.g., *Chaetomorpha, Ectocarpus,* and *Polysiphonia*), blue-green algae, and small epifaunal organisms, benthic diatoms form a fur-like covering on the leaves of macrophytes, (e.g., seagrasses: *Cymodocea, Halodule, Ruppia, Posidonia, Thalassia,* and *Zostera;* and macroalgae: *Enteromorpha* and *Ulva*). This fur-like covering, termed periphyton or Aufwuchs, furnishes grazing organisms with a rich food supply.[10] Addressing the food webs of seagrass meadows in the northwest Gulf of Mexico, Kitting et al.[57] reveal that epiphytic algae have high productivity and excellent palatability for consumers. These flora possibly play a more critical trophic role than the macrophytes themselves. Invertebrates of seagrass meadows primarily assimilate epiphytes rather than seagrasses, as evidenced by comparisons of stable carbon isotopes (δ^{13}C). Detritivores ingest epiphytic algae with seagrass detritus and associated bacteria.[58,59] Benthic diatoms (e.g., *Cocconeis, Licmophora,* and *Nitzschia*) may be key components of the epiphytic communities.[26]

Epiphytes occasionally cover an appreciable portion of vascular plant leaves and constitute a substantial fraction of the leaf biomass. As many as 100 species of epiphytic algae have been identified on seagrasses, although only a few species may outnumber all others.[60] For example, in the Choptank River estuary, the dominant species of epiphytic algae affixed to the two predominant submerged vascular plants, *Potamogeton perfoliatus* and *Ruppia maritima,* are *Nitzschia closterium* and *N. paradora,* other pennate diatoms (e.g., *Amphora, Amphipora,*

Cocconeis, Gyrosigma, Navicula, and *Pleurosigma), Melosira nummuloides,* and a few dinoflagellates (dominated by *Peridinium*).[61]

Epiphytic populations undergo seasonal and annual variations in production that can be large depending on geographical location. Marshall[62] obtained epiphytic production values of 20 g C per square meter per year on *Zostera marina* in Massachusetts. Epiphytes living on turtlegrass, *Thalassia testudinum,* in Florida waters have a much higher total annual production of 200 g C per square meter per year,[63] which is estimated to be 20% of the average net production of the turtlegrass of the area.[60] The diatom *Isthmia nervosa* periodically increases markedly in numbers on *Zostera* leaves in Alaska, resulting in an abrupt rise in biomass of nearly 50% of the total *Zostera* leaf and epiphyte dry weight.[60] Epiphytes comprise an average of 18 to 50% of the combined *Z. marina* leaf and epiphyte production;[64,65] in a *T. testudinum* bed, Jones[66] discerned that they accounted for 22% of the total production.[67]

Because of their association with submerged vascular plants, epiphytes reap several benefits. First, they have a greater proximity to light, and water currents transport nutrients to them while removing growth-inhibiting substances and sediments, thereby enhancing their growth.[67,68] Second, the transfer of nutrients and their release through seagrass leaves bathes epiphytes in high nutrient concentrations. Various metabolic exudates, although poorly understood, may influence the colonization of marine phanerogams. Some extracts from recently dead *Z. marina* leaves inhibit growth and have lethal effects on microalgae and bacteria.[69]

The size of the plant substrate for attachment governs to some degree the types of microalgae colonizing the surfaces. For instance, the pennate diatom, *Cocconeis scutellum,* has a large surface area for attachment and prefers to affix to more thickly branched species.[70] Orth and Van Montfrans[67] surmise that broad-bladed seagrass genera (e.g., *Posidonia* and *Zostera*) could favor floral species with a larger surface area for attachment than narrow-bladed seagrass genera (e.g., *Ruppia* and *Syringodium*). Sullivan[71] observed that one third of the diatoms on *R. maritima* blades consisted of the narrow form, *Navicula pavillardi*. Blades on *Z. marina* and *T. testudinum* are principally covered by diatoms of the genus *Cocconeis*.[72-74]

The list of organisms depending heavily on epiphytes as a food source is extensive.[57,75] Protozoans consume epiphytic diatoms as is apparent in studies of salt marsh Aufwuchs (= periphyton) communities.[76,77] Foraminiferans subsist on bacteria, diatoms, and chlorophytes.[78] Copepods (e.g., *Nitocra typica*) also feed on the periphyton.[67,79]

Microfauna and meiofauna living in epiphytic-felt flora (i.e., protozoans — ciliates, flagellates, foraminifers — free-living nematodes, polychaetes, rotifers, tardigrades, copepods, and ostracods) utilize the microorganisms (e.g., diatoms and bacteria) as food while serving as ration themselves for macrofauna. Small algae, such as species of *Cladophora, Enteromorpha, Hypnea,* and *Polysiphonia,* are common algal-felt constituents. Gastropods, including those in the families Cerithidae, Rissoidae, and Trochidae, graze on algal felt.[80] Other invertebrates known to rely on epiphytic algae as a dietary component are the sea hare, *Aplysia* spp., the tropical gastropod, *Strombus gigas,* the gastropods, *Bittium alternatum, B. varium, Littorina saxatilis,* and *Ilyanassa obsoleta,* the small prosobranch, *Modulus modulus,* the caprellid amphipod, *Caprella laeviuscula,* the isopods, *Idotea* spp., the decapod crustacean, *Palaemonetes pugio,* in addition to many other epifaunal species.[67]

In addition to growing on the surfaces of many types of macrovegetation and building brown or green patches (often composed of blue-green algae, diatoms, and euglenophytes) on sediment surfaces,[81,82] a diverse assemblage of microflora settle on man-made structures in marine and estuarine environments, affecting subsequent macrofouling of these structures.[83,84] The research of intact microflora on surfaces of man-made materials has been greatly facilitated by scanning electron microscopy (SEM) which, according to Novak (p. 145),[75] "enables the *in situ* study of undisturbed attached microorganismic assemblages, as well as the simultaneous study of organisms largely differing in size, such as macroalgae, diatoms, and bacteria". Other techniques require scraping or removing organisms by ultrasonic treatment; these methods destroy the community. Recently, the SEM has been applied to quantitative assessments of microorganisms on artificial substrates.

The development of marine fouling communities on man-made surfaces is preceded by the sorption of high molecular weight polymeric substances from the sea that serve as conditioning films or interface conversion layers.[85] Subsequently, bacteria, diatoms, and other microorganisms settle on the surfaces, with rod-shaped chemoheterotrophic bacteria initially colonizing the surfaces and their organic secretions constituting the main fraction of the primary film. Secondary colonizers consist of stalked or filamentous bacteria.[84,86] Once the bacterial film is secreted, diatoms, microalgae, and protozoa begin to establish.[86] Species of diatoms colonizing man-made surfaces encompass a restricted group of organisms, and the genera frequently mentioned in the literature as colonizers are *Amphora, Cocconeis, Licmophora, Navicula, Nitzschia, Pleurosigma, Synedra,* and *Tropidoneis.*[87] The microorganisms proliferate within the fouling deposits on man-made surfaces, together with organic secretions, detritus, inorganic precipitates, and corrosion products.[86]

An irreversible ecological succession of microorganisms evolves at some point after the attachment of a primary population of bacteria on marine surfaces, proceeding to the development of a climax microbial community.[88-90] The biofilm, viewed microscopically, appears as minute organisms, principally bacteria, algae, and protozoa, enmeshed in a thick layer of bacterial polymer. Trapped debris and dead cells — a nutrient source — accumulate in the matrix.[90] In sum, five stages in the formation of the primary biofilm have been deduced. The first four stages take place rapidly, within a few hours time, whereas the fifth stage requires days or weeks to materialize. Mitchell and Kirchman[90] espouse the five stages as: (1) an initial conditioning stage, (2) chemical attraction of motile bacteria, (3) reversible adsorption of both motile and nonmotile bacteria, (4) an irreversible stage of attachment mediated by bacterial polymers, and (5) the development of secondary microflora. As noted by Mitchell and Kirchman,[90] a conditioning film of polymers on the marine surfaces precedes the establishment of the primary bacterial community, which progresses in stages involving chemotaxis of motile bacteria, reversible sorption, followed by the permanent attachment of the primary bacterial film by polymeric fibrils. Attachment of microalgae results from both positive chemoreception and attachment of specific proteins to bacterial polysaccharides. Techniques to prevent the buildup of bacterial slimes on marine surfaces entail the chemical treatment of the surfaces to induce negative chemotaxis. Similarly, negative chemotaxis has been an effective means of controlling algal fouling.[90] Repellent chemicals (e.g., benzoic and tannic acids), for example, seem to substantially mitigate algal fouling when surfaces are properly treated.[91]

2. Benthic Macroalgae

Reference has been repeatedly made in the literature to the species-poor condition of benthic macroalgae in estuaries.[10] Seaweeds in estuaries cover only a small area of the bottom, with many populations concentrating on rocks, bulkheads, piers, and other hard substrates. Hence, *Fucus* spp. often attach to rocks. *F. ceranoides* prefers lower salinity waters and is confined to estuaries. Other species of *Fucus* (e.g., *F. vesiculosus),* meanwhile, cannot tolerate reduced salinities and, therefore, are found on marine coasts.[49] On mudflats, however, filamentous green algae, such as *Enteromorpha* spp., can predominate, as in the Eden estuary, Scotland, where *E. prolifera* is an abundant species.[49]

High turbidity and sedimentation in estuaries limit light penetration and impinge on the macrofloral community. Thus, the benthic macroalgae inhabiting this environment belong to a small number of widespread genera.[10] Most of these species are classified in three major groups: (1) Chlorophyta (green algae), (2) Phaeophyta (brown algae), and (3) Rhodophyta (red algae).[38] Genera of Chlorophyta typically encountered include *Chaetomorpha, Cladophora, Codium, Enteromorpha,* and *Ulva. Ascophyllum, Fucus,* and *Pelvetia* are genera of Phaeophyta continually recorded in estuaries of the Northern Hemisphere and *Colpomenia, Dictyota,* and

Hormosira, in estuaries of the Southern Hemisphere. The red algae, *Bostrychia* and *Caloglossa,* live in salt marsh and mangrove biotopes, while *Ceramium, Lawrencia,* and *Polysiphonia* affix to the surfaces of seagrasses in shallow subtidal waters. *Gracilaria* is yet another constantly occurring red algal genus in estuaries.[10]

Benthic macroalgae usually require attachment to a solid substrate to prevent them from being dislodged and carried away by bottom currents and waves.[81] This attachment is accomplished by adhesive, root-like holdfasts or basal disks that enable the plants to live on shells, rocks, man-made structures, and other hard surfaces. In contrast to vascular plants, macroalgae do not have roots; consequently, they do not normally colonize soft, sedimentary shores.[92] On mudflats, benthic algal communities consist of epipelic, epipsammic, or drift populations.

Some estuaries contain high densities and biomasses of the macroalgae.[93] A number of taxa (e.g., *Blidingia minima* var. *subsalsa, Enteromorpha clathrata,* and *Vaucheria* spp.) attain peak abundances in these shallow coastal waters and serve as a major source of organic matter.[94,95] Benthic macroalgal assemblages have a distinctive spatial distribution in estuarine environments, with species richness and diversity characteristically declining inland.[95] Perennial red or brown algae are more prevalent at greater distances from land as opposed to annual species, especially green algae, which extend farther upestuary.[96] Salinity exerts a major control on the distribution patterns of these flora.[97]

The spatial patterns of seaweeds in estuaries are contingent upon the dispersal capabilities of the populations. Seaweeds disperse either as free-floating macroscopic forms or as microscopic propagules (e.g., spores, gametes, and zygotes);[98] currents, of course, play a crucial role in the magnitude of their dispersal.[99] Hydrographic conditions also affect the settlement and attachment of the seaweeds.[100-103] Propagule dispersal is, in part, a function of biological factors, notably the periods of viability, motility, and the potential time restrictions for adhesion.[98] Thus, Zechman and Mathieson[98] uncovered maximum counts of propagules during the spring-early summer months at Adams Point, NH, and these peak numbers just preceded the period of highest summer biomass on the shore.

Wilkinson and Rendall,[104] considering the usefulness of benthic macroalgal distribution in pollution assessment, differentiated two broad physiognomic types of flora in estuaries of England. In the upper reaches subject to greater pollution effects, mat-forming flora, composed largely of green macroalgae, diatoms, and xanthophytes, predominate. In the lower reaches bordered by more rocky shorelines, impoverished marine flora dominated by fucoids are conspicuous. While some botanists have formulated species lists of algae representative of different estuarine zones conforming to the Venice system of classification,[105] Wilkinson and Rendall[104] have argued that such lists are not universally applicable, as exemplified by the small estuaries of England with widely variable salinities at any given point and a macrofloral community broadly tolerant of environmental conditions. Consequently, the prospect of classifying estuaries by means of benthic algal species distribution has limited utility in many geographical regions.

a. Zonation

In intertidal zones, typically on rocky shores, seaweeds display a horizontal banding due to the dominance patterns of the populations.[105] A distinctive species zonation is manifested on many artificial embankments, such as those along the River Thames,[106] and it can be much more pronounced here than on natural marsh banks.[107] Green algae are clearly zoned on man-made structures in other tidal systems of England as well.[108] A similar vertical zonation generally exists in estuaries worldwide.[97]

Several factors regulate the vertical distribution of benthic macroalgal species. Research on intertidal fucoid algae suggests that the lower limits are determined by species competition, and the upper limits, by physiological stresses associated with emersion.[105,109] Competition for light among the macroalgae may be the deciding factor concerning the lower limits of their vertical distribution in the intertidal and shallow subtidal zones. Certainly grazing can impose upper limits to the distribution of some subtidal macroalgae.[110] However, stresses associated with emersion are key determinants to the upper limits of many seaweeds.[7] As emphasized by Dring (p. 133),[7] "... the exact physiological cause for a specific limit, or the precise environmental factor controlling it in the field, has not yet been established for any species at any site".

b. Type Examples
1. Grays Harbor, WA

In the Grays Harbor estuary, Thom[95] documented a highly diverse and abundant assemblage of macroalgae. He registered 29 taxa in drift communities, in mats on sand, and in epifloral assemblages attached to boulders, tree roots, logs, other algae, and vascular plants. The most abundant populations during the sampling period were *Enteromorpha clathrata* var. *crinita, E. linza, E. intestinalis, Fucus distichus* ssp. *edentatus, Polysiphonia hendryi* var. *deliquescens,* and *Porphyra sanjuanensis. E. intestinalis* and *F. distichus* ssp. *edentatus* had the most widespread distribution. Green algae were collected in many different habitats throughout the estuary. The unattached macroalga, *E. clathrata* var. *crinita,* occurred as thick, drift aggregations in subtidal waters, accumulating as dense masses on intertidal mudflats, particularly in late summer. *P. hendryi* var. *deliquescens,* a filamentous red alga, was also a common, free-floating form. In stable substrates of intertidal zones, the perennial brown alga, *F. distichus* ssp. *edentatus,* attained prominence. Together with *P. sanjuanensis,* it affixed to the vascular plant, *Salicornia virginica,* in midintertidal areas. The yellow-green alga, *Vaucheria* sp., developed as dense mats on mudflats, but generally had a patchy distribution. *Rhizoclonium riparium* likewise grew in dense mats in sand.

The species composition of the benthic macroalgae changed along the longitudinal axis of the estuary, from the exposed rocky coast near its mouth toward its head. The floral assemblage on the exposed rocky coast was dominated by large brown and red seaweeds. When advancing eastward (upestuary), workers recorded a decline in the number of taxa, and the assemblage near the head of the estuary had few strongly euryhaline forms. Although many of the species persisted year-round in the estuary, large seasonal variations in population abundances were apparent.

The mean net primary productivity of the macroalgae ranged from 0.09 to 1.79 g C per square meter per hour. Excluding the highly productive *Zostera* beds, the benthic macroalgae supplied the highest floristic production in the estuary, with an annual production estimate of 47.8×10^6 kg C per year accounting for about 22% of all the organic carbon fixed in the system. The most abundant macroalgal species had moderate to high levels of net productivity compared to other estuaries.

Salinity and the availability of a suitable substrate probably are the main factors controlling the distribution of most of the benthic macroalgae in Grays Harbor. The large seasonal fluctuations in macroalgal species abundances of estuaries, in general, have been attributed to variations in air temperature, water temperature, salinity, nutrients, light intensity, photoperiod, and precipitation.[95,97,111] The same factors seem to modulate seasonal macroalgal abundances in Grays Harbor.[95]

Thom[95] refers to investigations of macroalgae in other West Coast estuaries of North America. He comments on the work of Davis and McIntire[112] in Netarts Bay and Yaquina Bay, OR; Neushul,[113] Thom,[114,115] Thom et al.,[116] and Hodgson and Waaland[117] in Puget Sound, WA; and Pomeroy[118] and Pomeroy and Stockner[111] in the Squamish River delta, British Columbia. The findings of these studies illustrate that benthic macroalgae can be important in some estuaries in terms of their abundance, diversity, and productivity.

2. Barnegat Bay, NJ

The benthic macroalgae of Barnegat Bay, like those of the aforementioned West Coast estuaries, undergo pronounced seasonal changes in species composition, abundance, and productivity. At least 116 species of benthic algae have been recovered from the bay (Table 4),[119]

TABLE 4
List of Benthic Algal Species Identified in Barnegat Bay, NJ

Chlorophyta

Blidingia marginata (J. Ag.) P. Dang.
B. minima (Näg. ex Kütz.) Kylin
Bryopsis hypnoides Lamour.
B. plumosa (Huds.) C. Ag.
Chaetomorpha sp.
C. aerea (Dillw.) Kütz.
C. linum (O. F. Müll.) Kütz.
Cladophora sp.
C. albida (Huds.) Kütz.
C. crystallina (Roth) Kütz.
C. refracta (Roth) Kütz.
C. sericea (Huds.) Kütz.
C. vagabunda (L.) Hoek
Codium fragile (Sur.) Hariot
Endophyton sp.
Enteromorpha sp.
E. clathrata (Roth) Grev.
E. compressa (L.) Grev.
E. intestinalis (L.) Link
E. linza (L.) J. Ag.
E. plumosa Kütz.
E. prolifera (O. F. Müller) J. Ag.
Gomontia polyrhiza (Lagerh.) Born. & Flah.
Monostroma sp.
Percursaria percusa (C. Ag.) Rosenv.
Phaeophila viridis (Reinke) Burrows in Parke & Dix.
Pseudendoclonium submarinum Wille
Rhizoclonium sp.
R. riparium (Roth) Harv.
Ulothrix sp.
U. flacca (Dillw.) Thur. in Le Jol.
U. implexa Kütz.
Ulva lactuca L.
Ulvaria oxysperma (Kütz.) Blid.
Ulvella lens

Phaeophyta

Ascophyllum nodosum (L.) Le Jol.
Asperococcus fistulosus (Huds.) Hook.
Desmotrichum undulatum (J. Ag.) Reinke
Ectocarpus sp.
E. siliculosus (Dillw.) Lyngb.
Elachista fucicola (Vell.) Aresch.
Fucus sp.
F. spiralis L.
F. vesiculosus L.
Giffordia sp.
G. granulosa (Sm.) Hamel
Halothrix lumbricalis (Kütz.) Reinke
Leathesia difformis (L.) Aresch.
Myrionema strangulans Grev.
Myriotrichia sp.
M. filiformis Harv.
Petalonia fascia (O. F. Müll.) O. Kuntze

TABLE 4 (continued)
List of Benthic Algal Species Identified in Barnegat Bay, NJ

Phaeophyta (continued)

Pilayella littoralis (L.) Kjellm.
Punctaria latifolia Grev.
P. plantaginea (Roth) Grev.
Ralfsia clavata (Harv. in Hook) Crouan frat.
R. verrucosa (Aresch.) J. Ag.
Scytosiphon lomentaria (Lyngb.) Link
Sphacelaria sp.
S. cirrosa (Roth) C. Ag.
Spongonema tomentosum (Huds.) Kütz.
Stilophora rhizodes (Turn.) J. Ag.

Rhodophyta

Agardhiella subulata (C. Agardh) Kraft & Wynne
Anotrichium tenue (C. Ag.) Näg.
Antithamnion sp.
A. cruciatum (C. Ag.) Näg.
A. plumula (Ellis) Thur. in Le Jol.
Asterocystis ramosa (Thwaites in Harv.) Gobi ex Schm.
Audouinella sp.
A. secundata (Lyngb.) Dix.
Bangia atropurpurea (Roth) C. Ag.
Callithamnion sp.
C. baileyi Harv.
C. byssoides Arn. ex Harv. in Hook.
C. corymbosum (Sm.) Lyngb.
C. roseum (Roth) Lyngb.
Ceramium sp.
C. diaphanum (Lightf.) Roth
C. fastigiatum Harv.
C. rubrum (Huds.) C. Ag.
C. strictum Harv.
Champia parvula (C. Ag.) Harv.
Chondria sp.
C. baileyana (Mont.) Harv.
C. sedifolia Harv.
C. tenuissima (Good. et Woodw.) C. Ag.
Dasya baillouviana (Gmel.) Mont.
Erythrotrichia ciliaris (Carm. ex Harv. in Hook.) Thur. in Le Jol.
Fosliella lejolisii (Rosan.) Howe
Gelidium crinale (Turn.) J. Ag.
Goniotrichum alsidii (Zanard.) Howe
Gracilaria tikvahiae McLach.
Hypnea musciformis (Wulf.) Lamour.
Lomentaria baileyana (Harv.) Farl.
Polysiphonia sp.
P. denudata (Dillw.) Grev. ex Harv. in Hook.
P. harveyi Bail.
P. nigra (Huds.) Batt.
P. nigrescens (Huds.) Grev.
P. subtilissima Mont.
P. urceolata (Lightf. ex Dillw.) Grev.
Porphyra sp.
P. leucosticta Thur. in Le Jol.

TABLE 4 (continued)
List of Benthic Algal Species Identified in Barnegat Bay,
NJ

Rhodophyta (continued)

P. umbilicalis (L.) J. Ag.
Spermothamnion sp.
Spyridia filamentosa (Wulf.) Harv. in Hook.

Cyanophyta

Agmenellum quadruplicatum (Menegh.) Breb.
Calothrix sp.

Chrysophyta

Biddulphia pulchella Gray
Grammatophora marina (Lyngb.) Kütz.
Licmophora abbreviata Ag.
Navicula grevilleana Hendey
Rhabdonema adriaticum Kütz.
Striatella unipunctata (Lyngb.) Ag.
Synedra sp.

Xanthophyta

Vaucheria sp.

From Loveland, R. E., Brauner, J. F., Taylor, J. E., and Kennish, M. J.,
Ecology of Barnegat Bay, New Jersey, Kennish, M. J. and Lutz, R. A., Eds.,
Springer-Verlag, New York, 1984, 78. With permission.

many of them drifting passively on the bottom unattached to any substrate. Sixteen species have been examined for their epiphytes (Table 5).[120] Species of Chlorophyta, Phaeophyta, and Rhodophyta far exceed all others; the Chlorophyta and Rhodophyta taxa predominate in the spring and summer, whereas the Phaeophyta taxa are most conspicuous in the winter. Species diversity peaks in late spring. It is lowest from the late summer through the fall and rises with the establishment of the winter flora. Only six species occur in the estuary year-round, with 92 species being present for 6 months or less.[120]

The dominant species are similar from year to year. *Agardhiella subulata, Ceramium fastigiatum, Codium fragile, Gracilaria tikvahiae,* and *Ulva lactuca* typically rank among the top ten macrophyte species in the estuary (Table 6). *U. lactuca* usually has the greatest biomass of all macroalgal populations.[119] Moeller[121] estimates the standing crop of this species at 53.6 g C per square meter.

Biomass values of macroalgae in other East Coast estuaries of the U.S. support the contention that this floral group may contribute significantly to various estuarine processes.[96] In Great Bay, NH, for example, Chock and Mathieson[122] computed maximum macroalgal biomasses in the summer ranging from 480 to 590 g C per square meter. This estuary has an abundance of warm temperate or "mixed floras".[123] In the Niantic River, CT, the standing crop of *Codium* fragile purportedly exceeds 1200 g C per square meter.[124] Benthic macroalgal productivity frequently approaches or surpasses phytoplankton productivity in estuaries and coastal waters.[92,125]

TABLE 5
Substrate Algae and Their Epiphytes in Barnegat Bay, NJ

Substrate	Epiphytes
Chlorophyta	
Chaetomorpha linum	*Phaeophila viridis*
	Pseudendoclonium submarinum
	Ulvella lens
	Desmotrichum undulatum
	Ectocarpus siliculosus
	Halothrix lumbricalis
	Myrionema strangulans
	Myriotrichia filiformis
	Pilayella littoralis
	Ceramium rubrum
	Goniotrichum alsidii
	Polysiphonia denudata
	P. nigrescens
	Porphyra umbilicalis
Cladophora sericea	*Phaeophila viridis*
	Ulvella lens
Rhizoclonium sp.	*Audouinella* sp.
Ulva lactuca	*Bryopsis hypnoides*
	Phaeophila viridis
	Myrionema strangulans
	Champia parvula
Phaeophyta	
Fucus spiralis and *F. vesiculosus*	*Enteromorpha linza*
	Ulva lactuca
	Ectocarpus siliculosus
	Elachista fucicola
	Giffordia granulosa
	Pilayella littoralis
	Scytosiphon lomentaria
	Ceramium rubrum
	Erythrotrichia ciliaris
	Polysiphonia denudata
	P. harveyi
	Porphyra umbilicalis
Rhodophyta	
Agardhiella subulata	*Halothrix lumbricalis*
	Audouinella sp.
	Polysiphonia denudata
	P. nigrescens
	P. subtilissima
Ceramium diaphanum	*Phaeophila viridis*
	Audouinella sp.
C. fastigiatum	*Phaeophila viridis*
	Pilayella littoralis
	Audouinella sp.
C. rubrum	*Ulothrix implexa*
	Audouinella sp.
	Erythrotrichia ciliaris
Gelidium crinale	*Porphyra umbilicalis*
Gracilaria tikvahiae	*Cladophora albida*
	Sphaecelaria cirrosa
	Ceramium rubrum
	Polysiphonia denudata
Polysiphonia denudata	*Phaeophila viridis*

TABLE 5 (continued)
Substrate Algae and Their Epiphytes in Barnegat Bay, NJ

Substrate	Epiphytes
	Audouinella sp.
	Goniotrichum alsidii
P. harveyi	*Phaeophila viridis*
	Ulvella lens
	Punctaria latifolia
	Audouinella sp.
	Callithamnion roseum
	Goniotrichum alsidii
P. nigrescens	*Phaeophila viridis*
	Ulothrix flacca
	Ulvella lens
	Desmotrichum undulatum
	Audouinella sp.
	Erythrotrichia ciliaris
P. subtilissima	*Desmotrichum undulatum*
	Audouinella sp.
	Ceramium rubrum

From Loveland, R. E., Brauner, J. F., Taylor, J. E., and Kennish, M. J., *Ecology of Barnegat Bay, New Jersey,* Kennish, M. J. and Lutz, R. A., Eds., Springer-Verlag, New York, 1984, 78. With permission.

TABLE 6
The Top Ten Macrophyte Species Collected in Barnegat Bay between 1969 and 1973[a]

	Year				
Rank	1969	1970	1971	1972	1973
1	*Ulva lactuca*	*Ulva lactuca*	*Ulva lactuca*	*Ulva lactuca*	*Ulva lactuca*
2	*Codium fragile*	*Gracilaria tikvahiae*	*Zostera marina*[b]	*Zostera marina*[b]	*Gracilaria tikvahiae*
3	*Zostera marina*[b]	*Zostera marina*[b]	*Codium fragile*	*Gracilaria tikvahiae*	*Ceramium* sp.
4	*Gracilaria tikvahiae*	*Codium fragile*	*Gracilaria tikvahiae*	*Codium fragile*	*Enteromorpha intestinalis*
5	*Ceramium fastigiatum*	*Enteromorpha linza*	*Enteromorpha intestinalis*	Unidentified Ulvaceae	*Zostera marina*[b]
6	*Polysiphonia harveyi*	*Agardhiella subulata*	*Enteromorpha* sp.	*Enteromorpha intestinalis*	*Codium fragile*
7	*Cladophora* sp.	*Polysiphonia harveyi*	*Agardhiella subulata*	*Agardhiella subulata*	*Spyridia filamentosa*
8	*Agardhiella subulata*	*Ceramium* sp.	*Ruppia maritima*[b]	*Chaetomorpha aerea*	*Champia parvula*
9	*Ceramium* sp.	*Ceramium fastigiatum*	[c]	[c]	*Polysiphonia* sp.
10	*Chaetomorpha linum*	*Polysiphonia nigrescens*	[c]	[c]	*Polysiphonia nigra*

[a] Species ranked according to the percent dry weight summed over the entire year.
[b] *Zostera marina* and *Ruppia maritima* are vascular plants commonly found in algal samples.
[c] All other taxa contributed less than 0.1 g.

From Loveland, R. E., Brauner, J. F., Taylor, J. E., and Kennish, M. J., *Ecology of Barnegat Bay, New Jersey,* Kennish, M. J. and Lutz, R. A., Eds., Springer-Verlag, New York, 1984, 78. With permission.

c. Ecological and Economical Value

1. Ecological Significance

Seaweeds are a favorite source of food for some herbivores. Sea urchins consume considerable quantities of macroalgae and have completely decimated beds in some areas. Algal-free bottoms may be maintained by relatively low densities of sea urchins. Breen and Mann,[126] Chapman,[127] and Miller[128] calculate that sea urchin biomasses of 150, 250, and 140 to 250 g/m², respectively, keep the bottom nearly devoid of seaweed. Miller[128] outlined the following sequence of stages in the abundance of seaweeds and sea urchins based on data gathered in St. Margaret's Bay, Nova Scotia.

1. Sea urchins are abundant at the edge of dense algal beds and rare within the beds.
2. Sea urchins eat away the beds from the edge.
3. Only small refuges remain for the algal beds.
4. Sea urchins present in moderate abundance maintain the bottom free of algae.
5. Sea urchins experience mass mortality due to disease, but the refuges remain.
6. Sea urchin recruitment rises as larvae settle. The algal community develops through successional stages.
7. Sea urchin and algal biomass increase.

Benthic grazers other than sea urchins have the potential to regulate macroalgal populations. Some macroinvertebrates on mudflats control *Enteromorpha* blooms.[129] A few fishes, too, ingest macroalgae, especially their epiphytes.[128]

Benthic macroalgae not consumed while alive enter the detrital food web where they form an organic substrate rapidly colonized by microorganisms. The detritus generated from these plants, being less refractory than that derived from vascular plants, is decomposed rapidly. The macroalgae possess few or no resistant compounds, such as the cross-linked celluloses and lignins composing vascular plants. Therefore, in estuaries, they may be degraded within several weeks, whereas remains of vascular plants high in lignins may persist for months or years.[51,130] Although the removal of macroalgal detritus is comparatively rapid, the detritus can still constitute a substantial fraction of the particulate organic matter in these systems.[131] Decaying macroalgae, in fact, exacerbate water quality degradation — for example, in some sections of the Potomac River estuary[38] — by creating high nutrient loadings.

2. Economical Value

Seaweeds have been historically exploited by man for a host of domestic and industrial uses. They are a source of food, supplements in animal meals, medicine, fodder, fertilizer, fuel, and soil conditioners.[132] In addition, these plants have been utilized for their salts and phycolloids, or gums, and have been of value in paper production.[27,133,134] Table 7 presents the various uses of seaweeds in different parts of the world.

Most human consumption of seaweed takes place in the Far East. Green algae of the genera *Enteromorpha* and *Ulva* are prepared as salads. About 90 species of red algae and 70 species of brown algae have value as food. Of red algae, the genera *Chondrus* and *Gracilaria* are harvested for consumption. Harvesting of brown algae for food has concentrated on species of the orders Fucales and Laminariales.[27]

The utilization of seaweeds in industry centers around the production of phycolloid extracts (i.e., agar, alginate, and carrageenan). The value of these substances is a function of their gelling and stabilizing properties, accounting for multiple applications in the food and pharmaceutical industries.[132] Agar continues to be well known for its inclusion in microbiological media. Alginates have a wider range of uses in the food, paint, paper, and textile industries. Dawes[27] relates the subsequent applications of alginates in the food and pharmaceutical industries: (1) as emulsifiers and stabilizers in dairy products (e.g., sherbets, ice cream, cheese, chocolate, milk,

TABLE 7
Utilization of Seaweeds

Genus	Used for									Where used					
	Food					Industrial				Asia			North America	Europe	Other
	Utilization level	Cultivated	Human	Livestock	Agar-agar	Alginates	Paste	Chemical drugs	Manure	Japan	East	West and Southeast			
Green algae															
Caulerpa	x														Philippines
Chaetomorpha												x			South America
Codium			x	x				x		x	x	x			Oceania
Enteromorpha	x	x	x							x	x	x	E	x	Oceania
Monostroma	x	x	x							x	x	x		x	
Ulva	x	x	x	x					x	x	x		x	x	South America, Oceania
Brown algae															
Alaria	x	x	x	x	x					x			EW	x	
Ascophyllum		x		x									E	x	
Cladosiphon				x						x					
Cystophyllum										x		x			
Dictyota												x			Oceania
Durvillea		x										x			Australia/New Zealand, South America
Endarachne			x							x					
Ecklonia	x	x	x			x		x	x	x	x				
Eisenia	x	x	x			x		x	x	x	x				
Fucus			x	x					x				W		Oceania

Genus	1	2	3	4	5	6	7	8	9	10	Distribution
Analipus	xx	x		x		x					
Laminaria	x	x	x	x	x	x		x	x	E	x	Australia/North America, South Africa, Oceania
Macrocystis	x	x	x		x	x		x	x	W	x	
Mesogloia	x	x		x	x	x		x			
Nemacystis	x	x		x	x	x		x			
Nereocystis	x			x	x	x		x	W			Australia/New Zealand
Padina				x		x		x			
Pelagophycus				x		x		x	x	x	
Pelvetia	x	x		x	x	x		x	x		
Petalonia	x	x	x	x					x			South America
Phyllogigas				x							
Sargassum	x	x		x	x	x		x	x	x	
Tinocladia	x	x		x	x			x	x	x	
Turbinaria						x			x		
Undaria	xx	x	x	x	x	x		x	x		
Red algae											
Acanthopeltis			x		x			x	x		
Acanthophora	x					x		x	x		
Agardhiella	x								x		
Ahnfeltia	x		x	x	x	x		x	x	x		Oceania
Asparagopsis	x			x	x	x		x				Oceania
Carpopeltis	x	x		x			x	x	x		
Ceramium	x		x	x	x		x	x	x		
Chondrus	x	x	x					x	x		
Corallopsis											
Digenia					x						
Eucheuma	xx	x	x	x		x		x	x	E	x	
Furcellaria			x		x			x			
Gelidiopsis						x			x	x	
Gelidium	x	x	x	x	x	x		x	x	x	x	South America, Oceania
Gigartina	x	x	x	x	x	x	x	x	x	EW	x	
Gloiopeltis	x	x	x	x	x	x	x	x	x	x	
Gracilaria	x	x	x	x	x	x	x	x	x	x		Oceania
Grateloupia		x	x	x	x	x	x	x	x			Oceania
Griffithsia				x				x			
Gymnogongrus	x		x	x	x	x		x	x	x		Oceania

TABLE 7 (continued)
Utilization of Seaweeds

Genus	Utilization level	Cultivated	Human	Livestock	Agar-agar	Alginates	Paste	Chemical drugs	Manure	Japan	East	West and Southeast	North America	Europe	Other
			Food				**Industrial**			**Asia**			**North America**	**Europe**	**Other**
Hypnea	x	x	x		x					x		x			Oceania
Iridaea	x	x	x				x			x	x		W	x	Oceania
Laurencia			x								x	x		x	Oceania
Meristotheca			x	x						x	x				
Nemalion			x							x	x				
Pachymeniopsis							x			x					
Porphyra	xx	x	x							x	x		EW	x	South America/Australia/New Zealand
Pterocladia			x		x					x					Oceania
Rhodymenia	xx	x	x								x		EW	x	
Sarcodia										x		x			
Suhria												x			Africa
Turnerella							x			x					

From Dawes, C. J., *Marine Botany*, John Wiley & Sons, New York, 1981. With permission.

puddings, and toppings); (2) as thickeners in syrups, sauces, salad dressings, soaps, shampoos, toothpastes, shaving creams, medicines, lipsticks, insecticides, plastics, fireproofing fabrics, and polishes; and (3) as surgical threads in the medical field. Carrageenan, in turn, has uses in the food industry, largely in dairy products. It also has value in the manufacture of cosmetics and pharmaceutical products. Agar and carrageenan are obtained from red algae, and alginate derives from brown algae. For a comprehensive description of the industrial uses of seaweeds, the reader should consult the work of Dawes[27] and McLachlan.[132]

B. VASCULAR PLANTS
1. Salt Marsh Grasses
a. Species Composition

Along the margins of temperate estuaries, especially in sheltered areas, salt marsh communities — halophytic grasses, sedges, and succulents — develop on muddy sediments at and above the midtide level.[10] Salt marsh communities dominate the vegetation of intertidal zones in mid- and high-latitude regions (Figure 2).[2] Beeftink[135] identifies six types of salt marshes dependent on salinity and tidal range: (1) estuarine, (2) Wadden, (3) lagoonal, (4) beach plain, (5) bog, and (6) polderland. He gives examples of each type from the European coastline. Hence, estuarine-type salt marshes border the mouths of western European rivers. Salt marsh vegetation of the Wadden type fringes the Wadden Sea. The inner side of the "Nehrungen" on the south Baltic Coast and the Fleet landward of the Chesil on the Dorset Coast harbor the lagoonal type. The beach plain type lines Boschplaat, part of a Dutch Wadden island. The bog type can be perceived in South England, Southwest Ireland, and the Baltic. The Dutch estuarine polder area typifies the polderland salt marsh.

Differences between the salt marsh types have been ascribed to a number of factors, but chiefly: (1) the character and diversity of the indigenous flora; (2) the effects of climatic, hydrographic, and edaphic factors upon this flora; (3) the availability, composition, mode of deposition, and compaction of sediments, both organic and inorganic; (4) the organism-substrate interrelationships, including burrowing animals and the prowess of plants in affecting marsh growth; (5) the topography and areal extent of the depositional surface; (6) the range of tides; (7) the wave and current energy; and (8) the tectonic and eustatic stability of the coastal area.[136] Marked geographical disparaties highlight the species composition of the salt marshes. Along the Atlantic Coast of North America, for example, the diversity of the flora is lower than that along the North Atlantic Coast of Europe, with the cordgrass, *Spartina alterniflora,* dominating between mean sea level and mean high water and grading landward into species of *Juncus* and *Salicornia,* as well as *Distichlis spicata* and *S. patens.*[92] In Europe, marshes bordering the Atlantic Ocean, the English Channel, and the North Sea show distinct compositional variations. The Atlantic marshes are dominated by the genera *Festuca* and *Puccinella;* the South Coast of England, by *S. anglica* and *S. townsendii;* and the North Sea, by species of *Armeria, Limonium, Spergularia,* and *Triglochin,* in addition to the sea plantain, *Plantago maritima.*[92]

Figure 3 depicts the three principal groups of salt marshes in North America: (1) Bay of Fundy and New England marshes, (2) Atlantic and Gulf coastal plain marshes, and (3) Pacific marshes. More than 6×10^5 ha of salt marshes rim the Atlantic Coast of North America.[92] The Pacific Coast, a tectonically active region, is rather depauperate in salt marsh grasses, most being confined to relatively narrow fringes of lagoons and sheltered bays. The plant associations of the three major salt marsh groups are unalike and serve to define their respective areas.

Mitsch and Gosselink[18] and Frey and Basan[136] have compiled the characteristic flora of the three major salt marsh groups of North America. In New England salt marshes, lower and upper salt marsh plants are discernible. Two ecophenotypes (growth forms) of *Spartina alterniflora* exist in these habitats. The tall form of *S. alterniflora* covers the lower marsh on the intertidal zone, which lies adjacent to the estuary or tidal creek. As the upper marsh is approached, the

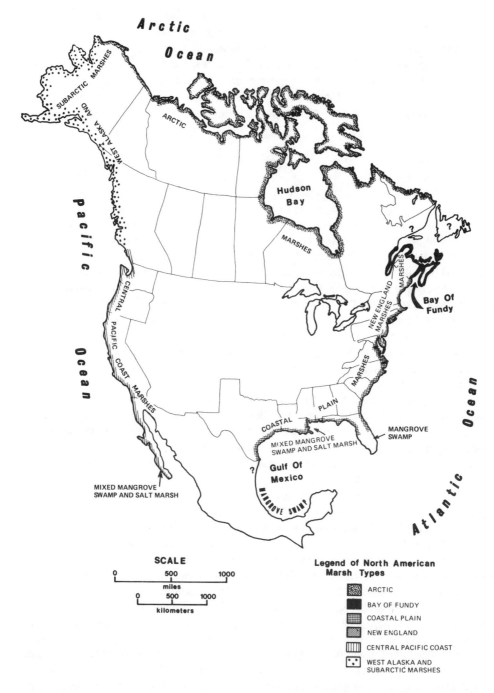

FIGURE 2.   Schematic distribution of the three major groups of salt marshes in North America: (1) Bay of Fundy and New England marshes, (2) Atlantic and Gulf coastal plain marshes, and (3) Pacific marshes. (From Frey, R. W. and Basan, P. B., *Coastal Sedimentary Environments,* 2nd ed., Davis, R. A., Ed., Springer-Verlag, New York, 1985, 225. With permission.)

stems of the species gradually shorten. The short form of *S. alterniflora* concentrates on the upper marsh together with *S. patens* (salt meadow grass), *Distichlis spicata* (spike grass), *Iva frustescens* (marsh elder), and forbs. Beyond the higher marsh, also termed the *S. patens* zone, pure stands of *Juncus gerardi* (black grass) grow at normal high tide. The upper limit of the tidal

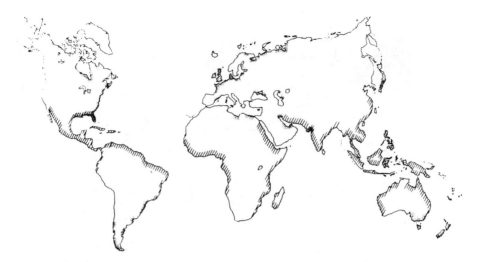

FIGURE 3.   Geographic distribution of salt marshes (stipled area) and mangroves (obliquely-hatched area). (From Ferguson, R. L., Thayer, G. W., and Rice, T. R., *Functional Adaptation of Marine Organisms,* Vernberg, F. J. and Vernberg, W. B., Eds., Academic Press, New York, 1981, 9. With permission.)

marsh is often fringed by the growth of *Panicum virgatum* (switch grass), *I. frustescens,* and other species of plants. Similar zones of salt marsh vegetation can be found on the East Coast of Canada.[22] Common high marsh floras of the Bay of Fundy region include *S. patens, Limonium nashi, Plantago oliganthos, Puccinellia maritima,* and *J. gerardi.*

On the Atlantic Coast south of Chesapeake Bay, the floral zonation pattern of coastal plain marshes parallels that of the marshes of New England. However, they differ by having (1) tall *S. alterniflora* frequently restricted to narrow bands along creeks, (2) a middle marsh zone typified by the diminutive form of *S. alterniflora,* and (3) an upper marsh in which *J. roemerianus* replaces *J. gerardi.*[18] Other plants generally occupying the upper marsh are *S. patens, D. spicata,* and *Salicornia* sp. (Figure 4).

The south Atlantic Coast possesses high marshes dominated by *S. alterniflora, Salicornia bigelovii, S. europaea, S. virginica, D. spicata,* and *J. roemerianus.* These plants form zones in mature marshes. Mangroves replace or intermix with salt marsh grasses in southern Florida and sections of the Gulf Coast.[136]

Depending on location, either *S. alterniflora* or *J. roemerianus* predominates in the coastal plain marshes of the Gulf states. The salt marshes of southern Alabama and Mississippi, for instance, are dominated by *J. roemerianus,* with *S. alterniflora* and *S. patens* acting as subdominants. In Louisiana, however, *S. alterniflora* is the dominant species and *D. spicata, J. roemerianus,* and *S. patens* play subdominant roles. A total of 347 species of halophytes have been recorded on the Atlantic and Gulf Coasts; 32 of these species live in all of the coastal states.[137] *S. alterniflora* is, by far, the dominant salt marsh plant in the intertidal zone of coastal regions.[2,138]

The diversity of salt marsh floras on the Pacific Coast of North America tends to be greater than that of the salt marsh floras on the Atlantic Coast; moreover, their zonation and succession patterns are more complex.[139] Approximately 54 species of plants have been chronicled in salt marsh biotopes south of the Canadian-U.S. border, and about 40 species, north of the border. *Puccinellia phryganodes* dominates the lower marsh of many areas along the northern Canadian and western Alaskan coasts, whereas *Spartina foliosa* is the most abundant plant of the lower marsh in many regions along the western North American Coast. Macdonald[140] points out five climatic groups of salt marshes on the Pacific Coast: (1) Arctic (northern Canada and northern and western coasts of Alaska), (2) subarctic (ranging from Anchorage, Alaska to the Queen

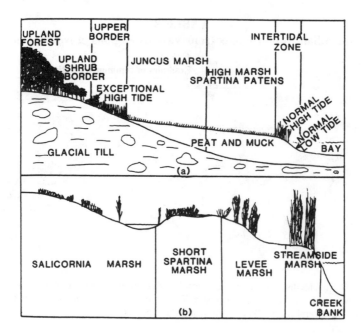

FIGURE 4. Plant zonation in typical salt marshes of North America. (a) New England salt marsh. (b) Georgia salt marsh. (a: From Teal, J. M., *Ecology*, 43, 614, 1962. b: From Miller, W. R. and Egler, F. E., *Ecol. Monogr.*, 20, 143, 1950.

Charlotte islands), (3) temperate (ranging from British Columbia to central California), (4) dry Mediterranean (southern California), and (5) arid (Baja, California and Mexico). Table 8 yields examples of salt marsh vegetation in various regions of North America.

b. Salt Marsh Formation

Salt marshes form in sheltered coastal areas where sedimentation is ensured, and erosion remains slight. The source of the sediment may be rivers, oceans, or the marsh itself. Establishment of the marsh is principally a function of the hydrographic and sedimentary regimes.[4,141] Salt marsh halophytes colonize mudflat surfaces between tidal levels of mean high-water neap (MHWN) and mean high water (MHW).[135] When sedimentation produces a surface above the MHWN, salt marsh formation can normally proceed since the MHWN usually sets a limit for the establishment of the halophytes.[17] Mud and silty sand provide substrates conducive to the early developmental stages of the marsh. A mat of protists, prokaryotes, and filamentous algae growing on the surface traps and binds sediment and promotes the progradation of the marsh. Accretion of the foreshore through sedimentation is necessary, but if sedimentation rates are too high, mature development of the marsh can be arrested. Sedimentation rates of 3 mm or less of accretion per year appear to be low, and 10 mm or more per year seem to be high. Rapid rates of sedimentation compromise species richness and can restrict or arrest successional development of the salt marsh system. Slow but gradual accretion enhances the maturation process.[13] Figure 5 illustrates the sequence of events in the development of a salt marsh.

In profile, salt marsh sediments generally have pedologic structures and exhibit soil types and horizons.[136] The fine sediments of the marsh may have multiple sources, being derived from rivers, the inner continental shelf, barrier washovers, erosion of cliffed headlands, alongshore drift of beach sediment, and marine organic aggregates. Tidal currents in the estuary transport the sedimentary particles to the marsh surface where they are deposited.[136] The action of sediment lag and scour lag associated with declining energy causes a lateral sorting of grain sizes, with particles progressively decreasing in size from the lower to the upper tidal marsh.

TABLE 8
Common Salt Marsh Plants from Various Regions of North America

Location	Examples of common vegetation	
	Lower marsh	Upper marsh
Eastern North America		
New England	*Spartina alterniflora*	*S. patens*
		Distichlis spicata
		Juncus gerardi
		S. alterniflora (dwarf)
Coastal Plain	*S. alterniflora*	*S. patens*
		D. spicata
		Salicornia sp.
		J. roemerianus
Bay of Fundy	*S. alterniflora*	*S. patens*
		Limonium nashii
		Plantago oliganthos
		Puccinellia maritima
		J. gerardi
Gulf of Mexico		
North Florida/South Alabama and Mississippi	Dominant *J. roemerianus* Subdominant *S. patens* *S. alterniflora*	
Louisiana	Dominant *S. alterniflora* Subdominant *S. patens* *D. spicata* *J. roemerianus*	
Arctic		
Northern Canada/Europe	*Puccinellia phryganodes*	*Carex subspathacea*
Western Alaska	*P. phryganodes*	*Puccinellia triflora*
		Plantago maritima
		Triglochin sp.
		P. maritima
Western North America		
Southern California	*S. foliosa*	*Salicornia pacifica*
		Suaeda californica
		Batis maritima

From Chapman, V. J., Salt Marshes and Salt Deserts of the World, Wiley-Interscience, New York, 1960, 392.

Turbidity maxima in proximity to the marshes probably contribute many fine grains to the adjacent marsh. Clay, silt, and floating detritus are carried landward on the marsh surface. In the marsh interior, this particulate matter admixes with biogenic sediments produced *in situ*. Plants and sediments of the marsh hasten sediment accumulation by damping waves, impeding current flow, and stabilizing the substrate. Bacterial and diatom films act as agents in the trapping of sediment, whereas roots of vascular plants anchor the substrate. All of these factors facilitate the buildup of thick accumulations of mud in most marshes while mitigating erosion of the marsh surface via wind-generated waves. When roots, stems, and leaves of salt marsh plants die and undergo microbial degradation, peat forms below the marsh surface and can attain appreciable

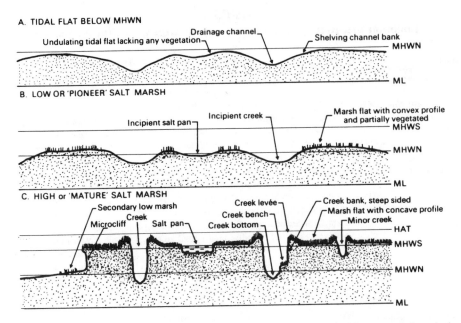

FIGURE 5.   Profiles of tidal flat and salt marsh habitats depicting physiographical features and a hypothetical sequence of salt marsh development. (A) High level tidal flat; (B) low level marsh: vascular plants colonize
the higher points of the tidal flat; and (C) high marsh: surface completely vegetated, creeks and pans sharply
defined. Secondary low marsh: develop where the surface has been eroded due to creek movement or the
formation of an erosion microcliff. (From Long, S. P. and Mason, C. F., *Saltmarsh Ecology*, Blackie & Son,
Glasgow, 1983. With permission.)

thicknesses, as reflected in some New England marshes.[142] In these marshes, changes in
physical-chemical conditions in sediments (e.g., substrate redox levels) with increasing marsh
elevation are determinants of pattern in the plant community.[143]

With continued sediment accretion, the salt marsh surface raises, and drainage improves.[4] A
sequence of changes in community composition — a succession — takes place in the salt marsh.
Stages of plant and animal succession have been related to the state of marsh maturation (i.e.,
youth, maturity, and old age). Frey and Basan (p. 238)[136] remark, "In geologic terms, marsh
maturation is a by-product of simple sediment progradation whereby the high marsh has
displaced the low-marsh by both vertical and lateral succession". In youth, the low-marsh habitat
nearly covers the entire marsh. In maturity, the low and high marsh occupy nearly equal areas.
In old age, the high marsh accounts for most of the total marsh area (Table 9). Rates of tidal
deposition are paramount to the development of the low marsh relative to the high marsh
regimes.[140] In a mature marsh, creeks or tidal drainage channels constitute distinctive physiographic features. Salt pannes also dot the surfaces of many of these systems.[17]

Dame[138] encapsulates the conceptual models formulated to describe salt marsh development.
The bi-directional model deals with the expansion of salt marshes from the accumulation of
sediments and submergence of upland areas. The deltaic or estuarine model describes the
formation of salt marshes at the mouth of major rivers. Finally, the emerging coast model treats
marsh development on tectonically active coasts, such as the West Coast of the U.S., where the
gradual emergence of the marsh results in the progressive replacement of the salt marsh plants
by upland species in a developmental sequence.

c. Ecological Importance
Five major ecological roles have been ascribed to salt marsh ecosystems.

TABLE 9
Idealized Stages in Marsh Maturation, Barrier Islands of Georgia[a]

Stage		Characteristics
Youth	A	Substantially more than 50% of area consists of low marsh; in early youth, total area may consist of low marsh, vegetated exclusively by *Spartina alterniflora*, initiated either as small marsh islands along higher parts of tidal flats or as narrow fringing marshes around sound or estuary margins; zonation patterns absent to very simple in early youth, giving way to a complex of repetitive low-marsh zones governed by the density and distribution of tidal drainages; high-marsh vegetation restricted mainly to terrestrial fringes
	B	Drainage systems well developed; pronounced meandering and crevasse splaying, with possible headward erosion of individual tributaries, during early youth; intensity of these processes declines progressively during middle and late youth and the positions of drainage channels become correspondingly more stable
	C	Relatively rapid sedimentation, especially during early youth; marsh substrates actively accrete, both vertically and laterally, until lateral growth is inhibited by margins of sounds or estuaries; during middle and late youth, accretion is most vertical, and the rate of deposition decelerates as the marsh increasingly approaches an equilibrium among topography, tidal hydraulics, and sediment supply
	D	Stratigraphic record consists predominantly of low-marsh environments, as modified by channel migrations; in many places the vertical sequence consists only of thin veneers of Holocene sediment spread over shallow basements of Pleistocene sand, rather than the fill of open lagoons or estuaries; these sand platforms are remnants of old barrier islands
Maturity	A	Low- and high-marsh areas approximately equal in size; low marsh and lower edge of high marsh vegetated by ecophenotypes of *S. alterniflora*; in early maturity the remaining high-marsh zones may consist of mixtures of *S. alterniflora*, *Salicornia* spp., or *Distichlis spicata*, followed by isolated stands of *Juncus roemerianus*, as is true in late youth, whereas in later maturity these plants typically occur either in mosaic clumps or in narrow, concentric zones
	B	Good drainage system remains, especially in low-marsh areas; but many tidal creeks are partially or totally infilled in the high marsh; in late youth and early maturity much lateral erosion and rotational slumping occur along creek banks; yet erosion in one place tends to be compensated by deposition in another, so that little net difference in channels results
	C	Relatively slow deposition; tidal sedimentation restricted mainly to low-marsh areas; extremely slow rates of deposition in the high marsh, except where barrier washovers occur or torrential rains sweep sands off adjacent Pleistocene or Holocene barrier island remnants; where supplies of washover or terrestrial sediments are not readily available and tidal hydraulics prevent significant deposition of clays and silts, the transition from mature to old-age marshes may be exceedingly slow
	D	Stratigraphic records are variable; those in low-marsh areas are similar to ones from youthful marshes, whereas ones in high-marsh areas may depict the succession from lowest to highest marsh, including numerous channel migrations and fills
Old age	A	Substantially more than 50% of area consists of high marsh; in late old age, virtually all of the area may consist of high marsh, including the encroachment of quasiterrestrial or terrestrial vegetation upon the marsh surface; in early and middle old age the zonation consists of concentric bands of short *S. alterniflora*, *Salicornia* spp., *Distichlis spicata*, and *Juncus roemerianus*, followed by or admixed with such plants as *Sporobolus*, *Borrichia*, and *Batis*; taller forms of *S. alterniflora* are restricted to the few drainage channels remaining
	B	Drainage mostly by surface runoff; most tidal channels are filled, and the marsh substrate is more or less uniform in elevation; aeolian processes are correspondingly more important in the distribution and reworking of sediments
	C	Extremely slow rates of deposition; tidal processes are largely ineffectual, and the transition to a terrestrial environment depends upon the availability of terrestrial sediments and mechanisms for their dispersal

TABLE 9 (continued)
Idealized Stages in Marsh Maturation, Barrier Islands of Georgia[a]

Stage	Characteristics
D	Stratigraphic records should show complete sequences, from underlying subtidal or low intertidal sediments, through the earliest marsh stages, to the oldest marsh or quasiterrestrial environments; old marshes probably are not obtained without the deposition of washover or terrestrial sands; hence, the final record would be a coarsening-upward sequence

[a] This model, which requires considerable testing, applies only to the more seaward marshes in Georgia, such as those associated with dissected barrier islands. More landward marshes, particularly those adjoining freshwater drainages, differ not only with respect to discrete modes of growth, but also by plant zonations. In brackish marshes, for example, *S. alterniflora* may be largely or totally replaced by *S. cynosuroides,* and *J. roemerianus* may occupy most of the marsh area.

From Frey, R. W. and Basan, P. B., *Coastal Sedimentary Environments,* 2nd ed., Davis, R. A., Ed., Springer-Verlag, New York, 1985, 225. With permission.

1.   They rank among the most productive ecosystems on earth.[144] Phytoplankton in tidal creeks, as well as benthic algae and vascular plants of the marsh, contribute a large quantity of organic matter. Above-ground production estimates of *Spartina* salt marshes range from approximately 200 to 3000 g C per square meter per year.[130] Below-ground production is even greater. Estimates of *S. alterniflora* root production range from about 220 to 2500 g C per square meter per year at creekside and 420 to 6200 g C per square meter per year inland.[145] Dame[138] stresses that the total net primary production of this plant on a global scale decreases as latitude increases, the highest value of 7800 g C per square meter per year being registered in Georgia[144] and the lowest reading of 1000 g C per square meter per year in Maine.[146,147] Recently, Dame and Kenny[148] computed net above-ground primary production figures of 2188 g C per square meter per year at creekside, 724 g C per square meter per year at midmarsh, and 1295 g C per square meter per year at high marsh for *S. alterniflora* marshes in the euhaline North Inlet estuary. Net below-ground primary production equaled 2363 g C per square meter per year at creekside and 5445 g C per square meter per year at high marsh. Production by mud algae on salt marshes in Massachusetts (105 g C per square meter per year), Delaware (160 g C per square meter per year), and Georgia (324 g C per square meter per year) amounts to 25 to 33% of the vascular plant production.[41,149-151] However, Zedler,[152] monitoring net primary production of mud algae in southern California, discovered levels between 320 and 588 g C per square meter per year, which equaled 76 to 140% of the vascular plant production. Measurements of the fraction of net primary production of salt marshes exported to adjacent estuaries range from 20 to 45%.[18] Although the salt marsh habitat is subject to widely variable temperatures, salinities, and alternating flooding and drying conditions, subsidies in the form of nutrient import, abundance of water, and tides foster high productivity.[18]

2.   Production of detritus in salt marshes is substantial, with up to 80% of the above-ground primary production and 100% of the below-ground primary production of *S. alterniflora* decomposing *in situ.*[138,153,154] With herbivorous grazing on salt marsh vegetation being slight, much of the organic carbon production passes to the detritus food web. Peat accumulates below-ground in the marsh at a rate of about 5% of the annual production. A three-phase decay of the litter of salt marsh grasses entails an initial phase of rapid weight loss, in which 5 to 40% of the litter is lost via leaching of soluble compounds.[154] This phase lasts less than 1 month. A second phase responsible for the loss of an additional 40 to 70% of the organic matter involves microbial degradation of the material and

subsequent leaching of hydrolyzed substances. This slower, second phase persists for as long as 1 year. The third phase, lasting perhaps another year, proceeds very slowly; only refractory substances (approximately 10% of the original material) remain. These phases of litter decay have been verified in a 2-year study of decomposition of four types of litter of *S. alterniflora*.[155] The extent and rate of litter decomposition, as shown by Wilson et al.,[155] depend largely on the chemical composition of the litter, especially the phenolic compounds. The interactions of the litter, microbes, and detritus-feeding organisms, along with the prevailing environmental conditions, control the rate of litter decomposition. Of the phenolic compounds composing *S. alterniflora* litter, the soluble phenolic acids act as herbivore deterrents and antimicrobial agents.[156] These stable compounds seem to inhibit detritivores.[157] Cell wall phenolics are principally hydroxycinnamic acids[158,159] that may hinder enzymatic hydrolysis of the cell wall polysaccharides.[158] The third group of phenolics — lignins — mechanically strengthens the cell walls of the plants. These heteropolymers provide for increased resistance of salt marsh plants to microbial degradation, thereby slowing the rate of decomposition. As salt marsh detritus accumulates and decays, it enriches in recalcitrant lignin and nitrogen. The sources of the nitrogen are microbial or fungal biomass or the condensation of reactive phenolics and carbohydrates with proteins and other products of microbial activity.[155,159-161] The export of salt marsh detritus into surrounding estuarine waters supports detritus-based production in the estuary that in many systems exceeds the phytoplankton-based production.

3. Salt marshes are a haven for many animals.[27] Leaves and stems serve as attachment sites for numerous epibiotic organisms.[18] Significant numbers of micro-, meio-, and macrobenthos live on or in the bottom sediments. Odum[162] postulated that salt marsh production nurtures as much as 95% of all recreational and commercial neritic fish caught in the fisheries.

4. Root systems of the vascular plants anchor the sediment, which stabilizes the substrate and mitigates erosion.

5. Salt marshes are sources or sinks of nutrients and trace metals.

d. Salt Marsh Fauna

Daiber[36,163] surveyed the literature on salt marsh faunal distributions, ecology, and life history. Teal[150] called attention to three major categories of invertebrate fauna which inhabit salt marshes, specifically: (1) terrestrial species (from the general marsh or from its upper limits), (2) aquatic species (from the seaward edge, from the creek sides, and from the general marsh), and (3) marsh-evolved species (either with or without a planktonic stage).[4] Protozoans (e.g., foraminifera) and meiobenthic populations (e.g., nematodes, harpacticoid copepods, oligochaetes, polychaetes, kinorhynchs, turbellarians, amphipods, and ostracods) account for a sizable fraction of the living biomass of surface sediments. Gerlach[164] conveyed that foraminifera and meiofauna are responsible for 12 and 30%, respectively, of the living biomass of the surface sediments. An estimated 20% of the food of deposit-feeding macrobenthos in subsurface sediments consists of these organisms.[164] Bacteria are an important food of protozoans and meiofauna, attaining densities as high as $10^9/cm^3$ in bottom sediments and $10^6/cm^3$ in overlying waters.[4]

Shelled mollusks include some of the most obvious inhabitants of the salt marsh. *Geukensia demissa,* the ribbed mussel, grows in clusters on the marsh surface. This bivalve can be seen in some areas nearly buried in sediment or attached to the stems of salt marsh grasses (e.g., *Spartina alterniflora)*. In the salt marshes of Sapelo Island, GA, the mussels have a highly clumped distribution, being most dense near the heads of small creeks.[165]

Other benthic macrofauna, such as gastropods, also are an abundant and common faunal component of salt marshes. *Melampus bidentatus,* the coffee bean snail, inhabits only salt marshes flooded by normal tides.[166] To avoid tidal inundation, these snails crawl up grass stems and debris.[36] Mud snails (*Hydrobia* spp.) and periwinkles (*Littorina* spp.) likewise utilize the salt marsh environment. *Hydrobia* is a deposit feeder that periodically retreats into the sediment in search of food and to avoid dessication.[4] The periwinkles consume plant matter,[167] grazing on salt marsh grasses and benthic algae.

Fiddler crabs, *Uca* spp., have been the subject of numerous studies in salt marshes, especially related to their reproductive and feeding behavior.[168] The burrowing activities of *Uca* may profoundly affect the marsh substrate and, secondarily, the production of vascular plants.[169-172] On the East Coast of North America from Massachusetts to Quintana Roo, Mexico, 15 species of fiddler crabs have been reported; their distributional patterns are strongly correlated with temperature, sediment type, and local or regional salinity.[173] As deposit feeders, fiddler crabs ingest detritus and attached microbes. However, at least one species, *U. pugnax*, may selectively feed on meiofauna.[4]

Teal[174] separated marsh crabs into three behavioral groups based on their responses to the stage of the tide. These groups are (1) *Eurytium limosum, Panopeus herbstii,* and *Sesarma reticulatum*, which become active at high tide or during cloudy meteorological conditions; (2) *U. minax* and *U. pugnax*, which are primarily active at low tide and feed at high tide underwater or at low tide in tidal pools; and (3) *S. cinereum* and *U. pugilator*, which remain active only in the air.[36] *Carcinus maenas*, the green crab, is the only European crab species recovered in U.S. salt marshes.

Less obvious but nevertheless of value to the overall ecology of salt marshes are smaller crustaceans — amphipods (e.g., *Orchestia gammarella* and *O. palustris)* and isopods (e.g., *Paragnathia maxillaris).* Many amphipod and isopod species undergo tidal migration, as do many copepod populations. The myriad of small burrows located along channel banks house marine amphipods and isopods.

Various populations of shrimp congregate in creek channels. Results of laboratory feeding experiments on postlarval brown shrimp *(Penaeus aztecus),* for example, indicate that planktonic diatoms (e.g., *Skeletonema costatum)* and epiphytes of *Spartina* are prime foods for this organism.[175] *P. aztecus* migrates into estuaries from the ocean during the postlarval stage, but returns to offshore waters as an adult. The postlarval penaeids require certain conditions, found in estuarine environments, for successful development.[176,177] They grow rapidly in these shallow coastal systems.[175]

Able et al.[178] determined that salt marsh peat reefs provide a habitat for small juvenile lobsters *(Homarus americanus)* (less than 40 mm in carapace length). Investigating salt marsh peat reefs at Nauset Marsh, Cape Cod, MA, Able et al.[178] advocated the peat reefs as potentially important nurseries for inshore lobster populations. All benthic stages of the lobster can use salt marsh peat reefs, rendering special significance to this habitat.

Tidal marsh creeks serve as a nursery habitat for many marine and estuarine fishes.[179,180] Gobies, menhaden, mullet, mummichog, and other ichthyofauna use creek channels.[4] The common mummichog, *Fundulus heteroclitus,* and the spotfin killifish, *F. luciae,* have life histories closely tied to salt marshes.[181] These two species consume various invertebrate taxa on vegetated marsh surfaces along much of the U.S. Atlantic Coast, while inhabiting the marsh environment year-round.[182-184] For both of these cyprinodontid species, the vegetated intertidal salt marsh represents a primary nursery habitat.[181]

Insects remove variable amounts of salt marsh grasses. Aphids, grasshoppers, moth larvae, and thrips graze heavily on salt marsh vegetation in certain areas. The grasshopper, *Orchelimum fidicinium,* feeds on leaves of *S. alterniflora*. Although Smalley[185] demonstrated that only 1% of the net aerial primary production of an *S. alterniflora* marsh was ingested by *O. fidicinium,* this insect may actually destroy far more plant material than it consumes.[186] Other grasshoppers sampled in salt marshes are *Chrothippus albomerginatus, Clinocephalus elegans, Conocephalus dorsalis, Nemobium sparselus,* and *Paroxua clavuliger*.[36] Another major grazing insect, the sap-sucking planthopper, *Prokelisia marginata*, subsists on material translocated through the vascular vessels of *Spartina* or on the contents of mesophyll cells.[138,186] The annual losses of *Spartina* grasses attributable to this insect may be much greater than the losses due to other herbivores.[186]

In a study of three estuarine marsh communities (i.e., *S. Alterniflora, Spartina-Distichlis-Salicornia,* and *J. roemerianus*), Barnes[187] addressed the ecological distribution of spiders. He observed highest spider densities and species numbers in the *S. alterniflora* zone. Limitations on web building in the higher two zones were deemed to be the overriding cause of the lower densities and species numbers. The sparsest populations existed in the *Juncus* zone which had little branching for spiders to build webs.[36]

Flies and mosquitos reach very high abundances in salt marshes. Efforts to control mosquito populations generally involve the filling of depressions and ditching of the marsh surface.[188] Salt marshes from New Jersey to North Carolina provide habitat for *Aedes, Anopheles,* and *Culex* mosquitos.[189-191] Larval greenhead flies (*Tabanus* spp.) and sheep flies (*Chrysops* spp.) distribute throughout areas of salt marshes of the mid-Atlantic Bight.[192-194] In addition to flies and mosquitos, some lepidopterans and midges are native to salt marshes.[4]

In Scott Head Island, Norfolk, England, Evans et al.[195] researched four species of beetles — *Bledius spectabilis, Cillenus lateralis, Dichirotrichus pubescens,* and *Heterocerus fenestratus.* Davis and Gray[196] examined ants, mainly in the high ground of the *S. patens* zone, and other insects in North Carolina marshes. Luxton[197,198] studied salt marsh acarines in England.

Salt marshes are ideal locations for many avifaunas. For example, avocets, herons, spoonbills, and other birds feed in marsh pools, while waders use saltings as roost areas.[4] Vegetational zonation, tidal flooding, and salinity influence bird distributions in this biotope.[36] Hence, the clapper rail, *Rallus longirostris crepitan,* is a resident of saline lower marshes, but the king rail, *R. elegans,* and the Virginia rail, *R. limicola,* prefer fresh and brackish marshes. Within the *S. alterniflora* zone of Georgia salt marshes, the resident clapper rail, *R. longirostris waynei,* has distributional preferences.[199] Burger et al.[200] and Burger and Olla[201,202] detail breeding, habitat selection, migration, and foraging behavior of marine birds and shorebirds, many of which frequent coastal marshes.

Coastal marshes are optimal foraging habitat for waterfowl. Stewart,[203] recounting research of the upper Chesapeake Bay, identified 13 major types of waterfowl habitat. Five of these habitats were in marshes differentiated by salinity distributions: (1) coastal embayed marsh, (2) salt estuarine bay marsh, (3) brackish estuarine bay marsh, (4) fresh estuarine bay marsh, and (5) estuarine river marsh. Table 10 inventories the ducks and geese associated with these marsh habitats. The fresh estuarine bay marsh had the most species (N = 13), and the salt estuarine bay marsh, the fewest species (N = 9). Intermediate numbers of species corresponded to the brackish estuarine bay marsh (N = 10), the coastal embayed marsh (N = 11), and the estuarine river marsh (N = 12). Despite not having the most species, the brackish estuarine bay marsh was considered by Stewart[203] to be the most valuable waterfowl habitat of the upper Chesapeake Bay. Of all waterfowl species, the black duck showed the widest distribution, being the only principal species in all five types of tidal marsh. Perry[204] has updated the work of Stewart,[203] supplying long-term data (1948—1986) on Chesapeake Bay waterfowl populations.

Ducks and geese consume salt marsh grasses, but only the greater snow goose, *Anser caerulescens atlantica,* has a noticeable impact on salt marshes. Smith[205] contended that geese in some North Carolina marshes can remove more than 50% of the plant biomass. Together with other terrestrial herbivores, therefore, these avifauna may clear broad areas of *Spartina.*[138]

Small mammals graze on salt marsh vegetation, although they have a much smaller effect than herbivorous insects and waterfowl. Included among this group are meadow mice (i.e., *Microtus californicus* and *M. pennsylvanicus*), rats, moles, muskrats, and shrews.[206,207] Mice other than meadow mice have been captured in salt marshes as well (e.g., house mice, *Mus musculus;* meadow jumping mice, *Zapus hudsonius;* and white-footed mice, *Peromyscus leucopus*).[36]

TABLE 10
Waterfowl Species Associated
with Marsh Habitats of Upper
Chesapeake Bay

Black duck
Ringed-neck duck
Wood duck
Mallard
Canada goose
Snow goose
Blue-wing teal
Green-wing teal
American widgeon
Common merganser
Hooded merganser
American coot
Gadwall
Pintail
Shoveler
Whistling swan

From Stewart, R. E., *U.S. Fish Wildl. Serv. Spec. Sci. Rep. Wildl.*, 65, 208, 1962. With permission.

2. Tidal Freshwater Marshes

Odum et al. (p. 1)[23] describe tidal freshwater marshes as a distinctive type of wetland "ecosystem located upstream from tidal saline wetlands (salt marshes) and downstream from nontidal freshwater wetlands (Figure 6). They are characterized by: (1) near freshwater conditions (average annual salinity of 0.5‰ or less except during periods of extended drought), (2) plant and animal communities dominated by freshwater species, and (3) a daily, lunar tidal fluctuation." While mesohaline and polyhaline estuarine marshes tend to be dominated by *Spartina alterniflora* and *S. patens* and oligohaline estuarine marshes by *S. cynosuroides,* tidal freshwater marshes are more diverse, with species of grasses, rushes, shrubs, broad-leaved plants, and herbaceous plants being represented. In North America, peak development of tidal freshwater marshes takes place on the East Coast of the U.S. from Georgia to southern New England. The most extensive tracts occur in the mid-Atlantic states.[23] New Jersey alone may have more than 8×10^4 ha of tidal freshwater marshes or about one half of the total amount of freshwater marsh coverage along the Atlantic Coast.[18]

a. Benthic Flora

Table 11 delineates the common species of vascular plants growing in the tidal freshwater marshes. High species diversity is characteristic, and species composition and community structure vary markedly from one geographical region to another.[22] According to Odum et al.,[23] tidal freshwater marsh plants typically consist of five floral assemblages: (1) broad-leaved emergent perennial macrophytes (e.g., spatterdock, arrow-arum, pickerelweed, and arrowheads); (2) herbaceous annuals (e.g., smartweeds, tear-thumbs, burmarigolds, jewelweed, giant ragweed, water hemp, and water-dock); (3) annual and perennial sedges, rushes and grasses (e.g., burbrushes, spike-rushes, umbrella-sedges, rice cutgrass, wild rice, and giant cutgrass); (4) grasslike plants or shrub-form herbs (e.g., sweetflag, cattail, rose-mallow, and water parsnip); and (5) hydrophytic shrubs (e.g., button brush, wax myrtle, and swamp rose).

Mitsch and Gosselink[18] correlated the distributions of plant tidal freshwater marshes with differences in elevation. In streams and permanent ponds, submerged vascular plants (e.g.,

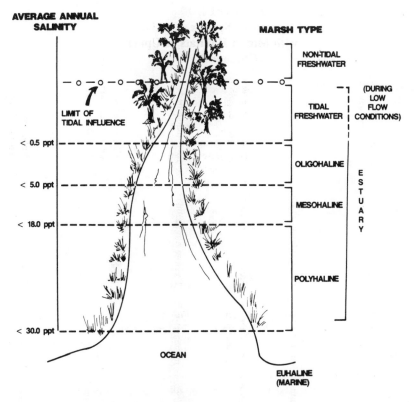

FIGURE 6. The distribution of tidal freshwater marshes in relationship to average annual salinity. (From Odum, W. E., Smith, T. J., III, Hoover, J. K., and McIvor, C. C., The Ecology of Tidal Freshwater Marshes of the United States East Coast: A Community Profile, FWS/OBS-83/17, U.S. Fish and Wildlife Service, 1984. With permission.)

Elodea spp., waterweed; *Myriophyllum* spp., water milfoil; *Nuphar advena,* spatterdock; *Potamogeton* spp., pondweeds) are usually encountered. Annuals — *Acnida cannabina* (water hemp), *Bidens laevis* (bur marigold), *Polygonum punctatum* (water smartweed), and other species — spread over the creek banks in summer, but disappear each fall. *Ambrosia trifida* (giant ragweed) populates the natural stream levee and gives way to broad-leaved monocotyledons on the low marsh behind the levee. Commonly inventoried on the low marsh are the flora *Peltandra virginica* (arrow arum), *Pontederia cordata* (pickerelweed), and *Sagittaria* spp. (arrowhead). A highly diverse group of annuals and perennials cover the high marsh. In the mid-Atlantic region, perennials dominate the high marsh early in the season, and annuals peak in this area somewhat later. *Acnida cannabina, Bidens laevis, Hibiscus coccineus* (rose mallow), *Pilea pumila* (clearweed), and *Polygonum arifolium* (tearthumb) are types of annuals living here. Nearly pure stands of *Spartina cynosuroides* (big cordgrass), *Typha* spp. (cattail), *Zizania aquatica* (wild rice), and *Zizaniopsis miliaceae* (giant cutgrass) frequently appear. Botanists have drawn generalizations regarding the plant associations in tidal freshwater marshes even though the species composition changes significantly with latitude.[18,23]

Primary production of tidal freshwater marshes parallels that of salt marsh systems, ranging from about 1000 to 3500 g C per square meter per year.[23] While much of the production in these freshwater marshes is due to vascular plants, phytoplankton and benthic algae also contribute organic carbon to the system. Mitsch and Gosselink[18] recognize three broad zones of primary production in tidal freshwater marshes: (1) a zone of low production on the low marsh bordering tidal creeks dominated by broad-leaved perennials; (2) a zone of highest production of freshwater species in areas of the high marsh dominated by perennial grasses and other erect,

TABLE 11
Common Species of Vascular Plants Occurring in the Tidal Freshwater Habitat

Species	General characteristics	Habitat preference	Salinity tolerance	Associated species
Acorus calamus (sweet-flag)	Grows in dense colonies propagating mainly by rhizome; stemless plants up to 1.5 m with stiff, narrow basal leaves; cylindrical inflorescence emerges from side of stem (open spadix); aromatic	Shallow water or wet soil; channel margins	Fresh	Peltandra virginica Polygonum spp. Impatiens capensis
Alternanthera philoxeroides (alligatorweed)	Hollow stems with simple branches bearing opposite, lance-shaped leaves; forms dense mats; flowers on long panicles; perennial	Extremely adaptable; often emersed	Fresh to oligohaline	—
Amaranthus cannabinus (water-hemp)	Erect, fleshy and stout; up to 2 m; leaves lanceolate with blades as long as 20 cm; not conspicuous until midsummer when it towers above other marsh forbs	Common to levee sections of the tidal marsh habitat; tolerates periodic inundation	Fresh to mesohaline	Peltandra virginica Polygonum spp. Bidens spp.
Asclepias incarnata (swamp milkweed)	Tall, leafy, pink-flowered herb growing solitary or in small, loose groups; lance-shaped, opposite leaves; reproduces via seeds or rhizomes	Cosmopolitan; grows in many wetland situations; high marsh species	Fresh to oligohaline	High marsh herbs
Bidens coronata B. laevis (burmarigold)	Annual plants up to 1.5 m tall, solitary or in small scattered groups; loosely branched above with opposite leaves; leaf shape variable but generally toothed or lanceolate; impressive yellow bloom late in the growing season	Cosmopolitan, growing in the upper two thirds of the intertidal zone on wet mud or in shallow water	Fresh	Polygonum spp. Amaranthus cannabinus Other Bidens spp.
Calamagrostis canadensis (reed-bentgrass)	Slender grass up to 1.5 m, generally forming dense colonies; long, flat leaves; loose, ovoid panicle with purplish color; perennial	Wet meadows and thickets	Fresh?	Typha spp. Acorus calamus

Species	Description	Habitat	Salinity	Associated species
Carex spp. (sedges)	Grasslike sedges, culms mostly three-angled, bearing several leaves with rough margins; up to 2 m tall and usually in groups; perennial from long, stout rhizomes	Low areas with frequent flooding or damp soil	Fresh	—
Cephalanthus occidentalis (buttonbush)	Branched shrub up to 1.5 m tall with leathery smooth opposite leaves and white flowers crowded into dense, spherical, stalked heads; flowers June through August; leaf petioles reddish	Upland margins and raised hummocks of tidal freshwater marshes; wet soil	Fresh to oligohaline	*Hibiscus* spp. *Cornus amommum*
Echinochloa walteri (water's millet)	Grass up to 2 m, solitary or in small groups; long, moderately wide leaf blades; flowers in a terminal panicle which is ovoid; greenish purple, and appears in July/August	Shallow water; moist areas, disturbed sites	Fresh to oligohaline	—
Hibiscus spp. *Kosteletzkya virginica* (mallows)	Shrubform herbs up to 2 m, scattered or in large colonies; leaves wedge-shaped or rounded and alternate; large, showy pink or white flowers appearing in midsummer; perennial	Freshwater marshes or the upland margin of saline marshes with freshwater seepage	Fresh to mesohaline	*Typha* spp. *Spartina cynosuroides* *Polygonum* spp. *Impatiens capensis*
Eleocharis palustris E. quadrangulata (spikerushes)	Perennials with horizontal rootstocks; culms stout, slender, and cylindrical or squarish with a basal sheath; flowers crowded onto terminus of spikelet; between 0.5 and 1.5 m	Channel margins or stream banks in shallow water; muddy, organic substrates	Fresh to oligohaline	*Pontederia cordata* *Scirpus* spp. *Juncus* spp. *Leersia oryzoides*
Impatiens capensis (jewelweed)	Annual plants up to 2 m with succulent, branched stems with swelling at the joints; colonial; leaves alternate and ovate or elliptic with toothed margins; flowers orange and funnel-like, appearing in July/August	Same as *Bidens* spp.; also grows in shaded portions of marshes	Fresh	*Bidens* spp. *Typha* spp. *Polygonum* spp.

TABLE 11 (continued)
Common Species of Vascular Plants Occurring in the Tidal Freshwater Habitat

Species	General characteristics	Habitat preference	Salinity tolerance	Associated species
Iris versicolor (blue flag)	Flat, swordlike leaves arising from a stout creeping rhizome; large, purplish-blue flowers emerge in spring from a stiff upright stem; perennial	High, shaded portions of the intertidal zone in damp soil; will not tolerate long inundations	Fresh	None in particular
Leersia oryzoides (rice cutgrass)	Weak slender grass growing in dense, matted colonies; leaf sheaths and blades very rough; emerges from creeping rhizomes and often sprawls on other vegetation	Midintertidal zones of marshes; high diversity vegetation patches	Fresh to oligohaline	Many; none in particular
Lythrum salicaria Decodon verticillatus (loosestrife)	Shrubform herbs forming large, dense colonies; aggressive; up to 1.5 m in height with lanceolate leaves opposite or whorled; upper axils branched with small purplish-pink flowers; terminal spikes pubescent; annual	Moist portions of marshes; high intertidal or upland areas	Fresh to oligohaline	*Hibiscus* spp. *Convolvulus* spp.
Mikania scandens (climbing hempweed)	Long, herbaceous vine forming matted tangles over other emergent plants; heart-shaped leaves; dense, pinkish flower clusters; slender stem; propagates by both seed and rhizome; perennial	Open, wooded swamps and marshes; shrub thickets	Fresh to oligohaline	—
Myrica cerifera (wax-myrtle)	Compact, tall, evergreen shrub with leathery alternate leaves; spicy aroma; waxy, berry-like fruits; forms extensive thickets	Most all coastal habitats; border between intertidal zone and uplands	Fresh	*Acer rubrum* *Nyssa* spp. *Taxodium distichum*
Nuphar lutecum (*N. advena*) (spatterdock)	Plant with floating or emergent leaves and flowers attached to flexible underwater stalks; rises from thick rhizomes imbedded in	Constantly submerged areas up to 1.5 m depth, or, if tidal, near or below mean low water in deep organic muds	Fresh	Usually in pure stand

Species	Description	Habitat	Salinity	Associated species
	benthic muds; flowers deep yellow, appearing throughout the summer			
Nyssa sylvatica *N. aquatica* (gum)	Medium-sized tree (10 m) with numerous horizontal, crooked branches; leaves crowded at twig ends turning scarlet in fall; flowers appear in April/May	Marsh/upland borders	Fresh	*Acer rubrum* *Myrica cerifera* *Alnus* spp.
Panicum virgatum (switchgrass)	Perennial grass 1—2 m in height in large bunches with partially woody stems; nest of hairs where leaf blade attaches to sheath; large, open, delicately branched seed head produced in late summer; rhizomatous	Dry to moist sandy soils or the midintertidal portions of tidal freshwater marshes; disturbed areas	Fresh to mesohaline	*Hibiscus* spp. *Scirpus* spp. *Eleocharis palustris*
Peltandra virginica (arrow-arum)	Stemless plants, 1—1.5 m tall, growing in loose colonies; several arrowhead-shaped leaves on long stalks; emerge in rather dense clumps from a thick subsurface tuber; flowers from May to June	Grows predominantly as an emergent on stream margins or intertidal marsh zones on rich, loose silt	Fresh to oligohaline	*Pontederia cordata* *Zizania aquatica* Many other species
Phragmites australis (common reed)	Tall, coarse grass with a feathery seed head; 1—4 m in height; grows aggressively from long, creeping rhizomes; perennial; flowers from July to September	Extremely cosmopolitan, growing in tidal and nontidal marshes and often associated with disturbed areas	Fresh to mesohaline	*Spartina cynosuroides* *Zizania aquatica*
Polygonum arifolium *P. sagittatum* (tearthumbs)	Plants with long, weak stems up to 2 m tall, usually leaning on other vegetation; leaves sagitate in shape and alternate; leaf midribs and stems armed with recurved barbs; flowers small and appearing in late summer; annual	Shallow water or damp soil; middle to upper intertidal zone	Fresh to oligohaline	*Bidens* spp. *Hibiscus* spp. *Impatiens capensis*
Polygonum punctatum *P. densiflorum*	Upright plants growing from a fibrous tuft of roots; narrowly to	Upper three quarters of intertidal zone in freshwater marshes on wet	Fresh to oligohaline	Many species

TABLE 11 (continued)
Common Species of Vascular Plants Occurring in the Tidal Freshwater Habitat

Species	General characteristics	Habitat preference	Salinity tolerance	Associated species
P. hydropiperoides (smartweeds)	widely lanceolate leaves with stalks basally enclosed within a membranous sheath; up to 1 m; flowers at spike at end of stalk	or damp soil		
Pontederia cordata (pickerelweed)	Rhizomatous perennial growing in dense or loose colonies; plants up to 2 m tall; fleshy, heart-shaped leaves with parallel veins and emerging from spongy stalks; flowers dark violet-blue, appearing June to August	Lower intertidal zone of tidal freshwater marshes	Fresh to oligohaline	*Nuphar luteum* *Peltandra virginica* *Sagittaria latifolia*
Rosa palustris (swamp rose)	Shrub up to 2 m growing in loose colonies; stems lack prickles except for those occurring at bases of leaf stalks; pinnately compound leaves with fine serrate margins; showy, pink flowers appearing July/August	High intertidal zones or wet meadows	Fresh to oligohaline	*Cephalanthus occidentalis*
Rumex verticillatus (water dock)	Erect, robust annual with dark-green, lance-shaped leaves; stem swollen at nodes; attains heights over 1.5 m and grows solitary or in loose colonies; flower head is evident in late spring and can be 50 cm in length	Wet meadows or pond margins on mud or in shallow water	Fresh to oligohaline	—
Sagittaria latifolia (duck-potato) *S. falcata* (bultongue)	Perennial herbs; stemless, up to 2 m in height and emerging from fibrous tubers; leaves arrowhead-shaped or lanceolate with white flowers in whorls appearing on a naked stalk in July/August	Borders of rivers or marshes in low intertidal zones on organic, silty mud	Fresh to oligohaline	*Peltandra virginica* *Pontederia cordata*

Species	Description	Habitat	Salinity	Other associates
Scirpus validus (soft-stem bulrush) *S. cyperinus* (woolgrass) *S. americanus* (common three square)	Medium to large rushes with cylindrical or triangular stems; inconspicuous leaf sheaths; usually grow in small groups; bear seed clusters on end or side of stem; perennial	Brackish to fresh shallow water or low to middle intertidal zones on organic clay substrates	Fresh to mesohaline	Other rushes *Typha* spp.
Sparganium eurycarpum (great burreed)	Stout upright forbs up to 1 m with limp, underwater, emergent leaves attached basally and alternating up the stem; toward the terminus, stems zig-zag bearing sphere-like clusters of pistillate and staminate flowers	Partially submerged, shallow water marsh areas; lower to middle intertidal zones	Fresh	*Zizania aquatica* *Leersia oryzoides* *Polygonum* spp.
Spartina cynosuroides (big cordgrass)	Perennial grass attaining heights in excess of 3 m, having long, tapering leaves and growing from vigorous underground rhizomes; found in dense monospecific or mixed stands	Channel and creek margins in tidal oligohaline marshes	Fresh to mesohaline	*Phragmites australis* *Typha* spp.
Taxodium distichum (bald cypress)	Tall tree with straight trunk (40 m), conifer-like but deciduous; light porous wood covered by stringy bark; unbranched shoots originating from roots as knees	Marsh/upland borders	Fresh?	*Nyssa* spp. *Acer rubrum*
Typha latifolia *T. domingensis* *T. angustifolia* (cattails)	Stout, upright reeds up to 3 m forming dense colonies; basal leaves, long and sword-like, appearing before stems; yellowish male flower disintegrates leaving a thick, velvety-brown swelling on the spike; rhizomatous; perennial	Very cosmopolitan, occurring in shallow water or upper intertidal zones; some disturbed areas	Fresh to mesohaline	Many associates
Zizania aquatica (wild rice)	Annual or perennial aquatic grass, 1—4 m tall, usually found in colonies; short underground roots, stiff hollow stalk, and long, flat, wide leaves with rough edges; male and female flowers separate along a	Fresh to slightly brackish marshes and slow streams, usually in shallow water; requires soft mud and slowly circulating water	Fresh to oligohaline	*Peltandra virginica* Many other species

TABLE 11 (continued)
Common Species of Vascular Plants Occurring in the Tidal Freshwater Habitat

Species	General characteristics	Habitat preference	Salinity tolerance	Associated species
	large terminal panicle in late summer			
Zizaniopsis mileacea (giant cutgrass)	Perennial by creeping rhizome; culms 1—4 m high; long, rough-edged leaves geniculate at lower nodes; large, loose terminal panicles appearing in midsummer; aggressive	Swamps and margins of tidal streams	Fresh?	—

From Odum, W. E., Smith, T. J., III, Hoover, J. K., and McIvor, C. C., The Ecology of Tidal Freshwater Marshes of the United States East Coast: A Community Profile, FWS/OBS-83/17, U.S. Fish and Wildlife Service, Washington, D.C., 1984.

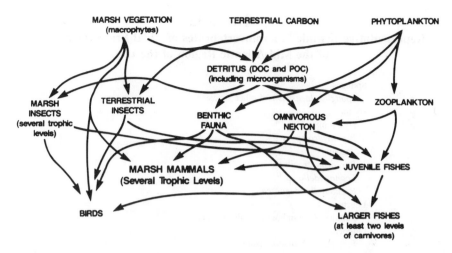

FIGURE 7. Hypothetical energy flow diagram for a tidal freshwater marsh. (From Odum, W. E., Smith, T. J., III, Hoover, J. K., and McIvor, C. C., The Ecology of Tidal Freshwater Marshes of the United States East Coast: A Community Profile, FWS/OBS-83/17, U.S. Fish and Wildlife Service, 1984. With permission.)

tall species; and (3) a zone of high production late in the season attributable to the high marsh mixed annual association. At least five factors affect the production estimates: (1) species composition and density of plants; (2) heterogeneity in plant community composition; (3) seasonal patterns of biomass allocation; (4) seasonal biomass turnover ascribable to leaf mortality, decomposition, and herbivory; and (5) inherent problems measuring background production.[23]

Figure 7 depicts an energy flow diagram for a tidal freshwater marsh; the arrows transcribe energy pathways. Major groups of organisms are labeled in the diagram. The marsh ecosystem has three principal sources of organic carbon, the largest source being vascular plants followed by terrestrial flora and phytoplankton. It is primarily a detritus-based system.[23]

Table 12 compiles representative benthic macrofauna of mid-Atlantic, tidal freshwater marshes. The benthic fauna of the marshes provide ration for many species of fishes and birds. Ichthyofauna sampled in this habitat include freshwater and estuarine species, anadromous populations, and a few marine forms. Odum et al.[23] formulated a list of the fishes collected in this environment.

Tidal freshwater marshes are ecologically and economically important because they serve as valuable nursery grounds for recreationally and commercially important fishes.[208] In addition, birds and wildlife populations utilize the marsh area, and some endangered species have been observed here as well. Finally, this system may be a source or sink of materials (e.g., nutrients) for estuaries.[209]

3. Seagrasses

a. Species Composition

An extremely widespread group of vascular plants, the seagrasses, occurs in estuarine and coastal marine environments of all latitudes except the most polar. These monocotyledonous angiosperms are highly productive and form a habitat for faunal assemblages. A well-developed root and rhizome system, strap-like leaves, and stems create subhabitats for a rich faunal community. Kikuchi and Peres[80] reveal four subhabitats associated with seagrasses: (1) sediment fauna, (2) rhizome and stem biota (e.g., amphipods, bivalves, and polychaetes), (3) leaf epiphyton (e.g., microbiota, anemones, hydroids, sea-mats, crustaceans, echinoderms, nema-

TABLE 12
Representative Benthic Macroinvertebrates of Tidal Freshwater Marshes
of the Mid-Atlantic Region of the U.S.

Sponges
   Spongilla lacustris and other species
Hydra
   Hydra americana
   Protohydra spp.
Bryozoans
   Barentsia gracilus
   Lophopodella sp.
   Pectinatella magnitica
Leeches
   Families Glossiphoniidae, Piscicolidae
Oligochaetes
   Families Tubificidae, Naididae
Insects
   Dipteran larvae (especially family Chironomidae)
   Larvae of Ephemeroptera, Odonata, Trichoptera, and Coleoptera
Amphipods
   Hyallela azteca
   Gammarus fasciatus
   Lepidactylus dytiscus (southeastern states)
Crustaceans
   Crayfish
   Blue crab, *Callinectes sapidus*
   Caridean shrimp, *Palaemonetes paladosus*
Mollusks
   Fingernail clam, *Pisidium* spp.
   Asiatic clam, *Corbicula fluminea* (formerly *C. manilensis*)
   Brackishwater clam, *Rangia cuneata*
   Pulmonate snails (at least six families)

From Odum, W. E., Smith, T. J., III, Hoover, J. K., and McIvor, C. C.,
The Ecology of Tidal Freshwater Marshes of the United States East
Coast: A Community Profile, FWS/OBS-83/17, U.S. Fish and Wildlife
Service, Washington, D.C., 1984.

todes, polychaetes, and gastropods), and (4) nekton which swim among the leaves (e.g., cephalopods, crustaceans, and fish).[4]

Forty-nine species of seagrasses belonging to 12 genera have been identified worldwide, as noted in Section II.A. Three of the genera are classified in the family Hydrocharitaceae (i.e., *Enhalus, Halophila,* and *Thalassia*) and nine in the family Potamogetonaceae (i.e., *Amphibolis, Cymodocea, Halodule, Heterozostera, Phyllospadix, Posidonia, Syringodium, Thalassodendron,* and *Zostera*). These vascular plants have adapted to saline conditions, living submerged in estuaries or shallow coastal waters where they withstand waves and tidal currents.[24] Although attaining maximum biomass in subtidal waters, these hydrophytic monocots may extend into the lower intertidal zone, remaining independent of any salt marsh vegetation.

The geographical distribution of seagrasses is a function of climate, salinity, light, and turbidity. On a global scale, more genera of seagrasses live in tropical (seven) than temperate (five) waters.[27] Eleven species of seagrasses belong to the genus *Zostera,*[4] the most important taxum of midlatitude estuaries. *Z. marina* has the widest distribution of all the species of *Zostera,* proliferating in muddy and sandy sediments of temperate and boreal waters of both the Atlantic and Pacific Oceans.[210] In tropical climates, *Cymodocea, Halodule,* and *Syringodium* inhabit mud, sand, and coal-debris bottoms along with *Thalassia. Cymodocea* and *Posidonia* are common in the Mediterranean. In Australia, warm temperate waters support concentrations of *Amphibolis* and *Posidonia.*

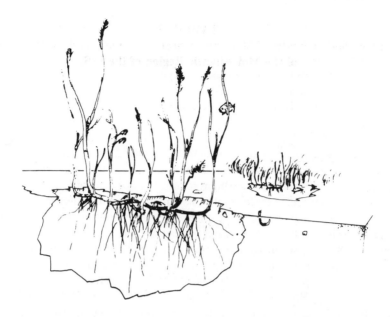

FIGURE 8. *Zostera marina* meadow displaying leaves, shoots, rhizomes, and roots. Network of rhizomes and roots below the sediment surface supply substantial amounts of organic detritus to estuaries. (From Kenworthy, W. J. and Thayer, G. W., *Bull. Mar. Sci.*, 35, 364, 1984. With permission.)

The majority of seagrass species (approximately 75%) occurs in subtidal waters of the Old World. The Indian Ocean and waters of the western Pacific serve as the center of distribution of the Old World forms. The Caribbean represents the center of distribution of the other seagrass species (approximately 25%).[27] The genera *Halodule, Phyllospadix, Thalassia,* and *Zostera* contain the most abundant species in North America.

Within an estuary, salinity, light, and turbidity are major environmental factors limiting the distribution and growth of seagrasses. While *Halodule* and *Zostera* inhabit waters with salinities as low as 10‰, *Cymodocea, Posidonia, Thalassia,* and *Thalassodendron* prefer higher salinities in excess of 20‰.[10] The observed distribution of seagrass species in some systems, although largely due to salinity effects, may be confounded by other factors, such as turbidity. In highly turbid estuaries, the distribution of seagrasses is restricted as a result of the diminution of light penetration in the water column. A gradient in light availability with increasing depth from intertidal to shallow subtidal zones causes distinct banding of seagrass populations, as light attenuation rises with elevated seston concentrations.[211] For example, in Barnegat Bay, NJ, *Z. marina* abuts the shoreline in waters less than 1 m deep.[119] *Zostera* lives at much greater depths in clearer systems, having been collected 6 m or more below the sea surface.[10] It extends to depths up to 30 m on the California Coast.[212]

Seagrasses are well adapted to the physical stresses of the estuarine environment, and establish thick meadows on muddy sediments and other soft substrates. As colonization proceeds, the seagrasses reduce water flow, thus increasing sedimentation rates within the beds. In addition to roots which stabilize the plants in the bottom, horizontal stems or rhizomes branch below the sediment surface to anchor the organisms (Figure 8). Roots project downward from the base of each shoot as well as from the rhizomes. They penetrate anoxic sediment layers and act as a pathway for the release of nutrients (e.g., ammonia and phosphate) from the sediments through the plants.[213,214] The extensive root and rhizome systems, together with dense strap-like leaves, trap sediments and mitigate erosion.[215-217, 644]

Under ideal conditions, seagrasses have very large biomasses and high rates of primary production. They form dense meadows of a single species or mixtures of two or more species.

In less productive areas, scattered clusters or isolated individuals rise above the sediment-water interface. Production figures range from 58 to 1500 g C per square meter per year, surpassing maximum biomass values by about 2 to 2.5 times.[130] Highest seagrass production takes place in subtropical and tropical waters. The biomass of epiphytes attached to seagrasses parallels that of the vascular plants themselves.[92,218,219]

b. Benthic Fauna

Highly diverse assemblages of invertebrates and fishes use seagrass biotopes as a nursery, as a feeding ground, and as an area for reproduction. Summerson and Peterson (p. 63)[220] add that "In temperate and subtropical embayments, benthic invertebrates sampled by corers, grabs, or bottom dredges characteristically exhibit higher densities and species diversities within seagrass beds than in nearby unvegetated sediments". A similar pattern operates in tropical seagrass systems.[220,221] Two hypotheses possibly explain the higher densities and species diversities within seagrass beds: (1) the baffling action of seagrass leaves traps invertebrate larvae and suspended food; and (2) seagrass plants inhibit predation, enabling greater densities and diversities of benthic fauna within the seagrass beds compared to unvegetated sediments. An overall greater habitat complexity afforded by the seagrasses perhaps accentuates these local differences.[222] Hence, in Back Sound, NC, Summerson and Peterson[220] divulged substantially greater densities of epifauna (52 times greater) and infauna (3 times greater) in a seagrass bed composed of a mixture of eelgrass, *Z. marina,* and shoalgrass, *Halodule wrightii,* than on an adjacent sandflat. Quantitative differences of benthic fauna have also been ascertained in temperate eelgrass *(Zostera)* beds in North Carolina,[223] Maryland,[215] southern California,[224] the Netherlands,[225] and Japan,[226] as well as in shoalgrass (*Halodule* spp.) and subtropical turtlegrass (*Thalassia* spp.) systems.[227-230]

Kikuchi[26] offers five reasons why the seagrass bed is a favorable habitat or shelter for associated fauna. First, seagrass vegetation effectively increases the substrate surface area for epiphytic flora and fauna. Second, the seagrass bed mitigates wave and current action, generating calm underwater space preferred by many animals. Third, because of reduced water movement, suspended mineral and organic particles sink more easily to the seagrass bed, thereby building up the bed. Some of the particles supply food for the biota. Fourth, the canopy of dense leaves shades the estuarine seafloor, causing lower insolation than in surrounding unvegetated areas. Protected from excessive illumination and insolation, this shaded habitat appears to be beneficial to the benthos. Fifth, conditions of shelter and high food supply make seagrass beds attractive to juvenile and small-sized nekton.

Controversy abounds as to whether the abundance of macrofauna in dense seagrass meadows is due to habitat preference or predation. Caging experiments lend credence to the differential-predation hypothesis.[220] These experiments verify that the exclusion of large, mobile predators (e.g., crabs and fishes) has little effect on benthic organisms in seagrass beds, but a substantial impact on benthic communities in unvegetated habitats.[231-236] However, past research has not clearly established predation as the proximate cause of correlations between abundance of prey and structural complexity of all seagrass beds.[237] For instance, Bell and Westoby,[237] in field experiments on a subtidal bed of *Z. capricorni* in Botany Bay, New South Wales, Australia, demonstrated how correlations of the abundance of prey and density of shoots within a seagrass bed are proximately caused by habitat preference of prey.

1. Microfauna and Sessile Fauna

As mentioned previously, the faunal community of a seagrass bed can be subdivided into four categories based on the mode of life of the constituent fauna and the microhabitat structure (see Section II.A). These four categories are the epiphytic felt flora and associated fauna, sessile fauna, mobile epifauna, and swimming fauna. As an example, three groups of epiphytic algae live on *Posidonia* leaves: (1) diatoms, tiny phaeophytes (e.g., *Ectocarpus* sp. and *Ascocyclus orbicularis),* and other small species which form a felt-like coating; (2) encrusting calcareous populations (e.g., *Melobesia lejolisi);* and (3) larger more or less erect soft algae (e.g., phaeophytes: *Castagnea mediterranea, C. irregularis, C. cylindrica,* and *Gyraudia sphacelarioides;* and rhodophytes: *Ceramiales* sp. and *Nemalionales* sp.). The epiphytic felt harbors several groups of microfauna, namely, protozoans (e.g., ciliates and flagellates), free-living nematodes, polychaetes (e.g., *Pionosyllis pulligera* and *Polyophthalmus pictus),* rotifers (e.g., *Colurus leptus* and *Notommata naias),* tardigrades, copepods (e.g., *Dactylophusia tisboides, Idiaea furcata,* and *Laophonte stromi),* and amphipods (e.g., *Amphithoe rubricata).* The dominant sessile species, comprising more than 95% of the existing sessile invertebrates, are hydroids (i.e., *Campanularia (=Orthopyxis) asymmetrica, Monotheca posidoniae,* and *Setularia perpusilla)* and bryozoans (i.e., *Electra posidoniae* and *Microporella johannae).* Other commonly occurring sessile invertebrates include foraminiferans (e.g., *Iridia serialis, Rhizonubecula adhaerans,* and *Webbinella crassa)* and actinians (e.g., *Parastephanauge paxi).*

2. Mobile Epifauna (Creeping, Crawling, or Walking Forms)

Small opisthobranch gastropods (e.g., *Elysia viridis, Glossodoris gracilis, Hancockia uncinata, Loviger serradifalci,* and *Polycera quadrilineata)* and prosobranch gastropods (e.g., *Alvania, Bittium, Marginella, Persicula, Rissoa,* and *Tricolia)* account for the majority of the creeping fauna on *Posidonia. Asterina pancerii,* an asteroid, *Eleutheria dichotoma,* a medusa, and *Propeamussium hyalinum,* a bivalve, represent other characteristic creeping forms. The walking fauna are predominantly crustaceans, such as the amphipod, *Cymadusa crassicornis,* the harpacticoid copepod, *Porcellidium fimbriatum,* the isopods, *Astacilla mediterranea, Idothea hectica,* and *Synisoma appendiculata,* and the pagurids, *Catapaguroides timidus, Eupagurus anachoraetus,* and *E. chevreuxei.*

3. Swimming Fauna

Copepods, ostracods, mysids, and other small animals which rest by clinging to *Posidonia* leaves via their claws embody most of the members of this group. Characteristic species are *Palaemon xiphias,* mysids (e.g., *Siriella clausi* and *S. jaltensis),* as well as the hippolytid shrimps (e.g., *Hippolyte inermis, H. gracilis,* and *Thoralus cranchi).* Some fauna rest on *Posidonia* leaves by employing suckers, notably, the Anthomedusae (e.g., *Cladonema radiatum),* Limnomedusae (e.g., *Gonionemus vertens, Olindias phosphorica,* and *Scolionema suveense)* and Chaetognatha (e.g., *Spadella cephaloptera).*

The bulk of the swimming species below the leaf canopy are fishes, some of the more important taxa being the Labridae (e.g., *Crenilabrus, Ctenolabrus,* and *Julis),* Mullidae (e.g., *Mullus),* and Sparidae. Invertebrates worthy of mention are *Octopus vulgaris* and *Sepia officinalis.* Various decapod crustaceans may also be placed in this category; they commonly inhabit the "matte" layer or live at the base of the leaves (e.g., *Alpheus, Athanas, Macropipus, Processa,* and *Scyllarus).*

c. Ecological Significance

In addition to the high biological productivity typified by seagrasses, five important ecological roles have been assigned to these marine angiosperms. They are a source of food for grazing invertebrates and vertebrates, chiefly sea urchins, fishes (Acanthuridae and Scaridae), turtles, and waterfowl (ducks and geese).[238] Fry and Parker[25] confirmed that seagrasses and associated algae in Texas bays were primary nutrient sources for shrimp and fish. More than 340 animals consume seagrasses in the bays. Since few invertebrate species graze directly on live seagrasses, most of the production passes to the detrital pool. In the last decade, researchers have applied stable carbon isotope ratios as a tracer of food webs involving seagrasses.[239-242] Part of the eelgrass detritus may be retained or recycled in the sediments, and part exported as particulate organic matter or dissolved organic matter.[243] However, the relative amount of detrital material exported from the eelgrass bed, as well as the exact type of material exported, has not been established.[244,245] Seagrass detritus exported to unvegetated areas of the estuary supplies nutrition to microbes and larger detritivores.[243]

Having great potential as a stable food source, seagrasses resist microbial decay, being more recalcitrant to microbial breakdown than even *Spartina*.[90,246] Decomposition rates also seem to depend on where the detritus is deposited. For example, the leaves of turtlegrass purportedly decay more rapidly when deposited in intertidal zones than when they remain in submerged grassbeds.[247]

Seagrass beds shelter many aquatic populations, offering the organisms a refuge from predators. The beds are nursery areas for a large number of invertebrates and fishes,[248] and some economically important finfish and shellfish species rely heavily on them.[249,250] The decline of the commercial shrimp, *Penaeus duorarum*, in Florida has been ascribed to the dredging of seagrass beds.[251]

Because of their significance as attachment sites for a multitude of epiphytes that support food webs in estuaries, seagrasses play a vital role in the energy flow of these coastal systems.[252,253] The diet of numerous invertebrates includes epiflora attached to seagrasses.[58,67,254-258] Harlin[68] published a list of algal and invertebrate epiphytes of seagrass blades. More than 180 invertebrate epiphytes, 150 microalgal epiphytes, and 450 macroalgal epiphytes were recorded.[27,68] Seagrass blades may promote habitat complexity which increases the density and/or diversity of motile seagrass epifauna.[258]

Seagrasses remove nutrients from sediments and overlying waters, and due to this activity, they are important to nutrient cycling processes in estuaries.[259] They act as nutrient pumps; consequently, their presence assists in the regulation of water quality of shallow waters.[260] The removal of nutrients (e.g., phosphorus) from bottom sediments by seagrasses and their subsequent release to surrounding waters enhance the productivity of estuaries.

By trapping sediments and protecting the seafloor from the devastating impact of major storms, such as hurricanes, seagrasses have great potential value in coastal protection.[261] Hence, their ability to stabilize bottom sediments and mitigate erosion[2,130,262-267] not only has ecological ramifications, but economical implications as well. Seagrasses prevent millions of dollars worth of shoreline losses annually, especially in the Gulf of Mexico and contiguous waters which periodically experience severe storms in late summer and fall with a concomitant translocation of large volumes of sediment.

C. MANGROVES

Along subtropical and tropical coasts, associations of halophytic trees, shrubs, palms, and creepers form dense tickets or forests classified as mangroves or mangals. As explicated by Mann,[92] the term mangal is often used by ecologists to define the total community of plants and the term mangrove, to denote an individual shrub or tree, although both terms have been applied to either. These plant systems are well established in protected embayments, tidal lagoons, and estuaries between 25° N and 25° S latitudes.[4,268] The mangrove forest ranges locally from the highest-tide mark down nearly to mean sea level.[27,92] Sedimentation is typically high. The mangrove trees are shallow rooted, having prop or drop-type roots that terminate only a few centimeters in the ground; in some cases, cable roots extend horizontally from the stem base and support air roots (i.e., pneumatophores) projecting vertically upward through the sediment surface. Anchoring and feeding types of roots develop on prop, drop, and cable roots.[27]

1. Species Composition

The global distribution of mangroves has been related to continental drift.[268] The taxonomic list of these organisms incorporates monocots and dicots (primarily) (Table 13) which may be arranged into two main groups — the Old World mangal containing approximately 60 species

TABLE 13
A List of Families and Genera of Mangroves

Families and genera	Number of species	Indian Ocean West Pacific	Pacific America	Atlantic America	West Africa
Dicots					
Avicenniaceae					
Avicennia	11	6	3	2	1
Bombacaceae					
Camptostemon	2	2	0	0	0
Chenopodiaceae					
Suaeda[a]	2	0	0	1	1
Combretaceae					
Conocarpus[b]	1	0	1	1	1
Languncularia	1	0	1	1	1
Lumnitzera	2	2	0	0	0
Euphorbiaceae					
Exoecaria[b]	1	1	0	0	0
Leguminosae					
Machaerium[b]	1	0	1	1	1
Meliaceae					
Xylocarpus	10	8	?	2	1
Myrsinaceae					
Aegiceras	2	2	0	0	0
Myrtaceae					
Osbornia	1	1	0	0	0
Plumbaginaceae					
Aegiatilis	2	2	0	0	0
Rhizophoraceae					
Bruguiera	6	6	0	0	0
Ceriops	2	2	0	0	0
Kandelia	1	1	0	0	0
Rhizophora	7	5	2	3	3
Rubiaceae					
Scyphiphora	1	1	0	0	0
Sonneratiaceae					
Sonneratia	5	5	0	0	0
Sterculiaceae					
Heritiera[b]	2	2	0	0	0
Theaceae					
Pellicera	1	0	1	0	0
Tiliaceae					
Brownlowia[b]	17	17	0	0	0
Monocots					
Arecaceae					
Nypha[a]	1	1	0	0	0
Pandanaceae					
Pandanus[b]	1	1	0	0	0
Total	80	65	9	11	9

[a] *Suaeda* typically is a small- to medium-sized bush, but can become a small tree.
[b] At least some of the species in these genera are more typical of freshwater swamps behind the mangrove coastal swamp and are not considered to be true mangroves by some botanists.

From Dawes, C. J., *Marine Botany*, John Wiley & Sons, New York, 1981. With permission.

and the New World mangal containing about 10 species.[92] Only a few genera occur in both vegetative groups.[4]

Mangroves fringe as much as 60 to 75% of the tropical coastlines of the world.[269] Although principally lining coastal zones between 25° N and 25° S latitudes, they occasionally extend somewhat beyond these points. For example, the species *Avicennia marina* occupies coastal regions in Japan and Bermuda in the Northern Hemisphere and Australia and northern New Zealand in the Southern Hemisphere.

2. Physical and Biological Factors Limiting Distribution
a. Temperature

Of the regime of factors limiting the spatial distribution of mangroves, air temperature is of paramount importance. Mangroves thrive under tropical conditions where the air temperature exceeds 20°C and the seasonal temperature range is less than 5°C. Air temperatures below –4°C are fatal; hence, these communities cannot tolerate hard frosts.[270] Temperature tolerance varies somewhat among species. For instance, *A. marina* resists low temperatures better than other species (e.g., *Rhizophora mangle),* and ranges into higher latitudes than other mangrove populations. This does not imply, however, that *A. marina* is immune to freezing temperatures and hard frosts. As signified by Sherrod and McMillan,[271] the abundance and distribution of black mangroves in the extreme northern Gulf of Mexico depend to a large degree on the frequency, duration, and/or severity of freezing temperatures.

b. Salinity

Mangroves are facultative halophytes. Although mangrove species vary in their salinity tolerances,[272] contributing to the zonation patterns commonly observed, they outcompete terrestrial plants (e.g., tropical rain forests) in estuarine and stable, high salinity coastal environments of tropical and subtropical regions. The salinity of bottom sediments is a function of local precipitation, subterranean seepage, terrestrial runoff, evaporation, and tidal flushing. Because these factors vary considerably in many regions, the salt concentration in most mangrove swamp soils fluctuates markedly.[4] The growth of mangroves is affected by soil salinity, with stunting resulting from hypersaline levels. Despite the potential constraints imposed by soil salinity, mangroves inhabit coastal zones with a surprisingly wide range of salinities.

c. Tides

Plant communities of mangrove systems are often most extensive on gently sloping shorelines with a large tidal range.[273] Sediment accumulation in these areas facilitates seedling development which fosters community expansion. Nonetheless, a large tidal range is not a prerequisite for the success of mangroves; thus, in Puerto Rico, where the tidal range remains less than 0.5 m, mangrove swamps thrive,[274] and in the Everglades of Florida, mangroves grow along the upper reaches of rivers having only a slight tidal range.[270]

Tidal action transports mangrove seeds, influencing local as well as regional distribution of the vegetation. Tidal flushing affects the salt concentration of the substrate. At low tide, the soil salinity may rise dramatically due to evaporation during aerial exposure. Upon return of the tide, the soil becomes saturated once again, which contributes to its anaerobic condition.[270]

Tidal flow also transports oxygen and nutrients that enhance production of the mangroves. These plants obtain much of their nutrients from freshwater runoff,[272,275] but the tides redistribute substantial concentrations of them. Additionally, mangrove swamps may trap freshwater masses at high tide, largely controlling the flushing of freshwater runoff from a river. This trapping phenomenon has been manifested in a mangrove-fringed tidal river, the Ducie-Wenlock River in Northeast Australia.[276,277] The process allows more time for nutrient uptake and settlement of fine sediments,[277] thereby favoring the growth of mangroves and explaining, in part, the large size of the trees lining Australian rivers that drain broad watersheds.[272,275]

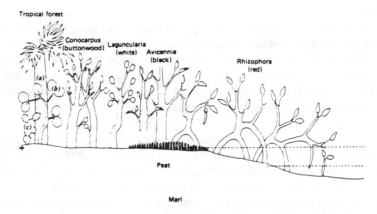

FIGURE 9. Mangrove swamp in southern Florida showing a typical zonation pattern. Red mangroves form the outer fringe in shallow subtidal waters and are replaced landward by black mangroves, white mangroves, buttonwoods, and a tropical rain forest. (From Dawes, C. J., *Marine Botany,* John Wiley & Sons, New York, 1981. With permission.)

d. Other Factors

The local geomorphology of an area is significant in the development of mangroves principally because it affects the physical and chemical factors (e.g., tidal inundation, soil salinity, and sediments) that control plant growth. Anthropogenic stresses from excessive logging, together with stochastic events (e.g., catastrophic storms and lightning fires), have strongly modified some mangrove communities. Not to be discounted are biological factors, especially interspecific competition. The relative competitive ability of the constituent flora of the community can lead to the exclusion of a species.

3. Zonation

Mangrove vegetation generally grows in a zoned pattern in which a single species or a group of species dominate specific bands. This zoned pattern, as viewed by Kuenzler (p. 347),[270] "... results from differences in rooting and growth of seedlings and from various competitive advantages which each species has in the several gradients present from below the low water to above the high water lines". The aforementioned physical and biological factors are critical to mangrove zonation. Despite being a common feature of mangrove systems, zonation can be absent or indistinct in some regions, often in areas having a small tidal range.

Watson,[278] appraising the zonation patterns of mangrove trees on the Malayan West Coast, designated five zones based on the frequency of inundation. Mann (p. 47)[92] embellished upon the description of these zones, proceeding from the lowest to the highest level: "(1) Species growing on land flooded at all tides: no species normally exists under these conditions, but *Rhizophora mucronata* will do so exceptionally; (2) species on land flooded by medium high tides: species of *Avicennia, Sonneratia griffithii,* and, bordering rivers, *Rhizophora mucronata;* (3) species on land flooded by normal high tides: most mangroves, but *Rhizophora* tends to become dominant; (4) species on land flooded by spring tides only: *Bruguiera gymnorhiza* and *B. cylindrica;* (5) species on land flooded by equinoctial or other exceptional tides only: *B. gymnorhiza* dominant, but *Rhizophora apiculata* and *Xylocarpus granatus* survive".

The mangroves of Puerto Rico have a strong zonal arrangement. Here, as in southern Florida (Figure 9) (see Section II.A), the red mangrove, *R. mangle,* fringes the outer margin of the community in shallow subtidal waters. Immediately landward, the white mangrove, *Laguncularia racemosa,* replaces *R. mangle.* Upland of the white mangrove is the black mangrove, *Avicennia germinans.* On the West Coast of Puerto Rico, the buttonwood, *Conocarpus erecta,* occurs landward of the white mangrove.

4. Succession

There is no consensus as to whether the species zonation in mangroves represents a true succession or if the mangrove vegetation reflects a steady state community. Many investigators infer that mangroves constitute either pioneer or seral successional stages in which the mangrove swamp evolves from an estuarine/mudflat community to a tropical rain forest. Based on this concept, the mangroves are evolving toward a freshwater, terrestrial plant community via modification of the habitat by the plant populations themselves (e.g., through soil formation). One species, therefore, alters the habitat where it grows, and, ultimately, another species supplants it. Gradually, the system shifts seaward, with the climax community established as an emergent tropical forest. This hypothesis of mangrove succession was championed by Davis[279] and supported by Kuenzler,[270] Chapman,[280] and others. Kuenzler[270] and Chapman[280] tout the mangrove swamps of Florida as an ideal example of this type of succession.

Alternatively, the mangrove populations do not induce vegetational changes themselves in a temporal sequence, but may, in fact, be under the control of external environmental factors.[27] This viewpoint envisions mangrove vegetation as a set of plants responsive to external forces and not directly involved in succession of a tropical rain forest. Changes in environmental conditions, such as sediment composition, wave action, currents, and sea level, would be expected to modify the zonal pattern.

5. Associated Flora and Fauna
a. Benthic Algae

Research on the productivity of mangrove systems has mainly dealt with organic material produced by the woody plants themselves. While the mangrove trees and shrubs certainly contribute the bulk of organic carbon to each system, benthic algae also are important primary producers. The roots of mangroves serve as ideal sites of attachment for benthic algae. Dense growths of the flora spread over the permanently submerged portions of red mangrove roots as, for example, in Puerto Rico, where species of *Acanthophora, Caulerpa, Hypnea, Lawrencia, Spyridia, Wrangelia,* and *Valonia* flourish. In this region, species of the genera *Centroceras, Enteromorpha, Murrayella, Polysiphonia,* and *Rhizoclonium* blanket portions of the intertidal zone. At the upper limit of high tide, *Bostrychia, Caloglossa,* and *Catenella* frequently appear.[270,281] Mangrove roots worldwide are normally associated with the following benthic algal genera: *Bostrychia, Caloglossa, Catenella,* and *Murrayella.*[270] The intertidal muds of Florida mangroves often support high densities of *Cladophoropsis* and *Vaucheria.* Below mean low water, tropical forms (e.g., species of *Acetabularia, Batophora, Caulerpa, Gracilaria, Halimeda, Penicillus,* and *Sargassum)* cover open shoal areas.[270,282,283] Unicellular and filamentous blue-green and green algae — a subterranean flora — also inhabit the mangrove environment of Florida.[270,284] The algal flora supply forage for benthic fauna of the mangroves, which prefer this plant material over that of the woody monocots and dicots.

Dawes[27] discussed the disparities that exist among the algal floras of mangrove swamps, contrasting the highly developed populations of the Caribbean systems (e.g., Puerto Rico) with the more depauperate floras of the tropical Pacific. The benthic algae of a Caribbean community may be subdivided into two major types; these are the filamentous and leafy forms. The filamentous forms produce felt-like mats or carpets on the sediment surface where they actively trap particles. Belonging to this group are the genera *Boodleopsis, Cladophora, Cladophoropsis,* and *Vaucheria.* The leafy forms tend to attach directly to mangrove roots — the pneumatophores, drop, and prop roots. They display microzonation. Species of *Bostrychia, Caloglossa,* and *Catenella* are examples. Bostrychia occurs in a zone above the two red algal genera, *Caloglossa* and *Catenella;* immediately above *Bostrychia* in the upper zone is the green algal genus, *Rhizoclonium.* These benthic algae like shaded, cool mangrove habitats. Consequently, they do not live in the mangrove forests of arid regions (e.g., those of the tropical Pacific).

b. Benthic Fauna

The benthic invertebrates of mangrove swamps are mostly filter and deposit feeders. In many mangroves, crustaceans and mollusks dominate the benthic faunal community. Crabs comprise a major portion of the faunal biomass of mangroves in the Caribbean. The intertidal flats of the mangrove islands of Florida Bay are sites of vigorous activity by the crabs *Uca pugilator, U. speciosa, U. thayeri,* and *Eurytium limosum.* Above the high water mark, *Aratus pisonii, Sesarma curacaoense,* and *S. reticulatum* reach high densities. The mud crab, *Rhithropanopeus harrisii,* as well as various amphipods and isopods, also attains high abundances in this zone.[285] Golley et al.[286] disclosed that fiddler crabs in the mangroves of Puerto Rico often dominate the benthic fauna in terms of biomass. These crabs feed on surface deposits at times of low tide. Gastropods ingest material deposited on the mud surface, and they consume mangrove roots. Species of *Cerithium, Cypraea, Littorina,* and *Melogena* are members of this group. The invertebrates manipulate large volumes of sediment in search of food and habitat. For instance, fiddler crabs *(Uca),* ghost crabs *(Dotilla),* tropical land crabs *(Cardisoma),* and prawns *(Upogebia)* construct burrows, enhancing the mixing and aeration of sediments that enable oxygen to penetrate to deeper layers. Bioturbation continually alters the physical and chemical characteristics of the substrate.[287]

Current velocity and turbidity may govern the relative importance of filter and deposit feeders in mangrove communities.[270] Waters having high current flow and low turbidity foster larger filter-feeding populations. Barnacles (e.g., *Balanus eburneus)* and coon oysters (e.g., *Ostrea frons)* exist in large densities attached to the roots or stems of mangroves in Florida. In Puerto Rico, the filter-feeder, *Crassostrea rhizophorae,* is abundant on the prop roots of red mangroves.[288] Other benthic macrofauna (e.g., barnacles, bryozoans, clams, mussels, crabs, hydroids, sea urchins, snails, shrimp, tunicates, and annelid worms) compete for food and space on the roots.

McConnaughey and Zottoli[289] summarize the invertebrate populations associated with the Florida-West Indies mangrove swamps. The most prominant species located here is *O. frons,* which clings to the roots. The coffee bean shell, *Melampus bidentatus,* affixes to the mangrove roots and is conspicuous along the upper margins of the swamp. Scattered in the muddy sediments are fiddler crabs (e.g., *Uca pugilator)* and the rose tellin (i.e., *Tellina lineata).* In addition, the periwinkle, *Littorina angulifera,* and crown conch, *Melongena corona,* disperse among the mangroves. The conch preys heavily on oysters. At the upper margin of the mangroves, land-dwelling crabs — the purple-clawed hermit crab *(Coenobita clypeata)* and the white land crab *(Cardisoma guanhumi)* — invade the area.

Macnae[267] evaluated the associations of marine invertebrates in mangroves of the Indo-West Pacific, where species of crabs (i.e., *Sesarma* spp. and *Uca* spp.) are numerous, along with gastropods. *Metopograpsus* sp., a blue crab, swarms over the prop roots. Bivalves, crabs, and gastropods colonize the seaward edge.[92] Conditions regulating the distribution of the fauna encompass both biotic and abiotic factors, namely: (1) the level of the water table, (2) the degree of substrate consolidation, (3) food availability (i.e., microflora, microfauna, and organic detritus), (4) resistance of the organisms to water loss, and (5) protection from the sun.

6. Ecological Importance

a. Production

Mangroves have several major ecological functions in estuaries and coastal oceanic waters. They supply significant quantities of primary production and provide a habitat for many populations of invertebrates and fishes. The bulk of their primary production enters detritus-based food webs, and a portion of it is exported as detritus to neighboring systems, thus stimulating productivity of contiguous waters. The outwelling of organic matter from mangrove forests promotes offshore production. The prop roots and pneumatophores, similar to the root systems of seagrass meadows, help to trap sediments and debris while mitigating erosion. These

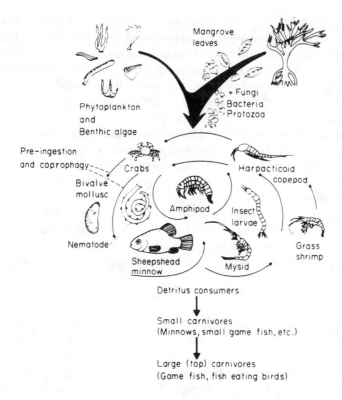

FIGURE 10.   Schematic diagram of the food web in the North River estuary, Florida, illustrating the flow of energy and materials through detritus-consuming omnivorous organisms. The cyclical nature of the diagram depicts the utilization and reutilization of detritus particles in the form of fecal material. Note that mangrove leaves are the principal energy source. (From Odum, W. E. and Heald, E. J., *Estuarine Research,* Vol. 1, Cronin, L. E., Ed., Academic Press, New York, 1975, 265. With permission.)

processes foster land accretion. The massive structure of the mangroves affords protection of the shoreline and stabilizes banks. Finally, mangroves filter land runoff, removing terrestrial products and retaining them within the system.

Estimates of the primary production of mangroves are consistently high,[290] with net primary production ranging from approximately 350 to 500 g C per square meter per year.[92] High primary production figures have been documented for mangroves in Australia,[291-295] Southeast Asia,[296] Puerto Rico,[286] Florida,[297,298] and Mexico.[299] The detritus generated in these systems yields energy for detrital based food webs;[300,301] the tidal regime largely determines whether the detritus production is primarily recycled *in situ* or exported.[292]

b. Detritus

Nearly 75% of the vegetation in the intertidal zone of the tropics consists of mangroves.[302] While terrestrial grazing removes a small portion of the live plant matter, most of the total production passes to litter which, subsequent to partial decomposition, serves as food for detritivores in adjacent estuarine waters. Hence, in red mangroves of Florida, Heald[303] determined that terrestrial grazers ingest only 5.1% of the total leaf production, the remaining litter being a detrital source useful as food. Grinders, deposit feeders, and filter feeders — three functional groups of detritivores — consume detritus, but probably derive most of their nutrition from the associated microbes.[238] In the North River estuary, Florida, mangrove leaf fall is the principal energy source for the system, supporting a complex web of organisms (Figure 10). The multiple reuse strategy of the detritus is manifested in the cyclical pattern shown in Figure 10.[4]

Substantial amounts of mangrove detritus can be exported to contiguous estuarine waters,[301] particularly when tidal exchange is intense. Odum and Heald[297] and Heald[303] observed a significant export of mangrove detritus in Florida. Boto and Bunt[292] noted a comparable effect in Queensland, Australia. However, more precise estimates must be made of the concentration of detritus degraded and recycled *in situ* relative to that exported from the system.[301,304]

c. Habitat Former

As a habitat, mangroves harbor diverse assemblages of epiflora, epifauna, benthic infauna, fishes, and birds. Insects and terrestrial fauna have also successfully adapted to the forest environment. In contrast to the low floral diversity in mangroves, faunal diversity is high.

d. Sediment Stabilizers

The roots and pneumatophores of mangroves accumulate sediments, trap litter, and aid in soil formation. Assisted by algal populations which build felty masses over the substrate, mangroves mitigate the impact of erosive forces.[27] Additionally, they protect the shoreline from the devastating impact of major storms incurred in tropical and subtropical regions.

IV. BENTHIC FAUNA

Based on their size, benthic invertebrates are classified as micro-, meio-, and macrofauna (see Section II.B). The microfauna embody those animals smaller than 0.1 mm in size, and the meiofauna, those forms greater than 0.1 mm but less than 0.5 mm in size. The macrofauna include individuals larger than 0.5 mm.[8]

Benthic fauna are separated into two broad categories according to their life habits: (1) infauna, which burrow and live in bottom sediments; and (2) epifauna, which live on the surface of sediments and on hard substrates (e.g., pilings, concrete structures, and rocks) attached to them either temporarily or permanently. These two groups comprise most of the benthic fauna of estuaries, although four additional groups are recognized (i.e., interstitial boring, swimming, and commensal-mutualistic forms).[8]

Multiple feeding strategies exist among the aforementioned faunal groups. At least five feeding modes are evident among the benthos. That is, the benthic fauna obtain their food as suspension feeders, deposit feeders, herbivores, carnivores-scavengers, and parasites.

A. MICROFAUNA

"Small animals" of the benthos are classified either as micro- or meiofauna, terms used in the practical sense to designate individuals passing through a sieve with a mesh size of 0.5 mm. The meiofauna encompass those metazoans retained on a sieve with a mesh size of 0.1 mm.[305] The term microfauna, therefore, is largely reserved for the protozoans of minute size (e.g., ciliates and foraminifera) which pass through sieves with mesh openings of 0.04 to 0.1 mm.

Some workers group foraminiferans with the meiofauna, whereas others place them with the microfauna on the basis of their weight.[34,306] The smallest protozoans, the zooflagellates, weigh about 10^{-11}g; the ciliates range from 10^{-10} to 10^{-6}g. Meiofauna, in turn, weigh less than 10^{-4}g (wet weight).[35]

The ciliates, present on the surfaces of grains and within capillary sediments, are the best-studied protozoan group, both taxonomically and ecologically.[34,307-311] Their abundance peaks in fine sediments and in mats of sulfur bacteria.[35] More than 300 species of interstitial forms are known,[312] but the actual number residing in a given habitat may be significantly greater. Aside from ciliates, other important protozoans living in or on marine sediments include amoebas, heliozoans, foraminiferans, and zooflagellates. These protozoans, however, have not been adequately investigated in estuarine environments.

The microfauna graze heavily on bacteria. Fenchel[313] discovered in samples of detritus, for example, that zooflagellates seem to feed exclusively on bacteria, whereas ciliates are omnivorous, feeding on bacteria as well as zooflagellates, other ciliates, and microalgae. The number of small zooflagellates bound to 1 g dry weight of detritus generally varies between 5×10^7 and 5×10^8, compared to 10^4 to 10^5 ciliates per gram dry weight.[314,315] The average biomass of zooflagellates and ciliates equals about 5×10^{-11} and 5×10^{-8}g, respectively, which in 1 g dry weight of detritus amounts to 5×10^{-3}g.[313] This biomass value approximates that of bacteria in 1 g of detritus, estimated to be 4×10^{-3}g. Consequently, the microfauna are deemed to be valuable nutritive sources for meiofauna and larger detritus feeders. Moreover, protozoan grazing on bacteria can modulate bacterial abundance in detritus and sediments.[313]

Protozoans, along with bacteria, fungi, and blue-green algae, are the most numerous biota in estuarine sediments.[38,316] These microbiota strongly influence the chemistry of the interstitial waters.[317] Perhaps most importantly, bacteria and fungi effectively decompose detrital material. Bacteria, in particular, convert organic matter into inorganic nutrients via decomposition processes (see Chapter 1),[81] and autotrophs may assimilate these nutrients if available.[130,318]

B. MEIOFAUNA

1. Taxonomic Considerations

Meiofaunal populations can be separated into two categories: (1) temporary meiobenthos, usually macrofaunal larvae, which spend only a portion of their life in sediments as part of the meiobenthos; and (2) permanent meiobenthos, which are meiobenthic throughout their life cycle. The second category contains members of Rotifera, Gastrotricha, Nematoda, Archiannelida, Tardigrada, Copepoda, Ostracoda, Mystacocarida, Turbellaria, Acarina, Gnathastomulida, and some specialized individuals of Hydrozoa, Nemertina, Bryozoa, Gastropoda, Soelenogastres, Holothuroidea, Tunicata, Priapulida, Polychaeta, Oligochaeta, and Sipunculida.[305,319] They are mostly interstitial forms moving between sedimentary particles, burrowers displacing sedimentary grains, and epibenthic or phytal organisms.[319]

2. Abundance

In sediments on the continental shelf, meiofaunal densities normally range from 10^5 to $10^7/m^2$.[320-322] The density and biomass of this group of organisms increase in shallow water environments, with very high densities ($2.6 \times 10^7/m^2$) recorded in intertidal mudflats.[319] The standing crop dry weight biomass of meiofauna averages about 1 to 2 g/m^2.[319] Typically, harpacticoid copepods and nematodes are the most abundant taxa. For instance, Bell[323] calculated that 73% of the meiofauna in a South Carolina benthic community consisted of nematodes. According to Dye and Furstenburg,[324] more than 80% of the total meiofauna in estuarine sediments are usually nematodes. The large numbers of nematodes quoted in the literature are not surprising because, as a group, the nematodes probably represent the most abundant metazoans in the biosphere.[325,326] Table 14 illustrates the high abundances of meiofauna in estuarine and marine sediments. Commonly occurring nematodes in European estuaries are registered in Table 15.

In a profile of seafloor sediments, the meiofaunal counts peak in the top few centimeters of the near-surface layers, gradually decreasing with increasing depth to the redox potential discontinuity (RPD) layer. In some cases, more than 90% of the meiofauna concentrate in the upper 1 cm of estuarine bottom sediments. Coull and Bell,[319] assessing the meiofauna of subtidal environments, stated that more than 95% of the fauna inhabit the top 7 cm of sediments and 60 to 70% of the fauna, the upper 2 cm of sediments.

As sheltering increases in estuaries, the abundance of meiobenthos also rises (Table 14).[3,327] The greater degree of sheltering afforded by estuaries combined with their high food supply enhances the development of a rich meiofaunal community. Estuarine areas, therefore, oftentimes support more meiofauna than other habitats.[324] Within an estuary, the abundance of

TABLE 14
Density of Intertidal Meiofauna[a]

Deposit	Location	Nematoda	Copepoda	Ostracoda	Oligochaeta	Total Meiofauna
Sand	S. of Stockholm, Locality "A", Sweden	38—169	31—833	0	214—498	391—1,529
	S. of Stockholm, Locality "B", Sweden	48—88	24—129	0	92—109	173—402
	Gothland, Sweden	8—14	0	0	17—112	225—1,048
	Scanian East Coast, Sweden	1	2—45	0	16—160	49—318
	Kattegat, Sweden	6—106	1—52	0	1—243	11—845
	Øresund, Denmark	115—497	1—10	0	1—375	249—764
	Hard Wadden, W. Denmark	10—1,050	<1—840	0—61	+	13—1,914
	Soft Wadden, W. Denmark	31—367	13—389	4—12	+	50—768
	Boulogne, France	3—367	3—192	0—<1	0	19—389
	Acachon, France	67—204	5—331	<1—18	0—12	83—591
	Eden estuary, St. Andrews, Scotland[α]	3,163	17	13	+	3,193
	Blyth estuary, England	300—1,320	*	*	*	—
	Whitstable, England	1,136—5,220	13—486	21—742	0—7	1,264—5,817
	Miami, FL	0—110	0—260	0	0	14—872
	Porto Novo, S. India	594—1,150	186—448	0	0—2	968—1,960
	E. Coast of Malaya	346—8,068	86—2,416	0	0	700—10,212
Coral sand	E. Coast of Malaya	32	144	0	12	244
Mud	Bristol Channel (Wales)	70—10,440	0—500	0—790	0—780	90—11,820
	Southampton Water, England	*	81—1,021	*	*	—
	Blyth estuary, England	228—2,830	*	*	*	—
	Salt marsh, Massachusetts	1,440—2,130	*	*	*	—
	Salt marsh, Georgia	460—16,300	*	*	*	—
	Vellar estuary, S. India	307—3,240	5—490	0—63	0	420—3,815

Note: * = not studied.

a   Numbers below 10 cm² of surface.

From McIntyre, A. D., *Biol. Rev.*, 44, 245, 1969. With permission.

TABLE 15
Nematodes Commonly Found in Estuarine and Brackish-Water Systems of Europe. Salinity Tolerances of Species also Indicated

Species	Salinity (‰)
Adoncholaimus fuscus	0.5
A. thalassophygas	0.5
Aegialoalaimus elegans	26.5
Anoplostoma viviparum	0.5
Antomicron elegans	2.1
Aponema torosus	26.0
Ascolaimus elongatus	0.5
Atrochromadora microlaima	0.5
Axonolaimus paraspinosus	0.9
A. spinosus	0.5
A. typicus	24.5
Axonolaimus sp. (aff. *spinosus*)	0.5
Bathylaimus assimilis	5.3
B. filicaudatus	30.5
B. longisetosus	0.5
B. stenolaimus	5.3
B. tenuicaudatus	0.5
Calomicrolaimus honestus	0.5
Calyptronema maxweberi	0.5
Camacolaimus barbatus	5.3
Campylaimus inaequalis	26.5
Choanolaimus psammophilus	0.5
Chromadora nudicapitata	17.0
Chromadora sp. (aff. *nudicapitata*)	17.0
Chromadorina erythropthalma	11.0—1.0
C. germanica	9.0
C. microlaima	24.0
Chromadorina sp. (aff. *germanica*)	24.0—9.0
Chromadorina sp. (aff. *viridis*)	5.0
Chromadorita fennica	15.0—4.0
C. guidoschneideri	25.0—2.0
C. nana	2.0
Chomardorita tentabunda	26.0—0.5
C. tenuis	25.0—4.0
Cyatholaimus punctatus	10.0
Daptonema leptogastrelloides	0.5
D. normandicum	5.3
D. oxycerca	0.9
D. procerum	26.0—11.0
D. setosum	0.5
D. tenuispiculum	30.5
D. trabeculosum	26.0—0.5
Daptonema sp. (aff. *biggi*)	0.5
Desmodora communis	5.3
Desmolaimus fennicus	0.9
D. zeelandicus	0.5
Desmoscolex falcatus	26.0
Dichromadora cephalata	0.5
D. geophila	0.5
D. hyalocheile	18.0
Diplolaimelloides altherri	22.0
D. islandicus	0.5
Eleutherolaimus oschei	27.5
E. stenosoma	0.5

TABLE 15 (continued)
Nematodes Commonly Found in Estuarine and Brackish-
Water Systems of Europe. Salinity Tolerances of Species
also Indicated

Species	Salinity (‰)
Eleutherolaimus sp. (aff. *stenosoma*)	0.5
Enoploides caspersi	18.0—0.5
E. labiatus	10.0
E. spiculohamatus	5.3
Enoplolaimus balgensis	18.0—0.5
E. litoralis	2.1
E. propinquus	2.1
E. vulgaris	0.5
Enoplus brevis	12.0
E. schulzi	5.3
Eurystomina terricola	5.3
Halalaimus gracilis	7.5
Halichoanolaimus robustus	11.0
Hypodontolaimus balticus	0.5
H. geophilus	0.5
H. inaequalis	15.0
H. setosus	0.5
Leptolaimus elegans	0.5
L. limicolus	11.0
L. papilliger	0.5
L. setiger	22.0
Metachromadora remanei	5.3
M. suecica	12.0
M. vivipara	0.9
Metalinhomoeus filiformis	11.0
Metaparoncholaimus sp. (aff. *campylocerus*)	0.5
Microlaimus globiceps	0.5
M. marinus	2.1
M. robustidens	5.4
Molgolaimus demani	24.5
Monhystera disjuncta	0.5
M. microphthalma	25.0—11.0
M. parva	0.5
M. paramacramphis	22.0
Monhystera sp. (aff. *filiformis*)	0.5
Monhystera sp. (aff. *parasimplex*)	3.5—2.0
Monhystrella parelegantula	11.0
Monoposthia costata	25.0
M. mirabilis	24.0
Nemanema cylindraticaudatum	10.0
Neochromadora izhorica	18.0—10.0
N. poecilosoma	8.6
N. poecilosomoides	18.0
N. trichophora	24.0
Odontophora armata	12.0
O. rectangula	24.0
O. setosa	5.0
Oncholaimus brachycerus	5.0
O. conicauda	18.0—10.0
O. oxyuris	0.5
Oxystomina elongata	11.0
O. unguiculata	31.0
Paracanthonchus caecus	0.5

TABLE 15 (continued)
Nematodes Commonly Found in Estuarine and Brackish-Water Systems of Europe. Salinity Tolerances of Species also Indicated

Species	Salinity (‰)
P. bothnicus	18.0—0.5
P. elongatus	25.0
P. sabulicolus	24.0
P. tyrrhenicus	5.0
Paracanthonchus sp. (aff. *bothnicus*)	18.0—0.5
Paracyatholaimus intermedius	30.0—0.5
P. proximus	0.5
P. ternus	0.5
P. truncatus	0.5
Paralinhomoeus lepturus	11.0
Pomponema sedecima	24.0
Praeacanthonchus punctatus	0.9
Prochromadorella longicaudata	24.0
Ptycholaimellus ponticus	0.5
Quadricoma scanica	26.0
Sabatieria longispinosa	24.0
S. pulchra	0.5
S. punctata	0.5
S. vulgaris	0.9
Setoplectus riemanni	0.5
Sigmophoranema rufum	18.0
Southernia zosterae	26.0
Sphaerolaimus balticus	22.0
S. gracilis	0.5
S. hirsutus	0.9
Spilophorella paradoxa	2.1
Spirinia parasitifera	22.0
Syringolaimus striaticaudatus	10.0
Terschellingia communis	0.9
T. longicaudata	0.9
Thalassoalaimus tardus	22.0
Theristus acer	5.3
T. blandicor	7.5
T. ensifer	18.0
T. flevensis	0.5
T. metaflevensis	0.5
T. meyli	0.5
Theristus pertenuis	11.0
T. scanicus	0.5
Trefusia conica	0.5
T. longicauda	5.3
Trichotheristus mirabilis	2.1
Tripyla cornuta	18.0—0.5
Tripyloides amazonicus	0.5
T. gracilis	5.3
T. marinus	0.5
Tripyloides sp. (aff. *marinus*)	0.5
Tubolaimoides tenuicaudatus	12.0
Viscosia rustica	24.0
V. viscosa	0.9

From Heip, C., Vincx, M., and Vranken, G., *Oceanogr. Mar. Biol. Annu. Rev.*, 23, 399, 1985. With permission.

meiofaunal populations usually is maximum in the intertidal zone, frequently in mudflats and salt marshes. The mean density of meiofauna in the mudflats of estuaries equals about $3 \times 10^5/m^2$, but can be as high as $6 \times 10^5/m^2$. In salt marshes, the mean meiofaunal density is approximately $5 \times 10^5/m^2$, although densities as great as 3 to $6.5 \times 10^6/m^2$ have been recorded. Fewer meiofauna populate sandflats where their density approaches $1 \times 10^5/m^2$.[324] The range of meiofaunal densities in the intertidal zone, as specified by Kerby,[328] amounts to 1.1×10^4 to $1.6 \times 10^7/m^2$; that in the subtidal zone, 4.0×10^3 to $3.2 \times 10^6/m^2$.

The species diversity of meiofauna in estuarine sediments is characteristically high. The sediment composition and texture exert some control on the type of meiofauna found. Hence, sandflat meiofauna, which dwell in bottom sediments to depths greater than 10 cm, are vermiform organisms that maneuver between sedimentary grains, utilizing interstitial spaces in the process.[12] Muddy sediments harbor stockier, burrowing species.[327] These meiofauna occupy near-surface sediments, concentrating in the upper centimeter of the sediment profile. Burrowing meiofauna prefer sediments with a median particle diameter of less than 125 μm. Coull and Bell (p. 194)[319] state:

> Interstitial groups, e.g., Gastrotricha (with one genus expected, *Musellifer*) and Tardigrada, are excluded from muddy substrates where the interstitial lacunae are closed. Obviously the converse is true in that a burrowing taxon, e.g., Kinorhyncha (one exception, *Cateria*), is excluded from the interstitial habitat. In those taxa that have both interstitial and burrowing representatives, e.g., Nematoda, Copepoda, Turbellaria, there is a difference in the morphology of mud and sand dwellers.

Nematodes penetrate to the greatest depths, and below 10 cm are the dominant members of the group. Dye and Furstenburg[324] sampled large numbers of nematodes at depths as great as 80 cm in the salt marshes of the Swartkops estuary.

3. Distribution

When traversing from intertidal to subtidal zones of an estuary, a zonation of meiofauna can be perceived,[3,329] even though horizontal patchiness along physical gradients periodically confounds results.[330] The distribution patterns of meiofauna depend on the physical-chemical conditions (e.g., temperature, salinity, dessication, and sediment grain size) of the environment and biotic interactions (e.g., competition and predation).[322,331] Salinity is a major factor affecting the spatial distribution of the meiofauna. Perkins[3] calls attention to the influence of gradual salinity changes in the Elbe estuary over a distance of 292 km, from the North Sea to the freshwater zone upriver. Of the 258 meiofaunal species identified in the estuary, 35% are restricted to waters greater than 18‰. Many meiofaunal species have adapted to euryhaline conditions in estuaries, such as in undersaturated beach sediments.[305,332] In general, the meiofauna residing in high intertidal zones are more tolerant of low salinities than the meiofauna living in low intertidal zones.[333] Meiofaunal species richness and density increase with increasing salinity from the head to the mouth of an estuary.[305,334-336] However, various interactive effects on meiofaunal dispersal, notably behavioral emergence, passive erosion, and passive transport, may obfuscate these longitudinal gradient patterns.[337]

The local and geographical distributions of meiofauna have been partially ascribed to temperature effects. Meiofaunal populations adjust their position in bottom sediments in response to prevailing temperature levels. Thus, the meiofauna of intertidal zones, having broader temperature tolerance limits, display peak densities in the upper centimeter of sediments where temperatures are most variable. Species occupying sediments below 1 cm tend to have lower thermal tolerance limits than the surface-dwelling forms.[332] Populations within the lower intertidal zone characteristically have lower thermal tolerance limits than those within the upper intertidal zone.[338]

Temporal variations in meiofaunal species composition and abundance correlate with seasonal temperature changes.[324,339-342] Dissolved oxygen, in combination with temperature,

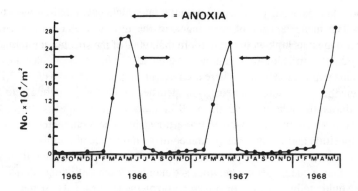

FIGURE 11.   Density of meiofauna in Harrington Sound, Bermuda from 1965 to 1968. Meiofauna disappear during periods of anoxia and do not repopulate sediments until 3 to 4 months following reoxygenation. (From Coull, B. C., *Limnol. Oceanogr.*, 14, 953, 1969. With permission.)

limits meiofaunal abundance. In the Swartkops estuary, for example, meiofaunal abundance peaks in the spring and fall when temperature and oxygen values are in the middle of their ranges. However, in the summer and winter, the number of meiofauna declines to a minimal figure due to low oxygen and temperature levels, respectively.[324,342]

Vertical zonation of meiofauna in subtidal sediments correlates with chemical zonal patterns. The meiofauna vary in their ability to tolerate low oxygen concentrations.[343,344] A well-oxidized, surface sediment layer is displaced at depth by a zone of low oxygen tension and reduced microniches and, ultimately, by an anoxic sulfide layer. In response, populations of meiofauna congregate near the sediment surface. Although the number of meiofauna approaches zero in the redox (RPD) layer, some taxa may live as true anaerobes below this layer.[343]

In Harrington Sound, Bermuda, recurring episodes of anoxia cause the meiofauna to disappear. Repopulation only takes place during periods of reoxygenation (Figure 11).[345] Factors other than oxygen availability affect the zonation of meiofauna in Harrington Sound as well as elsewhere, notably the amount of organic matter and food.[35,346] The interaction of multiple factors can obfuscate the effect of a single factor and may be responsible for a more drastic effect on the organisms than a single factor when nearing tolerance limits.[305]

Biotic factors contribute to the patchy distribution of meiofauna. The decomposition of organic matter, for instance, attracts meiofauna, and can create a patchy distribution.[347] The clustering of bacteria and microflora — meiofaunal food — likewise fosters localized concentrations of meiofauna.[348,349] Selective predation of meiofauna occasionally eliminates a taxon from a localized area, generating a discordant distribution.[305,350] The lack of a pelagic larval phase in the life history of the vast majority of meiofaunal species accounts for clumped patterns of juveniles.[319] The aggregated distribution of adults, in turn, is a strategy that ensures the reproductive success of the group.[351]

4. Reproduction and Growth

In contrast to most macrofaunal populations (except those in the high Arctic and the deep sea) which have high fecundities and a planktonic larval phase to maximize dispersal, the bulk of the meiofaunal forms produces only a few gametes, with individuals usually releasing from one to ten eggs at a time directly into the water. Alternatively, the meiofauna deposit their eggs in a cocoon attached to sand grains.[35] Brood protection and viviparity are common.[352] Benthic larvae emerge from hatched eggs, or young individuals in a rather advanced stage of development emerge from eggs in protective cocoons. Only species of the archiannelid genus *Polygordius* and several species of *Protodrilus,* according to Fenchel,[35] have higher fecundities among the meiofauna, producing as many as 200 eggs at a time. The larvae of these meiofauna are pelagic.

Most meiofauna undergo direct development; relatively few populations have larval stages. While the crustacean elements of the permanent meiofauna (e.g., copepods and ostracods) pass through larval stages during ontogeny, in more than 90% of the species, the larvae develop in the sediment rather than in the water column.[305] Many of the eggs of other meiofauna are "sticky" and remain clustered in concentrated areas. These adaptations ensure that relatively few eggs and larvae are lost from a system.

The probability of successful fertilization in the meiobenthos is increased via several mechanisms, namely, copulation, oviposition, and spermatophore transfer. Hermaphroditism and parthenogenesis raise the reproductive potential of many meiobenthic taxa. Although the reproductive output in terms of the total fecundity is small, several behavioral processes minimize reproductive losses. First, meiofaunal taxa may reproduce continuously through the year.[352,353] Second, relatively short generation times characterize numerous species. Third, life-history development can be delayed by some populations through resting eggs or larval-stage delay.[352-354] These reproductive adaptations promote large population densities, enabling the meiofauna to be highly successful in their habitat.

Vernberg and Coull[305] examine the environmental conditions that regulate reproduction and development of the meiofauna. A key factor is temperature which, when increased, normally decreases the time of egg development,[355] the time of postembryonic development,[356,357] and the generation time,[355,358] while increasing reproductive potential.[357,359] Salinity, crowding, and the quality of food also affect life-history development of these diminutive organisms.

5. Production

Meiofaunal production values published in the literature are based on indirect estimates of production either by respiration or by using an annual production/biomass ratio (P/B). In estuarine and marine sediments, nematodes are the most abundant metazoans, with densities of several million animals per square meter of seafloor in shallow subtidal environments.[360] Despite the significance of nematodes in the secondary production of estuaries and coastal oceanic waters, no estimates of marine nematode production have been derived from field research.[326] Energy-flow analyses on nematodes have relied on laboratory experiments and the extrapolation of these results to field populations. Relationships between respiration or production and biomass are frequent topics of study. However, practical limitations imposed by continuous reproduction of adults and the strong overlap of generations in the field relegate production estimates nearly exclusively to laboratory settings. Nevertheless, *in situ* production estimates are under investigation; most of the work is still in its infant stages.[319]

Laboratory observations have yielded valuable data on life-cycle turnover. By multiplying the standing crop biomass of the meiofauna by its turnover, estimates of production have been advanced.[324] In the Lynher estuary, England, Warwick et al.[361] computed an annual meiofaunal production value of 20.17 g C per square meter per year. This figure approximates that of the Ythan estuary, Scotland, which amounts to 19.5 g C per square meter per year.[362] Much lower meiofaunal production has been recorded in the Swartkops estuary, South Africa. Here, somewhat lower readings correspond to muddy sediments ($\bar{P} = 0.24$ g C per square meter per year) than sands ($\bar{P} = 0.46$ g C per square meter per year).[324,342]

Mean production of the harpacticoid copepod, *Huntemannia jadensis,* in the sandy beaches of Puget Sound, WA, equaled 1.0 g C per square meter per year.[363] A similar value was determined for the harpacticoid copepod, *Microarthridion littorale* ($\bar{P} = 0.87$ g C per square meter per year), in the salt marshes of South Carolina.[319] For a brackish water pond in northern Belgium, Heip[364] presented mean annual production measurements of the dominant copepod species, *Halicyclops magniceps* ($\bar{P} = 52$ mg C per square meter per year), *Tachidius discipes* ($\bar{P} = 105$ mg C per square meter per year), *Paronychocamptus nanus* ($\bar{P} = 279$ mg C per square meter per year), and *Canuella perplexa* ($\bar{P} = 281$ mg C per square meter per year).

Laboratory results on meiofaunal production may be unrealistically high. Conditions for meiofaunal growth and turnover are much more favorable in the laboratory than in nature. Consequently, life cycles appear to be more protracted in nature, and growth rates and generation times, lower than those evaluated under optimal laboratory conditions. Warwick[352] deals with some of the drawbacks of the laboratory culture of meiofaunal organisms.

C. MACROFAUNA

Benthic macroinvertebrates in estuaries have wide temporal and spatial variations in abundance related to both natural and anthropogenic perturbations.[37,365-372] Because of their generally restricted mobility, benthic macroinvertebrates have been employed to ascertain the impacts of pollutants in estuarine and coastal marine waters. For example, elaborate studies have been designed to evaluate the effects of organic loading and eutrophication on benthic communities.[373-377] San Francisco Bay has been the site of much research on the effects of wastewater discharges on macroinvertebrate populations and communities.[369,378] Workers have also successfully utilized these organisms in environmental impact assessments of thermal discharges[379] and heavy metals.[380] Surveys of macrofaunal communities subsequent to catastrophic oil spills are a common practice.[381,382] Populations of benthic macroinvertebrates have been recommended as key taxa for assessing water quality conditions in general.[383,384]

1. Distribution

The Crustacea, Mollusca, and Polychaeta are well represented in estuarine benthic communities. A characteristic feature of benthic macroinvertebrate populations in estuaries is their large temporal and spatial variations in abundance in both intertidal and subtidal zones. Fluctuations in density are most pronounced in opportunistic species (e.g., *Mulinia lateralis, Capitella capitata, Cistenides (=Pectinaria) gouldii,* and *Polydora ligni*). For some species, large variations in abundance reflect within-year periodicities of reproduction, recruitment, and mortality. Random environmental events (e.g., storms), however, often precipitate aperiodic density changes that exceed those caused by within-year fluctuations.[368] Heterogeneity of the benthic habitat leads to patchiness of species distributions.[365] The literature is replete with references to large year-to-year fluctuations in the abundance of macrobenthic fauna.[37,365-367,385-390]

a. Small-Scale Distribution Patterns

Wolff[391] summarized three levels of distribution patterns of estuarine benthos: (1) small-scale distribution patterns, (2) distribution patterns on the scale of the estuary, and (3) distribution patterns on a world scale. He related small-scale distribution patterns to depth, current velocity, sediments, and other factors. Water movements and sediment composition are deemed to be most important in heterogeneous environments. In intertidal zones, wave action is considered to be a major force structuring the benthic community.[392-394] Variations in boundary layer flow and strength influence recruitment and succession of macrofaunal populations.[395]

Sediment type affects the local distribution of the benthos.[366,396] The meroplankton of benthic populations search the estuarine seafloor for a favorable substrate. The preference of adults for specific sediments reflects, in part, the feeding mode of the taxa. For example, the bivalve, *Scrobicularia plana,* and the polychaete, *Polydora ligni,* inhabit muddy sediments, whereas the polychaetes, *Ophelia* spp., and amphipods, *Haustorius* spp. and *Bathyporeia* spp., live in sandy bottoms.[391] The fine sediments and high turbidities of muddy environments clog the feeding apparatus of suspension feeders, thereby precluding them from colonizing these areas. Deposit feeders are ideally suited to a lifetime of feeding in muddy substrates.[8] The distribution and niche space of soft-bottom polychaetes are dependent on sediment type.[397]

Benthic research during the past several decades has coupled the distribution of the benthos to sediment grain size and associated physical-chemical conditions.[8,287,398] The amount of *in situ* organic matter also correlates with the density levels of benthic populations in the sediments, as has been demonstrated for the lugworm, *Arenicola marina*,[399] the mudsnail, *Hydrobia ulvae*, and the tellinid bivalve, *Macoma balthica*.[400] The biomasses of these species also have been positively correlated with the content of organic matter in the sediments.

The infauna are responsive to vertical environmental gradients (e.g., dissolved oxygen). The macrofauna concentrate in the oxygenated surface layers of sediment, declining with increasing depth and decreasing oxygen supply. Some representatives of the microfauna and meiofauna, in contrast, peak in abundance in the deeper layers.[401] Most thiobios, for instance, are believed to be anaerobes preferring the chemocline and deeper reducing layers of sediment,[402] although the exact location of the species population maxima is controversial.[403]

Populations of infauna have a propensity for changing position in the sediment column during ontogeny, extending deeper into the sediment as the organisms age and grow larger in size. Relative position in the substrate is indicative of the ecological function of the benthos, that is, whether the populations are subsurface deposit feeders, suspension feeders, or another feeding type.[404] Rhoads and Boyer[405] compared the location of the infauna to the stage of successional development of the benthic community. Pioneering species of infaunal assemblages, according to this study, typify areas of physical disturbances and tend to be dominated by tubicolous or otherwise sedentary organisms that live in near-surface sediments. They feed in proximity to the sediment surface or from the water column.[406] Higher-order successional stages — termed equilibrium stages — occur in habitats removed from major physical disturbances. Bioturbating infauna, which feed at greater depths within the sediment, dominate these stages.

Vertical partitioning of sediments can mitigate competitive interactions among species in soft-bottom communities.[407] Organismal feeding and burrowing, in addition to the accessibility of epifaunal predators, may govern the vertical distribution and abundance of infaunal populations.[408-410] Constraints on burrowing depth ascribable to body size rather than resource partitioning of competitors have been shown to be critical in regulating the position of the infauna in sediments of the Rhode River, a subestuary of Chesapeake Bay.[407] Similar controls probably operate in other mid-Atlantic estuaries.

Both physical and biological disturbances play paramount roles as determinants of species abundance, distribution, and diversity in estuarine benthic communities. While physical disturbances of the benthos are caused by currents, waves, sediment deposition, sediment erosion, and so forth, biological disturbances are incurred principally from grazing, predation, and competition.[405,406,411-414] Rhoads and Germano (p. 293)[406] describe disturbance as "natural processes, such as seafloor erosion, changes in seafloor chemistry, and foraging disturbances which cause major reorganization of the resident benthos, or anthropogenic impacts, such as dredged material or sewage sludge dumping, thermal effluents from power plants, deposition of drilling muds and cuttings, pollution impacts from industrial discharge, etc". The successional pattern of a benthic community, moreover, hinges on the frequency and nature of disturbances.[415] Succession is defined by Gallagher et al.[416] as the temporal pattern of changes in the specific composition of a community. Sousa[417,418] and Connell and Keough[419] scrutinize the role of disturbances in natural ecosystems.

Along the estuarine seafloor, many benthic communities display patchy distribution patterns. Clustering of macrofaunal populations arises even over very small tracts of a few centimeters.[420] Patchiness on a scale of 50 m or less has been discerned in the Lough Hyne, Ireland.[398] Rogal et al.[421] discovered variations in macrobenthic community structure within sampling areas of 100 m^2 in Lubeck Bay (western Baltic Sea). Mechanisms responsible for inducing patchy distributions are linked to localized disturbances that alter faunal abundance and community structure. Benthic disturbances account for the spatial mosaic pattern of bottom communities in which parts of the community exist at different levels of succession.[415] An organism-sediment successional paradigm has been formulated, aiding in the explanation of benthic disturbances.[406] In general, pioneering species replace high-order assemblages (equilibrium species) in habitats

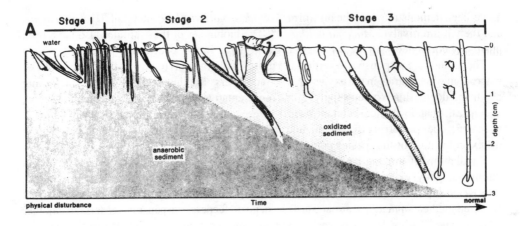

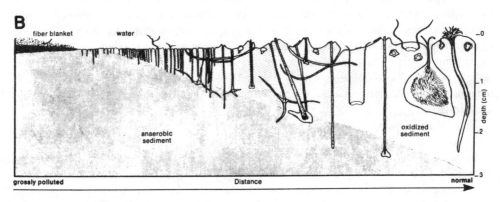

FIGURE 12.   Development of organism-sediment relationships over time following a major physical disturbance (A) and along a pollution gradient (B). Pioneering species (left side) tend to be tubicolous or otherwise sedentary organisms that live near the sediment surface and feed near the surface or from the water column. Particle bioturbation rates are low, and the aerated sediment layer is thin. High-order successional stages (right side) tend to be dominated by bioturbating infauna that feed at depth within the sediment. Particle bioturbation rates are high, and the aerated sediment layer is thick. (A) shows Stage I, II, and III seres. (From Rhoads, D. C. and Germano, J. D., *Hydrobiologia,* 142, 291, 1986. With permission.)

subjected to continual disturbances. Rhoads and Boyer[405] and Rhoads and Germano[406,422] give complete coverage of the population dynamics accompanying changes in organism-sediment relations.

Figure 12A illustrates the development of organism-sediment relations over time subsequent to a major physical disturbance, and Figure 12B depicts the unfolding of organism-sediment relations along a chronic pollution gradient. The significance of Stage I, II, and III seres and the importance of environmental disturbances are evident in this figure. The intensity and frequency of disturbance help to establish the successional stages present in the infaunal community.[405,423] In a new or newly disturbed bottom, pioneering species (Stage I sere) predominate; these forms tend to be small, opportunistic, tube-dwelling polychaetes with high densities (about $10^5/m^2$) only days or a few weeks after disturbance.[412,424] As expressed above, pioneering species feed near the sediment surface or from the water column. Hence, sedimentary effects attributable to these organisms are (1) subsurface deposit feeding which blankets the substratum with fecal pellets, (2) fluid bioturbation which pumps water into and out of the bottom through vertically oriented tubes, and (3) construction of dense tube aggregations which may influence microtopography and bottom roughness.[405]

A transitional stage (Stage II sere) may also be apparent, consisting of a diverse assemblage of tubicolous amphipods, mollusks, and polychaetes. Most members of this assemblage feed at, or near, the sediment surface, although some "head-down" feeders probe more deeply. The thickness of the aerated sediment layer is, in most cases, transitional between the early pioneering polychaete Stage I and the Stage III assemblage (Figure 12A).[406]

Expanses of the seafloor rarely subjected to physical disturbances are dominated by high-order seres (Stage III seres), primarily composed of infaunal deposit-feeders, and deeply burrowing errant or tube-dwelling forms which feed head down (i.e., conveyor-belt species) (sensu Rhoads[318]). Predominant members of Stage III assemblages include maldanid, pectinid, and orbinid polychaetes, caudate holothurians, protobranch bivalves, infaunal ophiuroids, and irregular urchins.[406] Organisms of the Stage III seres have longer life spans and larger body sizes than most infauna of Stage I and II seres.

Sedimentary effects associated with equilibrium species of Stage III seres are (1) water transfer and particle advection over vertical distances of 10 to 20 cm; (2) homogeneously mixed fabrics, with fecal pellets commonly found below the sediment surface; (3) void spaces (feeding pockets) at depths within the bottom; (4) redox discontinuity layers generally located at depths of 10 to 20 cm; and (5) surface microtopography that may be featureless and planar or covered with numerous feeding pits and fecal or excavation mounds.[405] High-order successional stages have been described in the *Maldanid-Nucula-Syndosmya* (polychaete-bivalve) community of the Clyde Sea,[425] the *Nucula-Nephtys* (bivalve-polychaete) assemblage of Buzzards Bay,[426,427] the *Molpadia-Euchone* (holothurian-polychaete) community of Cape Cod,[428,429] the *Maldane-Amphiura* (polychaete-ophiuroid) community of Ria de Muros, Spain,[430] and the *Lumbrimeris-Alpheus-Diolodonta* (polychaete-shrimp-bivalve) community of Kinston Harbor, Jamaica.[431] The dominance of equilibrium species does not prevent the occurrence of pioneering forms. Both may coexist if physical disturbances only involve near-surface sediments.[405]

The successional history of a benthic community can be ascertained in a vertical profile of the upper few centimeters of undisturbed seafloor. A rapid and accurate method of deciphering the stage of successional development of the benthos is by means of sediment-profile imaging. By employing a sediment-profile camera, up to 200 images of the benthos can be gleaned in one survey day covering several square kilometers. Reconnaisance maps of an area generated by a sediment-profile camera survey have great practical value in the formulation of management decisions concerning estuarine environments. Successional mosaics identified via sediment-profile imaging can be used to track long-term changes in benthic community structure and associated biogenic processes. The sediment-profile camera not only yields data of value for reconnaisance maps of successional seres, but also provides information on sediment modal grain size, depositional and erosional regimes, depth of the apparent redox boundary, sedimentary methane, bedforms, and disturbance gradients.[406]

Stochastic events or random perturbations of the benthic habitat, often associated with physical disturbances of the estuarine seafloor, create conditions conducive to major restructuring of the benthic community of some systems.[369] While abiotic factors, such as sediment erosion or deposition, precipitate aperiodic density changes of benthic populations during anomalous perturbations, biotic interactions likewise play a role in the structure of the faunal assemblages. For example, predators regulate the abundance of prey directly by consuming larvae, juveniles, and adults; alternatively, they indirectly influence the survivorship of their prey by burrowing through sediments, disturbing the sediment surface, and reducing larval settlement.[410] The predatory effects of macrobenthos on the structure of the benthic community have been well established in recent years (see Section IV.C.6).[432-436]

The significance of biological disturbances in modulating the organization of benthic communities can be easily overlooked due to confounding physical factors prevalent in estuarine ecosystems. Because observed community patterns may result from responses of organisms to spatial and temporal changes in the environment,[437,438] to competition,[439-443] to predation,[444-449] to anomalous events, or to the synergistic effect of interacting processes,[450]

TABLE 16
Classification of Approximate Geographic Divisions, Salinity Ranges, Types, and Distribution of Organisms in Estuaries

	Venice system		Ecological classification				
Divisions of estuary	Salinity ranges (‰)	Zones	Types of organisms and approximate range of distribution in estuary, relative to divisions and salinities				
River	0.5	Limnetic		Limnetic			
Head	0.5—5	Oligohaline		Oligohaline			
Upper Reaches	5—18	Mesohaline	Mixohaline				
Middle Reaches	18—25	Polyhaline			True estuarine		
Lower Reaches	25—30	Polyhaline					
Mouth	30—40	Euhaline		Stenohaline marine		Euryhaline marine	Migrants

From Carriker, M. R., *Estuaries*, Lauff, G. H., Ed., Publ. 83, American Association for the Advancement of Science, Washington, D.C., 1967, 442. With permission.

detection of the forces largely responsible for shaping the community structure is often obscured.[451] Recent research has emphasized the function of the dominant species in molding the structure of the community. Studies have been undertaken to evaluate their resistance to physical disturbances, their competitive interactions, and their influence on the biogenic structure of other populations.[452] The combined effect of biological and physical disturbances controls the distribution of benthic infauna. Thus, Gee et al. (p. 430)[447] remark that "... predation and physical disturbance play a major role in maintaining the community species diversity by reducing the abundance of the dominant species to a level at which competition for space is less intense and by making available free space for colonization by newly settled larvae". Species abundance and richness generally increase in the absence of predators.[232,408,447,453] Predatory infauna, in addition to epibenthic predators, act as regulatory agents in some habitats, being determinants of prey densities.[454-457] Investigations conducted in tidal flats, especially, reveal epibenthic predators as major determinants of species abundance patterns.[42,234,454,458] However, large variations in abundance of macrobenthos in intertidal zones frequently reflect other causative factors, notably recruitment failure or success, winter mortality, and storms.[377]

b. Distribution Patterns on the Scale of the Estuary

Environmental gradients along the longitudinal axis of an estuary impose constraints on the distribution of benthic macrofauna. When traversing from the estuarine head to its mouth, gradual shifts in salinity confine various macrobenthic populations to specific regions. Early workers correlated the salinity distribution in estuaries to zones of similar biotic composition. The Venice System classification, adopted by scientists at the Symposium on the Classification of Brackish Waters in 1958, subdivided estuaries into zones based on salinity ranges (i.e., limnetic, oligohaline, mesohaline, polyhaline, and euhaline zones). This classification scheme was later expanded to incorporate physiographic subdivisions and distributional classes of organisms (Table 16).[396] The term coenocline was coined to denote biotic change along the estuarine complex-gradient.[459,460] Areas of more abrupt change of biota coincided with salinity limits of the zones at 0.5, 5, 18, and 30‰, whereas within the zones, the biota were more homogeneous.[461] Zonation of the estuarine ecocline has been based mainly on patterns of species ranges rather than similarities of community composition, structure, and function within

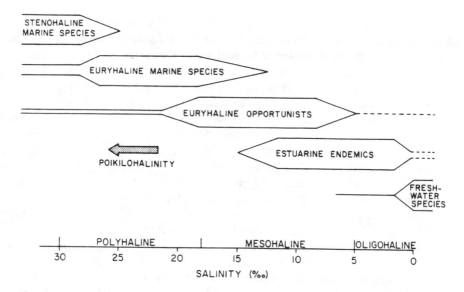

FIGURE 13.   General model of the distributional classes of species in a homoiohaline estuary. (From Boesch, D. F., *Ecology of Marine Benthos,* Coull, B. C., Ed., University of South Carolina Press, Columbia, 1977, 245. With permission.)

zones.[460] The Venice System classification has been applied most extensively by benthic ecologists in northwestern Europe, but it has also been used in North American estuaries.

Boesch[460] examined the macrobenthic coenocline of Chesapeake Bay, a large and relatively homoiohaline system similar to the brackish water bodies of northwestern Europe. Conclusions of his study indicate gradual and relatively uniform faunal changes along the homoiohaline gradient of the estuary, with zones of more accelerated alteration in the macrobenthos at salinity ranges of 3 to 8 and 15 to 20‰. Hence, the Venice System classification extrapolates reasonably well to this large, homoiohaline system. However, this classification scheme does not relate well to the faunal zonation patterns observed along other estuarine gradients, such as that of the Brisbane River estuary, Australia, which has a seasonally poikilohaline gradient with a much more abrupt coenocline change than that of Chesapeake Bay. With reference to species distributional classes — stenohaline marine species, euryhaline marine species, euryhaline opportunists, estuarine endemics, and freshwater species — Boesch[460] devised a general model of estuarine zonation in a homoiohaline estuary (Figure 13).

Salinity gradients vary greatly in estuaries. In large systems, they change more gradually than in small, tidally mixed estuaries, but seasonal freshwater pulses can have a marked impact on the larger bodies of water.[4] While seasonal and tidal fluctuations in salinity are natural events, anthropogenic effects of river impoundment, diversion of freshwater runoff, and tidal inlet modification may substantially alter the hydrodynamics, salinity regime, and benthic biota of impacted systems.[462]

A strong correlation has been established between the salinity tolerance limit and distribution of benthic fauna.[463] In the Newport River estuary, for instance, Wells[464] found that 303 species of epifauna declined upestuary with decreasing salinity. Laboratory experiments performed on a representative number of these species showed that the upestuary limit in their distribution was directly coupled to their tolerance of low salinities.

Estuarine benthic fauna respond to osmotic stress either through physiological or behavioral adjustments.[465-467] Osmoconformers, defined by Lincoln et al. (p. 177)[1] as "organisms having body fluids of the same osmotic concentration as the surrounding medium", depend largely on intracellular isosmotic regulation.[468] They cannot regulate cell volume and ionic composition. Marine species which are osmoconformers do not penetrate too far upestuary, but usually inhabit areas periodically exposed to full-strength salinities.[465] The distance that osmoconformers penetrate within estuaries is principally a function of the tolerance level of the species to wide variations in the concentration of body fluids.[4] Mollusks are osmoconformers whose successful adaptation to estuarine habitats is primarily ascribed to their ability to regulate cellular osmotic pressure, since they exhibit little anisosmotic extracellular regulation.[467] These organisms rely on behavioral mechanisms to survive salinity oscillations, either burrowing or closing their valves during unfavorable salinity perturbations. The Echinodermata, another group of osmo-conformers, is essentially a stenohaline marine phylum.[469] In contrast to the mollusks and echinoderms, many higher crustaceans are effective osmoregulators, that is, each organism "maintains the osmotic concentration of its body fluid at a level independent of the surrounding medium" (Lincoln et al., p. 177).[1] Because of osmoregulation, these crustaceans have adapted most successfully to the estuarine environment.

Two examples may be drawn from the literature to demonstrate the importance of salinity in controlling the distribution of benthic macroinvertebrates in estuaries. Flint and Kalbe,[470] conducting research on the structure and function of the benthic macroinvertebrate community of Corpus Christi Bay, distinguished between distinct community structures of riverine-influenced and oceanic-influenced stations. Differences were reported in species composition, taxa number, abundance, biomass, and species diversity of the benthic community among stations. Using cluster analysis on the species assemblages, Flint and Kalbe[470] confirmed that the total abundance decreased and the taxa number increased when moving away from the fluvial source. The mid-estuarine region had the maximum macroinvertebrate biomass.

Richter[471] performed an extensive quantitative survey of the macrofauna at 20 stations in the Rio Lingue estuary of southern Chile. Of the 16,775 animals collected, 83% consisted of oliochaetes, polychaetes, and ostracods. Bivalves (43.7%) and polychaetes (33.6%) accounted for most of the biomass. Through factor analysis of the data, Richter[471] differentiated three groups of infauna. One group was distributed in the upper estuarine zone in a region of low salinity. A second group occurred in the central part of the system; it spread into the upper or lower zones of the estuary. In the sandy, high salinity outlet area of the estuary, a third group dominated by marine species was identified. According to Richter,[471] sediment composition and turbulence, as well as salinity, probably controlled the distribution of the benthos.

Even though salinity is generally regarded as the most important factor governing the distribution of the benthic fauna on the scale of the estuary, gradients in other environmental factors elicit responses from the bottom fauna that affect community structure. Examples are gradients in turbidity, oxygen saturation, wave action, and pollution.[391] Interactive parameters at times frustrate explanations of the observed patterns.

1. Larval Dispersal

The distribution of numerous benthic macroinvertebrates is inextricably linked to dispersal during a planktonic larval stage. The life history of most major taxa of benthic invertebrates in estuaries includes a period of planktonic existence. The duration of the planktonic larval stage varies from species to species depending on the mode of development, environmental conditions, and chemical or physical cues that cause the larvae to settle and metamorphose. Thorson[472,473] organized planktonic larvae into three classes: (1) long-life planktotrophic forms in which free-swimming larvae feed on other planktonic organisms and remain in the plankton for 1 week to 2 or 3 months; (2) short-life planktotrophic forms which also feed while in the plankton, but the duration of larval life is 1 week or less; and (3) lecithotrophic forms which include nonfeeding pelagic larvae, with the yolk providing all nourishment for the larvae until metamorphosis.[474] Two additional developmental modes of marine bottom invertebrate larvae were likewise proffered by Thorson[472] (i.e., larvae undergoing direct development completely omitting a pelagic phase and larvae developing within the parental organism). Other individuals have extended Thorson's classification to incorporate demersal development,[475] and a new type of larvae — teleplanic larvae — which are planktotrophic forms transported great distances by transoceanic currents.[476,477] Further treatment of this subject is contained in Chapter 3, Section II.B.1 and other reviews.[477-482]

Jablonski,[483] citing the confusing terminology applied to the reproductive strategies of marine invertebrates, championed the two primary categories of larval development expounded by Jablonski and Lutz,[484] namely, planktotrophy and nonplanktotrophy. Spending considerably longer periods of time in the plankton than nonplanktotrophic forms, planktotrophic taxa are susceptible to passive transport by currents and may be swept out of estuarine environments to nearshore oceanic areas. The position of the larvae in the water column is critical to their dispersal because the magnitude and direction of currents change markedly with depth over a tidal cycle and, indeed, over very short time intervals.[485] While much speculation has been invoked concerning the ability of the larvae to adjust their vertical position by swimming in response to environmental cues, Scheltema[485] infers from available field observations that many larval forms act as either inert particles or neutrally buoyant objects. Having restricted swimming capacity, these larvae are likely to be translocated by passive transport.

The magnitude of larval dispersal is a function of hydrodynamic processes, larval behavior, and duration of the pelagic stages.[486-489] Day and McEdward[475] maintain that the movement of water masses and longevity of larvae dictate the potential for dispersal, but the actual degree of dispersal is determined in many cases by larval behavior. Horizontal advective transport of planktonic larvae can be maximized or minimized in estuaries by the swimming behavior of the organisms within constraints set by the duration of the planktonic period and strength of the water movement.[475] Bousfield,[490] Sandifer,[491,492] and Sulkin and Van Heukelem[493] note examples of estuarine species whose larval behavior govern their dispersal. Active control of larval distribution in the water column is most evident among the strongest swimmers, such as crustacean larvae.

Much of the literature on larval dispersal in estuaries supplies information on select groups, especially finfishes and decapod crustaceans (e.g., crabs and shrimps.)[494,495] Stancyk and Feller[495] specify several deficiencies that hinder the study of nondecapod larval transport in estuaries: (1) poor taxonomic knowledge of larvae, (2) problems with sampling and processing, (3) a perception of the lack of commercial value of the organisms, and (4) inadequate understanding of the physical and biological environments and extreme variation in physical characteristics between estuaries. Only a few estuarine invertebrate groups, other than the decapod crustaceans, have been investigated in sufficient detail to enable their dispersal patterns to be understood, and much of this knowledge is derived from descriptive or correlative field studies. Laboratory research on larval behavior yields additional useful data to elucidate the spatial and temporal distributions of larvae in estuaries.[496-498]

Knowledge of estuarine circulation is necessary to assess the transport of larval forms. Even with reasonably good data on the circulation of a particular system, however, evaluation of the larval dispersal patterns may be confused by the complexities of larval behavior. Scheltema,[485] for example, recounts the complex and strikingly different larval behavior of the American oyster *(Crassostrea virginica)* and the blue crab *(Callinectes sapidus)* in several mid-Atlantic estuaries (e.g., Delaware Bay and Chesapeake Bay).

The number of planktotrophic larvae of a species far exceeds that of lecithotrophic larvae. The immense numbers of planktotrophic larvae increase the probability of widespread dispersal. Predation, inadequate food sources, and the general vagaries of environmental conditions raise mortality rates and can reduce dispersal capability. Lecithotrophic larvae remain essentially free of the hazards associated with fluctuating food supplies in the plankton, attaining their energy from the yolk. Fewer lecithotrophic larvae are produced, however, with decreases in fecundity correlating with increased egg size in lecithotrophy.[477] In terms of the energy cost of reproduction, Chia[480] considers lecithotrophic development to be less expensive than planktotrophic development. In regard to caloric output by adults, planktotrophy presumably is cheaper than lecithotrophy.[475,493]

Planktonic larvae exercise some control in selecting suitable adult habitats. When unable to find a favorable substrate, some larvae delay metamorphosis until a more suitable site is located. In response to the proper chemical or physical cues, however, metamorphosis and settlement often take place rapidly. Multiple factors may induce larvae to metamorphose; they therefore affect the ultimate distribution patterns of adults on the estuarine seafloor. Settlement of larvae is influenced by algal substrates, bacterial coatings, organic matter, sediment size distributions, and chemicals released by individuals of the same species (e.g., pheromones). Meadows and Campbell,[500] Crisp,[481,501] Chia and Rice,[502] and Morse[503] explore the environmental factors and settlement-inducing substances modulating larval settlement and metamorphosis. Light,[497,504,505] gravity,[475] and hydrostatic pressure[506] elicit responses from marine invertebrate larvae which may be manifested as directional swimming (e.g., geotaxis and phototaxis) and changes in swimming velocity (e.g., barokinesis).[475]

Scheltema[485] groups the stimuli for larval settlement into two categories: (1) those requiring contact with the bottom (e.g., *Balanus*,[507] *Spirorbis*,[508] and *Haliotis*[509]); and (2) those resulting from modified behavior of the larvae in the water column (e.g., *Nassarius*,[510] *Phestilla*,[511] *Crassostrea*,[512,513] and *Capitella*[514]). Clearly, more research needs to be conducted on how larvae settle under conditions in the bottom boundary layer. Data collected from this type of research will be instrumental in comprehending the relative significance of settlement cues and larval behavior to hydrodynamics of the bottom boundary layer.[486]

Aggregations of some benthic populations reflect gregarious metamorphosis in response to chemical cues generated by members of the same species.[500,515,516] Larval recruitment (i.e., settlement, attachment, and metamorphosis) proceeds in a step-by-step fashion in which the organism initially detects specific biochemical cues that induce behavioral and developmental changes. Exogenous biochemical control of larval recruitment to benthic habitats has been documented among coelenterates, bryozoans, annelids, mollusks, echinoderms, arthropods, and ascidian chordates.[503] The role of pheromones in gregarious settlement of marine invertebrate larvae has been examined repeatedly during the past two decades. Pheromones are chemicals released from an organism that evoke behavioral, physiological or developmental responses in conspecifics.[517,518] A growing number of benthic invertebrate populations experience gregarious recruitment mediated by adult-derived chemical cues. Although many of these chemical cues remain uncharacterized, they have been implicated in patterns of settlement and metamorphosis of benthic larvae in the field. Burke[518] tabulated a list of benthic invertebrate larvae responding to cues associated with conspecific adults (Table 17).

Factors other than pheromones promote clumped distributions of some benthic populations. For example, larvae periodically undergo metamorphosis in aggregations due to the patchiness of suitable substrates.[518] This effect is manifested in the behavior of many fouling organisms such as *Hydroides dianthus*.[519]

The assessment of the factors responsible for inducing larval settlement and metamorphosis has been performed principally in the laboratory using still water.[520] However, at present, the observed infaunal distributions in the field can only be deduced from these laboratory investigations. As Butman (p. 114)[520] remarks, "Little is known about how initial patterns of larval settlement relate to the eventual distributions of adults." The role of larval settlement in determining the distribution and abundance of benthic populations remains largely enigmatic.

Planktotrophic and nonplanktotrophic modes of larval development lead to different magnitudes of larval dispersal which potentially affect the distribution patterns of adult benthic populations.[483] Lecithotrophic larvae usually disperse relatively short distances since they are pelagic for only several days to 1 week, or at most 2 weeks. These larvae likely concentrate near

TABLE 17
List of Marine Invertebrates for which the Presence of Conspecifics Influences
Settlement and Metamorphosis of Larvae

Phylum	Species	Evidence for chemical cue	Ref.
Cnidaria			
Anthozoa	*Ptilosarcus guerneyi*	Yes	Chia and Crawford (1973)
Hydrozoa	*Nemertesia antennina*	Yes	Hughes (1977)
	Clava squamata	Yes	Williams (1976)
	Kirchenpaueria pinnata	Yes	Williams (1976)
	Tubularia larynx		Pyefinch and Downing (1949)
Mollusca			
Bivalvia	*Crassostrea virginica*	Yes	Veitch and Hidu (1971)
	Ostrea edulis	Yes	Knight-Jones (1949); Bayne (1969)
	Mercenaria mercenaria	Yes	Keck et al. (1971)
Gastropoda	*Haliotis disces*	Yes	Seki and Kan-no (1981)
	Bittium reticulatum		Kiseleva (1967)
	Collisella strigatella		Wilson (1968)
	Rossoa splendida		Kiseleva (1967)
Annelida	*Sabellaria alveolata*	Yes	Wilson (1968)
	S. vulgaris	Yes	Eckelbarger (1975)
	Spirorbis borealis		Knight-Jones (1951)
	Phragmatopoma californica	Yes	Jensen and Morse (1984)
	Pomatoleios kraussi		Crisp (1977)
	Hydroides dianthus		Scheltema et al. (1981)
	Spirorbis pagenstecheri		Knight-Jones (1951)
	Sabellaria spinulosa	Yes	Wilson (1970)
	Polydora ligni		Blake (1969)
Arthropoda	*Aerceriella enigmata*		Straughan (1972)
	Balanus balanoides		Knight-Jones (1953)
	Elminius modestus	Yes	Knight-Jones and Stevenson (1950)
	Balanus crenatus	Yes	Knight-Jones (1953)
	B. amphitrite		Daniel (1955)
	B. tintinnabulum		Daniel (1955)
	Chthamalus stellatus		Daniel (1955)
Sipuncula	*Golfingia misakiana*	Yes	Rice (1978)
Echiura	*Urechis caupo*	Yes	Suer and Phillips (1983)
Echinodermata			
Crinoidea	*Florometra serratissima*		Mladinov and Chia (1983)
Holothuroidea	*Psolus chitonoides*		Young and Chia (1982)
Echinoidea	*Dendraster excentricus*	Yes	Highsmith (1982)
Chordata			
Urochordata	*Chelyosoma productum*		Young and Braithwaite (1980)
	Pyura haustor	Yes	Young (personal communication)

From Burke, R. D., *Bull. Mar. Sci.*, 39, 323, 1986. With permission.

the bottom of an estuary where they may be entrained in turbulent near-bottom flows. Dispersal capability of nonplanktotrophic larvae that show direct development is also limited relative to planktotrophs. They also have less capacity for delaying metamorphosis in the absence of a suitable substrate.[483,521,522] Olson[523] noted that the ascidian, *Didemnum molle*, which has a nonplanktotrophic mode of larval development, could delay settlement for up to 2 h. This species reportedly dispersed over several hundred meters of a sandflat. Few direct observations have been made of the absolute extent of larval dispersal of planktotrophs or nonplanktotrophs. Planktotrophic larvae, which typically occur in the plankton for 2 to 6 weeks prior to metamorphosis, can be transported large distances through an estuary until a suitable habitat is found for settlement.

Scheltema and Carlton[524] provide examples of marine fouling organisms with different modes of larval development. Nonplanktotrophic larvae of ascidians, spirorbinid polychaetes, and most of the common encrusting bryozoans have a short period of development lasting only a few hours. In contrast, planktotrophic larvae of attached barnacles, bivalves (e.g., mussels and oysters), serpulid polychaetes (e.g., *Hydroides*), and certain bryozoans belonging to the cheilostomatous genera *Membranipora, Electra,* and *Conopeum* exist in the plankton for several weeks to 2 months and, consequently, have a much greater dispersal capability.

c. Distribution Patterns on a World Scale

The cosmopolitan distribution of estuarine species has been studied previously.[396,525,526] Estuarine faunas, while resembling each other in various regions of the world, are not universally cosmopolitan. Wolff[391] compared the distribution of three groups of estuarine benthic faunas: (1) fouling species occurring in seawater and brackish water, (2) benthic species inhabiting coastal waters and estuaries, and (3) benthic species living almost exclusively in estuaries and other brackish waters. The first two groups of species indeed have a wide distribution, with many populations being cosmopolitan. Examples of fouling species distributed worldwide are the barnacles, *Balanus amphitrite* and *B. improvisus,* the ectoproct, *Barentia gracilis,* and the polychaete, *Mercierella enigmatica.* Some species of the coastal group along the Atlantic and Pacific Coasts of the U.S. have also been sampled in European waters.[461,527,528] The third group of exclusively estuarine species has fewer members with a worldwide distribution. It is not unusual for estuaries to harbor endemic faunas. Hence, the species composition of estuarine brackish-water faunas of South Africa embodies numerous endemic crustaceans.[529,530] Characteristic forms in the brackish waters of many estuaries and lagoons along the Atlantic and Gulf Coasts of North America are the blue crab, *Callinectes sapidus,* the mud crab, *Rhithropanopeus harrisii,* the isopod, *Cyathura polita,* and the polychaete *Streblospio benedicti.* On the Pacific Coast of North America, estuaries also have a variety of endemic faunas (e.g., the polychaete, *Nereis limnicola,* and the mysid, *Neomysis awatschensis*). Many endemic genera of crustaceans and mollusks inhabit waters of the Black and Caspian Seas.[391]

On a world scale, benthic invertebrates increase their distribution via adult migration, transport by rafting, and transport by human activities.[485] Only active and mobile epibenthic species increase their dispersal by means of adult migration. Faunal transport by rafting on floating materials represents a more likely method of dispersal. A most common way of dispersal is through human intervention. Since about 1850, man has introduced nearly 100 species of exotic marine invertebrates into San Francisco Bay.[531] This introduction has been accomplished primarily by the transport of fouling, boring, and ballast-dwelling organisms in ships and epizoic and nestling invertebrates in commercial oysters. The ballast water of ships, which frequently holds larvae and postlarvae of invertebrates, is a key element in the transoceanic and interoceanic dispersal of coastal marine invertebrates.[485,532]

2. Species Composition and Abundance

The severe time restrictions imposed on benthic ecologists to assess environmental impacts are not conducive to long-term investigations of soft-bottom invertebrate communities. Yet, long-term studies of the benthos are most valuable for establishing ecosystem alterations, for differentiating natural and anthropogenic changes in communities, and for generating and analyzing testable hypotheses.[533] Because of the wide temporal and spatial variations in the species composition and abundance of benthic macrofauna in estuaries, long-term studies yield the most insightful information on community structure and ecological patterns. However, before they contribute substantially to the explanation of these subjects, long-term benthic data sets collected in the field must fulfill certain minimum requirements, as connoted by Holland et al.[367] First, the data collection phase must exceed the mean life span of the longest-lived dominant species. Second, the sampling frequency must be adequate to quantify within year variation (e.g., seasonal patterns) for the short-lived species. Third, the data set must cover both extreme and average conditions. Fourth, field and laboratory experiments or modeling that test hypotheses about mechanisms and processes controlling long-term distributions must support the relationships suggested by the empirical data.

Few long-term, continuous monitoring programs of benthic communities in estuaries have been conducted to date because they are expensive, tedious, and do not guarantee definitive results.[368] Thus, short- or medium-length monitoring projects are much more prevalent today in surveying benthic communities and evaluating anthropogenic impacts. When long-term research has been implemented, estuarine scientists retain a better understanding of the mechanisms controlling system dynamics, the impacts of anthropogenic activities, as well as the relative magnitude of natural variation associated with seasonal dynamics, episodic events, and climatic trends.[534]

a. Long-Term Benthic Macroinvertebrate Studies
1. Barnegat Bay

From 1964 to 1968, Phillips[535] performed a qualitative assessment of the benthic macroinvertebrates of Barnegat Bay, NJ. Loveland et al.[536,537] conducted quantitative sampling of this fauna. Loveland and Vouglitois[37] reviewed these benthic faunal investigations.

Quantitative benthic sampling in the estuary was executed with a 0.05 m² Ponar grab. Between August 1969 and December 1973, 216 benthic invertebrate species were collected in Barnegat Bay (Table 18). The mean density of benthic macroinvertebrates in the bay during this 53-month period amounted to 2775/m² and ranged from 56 to 43,220/m². The numerically dominant species included the bivalve, *Mulinia lateralis,* the polychaete, *Cistenides (=Pectinaria) gouldii,* and the gastropod, *Acteocina canaliculata. M. lateralis* and *C. gouldii* are opportunistic species that experience large fluctuations in density.

In the mid 1960s, Phillips[535] recovered *M. lateralis* in densities ranging from 1 to 318/m², and *C. gouldii* in densities from 2 to 700/m². *A. canaliculata* was rare. By 1969, the densities of the three species increased dramatically; *M. lateralis* occurred in numbers as high as 36,840/m², with a mean value of 10,890/m².[538] The maximum and mean densities of *C. gouldii* in 1969 equalled 4980 and 1649/m², respectively. Somewhat less abundant, *A. canaliculata* appeared in densities up to 2540/m². The mean density of this species was 224/m².

Benthic macroinvertebrates had an unusually productive year in Barnegat Bay during 1969, but from 1969 to 1973, a marked decline in abundance ensued. The densities of the dominant forms — *M. lateralis, C. gouldii,* and *A. canaliculata* — dropped substantially during this 4-year period, although other common species (e.g., the gastropods, *Turbonilla interrupta* and *Mitrella lunata,* and the polychaete, *Scoloplos fragilis)* followed similar trends. Some macroinvertebrates, such as *Ampelisca* spp. and *Molgula manhattensis,* increased in abundance during this interval of generally diminishing densities.

The species composition of the benthic community of the estuary remained quite stable in space and time, while the absolute abundance of the species varied. The relative stability of the species composition of the benthic macrofaunal community of this estuary corroborates the findings in other systems. For example, Lie and Evans[645] documented persistent species composition during a 4-year study of the benthos of Puget Sound, as did Boesch et al.[539] in areas of Chesapeake Bay from 1960 to 1975.

Most of the benthic macroinvertebrates in Barnegat Bay are deposit feeders. From 1970 to 1973, for instance, 53 to 85% of the individuals in a sample were deposit feeders, 6 to 47% suspension feeders, and 3 to 23% carnivores. A preponderance of deposit-feeding organisms has

TABLE 18

Taxonomic List of Benthic Invertebrates Found in Barnegat Bay, NJ

Phylum Porifera
  Class Demospongiae

           Cliona celata (Grant)
           Halichondria bowerbanki (Burton)
           H. panicea (Pallas)
           Haliclona sp.
           Microciona prolifera (Ellis and Solander)

Phylum Cnidaria
  Class Hydrozoa
    Order Athecata

           Hydractinia echinata (Fleming)
           Pennaria tiarella (Ayres)
           Tubularia crocea (L. Agassiz)

    Order Thecata

           Campanularia sp.
           Obelia commissuralis (McCrady)
           Thuiaria argentea (Linnaeus)

  Class Anthozoa
    Order Actiniaria

           Diadumene leucolena Verrill
           Edwardsia elegans Verrill
           Halcampoides sp.
           Haliplanella luciae Verrill
           Haloclava producta (Stimpson)
           Metridium senile (Linnaeus)
           Actinothoe modesta (Verrill)

    Order Ceriantharia

           Cerianthus americanus Verrill

Phylum Platyhelminthes
  Class Turbellaria

           Euplana gracilis (Giard)
           Stylochus ellipticus (Girard)

Phylum Nemertinea
  Class Anopla

           Carinoma tremaphoros (Leidy)
           Cerebratulus lacteus (Leidy)

Phylum Sipunculida

           Golfingia improvisum Theel
           Golfingia sp.

Phylum Annelida
  Class Polychaeta
    Order Phyllodocida
      Family Phyllodocidae

           E. heteropoda Hartman
           Eteone lactea Claparede
           Eulalia viridis (Linnaeus)
           Eumida sanguinea (Oersted)
           Paranaitis speciosa (Webster)
           Phyllodoce arenae (Webster)
           P. maculata (Linnaeus)

      Family Polynoidae

           Harmothoe imbricata (Linnaeus)
           H. oerstedi (Malmgren)
           Lepidonotus squamatus (Linnaeus)

TABLE 18 (continued)
Taxonomic List of Benthic Invertebrates Found in Barnegat Bay, NJ

Family Sigalionidae

Sthenelais boa (Johnston)

Family Chrysopetalidae

Dysponetus pygmaeus Levinsen

Family Glyceridae

Glycera americana Leidy
G. capitata Oersted
G. dibranchiata Ehlers

Family Goniadidae

Glycinde solitaria Webster
Goniada maculata Oersted
Ophioglycera gigantea Verrill

Family Nephtyidae

Nephtys incisa Malmgren
N. picta Ehlers

Family Syllidae

Autolytus cornutus (A. Agassiz)

Family Hesionidae

Gyptis vittata Webster and Benedict
Podarke obscura Verrill

Family Nereidae

Nereis arenaceodonta Moore
N. pelagica Linnaeus
N. succinea (Frey and Leukart)
N. virens Sars
Platynereis dumerilii (Verrill)

Order Capitellida
Family Capitellidae

Capitella capitata (Fabricius)
Notomastus latericeus Sars

Family Maldanidae

Clymenella torquata (Leidy)
C. zonalis (Verrill)
Maldane sarsi Malmgren
Maldanopsis elongata (Verrill)

Order Spionida
Family Spionidae

Polydora ligni Webster
Scolecolepides viridis (Verrill)
Scolelepsis squamata (O. F. Muller)
Spio filicornis (O. F. Muller)
S. setosa Verrill

Family Chaetopteridae

Spiochaetopterus oculatus Webster

Family Sabellariidae

Sabellaria vulgaris Verrill

Order Eunicida
Family Onuphidae

Diopatra cuprea (Bosc)

Family Eunicidae

Marphysa sanguinea (Montagu)

Family Arabellidae

Arabella iricolor (Montagu)

Family Dorvilleidae

Stauronereis rudolphi (Della Chiaje)

TABLE 18 (continued)
Taxonomic List of Benthic Invertebrates Found in Barnegat Bay, NJ

Order Ariciida
  Family Orbiniidae

Orbinia norvegica (Sars)
Scoloplos fragilis (Verrill)
S. robustus (Verrill)

Order Cirratulida
  Family Cirratulidae

Cirratulus grandis Verrill
Tharyx acutus Webster and Benedict

Order Terebellida
  Family Pectinariidae

Pectinaria gouldii (Verrill)

  Family Ampharetidae

Asabellides oculata (Webster)

  Family Terebellidae

Amphitrite cirrata O. F. Muller
A. johnstoni Malmgren
A. ornata (Leidy)
Pista cristata (O. F. Muller)
P. palmata (Verrill)
Polycirrus eximius (Leidy)
P. medusa Grube
Terebellides stroemi Sars

Order Flabelligerida
  Family Flabelligeridae

Pherusa plumosa (O. F. Muller)

Order Sabellida
  Family Sabellidae

Sabella crassicornis Sars
S. microphthalma Verrill

  Family Serpulidae

Hydroides dianthus (Verrill)

Phylum Arthropoda
  Class Xiphosurida

Limulus polyphemus Linnaeus

  Class Pycnogonida

Callipallene brevirostris (Johnston)
Tanystylum orbiculare (Wilson)

  Class Crustacea
    Subclass Cirripedia
      Order Thoracica

Balanus balanoides (Linne)
B. eburneus (Gould)
B. improvisus (Darwin)

    Subclass Malacostraca
      Order Mysidacea

Heteromysis formosa (Smith)
Neomysis americana (Smith)

      Order Cumacea

Oxyurostylis smithi (Calman)

      Order Tanaidacea

Leptochelia savignyi (Kroyer)

TABLE 18 (continued)
Taxonomic List of Benthic Invertebrates Found in Barnegat Bay, NJ

Order Isopoda

Cyathura polita (Stimpson)
Edotea triloba (Say)
Erichsonella attenuata (Harger)
E. filiformis (Say)
Idotea baltica (Pallas)
Lironeca ovalis (Say)

Order Amphipoda
  Family Lysianassidae

Lysianopsis alba Holmes

  Family Ampeliscidae

Ampelisca abdita Mills
A. macrocephala Liljeborg
A. vadorum Mills
A. verrilli Mills

  Family Calliopiidae

Calliopius laeviusculus (Kroyer)

  Family Gammaridae

Elasmopus laevis Smith
Gammarus lawrencianus Bousfield
G. mucronatus Say
Maera danae Stimpson
Melita nitida Smith

  Family Bateidae

Batea catharinensis Muller

  Family Pontogeneiidae

Pontogeneia inermis (Kroyer)

  Family Hyalidae

Hyale sp.

  Family Corophiidae

Cerapus tubularis Say
Corophium tuberculatum Shoemaker
Erichthonius sp.
Unciola irrorata Say

  Family Ampithoidae

Ampithoe longimana Smith 1873
A. rubricata Montagu 1813
Cymadusa compta Smith

  Family Ischyroceridae

Jassa falcata (Montagu) 1818
Ischyroceros anguipes Kroyer 1838

  Family Aoridae

Lembos smithi Holmes 1905
Microdeutopus gryllotalpa Costa 1853

  Family Caprellidae

Aeginella longicornis Kroyer
Caprella geometrica Say
C. linearis Say

Order Decapoda
Infraorder Caridea
  Family Hippolytidae

Hippolyte zostericola (Smith)

  Family Crangonidae

Crangon septemspinosa Say

  Family Palaemonidae

Palaemonetes pugio Holthuis
P. vulgaris (Say)

  Family Majidae

Libinia dubia Milne-Edwards

TABLE 18 (continued)
Taxonomic List of Benthic Invertebrates Found in Barnegat Bay, NJ

Family Cancridae

Cancer irroratus Say

Infraorder Brachyura
Family Xanthidae

Eurypanopeus depressus (Smith)
Neopanope texana (Smith)
Panopeus herbstii H. Milne-Edwards
Rhithropanopeus harrisii (Gould)

Family Portunidae

Callinectes sapidus Rathbun
Carcinus maenas (Linnaeus)
Ovalipes ocellatus (Herbst)

Infraorder Anomura
Family Paguridae

Pagurus longicarpus Say
P. pollicaris Say

Phylum Mollusca
Class Gastropoda
Subclass Prosobranchia
Order Mesogastropoda

Bittium alternatum (Say)
Crepidula convexa Say
C. fornicata (Linnaeus)
C. plana Say
Epitonium rupicola Kurtz
Littorina saxatilis (Olivi)
Polinices duplicatus (Say)
Triphora nigrocincta (Adams)

Order Neogastropoda

Anachis avara (Say)
Busycon canaliculatum (Linnaeus)
B. carica (Gmelin)
Eupleura caudata (Say)
Mitrella lunata (Say)
Ilyanassa obsoleta (Say)
Nassarius trivittatus (Say)
N. vibex (Say)
Urosalpinx cinerea (Say)

Subclass Opisthobranchia
Order Cephalaspidea

Acteon punstostriatus (C. B. Adams)
Haminoea solitaria (Say)
Turbonilla interrupta (Totten)
Acteocina canaliculata (Say)

Order Nudibranchia

Doridella obscura (Verrill)
Doridella sp.
Cratena pilata (Gould)
Cratena sp.
Cuthona concinna (Alder and Hancock)

Class Bivalvia
Subclass Prionodesmata
Order Protobranchia

Nucula proxima Say
Solemya velum Say
Yoldia limatula (Say)

TABLE 18 (continued)
Taxonomic List of Benthic Invertebrates Found in Barnegat Bay, NJ

Subclass Pteriomorphia
  Order Prionodontia

                Anadara ovalis (Bruguiere)
  Order Pteroconchida

                Argopecten irradians (Lamarck)
                Crassostrea virginica (Gmelin)
                Geukensia demissa (Dillwyn)
                Modiolus modiolus Linnaeus
                Mytilus edulis Linne
Subclass Telodesmata
  Order Heterodontida

                Chiona cingenda Dillwyn
                Ensis directus Conrad
                Gemma gemma (Totten)
                Laevicardium mortoni Conrad
                Macoma balthica (Linne)
                M. tenta (Say)
                Mercenaria mercenaria (Linne)
                Mulinia lateralis (Say)
                Mya arenaria (Linne)
                Petricola pholadiformis Lamarck
                Pitar morrhuana (Linsley)
                Spisula solidissima (Dillwyn)
                Tagelus divisus (Spengler)
                Tellina agilis Stimpson
                T. versicolor Dekay
  Order Eudesmodontida

                Lyonsia hyalina Conrad

Phylum Ectoprocta

                Amathia vidovici (Heller)
                Bowerbankia gracilis Leidy
                Bugula turrita (Desor)
                Electra hastingsae Marcus
                Membranipora sp.

Phylum Echinodermata
  Class Asteroidea

                Asterias forbesi (Desor)
  Class Ophiuroidea

                Amphipholis squamata (Delle Chiaje)
  Class Echinoidea

                Arbacia punctulata (Lamarck)
  Class Holothuroid

                Cucumaria pulcherrima (Ayres)
                Leptosynapta tenuis (Ayres)
                L. roseola (Verrill)
                Thyone briareus (Lesueur)

Phylum Hemichordata

                Saccoglossus kowalevskyi (A. Agassiz)

Phylum Chordata
  Subphylum Urochordata
    Class Ascidiacea

                Botryllus schlosseri (Pallas)
                Molgula manhattensis (Dekay)
                Perophora viridis Verrill

TABLE 18 (continued)
Taxonomic List of Benthic Invertebrates Found in Barnegat Bay, NJ

Phylum Chaetognatha

Sagitta elegans Verrill

From Loveland, R. E. and Vouglitois, J. J., *Ecology of Barnegat Bay, New Jersey,* Kennish, M. J. and Lutz, R. A., Eds., Springer-Verlag, New York, 1984, 135. With permission.

been observed in other East Coast estuaries (e.g., Buzzards Bay, MA,[540] Barnstable Harbor, MA,[541] and waters at Hampton Roads, VA[542]). The numerical dominance of deposit feeders seems to be typical of higher-order successional stage, estuarine soft-bottom benthic communities.

2. Chesapeake Bay

Holland[366] and Holland et al.[367] surveyed the macrobenthos over a 14-year period in the mesohaline region of Chesapeake Bay (1971—1984) and the estuarine portion of the Potomac River (1980—1984). During this period, invertebrate samples were obtained with either a modified anchor dredge (1971—1976), a 0.12- or 0.10-m^2 hydraulically closing, van Veen grab (1977—1984), a 0.10-m^2 Smith-McIntyre grab (1977—1980), a 0.02-m^2 box core (1980—1984), or a 0.02-m^2 Ponar grab (1980—1984). Analyses of the Chesapeake Bay samples indicated the existence of five distinctive macrobenthic assemblages corresponding to major sediment types and salinity zones; these are a nearshore sand, a transitional muddy sand, a deep-water mud assemblage in the mid-bay and lower Potomac River, a nearshore sand, and a deep-water mud assemblage in the mid-Potomac River.[367,543-545] The relative abundances of the macrobenthic species differed among the five assemblages, but the species composition was similar. The transitional assemblage, found between low (5 to 10‰) and high (10 to 18‰) mesohaline salinities, had the greatest variation in species abundances. Differences in sediment characteristics accounted for much of the within habitat station variation of abundance. According to Diaz,[546] only salinity is more important than sediment type as a controlling factor on the benthic invertebrates of the estuary.

Based on temporal and spatial distributions, the macrobenthos of the Calvert Cliffs area have been subdivided into three groups — residents, opportunists, and nonresidents. Resident species (e.g., *Glycinde solitaria, Heteromastus filiformis, Macoma balthica, M. mitchelli, Micrura leidyi, Mya arenaria, Paraprionospio pinnata, Retusa canaliculata,* and *Scolecolepides viridis*) did not undergo long-term increases or decreases in abundance during the 14-year study period. The relatively stable physical-chemical environment in the midregion of the Chesapeake and lower Potomac River was responsible for the overall consistency of abundance.[367] Recruitment pulses caused much of the total variation in abundance of the dominant resident species. However, recruitment patterns likewise remained stable through time, with the amplitude of the recruitment pulses being determined by predators.[367] Large seasonal changes in abundance, when observed, were mainly attributed to predation,[232,236] but population reductions in summer also correlated with low dissolved oxygen levels.[545,547]

Opportunists (e.g., *Mulinia lateralis, Nereis succinea,* and *Streblospio benedicti*) fluctuated markedly in abundance from year to year primarily due to high fecundity and large recruitment pulses. Macrobenthic opportunists, which exerted a strong influence on the seasonal dynamics of the deep-water mud habitat, displayed high mortality rates subsequent to recruitment. This aspect of their population dynamics is typical of opportunistic species in other estuarine and marine systems.[548]

The nonresident benthos consisted principally of eurytolerant marine or polyhaline species (e.g., *Glycera dibranchiata* and *Tagelus plebeius*). Abundance of nonresident species peaked in years when salinity concentrations were above average. Variation in their abundances due to seasonal cycles was slight.

Polychaetes numerically dominated the benthic community in the Calvert Cliffs area, while mollusks dominated the standing crop by an order of magnitude.[545] Among major environmental factors affecting the benthos here, salinity and dissolved oxygen were of paramount importance. Salinity ranged from 5 to 22‰, but usually averaged between 10 and 16‰. During periods of abnormally low or high salinity, adult macroinvertebrates responded rather strongly to salinity fluctuations. The abundance of *Glycera dibranchiata, Glycinde solitaria, Macoma mitchelli, Paraprionospio pinnata, Acteocina canaliculata,* and *Tagelus plebeius* peaked in 1981—1982 when salinity exceeded the average for the preceding 10 years. In the low salinity years of 1972 and 1979, these six species had their lowest abundances. In contrast, *Corophium lacustre, Edotea triloba, Heteromastus filiformis, Haploscoloplos fragilis, Leptocheirus plumulosus, Macoma balthica, Micrura leidyi, Nereis succinea, Scolecolepides virides,* and *Streblospio benedicti* occurred in highest numbers during low salinity years and lowest numbers during high salinity years.

Holland et al.[367] documented three classes of species along the mesohaline salinity gradient of Chesapeake Bay. In the lower reaches of the Potomac River and the middle segment of the mainstem bay, the following euryhaline marine species were most numerous: *Acteocina canaliculata, Gemma gemma, Glycinde solitaria, Eteone heteropoda, Haminoe solitaria, Heteromastus filiformis, Macoma mitchelli, Micrura leidyi, Monoculodes edwardsi, Mulinia lateralis, Mya arenaria, Nereis succinea, Paraprionospio pinnata, Streblospio benedicti,* and *Tagelus plebeius.* Few of these populations occupied oligohaline waters (less than 5‰). In contrast, estuarine endemic species (e.g., *Cyathura polita, Leptocheirus plumulosus, Macoma balthica, Rangia cuneata, Scolecolepides viridis,* and *Tubificoides* sp.) were often taken from oligohaline habitats. They attained peak abundance in the middle reaches of the Potomac estuary. Other conspicuous populations collected at oligohaline salinities included freshwater forms (e.g., *Limnodrilus hoffmeisteri, Gammarus* sp., and larval stages of Chironomidae).

Abundance of the euryhaline marine species was positively correlated with salinity from 0.5 to 20‰. A negative correlation of salinity with abundance of estuarine species occurred over the range of about 8 to 20‰, and a positive correlation over the range of approximately 0.5 to 8‰. Abundance of freshwater forms was negatively correlated with salinity.

Effects of low dissolved oxygen concentration in summer were most pronounced on the macrobenthic assemblage in the deep-water mud habitat. Dissolved oxygen levels acted as a modulator of seasonal and year-to-year variation of the macrobenthos.[545] As a result, year-to-year abundance patterns for the deep-water mud assemblage differed from those of the other four assemblages. When dissolved oxygen levels dropped below 2 ppm in summer, a significant decline in the abundances of all species developed in the deep-water mud habitat. Even in late spring, low dissolved oxygen stress may have contributed to high mortalities in this assemblage. Anoxia generated high benthic macrofaunal mortality in the high and low mesohaline regions of the midbay and lower Potomac River that strongly influenced seasonal population cycles. However, with the dissipation of anoxic conditions in the fall, mobile adults and early stages of fall spawners (e.g., *Paraprionospio pinnata*) recolonized the previously anoxic sediments of the deep-water mud habitat.[367] Populations with planktonic developmental stages tended to repopulate defaunated areas more quickly than those that brood early life stages and have less dispersal capability (e.g., *Gemma gemma*). In the past decade, anoxic waters have encroached into shallower habitats and may threaten nursery and feeding grounds of a host of estuarine organisms.[550]

In sum, abundances of the macrobenthos, although affected by various abiotic and biotic factors, are chiefly influenced by salinity and dissolved oxygen concentrations. Most species increase or decrease in abundance in a predictable fashion in response to salinity changes, especially those leading to salinity extremes. The most acute perturbations in abundance arise annually in summer in the deeper waters of Chesapeake Bay as anoxic waters increase mortality of all macrobenthic species.

In spite of the temporal and spatial variations in abundance of the macrobenthos, several factors enhance the persistence of the organisms in the estuary. First, predation by fishes (e.g., *Leiostomus xanthurus)* and crabs (e.g., *Callinectes sapidus)* following recruitment pulses lowers faunal abundances and reduces opportunities for competitive interactions. Second, varied feeding modes foster the partitioning of space and food resources; competition among species, therefore, may be significantly reduced. Third, reproductive cycles with a periodicity of several months mitigate the impact of short-term environmental fluctuations on recruitment. Fourth, physical or chemical disturbances (e.g., storms and anoxia) frequently favor estuarine organisms adapted to wide perturbations in environmental conditions.[366]

3. Corpus Christi Bay

Between September 1974 and February 1979, Flint and Younk[365] sampled bottom communities in Corpus Christi Bay, TX, to establish a data base to assess natural and anthropogenic-induced variations in community dynamics of the benthos. A 0.09-m^2 Peterson grab was employed to sample the benthic communities monthly over the 4^1/$_2$-year study period. Field sample preparation involved sieving macroinvertebrates through a 0.5-mm screen, and in the laboratory, a stereo dissecting microscope was used to examine the samples.

Corpus Christi Bay is a shallow, lagoon-type estuary located in a semiarid region along the Texas coast. Having salinities generally higher than those of Barnegat and Chesapeake Bays in the range of 25 to 35‰, Corpus Christi Bay has been classified as a polyhaline system. Muds principally underlie the interior portions of the estuary, with mud and shelly sand lining its perimeters.

Benthic sampling recovered 313 taxa of macroinvertebrates belonging to 13 phyla. Polychaetes (mainly) and bivalve mollusks predominated in the grab samples. The dominant polychaete species were *Aricidea jeffreysii, Mediomastus californiensis, Paraprionospio pinnata,* and *Streblospio benedicti. Abra aequalis, Lyonsia hyalina floridana,* and *Mulinia lateralis* — seasonally abundant bivalves — generally dominated the molluscan group.

The investigators delineated significant differences in the benthic community between two major habitats: a 15-m deep channel area and a 3.5-m deep shoal region. The shoal habitat displayed higher population densities, numbers of species, and species diversities, while exhibiting lower equitability than the channel habitat. In the shoal habitat, one or two species usually dominated the benthic samples. Less dominance by individual populations existed in the channel habitat. The mean number of species per sampling station in the shoal habitat equaled 55.5 compared to only 21.6 in the channel habitat. The mean density of the macrobenthos at sampling stations in the shoal habitat ranged from 180 to 1700 animals per 0.09 m^2. The sampling stations in the channel habitat, in contrast, had mean densities between 35 and 580 animals per 0.09 m^2. The mean species diversity at shoal and channel stations amounted to 3.76 and 2.96, respectively. The shoal and channel stations differed markedly in terms of equitability of benthic communities, with the grand mean of the shoal communities being 0.416 and that of the channel communities, 0.609.

Flint and Younk[365] ascertained persistent benthic community patterns during the 4^1/$_2$-year study period. The shoal stations, for example, always showed higher species numbers, total densities, and diversities than the channel stations. Moreover, the fauna of the shoal habitat had more stable population densities relative to the channel habitat. Populations appeared to be less persistent in the channel habitat as well.

Disturbances (i.e., dredging, ship traffic, and shrimp trawling activities) continually disrupted the benthic community in the channel habitat. The disruption was manifested in the lack of distinct species assemblages, lower community densities, and higher variability in population numbers of many dominant species at the channel sites. Despite higher variability in the benthic community structure, at no time were the sediments devoid of fauna. The benthos had a high degree of resilience regardless of the type of disturbance. Subsequent to a disruption, species repopulated an area within a short period of time.

4. Puget Sound

As an estuary significantly deeper than Barnegat Bay, Chesapeake Bay, and Corpus Christi Bay, Puget Sound is subjected to fewer bottom habitat disturbances. Benthic data collected over a 20-year period at 200-m depth in the main basin of Puget Sound with a 0.1-m² van Veen grab sampler indicate large shifts in numerical abundance of the common species in spite of a low frequency of physical disturbance.[368] The bivalve, *Macoma carlottensis,* consistently dominated the benthic invertebrate community in most years, with an average density of 600/m² and, on two sampling dates, with a density exceeding 2000/m². Of the more than 120 benthic invertebrate species identified in the main basin of Puget Sound, many are equilibrium species with a longevity of greater than 1 year. Nearly all of these species occur in densities much less than 100/m².

Gradual changes in the numerical dominance of benthic invertebrates in Puget Sound can be traced from 1963 to 1983. Numerical shifts in dominance of the common species appeared at irregular multiyear frequencies. Hence, the dominant form, *M. carlottensis,* declined sharply in abundance during 1969 and 1970, when it was replaced as the dominant species by the polychaete, *Pectinaria californiensis,* which reached a density of 1000/m² (Figure 14). The polychaete, *Ampharete acutifrons,* normally a minor component of the community, underwent pulsing to a mean abundance greater than 2,000/m² in fall 1979. In 1982 and 1983, the bivalve, *Axinopsida sericata,* increased abruptly in abundance to nearly 100/m²; the rise in abundance approached 1000/m² during the mid-1980s. Although *M. carlottensis* decreased acutely in abundance in the late 1960s, it began to return to numerical dominance in late 1970 and 1971,[368] attaining a mean abundance above 2000/m² in 1977 and 1982.

Between 1964 and 1978, community dominance changed rapidly (Figure 13). The combined abundance of *M. carlottensis, P. californiensis, Ampharete acutifrons,* and *Axinopsida sericata* — the four species contributing most to the total faunal abundance — progressively increased during this interval. In conclusion, Nichols[368] speculates that the benthic community, as well as the benthic habitat, may be changing in Puget Sound over a long-term time frame.

5. Upper Clyde Estuary

Henderson[376] analyzed the long-term changes in species composition, density, and dominance patterns of benthic macroinvertebrates in the upper Clyde estuary, England, between 1974 and 1980. Organic pollution plagues this estuary; consequently, the macrobenthic invertebrate community is dominated by brackish, pollution-tolerant oligochaetes and polychaetes. Both natural and anthropogenic factors are responsible for the temporal fluctuations in the constituent populations of the community.

The benthic monitoring program was subdivided into two time periods spanning from 1974 to 1976 and from 1977 to 1980. The dominant members of the benthic macroinvertebrate community of the upper Clyde estuary between 1974 and 1976 included the Tubificid oligochaetes *Tubificoides benedeni, T. costatus, T. pseudogaster,* and *Monopylephorus rubroniveus;* the Enchytraeid, *Lumbricillus lineatus;* and the Naidids, *Nais elinguis* and *Paranais littoralis.* These species primarily inhabited the midsegment of the upper estuary. Between 1977 and 1980 inclusive, additional marine representatives became more commonplace (i.e., polychaetes — *Manayunkia aestuarina, Polydora ciliata,* and *Streblospio shrubsolii*). Some species (i.e., *Carcinus maenas, Corophium volutator,* and *Hydrobia ulvae*) appeared for the first time.

The abundance of the benthic fauna increased with increasing distance downestuary. Highest densities occurred at three stations downestuary: (1) the Cart confluence (less than 100 to 68,000

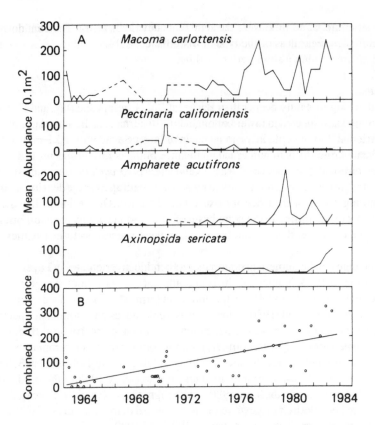

FIGURE 14.   Mean abundance of four numerically dominant, macrobenthic species in the main basin of Puget Sound over a 20-year period. (From Nichols, F. H., *Estuaries*, 8, 136, 1985. With permission.)

individuals per square meter), (2) Dalmuir (less than 500 to 230,000 individuals per square meter), and (3) Erskine (24,000 to 800,000 individuals per square meter). Upestuary of the Cart confluence, densities dropped substantially, often falling to 0 individuals per square meter between 1974 and 1976. In general, the densities increased at upestuary stations after 1976. They also remained high through 1980 at the Cart confluence and Dalmuir locations. At Erskine, however, a gradual diminution in density was evident, and by late 1980, it had fallen to only 1000 individuals per square meter.

Of the Tubificidae at Erskine, *T. benedeni, T. costatus,* and *T. pseudogaster* attained peak densities of less than or equal to 75,000, greater than 100,000, and 20,000 individuals per square meter, respectively, between 1974 and 1976. Subsequent to 1976, however, the abundance of these forms decreased, and by 1980, the densities of *T. benedeni, T. costatus,* and *T. pseudogaster* were 26,000, 3,000, and less than 200 individuals per square meter, respectively. Also at Erskine during the years 1974 to 1976, *L. lineatus* and *M. rubroniveus* reached highest densities, with *L. lineatus* having maximum counts up to 5,000 individuals per square meter and *M. rubroniveus,* greater than 10,000 individuals per square meter. Other species found in high densities during the entire survey were *P. littoralis,* exceeding 500,000 individuals per square meter at Dalmuir and Erskine, and *Capitella capitata* occasionally surpassing 100,000 individuals per square meter between 1974 and the end of 1977.

The dominance patterns of benthic macroinvertebrates changed temporally at the sampling sites. In some cases, a well-defined inverse relationship of dominance was maintained among two species. For example, *P. littoralis* and *T. benedeni* always showed inverse dominance patterns despite highly erratic fluctuations in dominance between 1974 and 1976. A seasonal dominance pattern of these species emerged in 1976 and 1977, with *T. benedini* dominating in summer and *P. littoralis* between October and March.

6. Dutch Wadden Sea

Beukema and Essink[370] and Beukema[551] monitored the benthic macroinvertebrates of the Dutch Wadden Sea tidal flats. Only six or seven macrobenthic species of the tidal flats add an average of more than 1 g/m² each to the total biomass.[551] The succeeding seven species accounted for most of the total biomass in 1971 and 1972: the blue mussel *Mytilus edulis* (23% of the total biomass), the lugworm *Arenicola marina* (19%), the sandgaper *Mya arenaria* (17%), the cockle *Cerastoderma edule (=Cardium edule)* (16%), the tellinid bivalve *Macoma balthica* (8%), and the polychaetes *Nereis diversicolor* (5%) and *Lanice conchilega* (3%). These seven species supplied more than 90% of the total biomass in a later study performed in 1977 when *C. edule* contributed the most biomass (26%). Other species, such as *Corophium volutator* and *Hydrobia ulvae,* have much higher numerical densities, but extremely low biomasses. A large fraction of the macrobenthic invertebrate biomass (66%) is contained in the mollusks and polychaetes (about 33%), with coelenterates, crustaceans, echinoderms, and other groups comprising a minor fraction.

Beukema and Essink[370] compiled data on the macrobenthic fauna at Balgzand during 17 successive years (1969—1985). Fifteen stations were sampled at Balgzand, a 50-km² tidal flat area in the western perimeter of the Wadden Sea, in an attempt to unravel the degree of similarity of population fluctuation patterns. Results of this long-term sampling program revealed that nearly 50% of the fluctuation patterns of the populations within the study area had a high degree of similarity. Table 19 chronicles important macrobenthic invertebrates at Balgzand.

Reise[434] has recently detailed macrofaunal assemblages on the tidal flats of Königshafen, a shallow, sheltered bay on the island of Sylt in the northern part of the Wadden Sea. The dominant macrofaunal species form dense assemblages on the tidal flats. This is the case for *C. volutator* and *Nereis diversicolor* on the upper tidal flats, *A. marina* on the sandflats, and *Lanice conchilega* and *M. edulis* near the low water line.

Polychaetes numerically dominate the benthic macrofauna in most areas of Königshafen (Table 20). *Corophium volutator* and *H. ulvae* may be dominant in zones where they grow in dense aggregations. On sandflats, *A. marina* has a mean density of approximately 40/m². The cockle, *C. edule,* also is a prominant member of the benthic community of sandy tidal flats. Varying greatly in abundance on muddy flats, benthic macrofauna are frequently dominated by spionid and capitellid polychaetes (e.g., *Tharyx marioni*) and the oligochaete, *Tubificoides benedeni.* The macrofaunal biomass averages about 15 g C per square meter on the tidal flat, but where dense beds of mussels grow, the mean annual biomass approaches 1000 g C per square meter.

The tidal flats of the Dutch Wadden Sea cover an area of 1300 km². McLusky[49] estimated the mean macrofaunal biomass of the flats to be 27 g/m² ash-free dry weight, with a range of 19 to 34 g/m² ash-free dry weight. At the highest and lowest levels of the intertidal zone, where environmental factors are extreme, a low absolute abundance and biomass of species exist. Maximum macrofaunal biomass, exceeding 50 g C per square meter, takes place between the midtide level and 50 cm below the midtide level in sediments having a silt content of 10 to 20%. The highest species diversity values, greater than 14/0.45 m², also are encountered at the midtide level, about 25 cm below the sediment-water interface in sediments having a silt content of 2 to 25%.[49,552]

3. Species Diversity

When proceeding from oceanic waters into estuaries, a gradient is evident in the diversity of macróbenthic fauna. Species richness, the number of species present, generally declines upestuary to a minimum number at a critical salinity which reflects the incapabilities of many

TABLE 19
Quantitatively Important
Macrobenthic Invertebrates
Collected during a 17-Year
Sampling Program at
Balgzand, Dutch Wadden Sea

Abra tenuis (Montagu)
Anaitides spec. div.
Angulus tenuis (Da Costa)
Antinoella sarsi (Kinb.)
Arenicola marina (L.)
Bathyporeia spec.
Carcinus maenas (L.)
Cerastoderma edule (L.)
Corophium volutator (Pall.)
Crangon crangon (L.)
Eteone longa (Fabr.)
Heteromastus filiformis (Clap.)
Hydrobia ulvae (Penn.)
Lanice conchilega (Pall.)
Littorina littorea (L.)
Macoma balthica (L.)
Magelona papillicornis F. M.
Mya arenaria L.
Mytilus edulis L.
Nemertini spec. div.
Nephtys hombergii Sav.
Nereis diversicolor (O. F. M.)
Scolelepis foliosa (A. & M.-E.)
Scoloplos armiger (O. F. M.)
Scrobicularia plana (Da Costa)

From Beukema, J. J. and Essink, K.,
Hydrobiologia, 142, 199, 1986. With
permission.

TABLE 20
Abundance of Macrofauna at Four Sites in April 1975, Königshafen, Island of Sylt[a,b]

Habitat % submergence	High flat 30	Grass bed 50	Sand flat 70	Mud flat 80
Anthozoa				
Sagartiogeton viduata (O. F. Müller)	0	19	0	0
Plathelminthes				
Pseudograffilla arenicola Meixner	0	350	0	0
Nemertini				
Tetrastemma melanocephalum (Johnston)	19	0	0	0
Amphiporus lactifloreus (Johnston)	6	25	0	0
Prostomatella arenicola Friedrich	0	6	13	0
Lineus viridis (Fabr.) Johnston	6	88	0	0
Polychaeta				
Harmothoe imbricata (L.)	0	6	0	0
H. sarsi (Klingenberg)	0	0	6	0
Anaitides mucosa (Oersted)	0	113	38	19

TABLE 20 (continued)
Abundance of Macrofauna at Four Sites in April 1975, Königshafen, Island of Sylt[a,b]

Habitat % submergence	High flat 30	Grass bed 50	Sand flat 70	Mud flat 80
Polychaeta (continued)				
Eteone longa (Fabr.)	13	44	50	6
Nereis diversicolor O. F. Müller	1,894	0	0	0
Microphthalmus aberrans (Webster & Bened.)	0	38	138	69
Nephtys hombergi Savigny	0	6	13	19
Arenicola marina (L.)				
0 − group	138	1,725	0	1,444
1 + group	0	13	50	0
Heteromastus filiformis (Clap.)	0	13	6	581
Capitella capitata (Fabr.)	0	2,506	456	119
Scoloplos armiger (O. F. Müller)	0	844	831	756
Magelona mirabilis (Jones 1977)	0	0	6	0
Spio filicornis (O. F. Müller)	0	0	25	6
Pygospio elegans Claparede	356	5,331	1,919	9,219
Polydora ligni Webster	63	13	0	44
Malacoceros fuliginosus (Clap.)	0	581	0	0
Tharyx marioni (Saint-Joseph)	0	50	88	1,775
Ampharete acutifrons (Grube)	0	431	6	38
Lanice conchilega (Pallas)	0	38	0	0
Fabrica sabella Ehrenberg	44	8,806	0	0
Oligochaeta				
Tubificoides benedeni (D'Udekem)	0	3,588	456	3,294
Bivalvia				
Mytilus edulis L.	6	25	0	19
Mya arenaria L.	0	25	6	13
Macoma balthica L.	13	156	6	13
Abra alba (W. Wood)	0	13	0	0
Tellina tenuis Da Costa	0	0	0	0
Cerastoderma edule (L.)	0	144	38	119
Gastropoda				
Littorina littorea L.	6	88	19	6
L. saxatilis (Montagu)	0	94	0	0
L. obtusata L.	0	0	0	6
Hydrobia ulvae (Pennant)	3,963	27,906	6	13
Retusa obtusa (Montagu)	0	6	0	0
Isopoda				
Idotea baltica (Pallas)	0	6	0	0
Amphipoda				
Corophium arenarium Crawford	0	0	63	0
C. volutator (Pallas)	7,725	0	0	38
Chaetogammarus marinus (Leach)	6	0	6	0
All individuals	14,258	53,097	4,245	17,622
Number of species	15	32	23	23

[a] Individuals below 1 m².

[b] From each site, 16 samples of 100 cm²/0—20 cm, where 0—2 cm was sieved with 0.25 mm, 2—5 cm was sieved with 0.5 mm, 5—20 cm was sieved with 1.0 mm.

From Reise, K., *Tidal Flat Ecology: An Experimental Approach to Species Interactions, Ecological Studies,* Vol. 54, Springer-Verlag, Berlin, 1985. With permission.

of the animals to tolerate salinity stress and to undergo extensive cell volume regulation. As the number of species diminishes, so does the interspecific competition, leading to niche expansion among species capable of inhabiting the estuarine seafloor.[8]

Levinton[8] deals at length with the concept of diversity. This parameter consists of two components: (1) species richness, S, described above; and (2) evenness, J, which measures the distribution of population sizes of the respective species. Various indexes of diversity have been advocated for marine ecological research. One of the most common indexes is the Shannon-Wiener measure, H'. This index combines the diversity components S and J and is determined by

$$H' = \sum_{i=1}^{S} p_i \ln p_i \tag{1}$$

where s is the number of species, and p_i is the proportion of species i. As S increases, so does H'. Levinton (p. 87)[8] acknowledges that, for a given S, H' is maximized (H'$_{max}$) when $P_1 = P_2 =$ ... $= P_s$, and H' equals ln S. Therefore, a measure of evenness is the following

$$J' = \frac{H'}{H'_{max}} = \frac{H'}{\ln S} \tag{2}$$

Only middle species strongly affect the measure, H', while abundances of the common or rare species do not.

A second commonly used index of diversity is the Simpson's index defined by

$$S_I = \sum_{i=1}^{S} \frac{n_i(n_i - 1)}{N(N - 1)} \tag{3}$$

where n_i represents the number of species i and N, the total number of individuals of all species. This index emphasizes the degree of dominance by one or several species.

A third measure, which Levinton (p. 88)[8] contends "... has biological meaning because it can be related to the diversity of resources available to the s species present when competitive interactions are considered", has been applied in diversity studies. This measure is given as

$$D_s = \frac{1}{\sum_{i=1}^{S} p_i^2} \tag{4}$$

a. Causes of Low Species Richness in Estuaries

Remane[553,554] formulated a species-salinity relationship based on his work in Baltic waters which shows a bimodality in species richness — one peak corresponding to freshwater and another to marine salinities (Figure 15). Findings from brackish lagoons in Denmark,[306,554] the Tay and Tees estuaries in England,[555] estuaries and brackish waters of the Delta area of the Netherlands,[391,461,556,557] and estuaries in Virginia[387] corroborate this relationship. The species richness of benthic invertebrates in estuaries is minimum within a salinity range of about 5 to 8‰, termed the critical salinity.[558,559] The number of freshwater and marine species gradually diminishes from the upper and lower reaches, respectively, into the very unstable brackish-water areas of most systems.

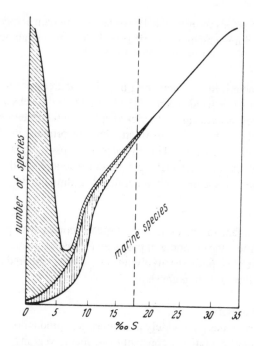

FIGURE 15. Relationship of salinity to species richness. Obliquely-hatched area: freshwater species. Vertically-hatched area: brackish-water species. White: marine species. (From Wolff, W. J., *Estuaries and Enclosed Seas,* Ketchum, B. H., Ed., Elsevier, Amsterdam, 1983, 151. With permission.)

While the species richness in estuaries may be quite low, the density of the taxa present can be great. Some species are well-adapted to environmental stresses in these coastal habitats. Carriker[396] attributes the high abundances of populations in estuaries, in part, to the reduction in the number of species competing for a limited food supply.

1. Stability — Time Hypothesis

In 1968, Sanders[560] postulated that the species diversity of the benthos is dependent on the stability of the environment as well as its geological history. For low-diversity communities in fluctuating environments, he coined the words "physically controlled", and for high-diversity communities in stable or predictable environments, he fashioned the term "biologically accommodated". Diversity tends to be greater in geologically ancient environments that are physically stable than in recent, variable environments. Because they are geologically young, unstable or unpredictable, and ephemeral, estuaries have a lower probability of speciation and a greater likelihood of extinction than more constant environments such as the deep sea. Organisms inhabiting estuaries must adapt to a highly fluctuating environment where perturbations in physical, chemical, and biological conditions are extreme. Thus, specialized species in estuaries typically do not survive as well as generalists with broad adaptability.

2. Environmental Stress

Another hypothesis formulated to explain fewer species in estuaries and other extreme environments is environmental stress. Accordingly, less species inhabit estuaries because fewer representatives of phyletic groups have evolved that are capable of successfully invading this environment.[8] Those that do penetrate estuarine waters often undergo ecological expansion as a result of reduced interspecies competition. However, the gradient in species richness of benthic invertebrates to a minimum number at the critical salinity range might reflect a stress gradient.

3. Resource Stability

Environments characterized by fluctuating trophic resources promote the proliferation of generalized rather than specialized species.[561,562] Complex food webs are less likely to evolve when trophic constancy is lacking. Estuaries display marked seasonal variations in food availability, as exemplified by the large fluctuations in primary productivity of temperate systems between summer and winter. This hypothesis helps to explain the higher number of specialized forms in offshore waters that have greater resource stability compared to inshore regions which support more species with generalized feeding habits.

4. Other Hypotheses

A high diversity of organisms arises in areas of greater competitive pressure, thereby forcing specialization.[8] In addition, the cropping of prey species fosters greater diversity by permitting more species to coexist.[563] Hypotheses involving competition and predation have received less attention than the stability-time hypothesis.

4. Biomass and Productivity

Benthic macroinvertebrates vary widely in biomass and production from estuary to estuary, but these parameters may be relatively constant from year to year in a given estuary. The mean biomass of the macrobenthos generally ranges from 10 to 25 g ash-free dry weight per square meter for an entire estuary and locally attains levels of 2000 g ash-free dry weight or more. The mean production for a whole estuary approaches 50 g ash-free dry weight per square meter per year. Organisms growing in dense aggregations (e.g., mussels or oysters) can reach production figures of 200 to 300 g ash-free dry weight per square meter per year. Highest production and biomass measurements have been recorded on suspension feeders such as cockles, mussels, and oysters, with deposit feeders yielding lower secondary production.[81]

Biomass and production estimates have been enumerated for a number of bivalve species. In England, Walne[564] computed biomass values of 200 g dry flesh weight per square meter for the hard clam, *Mercenaria mercenaria,* and the oyster, *Ostrea edulis.* Wolff[391] quoted biomasses for *Mytilus edulis* ranging from 400 to 1420 g dry flesh weight per square meter. Hibbert[565] calculated the biomasses of *Cardium (=Cerastoderma) edule, M. mercenaria,* and *Venerupis aurea* in Southampton water, producing estimates of 17.8 to 64.6, 7.7 to 50, and 0.6 to 1.14 g ash-free dry weight per square meter, respectively. Burke and Mann[566] registered a biomass value of 4.57 g flesh dry weight per square meter for *Mya arenaria* on the sandflats of the Petpeswick Inlet, Nova Scotia. In the Ythan estuary, Scotland, Chambers and Milne[567] disclosed a mean biomass figure of 4.86 g dry flesh weight per square meter for *Macoma balthica.* Wolff and de Wolf[568] published biomasses for *C. edule* in the Grevelingen estuary up to 46.7 g ash-free dry weight per square meter. Estimates have been made of the biomasses of *Crassostrea virginica* in Georgia (970 g ash-free dry weight per square meter)[569] and of *Mytilus edulis* in Delaware Bay (up to 2.2 g ash free dry weight per square meter).[570]

In the Lynher estuary, England, Warwick and Price[571] documented a mean biomass of 0.43 g dry weight per square meter for the polychaete, *Ampharete acutifrons.* A higher value equal to 4.22 g ash-free dry weight per square meter has been derived for *Nereis diversicolor* in the Ythan estuary.[567] For smaller oligochaete worms, McLusky et al.[572] demonstrated higher biomasses in organically enriched regions of the Forth estuary (27.97 g dry weight per square meter) than in more typical estuarine mudflats (6.30 g dry weight per square meter). The lugworm, *Arenicola marina,* in the Grevelingen estuary had a mean biomass of 8.75 g ash-free dry weight per square meter.

Similar to production figures, biomass values for macrofauna of temperate systems charac-

TABLE 21
Biomass Values of Macrobenthic Species Assemblages in Selected European and U.S. Estuaries

Area	Depth	Biomass (g m^{-2})	Ref. (sources from Ref. 391)
Wadden Sea, Denmark	Intertidal	174—497 (wet)	Smidt (1951)
Ringkøbing Fjord, Denmark	Intertidal	267—450 (wet)	Smidt (1951)
Long Island Sound	Subtidal	54.6 (dry)	Sanders (1956)
Barnstable Harbor	Intertidal	38—40 (dry)	Sanders et al. (1962)
North Carolina Estuary		8.1 (dry tissue)	Williams and Thomas (1967)
Puget Sound	12—195 m	8—9 (ash-free)	Lie (1968)
Firemore Bay, Scotland	Beach	1.3 (dry)	McIntyre and Eleftheriou (1968)
	Subtidal	3.7 (dry)	McIntyre and Eleftheriou (1968)
Biscayne Bay	Intertidal	30 (dry)	Moore et al. (1968)
Narragansett Bay	Intertidal musselbed	1852 (ash-free)	Nixon et al. (1971)
Lower Mystic River	15 m	1.0—50 (wet)	Rowe et al. (1972)
Balgzand, The Netherlands	Intertidal	9.6—25.1 (ash-free)	Beukema (1974)
Kiel Bay, Germany	>15 m	26.3 (ash-free)	Arntz and Brunswig (1975)
Orwell Estuary, England	Intertidal	46.0 (ash-free)	Kay and Knights (1975)
Stour Estuary, England	Intertidal	28.1—38.7 (ash-free)	Kay and Knights (1975)
Colne Estuary, England	Intertidal	7.8 (ash-free)	Kay and Knights (1975)
Dengie Flats, England	Intertidal	18.5 (ash-free)	Kay and Knights (1975)
Crouch Estuary, England	Intertidal	8.7 (ash-free)	Kay and Knights (1975)
Roach Estuary, England	Intertidal	8.7 (ash-free)	Kay and Knights (1975)
Foulness Flats, England	Intertidal	19.6 (ash-free)	Kay and Knights (1975)
Thames Estuary, England	Intertidal	18.8 (ash-free)	Kay and Knights (1975)
Medway Estuary, England	Intertidal	24.1 (ash-free)	Kay and Knights (1975)
Swale Estuary, England	Intertidal	32.8 (ash-free)	Kay and Knights (1975)
Tamar Estuary, England	Intertidal	13 (ash-free)	Warwick and Price (1975)
Southampton Water, England	Intertidal	90—190 (ash-free)	Hibbert (1976)
Wadden Sea, The Netherlands	Intertidal	27 (ash-free)	Beukema (1976)
Grevelingen Estuary, The Netherlands	Whole estuary	20.8 (ash-free)	Wolff and De Wolf (1977)
Byfjord, Sweden	Subtidal	4.5 (dry)	Rosenberg et al. (1977)
Delaware River	Intertidal fresh water	7.0 (dry)	Crumb (1977)
San Francisco Bay	Intertidal	10.1—30.5 (ash-free)	Nichols (1977)

From Wolff, W. J., *Estuaries and Enclosed Seas,* Ketchum, B. H., Ed., Elsevier, Amsterdam, 1983, 151. With permission.

teristically have wide seasonal variations. This seasonal pattern holds for individual species as well as for the total benthic fauna. Table 21 gives biomass estimates of macrobenthic species assemblages for a number of estuaries in Europe and the U.S.

Absolute abundance, rather than biomass, may be a better measure of benthic community dynamics.[12,37,540] Only a few randomly distributed animals often account for the major fraction of the total faunal biomass. In Buzzards Bay, MA, for example, Sanders[540] observed that only 0.15% of the species comprised 55.17% of the weight, and ten species constituted 95% of the total biomass. Benthic organisms with large exoskeletons such as bivalves, although present as sparse populations in many areas, frequently supply a disproportionately large amount of the biomass, even though smaller invertebrates (e.g., polychaetes) may numerically dominate the community.

In comparison to other ecosystems, the production of benthic macrofauna in estuaries is relatively high. This high production, as advanced by Wolff,[391] may be due to the large food supply and preponderance of opportunistic species which grow rapidly and have high turnover rates. Hence, on the Hamble Spit, Southampton, England, Hibbert[565] determined the annual production of *C. edule* (29.2 to 71.4 g ash-free dry weight per square meter per year), *M. mercenaria* (4 to 14 g ash-free dry weight per square meter per year), and *V. aurea* (0.7 to 1.25 g ash-free dry weight per square meter per year). Production of *Nereis diversicolor* in the Ythan estuary amounted to 12.78 g ash-free dry weight per square meter per year,[567] and that of the polychaete, *Ampharete acutifrons,* in the Lynher estuary equaled 2.32 g ash-free dry weight per square meter per year. Annual production of oligochaetes in the Forth estuary yielded 25 g ash-free dry weight per square meter per year. In the Grevelingen estuary, *C. edule* produced 10 to 120 g ash-free dry weight per square meter per year.[568] Also in the Grevelingen estuary, *A. marina* and *M. balthica* — both deposit feeders — supplied 3.3 to 6.3 and 0.3 to 7.9 g ash-free dry weight per square meter per year, respectively. Production estimates in this estuary likewise have been made for the grazing gastropods, *Hydrobia ulvae* (7.2 to 12.8 g ash-free dry weight per square meter per year) and *Littorina littorea* (6.1 g ash-free dry weight per square meter per year).

At Sapelo Island, GA, Bahr[569] calculated production figures of about 200 g ash-free dry weight per square meter per year for *C. virginica.* Sanders[573] compiled production estimates in Long Island Sound for the deposit feeders *Cistenides (=Pectinaria) gouldii* (1.7 g ash-free dry weight per square meter per year), *Nephtys incisa* (9.3 g ash-free dry weight per square meter per year), *Scrobicularia plana* (3 to 13 g ash-free dry weight per square meter per year), and *Yoldia limatula* (3.2 g ash-free dry weight per square meter per year). Burke and Mann[566] recorded production values for *L. saxatilis* (3.3 g ash-free dry weight per square meter per year), *M. balthica* (1.9 g ash-free dry weight per square meter per year), and *Mya arenaria* (11.6 g ash-free dry weight per square meter per year) at Petpeswick Inlet, Nova Scotia (Canada). Table 22 provides production measurements for macrobenthic assemblages in estuaries of Europe and the U.S.

5. Feeding Strategies

Benthic invertebrates have several modes of feeding, as mentioned briefly in Section II.B. Rhoads[318] describes five categories of benthic feeding strategies: (1) suspension feeders which consume seston suspended in the water column, (2) deposit feeders which consume deposited detritus, (3) herbivores which ingest plant matter, (4) carnivore-scavengers which eat live animals or recently dead animal tissue, and (5) parasites which obtain nutrition via the fluids of living organisms. The majority of benthic macroinvertebrates consists of suspension and deposit feeders. All types of organic matter in estuaries — both particulate and dissolved — are food sources, including bacteria and other microbes, benthonic and planktonic organisms, and dissolved organic matter.[391]

Suspension feeders usually inhabit sandy sediments, and deposit feeders, muddy sediments. In areas of high concentrations of suspended particles, deposit feeders supplant suspension feeders as the preponderant macrobenthic fauna because the particles clog the filtering structures of the suspension feeders. Sanders,[426] in an investigation of the benthic community of Buzzards Bay, MA, discovered the greatest proportion of deposit feeders in unstable muds. Here, the protobranch bivalves, *Nucula annulata* and *Yoldia limatula,* and the errant polychaete, *Nephtys incisa,* dominated the community. In muddy-sand bottoms, however, suspension feeders predominated. The spatial segregation of the deposit and suspension feeders denoted by Sanders[426] in Buzzards Bay also has been viewed on a worldwide scale.[318] Rhoads and Young[427] term the incompatibility of the two groups as trophic group amensalism.

Suspension feeders primarily strain phytoplankton from the water column as food, but also remove some detritus, although the value of the detritus in respect to energetic significance remains unclear. Other possible foods of these organisms are bacteria and zooplankton.[391] The

TABLE 22
Production Estimates for Estuarine Macrobenthic Assemblages in Selected Estuaries of Europe and the U.S.[a]

Area	Depth	Production	Ref. (sources from Ref. 391)
Long Island Sound	Subtidal	21.4 (infauna)	Sanders (1956)
	Subtidal	5.8 (epifauna)	Richards and Riley (1967)
Firemore Bay, Scotland	Whole bay	2.5 (g carbon)	McIntyre and Eleftheriou (1968)
Kiel Bight, Germany	15 m	17.9	Arntz and Brunswig (1975)
Tamar Estuary, England	Intertidal	13.3	Warwick and Price (1975)
Southampton Water, England	Intertidal	152—225 (?)	Hibbert (1976)
Grevelingen Estuary, The Netherlands	Whole estuary	50.3—57.4	Wolff and De Wolf (1977)

[a]   Values expressed as g ash-free dry weight per square meter per year.

From Wolff, W. J., *Estuaries and Enclosed Seas*, Ketchum, B. H., Ed., Elsevier, Amsterdam, 1983, 151.

barnacles, *Balanus balanoides* and *B. improvisus*, the bivalves, *C. virginica*, *M. mercenaria*, and *M. edulis*, and the bryozoans, *Bowerbankia gracilis* and *Conopeum seurati*, are examples of estuarine suspension feeders.

Deposit feeders have been subdivided into selective and nonselective types. Selective deposit feeders use various means to separate their food from sedimentary particles. Nonselective deposit feeders do not sort food particles from sediment, but ingest the sediment as a whole, digesting organic material and voiding the sediment and nondigestable particles. Bacteria, benthic microalgae, meiofauna, microfauna, and possibly detritus are important sources of food for nonselective deposit feeders, and bacteria and microalgae are favored food items of selective deposit feeders. The Tellinid, *M. balthica*, exemplifies selective deposit feeding and the lugworms, *Arenicola marina* and *Abarenicola claparedii*, nonselective deposit feeding.

Deposit feeders rework considerable quantities of sediment; consequently, they continually alter the substrate. As modifiers of the upper layer of sediment on the seafloor, the motile surface deposit feeders that ingest bulk sediment at high rates (e.g., Arenicolids, Enteropneusts, and Holothurians) are critical agents in the mixing and transport of particles, interstitial water, and dissolved gases.[574] Other sedentary deposit feeders, largely polychaetes, live in vertically oriented tubes, at times in dense aggregations. They transfer enormous volumes of grains from several centimeters depth to the sediment surface.[318]

Many motile macrobenthic fauna, especially crabs, echinoderms, and gastropods, are scavengers or predators. The horseshoe crab *Limulus polyphemus*, blue crab *Callinectes sapidus*, mud crab *Neopanope texana*, knobbed whelk *Busycon carica*, oyster drill *Urosalpinx cinerea*, lobed moon shell *Polinices duplicatus*, and sea star *Asterias forbesi* provide examples. *L. polyphemus* and *C. sapidus* are voracious predators; *L. polyphemus* digs through sediments, consuming clams, polychaetes, and other burrowing organisms,[12] whereas *C. sapidus* moves along the bottom in search of live or dead animals. *B. carica*, *U. cinerea*, and *P. duplicatus* have a more restricted diet, ingesting other mollusks, such as *Chione cancellata*, *M. mercenaria*, and *C. virginica*. Another predator, *A. forbesi*, also feeds heavily on mollusks. Xanthid crabs comprise an abundant and widely distributed group of commensals and scavengers.[575]

Other benthic fauna are classified as grazers. These species primarily subsist on benthic microalgae, (e.g., *Littorina)*; however, some eat macroalgae (e.g., *Strongylocentrotus*) or seagrasses (e.g., *Idotea chelipes*).[391] Valiela[51] remarks that grazers affect the abundance and distribution of attached microalgae. It is not unusual for entire algal patches to be removed via the activity of benthic grazers.

6. Trophic Interactions
a. Predator-Prey Experiments

Competitive and predatory-prey interactions have been investigated most successfully in the rocky intertidal zone, where manipulative experiments can be executed relatively easily. Research by Lubchenco and Menge[576] on the competitive hierarchy of *Balanus balanoides, Mytilus edulis,* and *Chondrus crispus* in the rocky intertidal zone of New England is one such example. Other classic work on competition in intertidal rocky shores is that of Paine[577] and Connell[578] on the Pacific Coast of North America.

In soft-bottom communities, the use of exclusion experiments has rapidly gained favor in studies of predation and competition. Exclusion experiments entail deploying cages to exclude predators from an area to assess their impact on epibenthic and infaunal organisms. When predators are removed from an area by cages or some other means of exclusion, an acute increase in their macroinvertebrate prey often takes place. In lagoons and estuaries, caging experiments on unvegetated soft bottom result in a two- to threefold increase in the density of macroinvertebrate prey in zones protected from predators.[432] Little evidence exists for competitive exclusion in these predator-free regions.[51] Caging experiments on vegetated soft bottoms lead to different conclusions. The exclusion of predators by cages in these habitats accounts for only very slight changes in the macroinvertebrate community.[432] The vegetation apparently serves a dual role to hide prey organisms and to exclude larger predators. These vegetated bottoms usually have communities with greater population abundances and species richness than unvegetated bottoms. In addition, they also show little evidence of competitive exclusion.[51]

In the intertidal zone, predators profoundly influence the structure of benthic communities in soft sediments. Reise,[225,234,455] examining the intertidal flats of European systems and using predator exclusion experiments, found that certain epibenthic predators (e.g., *Carcinus maenas, Crangon crangon,* gobiid fishes, flatfishes, and birds) substantially crop prey organisms. When protected by enclosures, benthic fauna on unvegetated mudflats increased in density by as much as tenfold that of unprotected areas. Again, no significant change in the benthos occurred in vegetated (seagrass) habitats when experimental enclosures were employed, demonstrating that the vegetation affords prey organisms natural protection from predators.

Differences in the benthic community composition within a caged area relative to natural control areas may arise from several factors.[233,579,580] According to Reise,[42] these are (1) chance (patchiness in the distribution of sediment-dwelling organisms), (2) physical artifacts (cages affect currents, sedimentation, erosion, and shading), (3) algae (trapping of drift algae by the cages), (4) attraction of larvae (netting of the cages, resembling vegetation on reef structures, attract pelagic and migrating stages of the benthos), (5) attraction of predators (cages act as a refuge for smaller predators), and (6) unwanted exclusions (in addition to predators, larger herbivorous fish and birds are excluded which might drastically reduce disturbance efforts). Partial cages have been constructed to differentiate between cage artifacts and effects of predator exclusion. Unequivocally, cage effects exist and can partially obscure changes caused by predator exclusion. Nevertheless, caging experiments are a powerful means of understanding predator-prey interactions in soft-sediment habitats of estuaries.

b. Predation, Competition, Amensalistic Interactions, and Community Structure

Under certain conditions, biological interactions (e.g., predation, competition, and amensalistic interactions) modulate local patterns of benthic community composition.[581] While direct competition and predation have been stressed by some investigators as the two principal biotic factors controlling distributions and abundances of organisms in benthic communities, recent studies suggest that indirect interactions may be equally important processes. Indirect interactions involve habitat modifications by a species that alters the density of another through some intermediary effect. A number of functional-group hypotheses have been proposed in the literature to explain observed distribution patterns of the benthos, including trophic-group amensalism, adult-larval interactions, and mobility-mode interactions. These hypotheses concentrate on density-dependent interactions and/or indirect interactions involving habitat modifications which may be of major significance in determining the composition of soft-bottom communities.[581]

Research during the past 25 years has verified the role of predation in structuring benthic communities.[51,446-450,453-457,577,582-585] Commito and Ambrose[458] argued that multiple trophic levels are discernible within the benthic infaunal communities of shallow coastal systems. Ambrose[410] advocated the application of a three-level interactive model of marine infaunal communities, composed of epibenthic predators, predatory infauna, and nonpredatory infauna. Based on this model, epibenthic predators (e.g., birds, crabs, and fishes) principally consume infaunal predators, but also feed to some extent on nonpredatory populations. Nonpredatory infauna are assumed to be primarily the prey of predatory infauna. Wilson,[449] reanalyzing the caging surveys discussed by Ambrose,[410] refuted the three-level interactive model and its application to soft-sediment communities.

Valiela[51] delved into the species interactions and predatory effects of fauna in salt marsh creek sediments of New England. The major macroinvertebrate taxa in this biotope are amphipods, an anemone *(Nematostella vectensis)*, annelids *(Capitella* spp., *Hobsonia florida, Manayunkia* spp., and *Streblospio benedicti)*, hydrobiid snails, and an oligochaete *(Paranaius litoralis)*. Crabs *(C. maenas)* and fishes (principally *Fundulus heteroclitus* and *F. majalis)* ingest benthic fauna and other prey during the warm months of the year. When cages are positioned to exclude predators, the biomass of benthic macroinvertebrates remains high. The biomass of the prey organisms diminishes sharply, however, when the predators have access to the benthos. Valiela[51] concluded that both the crabs and fishes remove much of the prey biomass, with the standing crop of macroinvertebrates dropping by an order of magnitude between late June and August. These predators, therefore, are key elements in structuring the benthic community.

Breitburg[50] performed studies on macroinvertebrate grazing of benthic communities that colonize plexiglass panels in subtidal waters. The most abundant grazers were the sea urchins, *Strongylocentrotus franciscanus* and *S. purpuratus,* and the omnivorous sea star, *Patiria miniata.* Other mobile benthic invertebrates included the sea urchin, *Lytechinus anamesus;* the sea stars, *Dermasterias imbricata, Orthasterias kohleri, Pisaster giganteus,* and *P. brevispinus;* the sea cucumber, *Parastichopus parvimensis;* the gastropods, *Astrea undosa, Cypraea spadicea,* and *Kelletia kellitii;* the crab *Cancer antennarius;* and the spiny lobster, *Panulirus interruptus.* Only grazer-resistant algal crusts, diatom/blue-green algal films, and short-lived filamentous algae grew abundantly on the plexiglass panels exposed to naturally high densities of sea urchins and sea stars. Other algae and sessile invertebrates were common on plexiglass panels protected from grazers. Temporal and small-scale spatial variation in the abundances of species arose from the plexiglass experiments. In sum, the macroinvertebrate grazers had a strong influence on species composition and population abundances of the epibenthic community.

Jensen and Jensen[448] evaluated epibenthic predation on juvenile benthic macrofauna of an intertidal flat in the Danish Wadden Sea. They determined that the shore crab, *Carcinus maenas,* was a key epibenthic predator of juvenile macrofaunal populations, preying heavily on the cockle, *Cerastoderma edule.* By precluding the development of cockle beds, *C. maenas* represented a major force in structuring the benthic community of the intertidal habitat of the Wadden Sea.

7. Animal-Sediment Relationships

Benthic organisms modify the physical and chemical characteristics of estuarine and coastal marine bottom sediments. Kennish,[130] Jumars and Nowell,[395] Brenchley,[404] Rhoads and Boyer,[405] Aller,[586,587] Yingst and Rhoads,[588] Taghon,[589] Eckman et al.,[590] Yingst,[591] Grant,[592] Alongi,[593] and Luckenbach[594] furnish details of animal-sediment relationships. These publications, while not exhaustive, give the reader overviews on the processes by which the benthos modify seafloor sediments. It is not the intention of this section of the chapter to summarize all that is known regarding animal-sediment interactions in estuarine ecosystems. However, the salient points of this subject area will be addressed.

The benthos alters seafloor sediments via feeding, burrowing, and physiological activities, and the biogenic alteration often promotes the restructuring of the original benthic community.[318,595] Probert (p. 893)[596] states, "Previous studies of marine soft-bottom communities have shown (1) that natural disturbances (especially biologically-mediated disturbances, which are usually localized and recur reasonably frequently) help maintain spatio-temporal heterogeneity of communities, and (2) that biogenic modification of sediment can affect sediment stability with respect to fluid forces and geotechnical properties and that this is an important factor in community organization, particularly in the trophic structure of the macrofauna". Moreover, biogenic alteration of sediment stability, along with natural disturbances, may sustain trophically mixed communities in areas where deposit feeders do not have a primary impact on sedimentary properties.[596] Bioturbation affects interparticle adhesion, water content of sediments, bed roughness, and geochemistry of interstitial waters. By the formation of fecal pellets, benthic macroinvertebrates modify grain-size distribution. The geotechnical properties of sediments are frequently changed in regions of intense bioturbation, which influences sediment stability with respect to fluid forces. The reworking of sediments by infaunal deposit feeders increases the water content and roughness of the bottom, thereby enhancing particle transport.

Wright[597] refers to a number of biogenic structures contributing to bed roughness, namely animal tubes, pits, and depressions ascribable to feeding and burrowing, excavation and fecal mounds, crawling trails and furrows, and individual organisms protruding above the sediment surface. These biogenic structures interact with fluid motions within the benthic boundary layer that erode, transport, and deposit sediments. The stabilization and destabilization of seafloor sediments have been linked to the construction of animal tubes.[594] For instance, at some critical density of tubes, a skimming flow develops due to the interactions of flow perturbations created by the individual tubes which protect the bottom sediments from erosion.[590] Arrays of seagrass blades also effectively stabilize the bed, but isolated roughness elements, such as single blades of seagrasses, animal tubes, biogenous mounds, and shell fragments, deflect fluid toward the bed, enhancing local scour and erosion.[590]

Bioturbation involves dual processes — biogenic particle manipulation and pore water exchange. It is best visualized in communities dominated by high-order successional stages where dense populations of errant infaunal deposit feeders rework sediments at depths of several centimeters and profoundly affect physical properties.[412,591] Conveyor-belt, deposit-feeding species, characteristic of equilibrium-stage benthic communities, are principal reworkers of estuarine bottom sediments.[405] Frequently feeding in proximity to the RPD, they can modulate microbial distribution and activity at the top and bottom of the conveyor belt.[598,599] In addition, the diagenesis of redox sensitive substances (e.g., organic matter and transition metals) is related to flow of materials along the conveyor belt.[600,601] In contrast, pioneering species (e.g., *Capitella capitata)* are largely tubicolous or sedentary forms that rework sediments most intensely in the upper two centimeters of the seafloor.[600,601]

By burrowing, ingesting, and defecating sediment, bioturbating organisms loosen the sedimentary fabric, accelerate nutrient mixing in the sediment, and increase the exchange of gases and other substances across the sediment-water interface. Bioturbation may be vital in reconstituting buried, polluted sediments back to the sediment-water interface.[602] Erodibility and sediment transport are affected by alteration of: (1) fluid momentum impinging on the bed, (2) particle exposure to the flow, (3) adhesion between particles, and (4) particle momentum.[395] Biota have long been implicated in the stabilization and destabilization of sediments. Sedimentologists have known for some time that sediment stabilization is enhanced via mucous binding of sediments by bacteria, microalgae, meiofauna, and macrofauna.[603-605] Activities of meiofauna and macrofauna may also destabilize sediments through direct sediment displacement, as well as through changes in surface microtopography, grain size, and grain exposure.[395,594]

Investigations of the interactions between benthic organisms and their sedimentary environment have progressed along two avenues. The first addresses how fluid flow in the vicinity of the bed and erodibility of surface sediments on the seafloor are impacted by irrigation, mucus, pelletization, and tube structure. The second concerns the measurement of rates of sediment reworking by benthos through burrowing, feeding, and other activities.[592]

Newly applied acoustic and photographic techniques have greatly facilitated field measurements of bed characteristics and macrofaunal processes. For example, side-scan sonography allays problems in data acquisition of the areal distribution patterns of bed roughness, sediment type, and morphology. Wright[597] covers the components of the side-scan system. The REMOTS™ sediment profiling camera, introduced by Rhoads and Germano,[606] yields data on the species composition of benthic fauna and the geochemistry, bottom micromorphology, and microstratigraphy of the upper few centimeters of the sediment column. Kennish,[130] Wright,[597] and Section IV.C.1.a of this chapter supply additional information on this unique camera.

Other field measurement techniques applicable to the study of the benthic regime, including the benthic boundary layer in estuaries, deal with gathering data on current profiles and turbulence. Instruments successfully designed and applied to this end are electromagnetic current meters, acoustic current meters, thermistors and heat transfer-type current meters, and impellors and laser-doppler current meters.[597] Where data acquisition of the seafloor or bottom boundary layer require long time series of several days or weeks, "tripods" have been effective. These three-legged frames support self-contained bottom-mounted instrumentation systems which rapidly sample and store high quality data on flow, suspended sediment concentration, bed micromorphology, and water quality (e.g., temperature, conductivity, and dissolved oxygen). Some of the more popular recording instrument packages are the U.S. Geological Survey GEOPROBE, the University of Connecticut systems, University of Washington tripods, the Virginia Institute of Marine Science tripods, and the BASS tripod.[597] Others are also available. The readers should consult the work of Holme and McIntyre[607] for additional literature concerning the practical methods used in the study of estuarine benthic macrofauna.

8. Fouling Organisms

Having long been the focus of ecological investigations, the boring and fouling communities of estuaries have received considerable attention due to the economic costs they inflict on society by damaging man-made materials. Macrofouling invertebrates consist of epifauna that settle and grow on the surfaces of natural and man-made structures. Their life processes generally reduce the longevity of these materials. The structural integrity and utility of wooden structures are often compromised by the growth of fouling populations, and the corrosion processes of metals commonly are altered by their presence. These organisms cause a myriad of problems to shipping and ocean engineering systems. Hence, boats and ships, navigational buoys, underwater sound equipment, saltwater piping systems, moored oceanographic instruments, as well as most submerged materials are susceptible to fouling infestation.[608] As a result, much basic research has been devoted to understanding the dynamics and structure of macrofouling populations and communities, respectively.

The macrofauna of marine fouling communities is highly diverse; more than 1000 species of macroinvertebrates have been identified in these communities. While most phyla are represented, the most frequently occurring ones include the Porifera (sponges), Coelenterata (hydroids), Annelida (tubeworms), Arthropoda (barnacles), Mollusca (mussels, clams, oysters, limpets), Bryozoa, Echinodermata (starfish), and Chordata (tunicates).[609] The predominant fouling populations live permanently attached to a substrate, but a number of important forms exhibit various levels of mobility (e.g., limpets, sea anemones, and starfish). Much of the early literature on macrofoulers is contained in a volume by Woods Hole Oceanographic Institution.[610] A more recent compendium *(Marine Biodeterioration: An Interdisciplinary Study),* which is the product of a symposium on marine biodeterioration held at the Uniformed Services University of the Health Sciences, Bethesda, MD from April 10 to 23, 1981, incorporates more up-to-date information.[611] Perkins,[3] Osman and Whitlatch,[423] Thorhaug and Marcus,[616] Field,[617] Sutherland,[618] Loveland and Shafto,[619] and Okamura[620] also present contemporary findings on macrofouling organisms.

Marine fouling communities experience large temporal changes in species composition and abundance, especially in temperate estuaries. Seasonal fluctuations in the fouling community reflect temporal variability in reproductive cycles, settlement, and recruitment. Despite the year-round reproduction of some species of the fouling community, the spawning and settlement of many forms are restricted to a specific season of the year, with the reproductive period differing among the fouling species. Such temporal variability in reproduction is largely related to seasonal changes in water temperature and salinity, and in some locations to seasonal changes in light, nutrients, and food availability.[621] Epifaunal communities undergo seasonal cycles both on the Atlantic Coast[615,622] and Pacific Coast[623] of the U.S.

Loveland and Shafto[619] and Shafto[624] describe the fouling community of Barnegat Bay, NJ, where seasonal changes in community structure are pronounced. Of the 38 species of fouling organisms they recorded on 143 exposure panels, 21 species were mobile and 17 sessile. Sessile species dominated the fouling community; *Balanus eburneus, Bowerbankia gracilis, Hydroides dianthus,* and *Membranipora* sp. were the most abundant forms at most sampling sites. Settlement of the dominant species occurred from May to October, but the maximum monthly rate of settlement of each species varied with location in the bay (Table 23). Settling rates of all dominants fluctuated markedly during the summer. *Hydroides dianthus* had the shortest settling period (May to October) and *B. eburneus,* the longest (December to October). Shafto[624] tabulated the fouling macroinvertebrates of the bay, together with their abundance and mode of life (Table 24).

Loveland and Shafto[619] distinguished between two types of fouling larval settlement in the estuary: (1) lower surface settlement and (2) upper surface settlement. *Balanus eburneus, Botryllus schlosseri, Bowerbankia gracilis, Hydroides dianthus, Membranipora* sp., and *Molgula manhattensis* settled most heavily on lower surfaces of sampling panels. *Corophium* sp., *Melita nitida, Polydora ligni,* and *Sabellaria vulgaris* settled more frequently on the upper surfaces of sampling panels. Little settlement took place on vertical surfaces. Differences in settling patterns of the species may be due to their photopositive or photonegative behavior, responses to mud and detritus accumulation, attraction to the same species, avoidance of interspecific competition for surface area, and attraction to microflora, bacteria, and substrate chemicals.

The fouling community at Beaufort, NC, is also extremely variable, and both reproduction and recruitment are highly seasonal. Some species (e.g., *Bugula* and *Balanus* spp.) recruit almost continuously; others (e.g., species of *Botryllus, Halichondria, Haliclona, Pennaria,* and *Tubularia*) recruit only periodically. Still other species of *Ascidia, Ostrea, Schizoporella,* and *Styela* recruit predictably each year.[625] Species assemblages that develop in the fouling community depend on the history of colonization, variations in larval recruitment, and exposure to predation.[621]

Larval dynamics and settlement strongly influence the temporal and spatial complexity of the adult fouling community. Most benthic macrofaunal populations of fouling communities have a free-swimming larval or pseudolarval stage that facilitates dispersal of the species. The larvae, although at times carried long distances by currents, are able to select particular substrates for settlement and metamorphosis, which increases the probability of their survival. Settlement is often in proximity to an adult. Gregarious settlement has been documented among barnacles, corals, oysters, and tubeworms, suggesting that chemoreception plays a role in the behavior of these larval forms.[609] Research on the ability of larvae to select a particular substrate has

TABLE 23

Dominant Species of the Fouling Community of Barnegat Bay, Showing Months of Maximum Settling Rates at Nine Sampling Sites

Species	Description	Site A	B	C	D	E	F	G	H	I
Balanus eburneus	Acorn barnacle; attached with a heavy calcareous shell	M/Jn	M/Jn	Jn	Jn	M/Jn	Jn	Jn	M/Jn	Jn
Botryllus schlosseri	Colonial tunicate; forms fleshy sheets	A/S	—	—	—	S	S/O	S	S	—
Bowerbankia gracilis	Colonial ectoproct; forms thin, connected stalks	S/O	Jy	Jy	S	S/O	S	A/S	S/O	A
Corophium sp.	Amphipod; lives in mud-detrital burrows	M	M/Jn	Jn	M/Jn	M/Jn	Jn/Jy	M/Jn	M/Jn	—
Hydroides dianthus	Polychaete worm; forms calcareous tubes	Jn	S/O	A/S	S/O	Jy/A	Jn/Jy	Jy	Jy/A/S	A
Melita nitida	Amphipod; lives in mud-detrital tubes	A/S	A	S	S/O	S	Jn/Jy	A	Jn/Jy	A
Membranipora sp.	Colonial ectoproct; forms calcareous sheets	Jn/Jy	A/S	S	—	Jn/Jy	Jn/Jy	Jy	Jy	Jn/Jy
Molgula manhattensis	Solitary, attached tunicate; fleshy body	Jy	S	—	—	S	S	A/S	S	S/O
Polydora ligni	Polychaete worm; lives in mud-detrital burrows	Jy	Jy	Jy	M/Jn	Jy	Jy	A	M/Jn/A	M/Jn/Jy/A

Note: M = May; Jn = June; Jy = July; A = August; S = September; O = October.

From Loveland, R. E. and Shafto, S. S., *Ecology of Barnegat Bay, New Jersey*, Kennish, M. J. and Lutz, R. A., Eds., Springer-Verlag, New York, 1984, 226. With permission.

TABLE 24
Abundance and Mode of Life of Fouling Organisms in Barnegat Bay, NJ

Phylum / Species	Frequency sampled (%)[a]	Number/ m²/sample[b]	Mode of life[c-e]
Porifera			
Haliclona canaliculata	3	0.51	E
Cnidaria			
Haliplanella luciae	7	2.96	E
Diadumene leucolena	3	1.12	E
Turbellaria			
Stylochus ellipticus	19	14.19	M
Euplana gracilis	5	1.12	M
Annelida			
Capitella capitata	4	26.15	M
Diopatra cuprea	1	0.10	T
Eteone lactea	1	0.10	M
Goniada maculata	1	0.41	M
Hydroides dianthus	30	2456.88	T
Nereis succinea	41	35.14	M
Phyllodocid sp.	1	0.20	M
Polydora ligni	37	310.81	T
Sabellaria vulgaris	4	2.86	T
Arthropoda			
Ampelisca sp.	2	0.72	T
Balanus eburneus	77	3155.30	E
Caprella penantis	6	35.14	M
Callipallene brevirostris	4	12.36	M
Cirolana concharum	1	0.10	M
Corophium sp.	26	129.82	T
Cyathura polita	2	7.66	M
Erichsonella sp.	1	0.10	M
Leptochelia savignyi	5	1.33	M
Melita nitida	35	90.59	T
Tendipedida sp.	4	21.76	M
Mollusca			
Bittium alternatum	9	2.15	M
Doridella obscura	10	10.93	M
Crepidula fornicata	12	22.88	M
Mitrella lunata	6	2.25	M
Ilyanassa obsoleta	1	0.10	M
Mytilus edulis	3	0.51	E
Solemya velum	1	0.20	M
Tellina agilis	1	0.10	M
Ectoprocta			
Bowerbankia gracilis	39	12.80%[f]	E
Electra hastingsae	14	2.60%[f]	E
Membranipora sp.	49	4.20%[f]	E
Chordata			
Botryllus schlosseri	12	1.10%[f]	E
Molgula manhattensis	20	244.42	E

[a]   Percentage of 143 total samples containing a species.
[b]   Mean number of individuals per species found in all 143 samples.
[c]   E = Encrusting or attached form.
[d]   T = Tube or burrow dwelling form.
[e]   M = Mobile form.

TABLE 24 (continued)
Abundance and Mode of Life of Fouling Organisms in
Barnegat Bay, NJ

f   Average percentage of board surface covered per sample.

From Loveland, R. E. and Shafto, S. S., *Ecology of Barnegat Bay, New Jersey*, Kennish, M. J. and Lutz, R. A., Eds., Springer-Verlag, New York, 1984, 226. With permission.

concentrated on the chemistry of inducing substances associated with the substrate. An effort has been made to characterize the chemical components of metamorphic inducers, but only a limited number of species have been examined. Most information on metamorphic inducers has been obtained on the abalone, *Haliotis rufescens,* the echinoid, *Dendraster excentricus,* the nudibranch, *Phestilla sibogae,* the polychaete genera, *Janua* and *Phragmatopoma,* barnacles, hydrozoans, and a sipunculan worm.[626]

Some species display a "larval choice" regarding settlement location. During site selection, fouling organisms scrutinize the surface of a substrate to test its acceptability. If the surface is favorable, a larva will attach to the substrate and eventually set up a permanent settlement on it. If the surface is not favorable, however, a larva retains the ability to leave the substrate, reenter the water column, and search for another site of attachment.

Fouling larvae attach either temporarily or permanently to a specific site. When exploring the suitability of a substrate, larvae may temporarily attach to the surface via a suction apparatus or a secreted sticky mucous substance.[609] Sea urchins, starfish, and fast-moving marine organisms successfully utilize suction devices, whereas gastropods, sea anemones, tunicates, and other slow-moving forms apply a mucous secretion for their temporary attachment. Adults that have some mobility (e.g., sea anemones and tunicates) generally use mucous secretion for permanent settlement. Barnacles, tubeworms, and those forms that settle for life produce a hardened or cured adhesive cement, sometimes reinforced with calcareous deposits.[609]

Little availability of bare substrate space for the attachment of fouling larvae at times poses a threat to settlement. Competition for space is intense in fouling communities, and competitive interactions in many regions govern the abundance and distribution of fouling organisms.[439,627] Space in marine hard-substrate communities becomes available through multiple pathways, that is, by the death of individuals or colonies of residents, by predation, and by abrasion (e.g., through the action of strong waves or currents).[621]

The chief competitive strategy of fouling species is overgrowth caused by relatively rapid growth rate, aggressive behavior, allelochemicals, and preemption of larval settlement by established individuals.[443,615,621,627,628] Jackson,[439] analyzing the strategies of solitary and colonial organisms in marine hard-substrate communities, iterated that solitary forms are superior competitors for space in rocky intertidal zones, but in most other locations, colonial fauna typically overgrow them. In subtidal zones of temperate and boreal climates, however, dense algal populations spread over the surfaces.

To increase the probability of survival, a competitively weak fouling population can escape in space or in time from disadvantageous competitive relationships. A species may flourish in a refuge zone beyond the spatial limits of competitive dominants, returning to colonize other areas when space becomes available. Alternatively, a fouling organism may find it advantageous to synchronize reproduction with times when a maximum amount of substrate space is available for colonization, or when competitive interactions are ameliorated. Other species reproduce almost year-round, ensuring that a cleared substrate is taken at some time.[621]

9. Boring Organisms

A specialized group of benthic fauna has adapted to boring into rock or wood. The piddock,

Pholas, lives in a variety of substrates from stiff clay to compact limestone.[3] Rock borers, (e.g., *Barnea candida, Zirphaea crispata,* and *Pholas* spp.) also have the capability of boring into wood.[3] Most woodborers belong to the genera *Bankia* and *Teredo;* of less importance are the genera *Bactronophorus* and *Zachsia.* In addition, some genera of the family Phoadidae include woodborers.

The destructiveness of the lamellibranch mollusks, *Bankia* spp. and *Teredo* spp. (shipworms), and the crustacean arthropods, *Limnoria* spp., is immense. Shipworms alone annually destroy millions of dollars worth of wooden piles, docks, bulkheads, boats, and barges by boring into the wood with their paired anterior shells. As they bore into the substrate, the shipworms grow in length and width, while building calcareous-lined burrows. The excavated wood is utilized as food.[629,639] Not only does wood serve as a food source, but also as a protective habitat. When left unabated, damage by woodborers weakens the internal structure of the wood until it collapses. Creosote-treated wood resists attack by shipworms and remains one of the most effective methods of preventing destructive action of the woodborers.

Shipworms can withstand adverse environmental conditions by plugging the burrow entrance with their pallets — elongated calcareous structures situated at the base of the siphons on either side beneath a muscular collar.[631] It is not uncommon for shipworms to live for more than a month sealed in their burrows. They utilize stored glycogen to survive extended periods of anaerobic conditions.[632]

The isopod, *Limnoria* sp., commonly called the gribble, does not construct elongate burrows into the wood structure, but burrows obliquely to the wood surface, leaving a tunnel approximately 0.1 cm in diameter. The tunnels usually penetrate no more than about 1.5 cm into the wood; they may reach a length of 4 to 5 cm.[3] Gribbles themselves do not exceed 5 mm in length, and wood pilings can contain more than 1×10^5 individuals.[12,633] Attack by *Limnoria* differs from that of shipworms because the tunnels intersect each other causing a breakdown of the superficial layers of wood. A friable, spongy mass results which is susceptible to mechanical destruction by waves, currents, and other physical forces. In many cases, *Limnoria*-infested wood has an appearance of an hourglass structure. Damage due to teredinids is often less overt. Shipworm burrows do not intersect, but the heart of infested wood may be riddled with them. The outward appearance of the wood frequently shows no signs of the attack, and the damage goes undetected until the structure collapses.[632]

Creosote-treated wood does not prevent *Limnoria* infestation. Yet, this wood tends to be less severely attacked than wood left untreated.[634] Peak abundance occurs between low water and about the half-tide level.[3] Pilings infested with *Limnoria* generally have a reduced diameter between the level of mean high tide and the sediment-water interface.[12]

10. Level-Bottom Community Concept

Petersen[635-639] championed the concepts of benthic communities and associations while working in the shallow coastal waters of Denmark during the early 1900s. Using a bottom grab, Petersen[635] attempted to decifer the statistical distribution of the fauna. It became evident to Petersen,[638] based on thousands of bottom samples, that broad areas of the seafloor contained a strikingly uniform combination of macrofaunal species, with a few species tending to outnumber all others. In other areas, another combination of benthic fauna occurred in a similar, uniform manner.[640] Few populations actually were common over broad segments of the seafloor. Ultimately, Petersen[638] accepted the statistical community concept.[6]

As a function of sediment characteristics and water depth, Petersen[636-638] identified nine different associations or communities of benthic fauna. Perkins[3] recounted them:

1. Shallow muds inhabited by *Macoma* communities, widespread in distribution, and composed principally of *M. balthica, Mya arenaria,* and *Cerastoderma (=Cardium) edule*
2. Shallow, muddy sands frequently found in sheltered creeks and characteristically inhab-

ited by *Syndosmya* (i.e. *Abra)* communities and composed principally of *A. alba, A. prismatica, Macoma calcarea,* and *Astarte* spp. *(Echinocardium cordatum* may be found with this association)

3. Shallow sandy bottom, on open coasts, inhabited by *Venus* communities, widespread in distribution, and composed principally of *V. gallina, Tellina fabula,* and *Montacuta ferruginosa (Echinocardium cordatum* frequently occurs in this association)

4. Sandy mud, at intermediate depth, inhabited by *Echinocardium-filiformis* communities, and composed principally of *E. cordatum, Amphiura filiformis, Turritella communis,* and *Nephtys* spp.

5. Soft mud, at greater depth than (4) inhabited by *Brissopsis-chiajei* communities, and composed principally of *B. lyrifera, Amphiura chiajei,* and *Abra nitida*

6. Soft mud, at greater depths than (5) inhabited by *Brissopsis-sarsi* communities, and composed principally of *B. lyrifera, Ophiura sarsi,* and *Abra nitida*

7. Mud, at depth in the Skagerrak, inhabited by *Amphilepsis-Pecten* communities and composed principally of *A. norvegica* and *Chlamys (=Pecten) vitrea*

8. Firm mud, deep in the Kattegat, inhabited by *Haploops* communities and composed principally of *H. tubicola* and *Chlamys septemradiata*

9. Coarse sands, at depth, inhabited by *Venus* communities which are widespread in occurrence and composed principally of *V. gallina, Spatangus purpureus, Echinocardium flavescens,* and *Spisula* spp.

Thorson[640] later introduced the concept of "parallel bottom communities" to help explain differences in benthic communities from different areas where physical conditions were very similar. He (p. 521)[640] perceived "... Parallel communities inhabiting the same type of bottom at similar depths, characterized by different species of the same genera, but replacing each other in accordance with the geographical regions". Thorson[640] named bottom communities on the basis of their species composition, for instance, *Amphiodia-Amphioplus, Amphiura, Macoma, Tellina,* and *Venus* communities.

Two examples of level bottom faunal communities in estuaries can be taken from the mid-Atlantic Bight region.[5] Amphipod crustaceans of the genus *Ampelisca* dominate the benthic communities of a number of southern New England bays and sounds, in shallow sand and silt-sand habitats. *Ampelisca abdita* and *A. vadorum* are two sibling species commonly found in these estuaries; maximum abundances have surpassed 35,000 individuals per square meter in some systems.[5] Mixed populations of the ampelicids, *A. macrocephala, A. vadorum,* and *A. verrilli,* concentrate on sandy bottoms in Buzzards Bay and Vineyard Sound. Narragansett Bay supports both *A. abdita* and *A. vadorum.* Sanders[641] documented an association of *A. abdita* with the polychaetes, *Nephtys incisa* and *Cistenides (=Pectinaria) gouldii,* and the bivalves, *Lyonsia hyalina, Mulinia lateralis,* and *Pitar morrhuana,* in Long Island Sound. *A. vadorum* occupied coarser sediments of the sound with other amphipods (e.g., *Erichthonius braziliensis)* and bivalves. In Great Bay, NJ, *A. abdita* also dominates extensive areas.[642] *Ampelisca* communities typically are productive, supplying forage for a host of organisms at higher trophic levels.

In addition to *Ampelisca* communities, a *Nephtys-Nucula-Yoldia* community lives in southern New England estuaries, usually on deeper muddy bottoms than *Ampelisca* communities which often surround them. Deeper regions of Buzzards Bay, Narragansett Bay, and Long Island Sound have been dominated by *Nephtys incisa, Nucula annulata,* and *Yoldia limatula* in the past. The polychaete, *N. proxima,* also has been a dominant species. Deposit-feeding organisms, which consume detritus on the surface (e.g., *Nucula* and *Yoldia)* and below the surface (e.g., *Nephtys),* characterize the *Nephtys-Nucula-Yoldia* community.

11. Zonation Concept

While the level bottom approach has garnered much interest over the years in studies of apparent patterns of the benthic fauna, a second approach — the zonation approach — has also advanced our knowledge of the benthos. The level bottom approach describes assemblages of benthic organisms which are basically constant, and relates them to the physical environment, especially the type of substrate. The zonation approach differs in that the benthic assemblages are related to a physical gradient or a combination of physical gradients.[6] This approach has been successfully applied to intertidal and shallow subtidal zones for obvious reasons. Investigations of zonation have been fruitful in mangrove and salt marsh systems, and particularly along rocky coastlines.[51] On rocky shores, organisms generally recur in regular patterns in distinct zones.

Some workers consider zonation to simply be the outcome of responses of benthic organisms to gradients of physical parameters. Biological interactions play a significant role in the observed distributions of macrofauna in certain habitats, however (e.g., rocky intertidal shores).[419,578,582,643] Biological agents aside, "zonations are now usually seen as the resultant pattern of distribution of organisms reacting to a series of gradients which may or may not overlap" (Erwin, p. 152).[6] Although much data on zonation have been derived from research on the intertidal zone, the development of new techniques and the utilization of SCUBA diving have extended sophisticated studies of zonation to the shallow subtidal realm. Here, data continue to be accumulated on the effects of physical factors, such as depth and light gradients, as well as biotic interrelationships, on the patterns or zones of benthic fauna.

V. SUMMARY AND CONCLUSIONS

The estuarine benthos consists of all organisms, both plant and animal, living on or in bottom sediments. Benthic flora and fauna are important in the energy budget of estuaries and may serve as habitat formers. Macrophytes (algae and vascular plants) constitute the major fraction of the benthic biomass of these coastal systems, with emergent and submergent vascular plants providing the most organic carbon. Bottom sediments are often devoid of benthic macroalgae which typically attach to hard surfaces, including man-made structures, shells, stones, and rocky shores. They often drift passively over the estuarine seafloor. Dense meadows of vascular plants (e.g., seagrasses) commonly grow in shallow subtidal areas. Salt marsh grasses and mangroves are conspicuous features of the intertidal zones of temperate and tropical regions, respectively.

Benthic fauna can be subdivided according to size into micro-, meio-, and macrofauna. They also can be grouped on the basis of life habits and adaptations into epifauna, which live on the estuarine seafloor or attached to a firm substrate, and infauna, which live in bottom sediments. Other (nonparasitic) species, however, are best classified as interstitial, boring, swimming, and commensal-mutualistic forms. Using feeding habits as criteria, five types of benthic fauna have been identified: (1) suspension feeders, (2) deposit feeders, (3) herbivores, (4) carnivores, and (5) scavengers. The distribution, function, and behavior of the benthos are clearly influenced by biological, chemical, and physical factors.

The benthic flora of estuaries can be separated into micro- and macrofloral components. The microflora are most extensively developed in intertidal habitats, where single-celled algae (i.e., diatoms, blue-green algae, and dinoflagellates) frequently discolor the sediments due to their large numbers. Single cells or filamentous colonies of microflora adhere to sediments and also attach to the surfaces of rocks, other plants, animals, and man-made materials. Motile forms migrate vertically in sediments in response to changing light intensities. Mats of green and blue-green algae underlie salt marshes and tidal flats. Blue-green algal mats are prominent in warm-water systems, for example, those estuaries located along the Gulf of Mexico.

In certain habitats (e.g., tidal flats), benthic microalgae supply a considerable amount of primary production. Annual production estimates typically range from about 25 to 250 g C per square meter per year. Benthic microalgae, which form mats on mudflats, bind sediment via the secretion of extracellular mucous films, thereby stabilizing the substrate and mitigating erosion. Many microalgal species grow as epiphytes on the leaves and stems of macrophytes, where they form a fur-like covering termed periphyton or Aufwuchs. Grazers utilize the periphyton as food. Grazing by mud snails, fiddler crabs, herbivorous fish, and other fauna controls the growth of epibenthic algae, limiting biomass and production. Epiphytic felt flora are habitation sites of micro- and meiofauna (i.e., protozoans, nematodes, rotifers, tardigrades, copepods, and ostracods) that consume associated microorganisms (e.g., bacteria and diatoms) and, in turn, supply ration for macrofauna. Species of *Cladophora, Enteromorpha, Hypnea,* and *Polysiphonia* are common components of the algal felt flora.

The attachment of epiphytes to submerged macroflora is beneficial for several reasons. First, the submerged plants act as support or anchorage for the epiphytes. Second, the greater proximity of the microflora to light enhances epiphytic growth, and water currents remove growth-inhibiting substances and transport nutrients.

Benthic macroalgae (seaweeds) cover only small areas of the estuarine bottom, usually attaining peak densities on hard surfaces. Because of strong currents and wave action, they generally require attachment to a solid substrate. Adhesive, root-like holdfasts or basal disks provide effective means of attachment for these plants.

Environmental conditions in estuaries are not conducive to luxuriant growth of benthic algae. High turbidity and sedimentation, for instance, mitigate light penetration. These factors account for the species-poor condition of benthic macroalgae in many of these complex systems. Most of the species belong to the Chlorophyta (green algae), Phaeophyta (brown algae), and Rhodophyta (red algae).

In some estuaries, benthic macroalgae are quite abundant; some taxa (e.g., *Blidingia minima* var. *subsalsa, Enteromorpha clathrata,* and *Vaucheria* spp.) reach maximum abundances in estuaries. The distribution of benthic macroalgae in these systems is distinctive along the longitudinal axis. The species number and diversity decrease when proceeding from the estuarine mouth to its head. Annual species, especially green algae, extend farther upestuary compared to perennial red or brown algae, which are more prevalent downestuary.

Seaweeds often display a zonation pattern, particularly on rocky shores, due to differences in dominance among populations. Species zonation may be more obvious on artificial embankments and other man-made structures than on natural banks. Factors contributing to zonation of macrofloral populations include species competition, grazing, and physiological stresses associated with emersion.

The ecological importance of seaweeds in estuaries is principally related to their value as a food source for herbivores. Live plant matter not ingested by benthic grazers ultimately enters the detritus food web where microorganisms colonize and decompose the material. Macroalgal detritus breaks down rapidly relative to vascular plant detritus which is much more refractory. It may be degraded in only a few weeks in estuaries, but the more recalcitrant vascular plant detritus requires several months or even years to decompose.

Seaweeds have several domestic and industrial uses. They are used by man as sources of food, supplements in animal feeds, medicine, fodder, fertilizers, fuel, and soil conditioners. In addition, they have value in the production of paper and phycolloid extracts (i.e., agar, alginate, and carrageenan).

The benthic macroflora are best represented in three major plant communities: salt marshes, seagrasses, and mangroves. Rooted vascular plants predominate in these communities, with benthic macroalgae being less important in terms of biomass and production. Globally, salt marshes range from mid-temperate to high latitudes and are replaced from suitable sites in the tropics by mangroves which generally lie between 25° N and 25° S latitude. Seagrasses have a very wide distribution; they can be found in shallow waters of all latitudes except the most polar.

Salt marsh plants dominate the vegetation of intertidal zones of estuaries in mid- and high-latitude regions. Consisting of halophytic grasses, sedges, and succulents, which develop on muddy sediments at and above the mid-tide level, salt marshes embody much of the 2×10^6 ha of estuarine wetlands in the lower 48 contiguous states. Most of the flora in salt marshes belong to a few cosmopolitan genera (e.g., *Spartina, Juncus,* and *Salicornia)* that are broadly distributed.

Six types of salt marshes have been distinguished (i.e., estuarine, Wadden, lagoonal, beach plain, bog, and polderland types). The formation of each type depends on salinity and tidal range. Three geographical units of salt marshes are recognized in North America: (1) Arctic, (2) eastern North America, and (3) western North America marshes. Salt marshes are well developed on the southern shore of Hudson Bay, the East Coast of Canada and the U.S., and the Gulf Coast of the U.S. Discernible differences occur in the species composition of the salt marshes of these three geographical units, although the predominance of *Spartina alterniflora* and *S. patens* has been noted in mesohaline and polyhaline marshes. The salt marsh grasses bordering estuaries frequently exhibit a well-defined zonation. For example, the coastal plain marshes along the Atlantic Coast south of Chesapeake Bay are characterized by a lower marsh of tall form *S. alterniflora* and an upper marsh of *S. patens, Distichlis spicata, Juncus roemerianus,* and *Salicornia* sp. Along the Atlantic Coast of North America, salt marsh communities tend to be less diverse and their zonation and succession patterns less complex than those along the Pacific Coast.

Sheltered coastal areas, where sedimentation is ensured and erosion remains slight, are ideal sites for salt marsh development. The halophytes initially establish populations on mudflat surfaces between the levels of MHWN and MHW. Growth of the salt marsh proceeds as sedimentation produces a surface above the MHWN, which forms the limit for the establishment of the halophytes. Maturation of the marsh is promoted by gradual accretion of sediments. Sedimentation rates of 3 to 10 mm/year enhance salt marsh maturation. Higher rates of sedimentation, however, may arrest successional development and restrict species richness. Sediment progradation plays a crucial role in the state of marsh maturation (i.e., youth, maturity, and old age), which has been related to the stages of plant and animal succession.

Salt marsh systems serve at least five significant ecological functions. They are highly productive; estimates of aboveground production of *Spartina* salt marshes range from 200 to 3000 g C per square meter per year, and below-ground production may be even greater. Much of this primary production enters the detritus food web upon the death of the plants, and forms an essential energy source for many estuarine organisms. Many animals utilize the salt marsh habitat for food, shelter, and reproduction. The rooted vegetation anchors the sediment and reduces erosion. The salt marshes also behave as sources or sinks of trace metals and nutrients.

The invertebrate fauna of salt marshes can be placed in three categories: terrestrial species, aquatic species, and marsh-evolved species. Of the smaller organisms, protozoans (e.g., foraminifera) and meiobenthos (e.g., nematodes, harpacticoid copepods, oligochaetes, poly-chaetes, kinorhynchs, turbellarians, amphipods, and ostracods) yield high biomasses in surface sediments. Some of the more important macrofauna occupying this biotope are bivalves (e.g., *Geukensia demissa),* gastropods (e.g., *Hydrobia* spp., *Littorina* spp., and *Melampus bidentatus),* and crustaceans (e.g., *Uca* spp.). In addition to the benthic invertebrates, numerous insects, fishes, and avifauna prefer the salt marsh habitat.

Tidal freshwater marshes replace tidal salt marshes as salinity diminishes inland. They develop in areas where the annual salinity averages 0.5‰ or less. Farther upstream, these marshes grade into nontidal freshwater wetlands.

Freshwater flora prevail in tidal freshwater marshes. Diversity of the flora is greater than in tidal salt marshes, with species of grasses, rushes, shrubs, broad-leaved plants, and herbaceous forms being found. The most extensive tracts of tidal freshwater marshes in the U.S. occur in the mid-Atlantic states, but well-developed marshes range from Georgia to southern New England. New Jersey alone accounts for approximately 50% of the total freshwater marsh coverage along the Atlantic Coast, approaching 80,000 ha.

As in the case of tidal salt marshes, tidal freshwater marshes show a horizontal zonation pattern. For example, on the Atlantic Coast of the U.S., lower and upper marsh zones are evident; broad-leaved, fleshy plants dominate the lower marsh and a higher diversity of forms (e.g., *Acorus calanus, Biden* spp., *Hibiscus moscheutos, Polygonums* spp., and *Typha* spp.), the upper marsh. Five floral groups typically comprise tidal freshwater marshes: (1) broad-leaved emergent perennial macrophytes; (2) herbaceous annuals; (3) annual and perennial sedges, rushes, and grasses; (4) grasslike plants or shrub-form herbs; and (5) hydrophytic shrubs. However, the species composition and community structure vary markedly across geographical regions and seasonally as well.

The ecological value of tidal freshwater marshes parallels that of tidal salt marshes. Primary production ranges from about 1000 to 3500 g C per square meter per year. Vascular plants are the main contributors to the production pool, while phytoplankton and benthic algae remain secondary in importance. Principally being detritus-based systems, these marshes also serve as sources or sinks of material (e.g., nutrients) for estuaries. Commercially important fishes use the marsh habitat as a nursery. Moreover, many insects, birds, and other wildlife inhabit these regions.

A recurrent biotope of exceptional value in estuaries worldwide is the seagrass community composed of monocotyledonous angiosperms that grow in shallow subtidal waters. Twelve genera of seagrasses (i.e., *Amphibolis, Cymodocea, Enhalus, Halodule, Halophila, Heterozostera, Posidonia, Phyllospadix, Syringodium, Thalassia, Thalassodendron,* and *Zostera*), incorporating about 50 species, have been identified worldwide. *Zostera* (eelgrass) is the dominant genus in temperate waters of North America, and *Thalassia* (turtlegrass) predominates in subtropical and tropical regions. Other abundant genera in North America include *Halodule* and *Phyllospadix.* Seven genera of seagrasses occur in tropical latitudes compared to five in temperate latitudes. *Zostera* has the broadest distribution of all seagrass genera in the temperate and boreal waters of the Atlantic and Pacific Oceans. It also has been the most intensely studied seagrass.

Locally, seagrasses are restricted to lower intertidal and subtidal zones in estuarine and coastal oceanic environments. Anchored to bottom sediments by a network of roots and rhizomes and structured with a dense arrangement of stems and leaves, seagrasses create a habitat for many other aquatic organisms. Sediment fauna, rhizome and stem biota, leaf epiphyton, and nekton form four productive subhabitats within seagrass systems. The thick foliage affords prey protection from predators and shelters organisms from strong currents and wave action.

Salinity, light, and turbidity strongly affect the distribution and growth of seagrasses. In turbid estuaries, where high seston concentrations restrict light transmission, seagrasses often are limited to waters less than 1 m deep. However, in regions of greater water clarity, seagrasses may extend to depths of 30 m.

Seagrasses are ecologically significant because of their high primary production and value as habitat formers. Estimates of their primary production range from 58 to 1500 g C per square meter per year. Numerous invertebrate and fish populations utilize seagrass habitats as nursery, feeding, and reproductive grounds. Large quantities of detritus generated from seagrasses also are important in the energy flow of many estuarine ecosystems. Additionally, these rooted vascular plants play a role in the cycling of certain critical nutrient elements.

Mangroves or mangals are assemblages of halophytic trees, shrubs, palms, and creepers that form dense thickets or forests in intertidal and shallow subtidal zones of subtropical and tropical waters. They thrive in protected embayment areas, tidal lagoons, and estuaries between 25° N and 25° S latitudes. Florida has the most elaborate mangrove communities in the U.S., dominated by the black mangrove *(Avicennia germinans),* red mangrove *(Rhizophora mangle),* and white mangrove *(Laguncularia racemosa).* Mangroves rim more than 170,000 ha of the Florida coastline; these plant associations are prominant along the Gulf Coast as well. Although extending northward as far as 32° N in the U.S., mangroves have established most successful communities south of 28 to 29° N latitude.

Worldwide, about 80 species of monocots and dicots belonging to 16 genera have been described in mangrove communities, with at least 34 species in nine genera believed to be true mangroves. Based on hydrodynamics (i.e., tidal action and freshwater hydrology) and topography, mangroves can be classified into six types of wetland systems: (1) overwash mangrove islands, (2) fringe mangrove wetlands, (3) riverine mangrove wetlands, (4) basin mangrove wetlands, (5) hammock mangrove wetlands, and (6) dwarf mangrove wetlands. Mangroves exhibit a zonal pattern related to the salinity tolerances of the plant populations, tidal inundation, and other factors. In South Florida, for example, an *R. mangle* fringe builds into shallow subtidal waters. *Avicennia germinans* occurs immediately landward of *R. mangle* from the lower to the upper intertidal zone. Upland of *A. germinans, L. racemosa* occupies sites in the middle to upper intertidal zone. Farther upland, *Conocarpus erecta* or buttonwood replaces *L. racemosa,* and, although occasionally inhabiting the upper intertidal zone, usually is part of the sand/strand vegetation. Some workers consider mangroves to be pioneer or seral successional stages evolving toward a freshwater, terrestrial plant community. According to this postulation, an estuarine/mudflat community can be transformed into a tropical rain forest.

The prop or drop-type roots of mangrove trees are shallow, and cable roots projecting horizontally from the stem base provide additional support. High rates of sedimentation typify mangrove systems. Submerged portions of mangrove roots frequently have dense growths of benthic algae attached to them (e.g., species of the genera *Bostrychia, Caloglossa, Catonella,* and *Murrayella).* A highly diverse assemblage of vagile and sessile faunal populations also makes home in the mangrove vegetation and tidal channels. Many of these benthic invertebrates are filter or deposit feeders; crustaceans and mollusks usually dominate the communities. Common benthic fauna encountered in mangroves are fiddler crabs (*Uca* spp.), snails *(Littorina),* barnacles *(Balanus),* and oysters (*Crassostrea* and *Ostrea).*

Like salt marshes, tidal freshwater marshes, and seagrass systems, mangroves are highly productive, having primary production estimates ranging from 350 to 500 g C per square meter per year. A substantial portion of this production supports detritus-based food webs. Mangrove swamps are a critical habitat for benthic invertebrates, fishes, and birds. Insects, reptiles, and terrestrial mammals have also successfully adapted to mangrove forests. The subtidal root systems of mangroves protect the shoreline, enhance bank stabilization, and promote land accretion, while mitigating erosion. Finally, mangrove trees are commercially valuable for water-resistent timber, charcoal, dyes, and medicines.

Benthic fauna can be separated into micro-, meio-, and macrofaunal components on the basis of size. Individuals which pass through sieves of 0.04- to 0.1-mm mesh constitute the microfauna. The meiofauna, metazoans weighing less than 10^{-4}g (wet weight), are retained on sieves of 0.04- to 0.1-mm mesh. The macrofauna are larger animals captured on sieves of 0.5- to 2.0-mm mesh.

The microfauna essentially consist of protozoans. The meiofauna can be subdivided into: (1) temporary meiofauna which are juvenile stages of the meiofauna and (2) permanent meiofauna (i.e., gastrotrichs, kinorhynchs, nematodes, rotifers, archiannelids, halacarines, harpacticoid copepods, ostracods, mystacocarids, and tardigrades as well as representatives of the bryozoans, gastropods, holothurians, hydrozoans, oligochaetes, polychaetes, turbellarians, nemertines, and tunicates). Whereas the absolute abundance of micro- and meiofauna far outnumbers the benthic macrofauna, the biomass of the macrobenthos is much greater.

In estuarine environments, meiofaunal densities average about $10^6/m^2$, and standing crop dry weight biomass figures range from approximately 1 to 2 g/m^2. The species composition of the estuarine meiofauna is a function, in part, of the sediment type. For instance, the meiofauna of sand flats are vermiform, interstitial species that negotiate crevices between sedimentary grains. In muddy sediments, burrowing forms predominate. The distribution of meiofauna is patchy when proceeding from intertidal to subtidal zones due to variability of the physical environment (e.g., temperature, salinity, dessication, and sediment grain size) and biological interactions (e.g., predation, competition, and bioturbation). Horizontal gradients in salinity strongly modulate the species composition of meiofauna. The species richness of this group generally increases with rising salinity from the head to the mouth of estuaries. Higher salinities also correspond to greater meiofaunal densities. Seasonal changes in the species composition and abundance of meiofauna have been correlated with their temperature sensitivity.

Chemical zonation in bottom sediments exerts some control on the vertical distribution of meiofauna. Vertical zonal patterns in subtidal sediments, for example, reflect the ability of the meiofauna to tolerate low oxygen concentrations. Other factors influencing the vertical zonation of fauna include food availability, amount of organic matter, sediment grain size, and selective predation.

The fecundity of the meiofauna is less than that of most macrofaunal species. These diminutive organisms usually release from 1 to 10 eggs at a time, although members of the archiannelid genus *Polygordius* and several species of *Protodrilus* produce up to 200 eggs per reproductive event. The majority of estuarine meiofauna undergoes direct development; brood protection and viviparity are common. The meiofauna constitute a highly successful group because of several reproductive adaptations, such as the year-round reproduction of some of the taxa, relatively short generation times, and the delay of life-history development via resting eggs or larval stage delay.

Meiofaunal production estimates have been determined for a number of estuaries. Annual meiofaunal production in the Lynher estuary, England, equals 20.17 g C per square meter per year, which compares favorably to that of the Ythan estuary, Scotland, which amounts to 19.5 g C per square meter per year. Production measurements are lower for the Swartkops estuary, South Africa, with values of 0.24 g C per square meter per year for muddy areas and 0.46 g C per square meter per year for sandy regions.

The species composition and abundance of benthic macrofauna have wide temporal and spatial variations. Fluctuations in abundance are conspicuous among the opportunistic species (e.g., *Mulinia lateralis, Capitella capitata, Cistenides (=Pectinaria) gouldii,* and *Polydora ligni*). Large changes in abundance of the macrofauna during the year may be ascribable to normal periodicities of reproduction, recruitment, and mortality. However, they may also be attributable to random environmental perturbations, where vagaries in physical or chemical conditions can precipitate large aperiodic density changes.

The spatial distribution of benthic macroinvertebrates has been assessed on three levels: (1) small-scale distributions within the estuary, (2) distribution patterns on the scale of the estuary, and (3) distribution patterns on a world scale. Local distributions of the benthos within estuarine habitats have been related to physical factors, notably waves and currents, sediment character-istics, and depth. Biological factors (e.g., predation and competition) and chemical factors (e.g., oxygen concentrations) also influence the distribution of the benthic macrofauna. The species composition of the benthic macrofauna within habitats is dependent on sediment type. The polychaete, *Polydora ligni*, and the bivalve, *Scrobicularia plana,* for example, prefer muddy sediments, while the amphipods, *Bathyporeia* spp. and *Haustorius* spp., and the polychaetes, *Ophelia* spp., live in sandy bottoms. The concentration of organic matter in seafloor sediments is important in regulating benthic community structure as well. The abundance of such forms as the lugworm, *Arenicola marina,* the mudsnail, *Hydrobia ulvae,* and the bivalve, *Macoma balthica,* has been coupled to the amount of organic matter in the bottom sediments.

As the concentration of dissolved oxygen diminishes with increasing depth in bottom sediments, the absolute abundance of the macrofauna likewise drops. Abundance of benthic macroinvertebrates peaks in the oxygenated surface layers. This relationship does not hold for various micro- and meiofauna, however, which may attain maximum numbers in deeper sediments (e.g., near the chemocline).

The distribution of the macrobenthos in the sediment column has also been related to the stage of successional development of the benthic community in addition to local physical disturbances. In habitats subjected to frequent physical disturbances, pioneering species of infauna — tubicolous or otherwise sedentary invertebrates living in near-surface sediments — tend to dominate the community. The pioneering forms feed near the sediment surface or from the water column. Habitats devoid of physical disturbances harbor higher-order successional stages or equilibrium stages of benthos dominated by bioturbating infauna which feed at greater depths within bottom sediments. Aside from physical disturbances caused by wave and current action, sediment deposition and erosion, as well as other factors, biological disturbances due to grazing, predation, and competition control the vertical distribution of the infauna. The proximity of epifaunal predators, elicits responses from infaunal prey that leads to their repositioning within the sediment. Disturbances of the estuarine seafloor have been enlisted by benthic ecologists as principal causative agents of the spatial mosaic pattern observed in bottom communities. Sediment profile imaging has proven to be a viable method of assessing vertical profiles of the estuarine seafloor and of determining the stage of successional development of the benthos. The sediment-profile camera has been a valuable tool in ascertaining the structure of the benthic community.

Larger-scale distributional patterns of the benthic macrofauna, from the head to the mouth of estuaries, develop from population responses to gradients in environmental conditions. Changes in salinity along the length of an estuary have a profound effect on the species distribution of the benthic fauna. Abrupt changes in biota have been discerned at salinities of 0.5, 5, 18, and 30‰. Along the homoiohaline gradient of the Chesapeake Bay, gradual and relatively uniform benthic macrofaunal changes occur except within the salinity ranges of 3 to 8 and 15 to 20‰, where the changes in the macrobenthos are more accelerated.

The distribution of benthic macroinvertebrates throughout an estuary is also a function of larval dispersal and success of recruitment. Many benthic faunal populations have a planktonic larval stage early in ontogeny, and at this time, the larvae experience high mortality due to predation and the vagaries of environmental conditions. Nevertheless, currents can transport the larvae estuarywide to populate favorable habitats. The length of the planktonic larval stage of benthic invertebrates varies depending on mode of development, environmental factors, and chemical or physical cues inducing the larvae to settle and metamorphose. Research has revealed three categories of planktonic larvae in estuaries: (1) long-life planktotrophic forms, (2) short-life planktotrophic forms, and (3) lecithotrophic forms. Two other modes of development of benthic macroinvertebrate larvae are direct development, omitting a pelagic phase, and viviparity in which the larvae develop within parental organisms. Some workers recognize only two primary categories of larval development, namely, planktotrophic and nonplanktotrophic development.

The degree of larval dispersal is not only contingent upon the duration of the pelagic stages, but also upon the behavior of the larvae and, perhaps most importantly, upon the hydrodynamic processes in the estuary. While horizontal advective processes and the longevity of the pelagic larval stage primarily determine the potential for dispersal, larval behavior is responsible in many cases for the actual degree of dispersal. Hence, the swimming behavior of the larvae in the water column can maximize or minimize horizontal advective transport. Strong swimmers (e.g., decapod crustacean larvae) maintain greater control over larval dispersal by the strength of their movements in the water column. Larval swimming behavior is responsive to environmental factors. Directional swimming (e.g., geotaxis and phototaxis) and changes in swimming velocity (e.g., barokinesis) of larvae have been linked to light, gravity, and hydrostatic pressure. Planktotrophic larvae, because of their higher abundance and longer pelagic existence, have greater dispersal capability than lecithotrophic larvae. Predation and harsh environmental conditions, however, cause elevated mortality rates that may limit their spatial distribution.

Meroplankton of many benthic macroinvertebrates delay metamorphosis until a favorable habitat is located. Certain chemical or physical cues trigger metamorphosis and settlement of these larvae. Chemicals released by adults of the same species purportedly induce larvae to metamorphose, which partially dictates population distributions on the estuarine seafloor. Additional factors possibly involved in regulating larval settlement are sediment size distributions and the presence of algal substrates, bacterial coatings, and organic matter.

Larval recruitment (settlement, attachment, and metamorphosis) of certain benthic populations (e.g., clams and oysters) reflects gregarious behavioral patterns mediated by adult-derived chemical cues. Pheromones, for example, elicit behavioral responses in the larvae that foster gregarious settlement. Biochemical control of larval recruitment to the benthos has been demonstrated among annelids, arthropods, bryozoans, ascidian chordates, coelenterates, echinoderms, and mollusks. Clumped distributions of benthic macroinvertebrates may arise from factors other than gregarious settlement of larvae to chemical cues. For instance, suitable substrates for larval settlement may themselves have a patchy distribution.

On a world scale, some estuarine macrobenthic species display a cosmopolitan distribution. However, many estuarine benthic macroinvertebrates are not universally cosmopolitan. Few benthic species which inhabit estuaries nearly exclusively exhibit a worldwide distribution. Fouling species living in marine and brackish water systems (e.g., *Balanus amphitrite* and *B. improvisus*) tend to have a wide distribution, with many cosmopolitan populations being represented. A third group — benthic species confined to coastal oceanic waters and estuaries — also contains representatives with a wide geographical distribution. The distribution of the benthos on a global scale can be increased by means of adult migration, transport by rafting, and transport by human activities.

The species composition and abundance of benthic macroinvertebrate communities in estuaries varies markedly through time. Consequently, to understand the structure and dynamics of these communities, long-term monitoring provides the most insight. Examples of long-term macrobenthic monitoring programs in the U.S. are presented for Barnegat Bay, Chesapeake Bay, Corpus Christi Bay, and Puget Sound, and in Europe for the Upper Clyde estuary (England) and the Dutch Wadden Sea.

Diversity is a useful parameter for assessing the stability of macrobenthic communities. It consists of two components — species richness and evenness. Two of the most widely employed diversity indices in estuarine and marine research are the Shannon-Wiener and Simpson's indices.

Estuarine benthic communities usually have low species richness, although the density of the taxa can be very high. The species richness of benthic invertebrates in estuaries is lowest at salinities between 5 and 8‰. The number of species decreases when proceeding from the lower reaches into unstable brackish-water areas, and the number of freshwater taxa declines from the upper reaches into these areas. Several hypotheses may explain the low species richness of benthic invertebrates in estuaries. According to the stability-time hypothesis, environments that are geologically young and unstable, such as estuaries, show low species diversity. Environmental stress, which is high in these coastal systems, also has been advanced to explain the low species numbers. Fluctuating trophic resources may contribute to reduced species richness as well.

The biomass and productivity of the macrobenthos vary widely in estuaries. For an entire estuary, the mean biomass of the macrobenthos generally ranges from 10 to 25 g ash-free dry weight per square meter, and the mean production approaches 50 g ash-free dry weight per square meter per year. Dense aggregations of animals, such as mussels or oysters, yield production values of 200 to 300 g ash-free dry weight per square meter per year. Locally, the mean biomass of these organisms attains levels of 2000 g ash-free dry weight per square meter or more. The secondary production of benthic macroinvertebrates in estuaries is rather high in comparison to other ecosystems.

Five types of benthic macrofaunal feeders are recognized: (1) suspension feeders, (2) deposit feeders, (3) herbivores, (4) carnivores-scavengers, and (5) parasites. Most benthic macrofauna are suspension and deposit feeders which inhabit sandy and muddy sediments, respectively. Environments with high concentrations of suspended solids cannot support suspension feeders because the particles clog their filtering structures. Suspension feeders mainly consume phytoplankton, but also remove some detritus, though the nutritional value of the detritus remains unclear. Deposit feeders are either selective or nonselective. Selective deposit feeders separate their food from sedimentary particles; nonselective deposit feeders ingest sediment and food particles together, digest the organic matter therein, and excrete the sediment and other nondigestable components. Bacteria, benthic microalgae, meiofauna, and microfauna comprise important food sources for nonselective deposit feeders, and bacteria and microalgae are excellent foods for selective deposit feeders. As a group, deposit feeders are ecologically significant because they continually rework bottom sediments, and in this process impact the biological, chemical, and physical characteristics of the estuarine seafloor.

Investigations of rocky intertidal zones have supplied much information on the biological interactions of benthic communities. In subtidal, soft-bottom environments, exclusion experiments using cages and other devices to preclude predators from killing prey organisms have been invaluable in evaluating the significance of predation on the structure of benthic communities. Results of exclusion studies on unvegetated soft bottoms of lagoons and estuaries indicate a two-to-threefold increase in the density of macroinvertebrate prey in areas protected from predators. Caging experiments on vegetated soft bottoms reveal less predation and only slight effects on the benthic community. The vegetation serves as protective coverage for the prey organisms. Vegetated soft bottoms generally harbor communities with greater abundance and species richness than unvegetated bottoms. Predation effects are pronounced in sediments of unvegetated tidal flats (e.g., Wadden Sea), where the density of benthic populations in protected enclosures increases by as much as tenfold that of unprotected areas.

Benthic fauna modify seafloor sediments via feeding, burrowing, and physiological activities. Animal tubes, pits and depressions, excavation and fecal mounds, crawling trails or burrows, and organisms protruding above the sediment-water interface exacerbate bed roughness, thereby affecting fluid motion and sediment erosion and transport in the benthic boundary layer. Bioturbation by benthic macrofauna (i.e., biogenic particle manipulation and pore water exchange) influences interparticle adhesion, water content of sediments, bed roughness, and geochemistry of interstitial waters. Both the stabilization and destabilization of bottom sediments have been attributed to biogenic activity of the benthos. These organisms also enhance nutrient mixing in sediments and the exchange of gases across the sediment-water interface. Among bioturbating organisms, pioneering species consisting principally of tubicolous or sedentary forms rework sediments most intensely in the upper two centimeters. In contrast, benthic communities dominated by high-order successional stages rework sediments at greater depths to several centimeters. Conveyor-belt, deposit-feeding species, for example, rework sediments well below the sediment-water interface.

Benthic research has been greatly accelerated by the use of newly developed acoustic and photographic techniques in the field. Side-scan sonography facilitates data collection on bed roughness, sediment type, and morphology. The REMOTS™ sediment profile camera is especially valuable in studying the species composition of the benthic community and the geochemistry, bottom micromorphology, and microstratigraphy of the upper part of the sediment column. Electromagnetic current meters, acoustic current meters, thermistors and heat transfer-type current meters, and impellors and laser-doppler current meters have supplied impressive amounts of data on current profiles and turbulence in the benthic boundary layer. Tripods have allayed difficulties in data acquisition of the bottom or benthic boundary layer over a long time series of several days or weeks. Types of tripods currently employed in field research are the U.S. Geological Survey GEOPROBE, the University of Connecticut systems, University of Washington tripod, the Virginia Institute of Marine Science tripod, and the BASS tripod.

Benthic macrofouling populations are of major economic significance in estuarine waters. These epifaunal organisms settle and grow on the surfaces of natural and man-made materials and generally reduce the longevity and structural integrity of the materials. Millions of dollars are spent annually on shipping and ocean engineering systems alone to deal with fouling problems. Marine fouling communities contain a highly diverse number of macrobenthos, with more than 1000 species represented. The most frequently occurring phyla of fouling communities are the Porifera, Coelenterata, Annelida, Anthropoda, Mollusca, Bryozoa, and Chordata. The temporal and spatial complexity of these communities depend heavily on larval dynamics and settlement. The larvae have the ability to select particular substrates for settlement and metamorphosis; gregarious settlement is related to the chemistry of inducing substances associated with the substrate. Although a number of macrofouling organisms display various levels of mobility, subsequent to settlement and metamorphosis, the majority of species comprising the community remains permanently attached to the substrate.

Competition for space is intense in fouling communities. In many estuaries, competitive interactions control the abundance and distribution of fouling populations. Competitive strategies of fouling organisms in hard-substrate communities include overgrowth caused by relatively rapid growth rate, aggressive behavior, allelochemicals, and preemption of larval settlement by established individuals. Solitary forms appear to be superior competitors for space in rocky intertidal zones, but the overgrowth of colonial fauna dominates in many other communities.

Boring organisms are a specialized group of benthic organisms adapted to burrowing into rock and wood. Two shipworm genera, *Bankia* and *Teredo,* embody most species of woodborers. The piddock, *Pholas,* bores into rock in addition to other substrates such as stiff clay. Some borers (e.g., *Barnea candida* and *Zirphaea crispata*) burrow into rock as well as wood.

Shipworms and gribbles *(Limnoria)* are of particular interest because of their tremendous destructiveness to wooden structures in estuarine and marine environments. Wooden piles, docks, bulkheads, boats, and barges incur millions of dollars worth of damage annually due to woodborer infestation. While creosote-treated wood resists attack by shipworms, it does not afford protection from *Limnoria* infestation. Shipworms bore into wood, weakening its internal structure. When left unabated, the damage progresses to the point where the wood collapses. *Limnoria* attack results in a breakdown of the superficial layers of wood, commonly leading to a friable, spongy mass with an hourglass appearance.

Studies of benthic communities have advanced along two lines: (1) the level-bottom community approach and (2) the zonation approach. The level-bottom community approach deals with the description of assemblages of benthic organisms which are basically constant, and relates them to the physical environment. The zonation approach, however, focuses on benthic assemblages along a physical gradient or a combination of gradients. This approach has been successfully applied to intertidal and shallow subtidal zones, whereas the level-bottom community concept has been most fruitful in investigations of subtidal zones of estuaries and shallow coastal marine environments.

REFERENCES

1. **Lincoln, R. J., Boxshall, G. A., and Clark, P. F.,** *A Dictionary of Ecology, Evolution and Systematics,* Cambridge University Press, Cambridge, 1982.
2. **Ferguson, R. L., Thayer, G. W., and Rice, T. R.,** Marine primary producers, in *Functional Adaptations of Marine Organisms,* Vernberg, F. J. and Vernberg, W. B., Eds., Academic Press, New York, 1981, 9.
3. **Perkins, E. J.,** *The Biology of Estuaries and Coastal Waters,* Academic Press, London, 1974.

4. **Boaden, P. J. S. and Seed, R.,** *An Introduction to Coastal Ecology,* Blackie & Son, Glasgow, 1985.
5. **Pratt, S. D.,** Benthic fauna, in *Coastal and Offshore Environmental Inventory, Cape Hatteras to Nantucket Shoals,* Mar. Publ. Ser. No. 2, University of Rhode Island, Kingston, 1973.
6. **Erwin, D. G.,** The community concept, in *Sublittoral Ecology: The Ecology of the Shallow Sublittoral Benthos,* Earll, R. and Erwin, D. G., Eds., Clarendon Press, Oxford, 1983, 144.
7. **Dring, M. J.,** *The Biology of Marine Plants,* Edward Arnold, London, 1982.
8. **Levinton, J. S.,** *Marine Ecology,* Prentice-Hall, Englewood Cliffs, NJ, 1982.
9. **Dauer, D. M., Ewing, R. M., and Rodi, A. J., Jr.,** Macrobenthic distribution within the sediment along an estuarine salinity gradient, *Int. Rev. Gesamten Hydrobiol.,* 72, 529, 1987.
10. **Day, J. H.,** The estuarine fauna, in *Estuarine Ecology: With Particular Reference to Southern Africa,* Day, J. H., Ed., A. A. Balkema, Rotterdam, 1981, 147.
11. **Dexter, D. M.,** Distribution and niche diversity of haustoriid amphipods in North Carolina, *Chesapeake Sci.,* 8, 187, 1967.
12. **Stickney, R. R.,** *Estuarine Ecology of the Southeastern United States and Gulf of Mexico,* Texas A & M University Press, College Station, 1984.
13. **Williams, R. B.,** The Ecology of Diatom Populations in a Georgia Salt Marsh, Ph.D. thesis, Harvard University, Cambridge, MA, 1962.
14. **Pomeroy, L. R., Darley, W. M., Dunn, E. L., Gallagher, J. L., Haines, E. B., and Whitney, D. M.,** Primary production, in *The Ecology of a Salt Marsh, Ecological Studies,* Vol. 38, Pomeroy, L. R. and Wiegert, R. G., Eds., Springer-Verlag, New York, 1981, 39.
15. **Odum, H. T.,** Biological circuits and the marine systems of Texas, in *Pollution and Marine Ecology,* Olson, T. A. and Burgess, F. J., Eds., Wiley-Interscience, New York, 1967, 99.
16. **Baretta, J. W. and Ruardij, P., Eds.,** *Tidal Flat Estuaries: Simulation and Analysis of the Ems Estuary, Ecological Studies,* Vol. 71, Springer-Verlag, New York, 1988.
17. **Long, S. P. and Mason, C. F.,** *Saltmarsh Ecology,* Blackie & Son, Glasgow, 1983.
18. **Mitsch, W. J. and Gosselink, J. G.,** *Wetlands,* Van Nostrand Reinhold, New York, 1986.
19. **Chapman, V. J.,** *Salt Marshes and Salt Deserts of the World,* Wiley-Interscience, New York, 1960.
20. **Chapman, V. J.,** Salt marshes and salt deserts of the world, in *Ecology of Halophytes,* Reimold, R. J. and Queen, W. H., Eds., Academic Press, New York, 1974, 3.
21. **Chapman, V. J.,** *Wet Coastal Ecosystems,* Elsevier, Amsterdam, 1977.
22. **Odum, W. E.,** Comparative ecology of tidal freshwater and salt marshes, *Annu. Rev. Ecol. Syst.,* 19, 147, 1988.
23. **Odum, W. E., Smith, T. J., III, Hoover, J. K., and McIvor, C. C.,** The Ecology of Tidal Freshwater Marshes of the United States East Coast: A Community Profile, FWS/OBS-87/17, U.S. Fish and Wildlife Service, Washington, D. C., 1984, 177 pp.
24. **den Hartog, C.,** *The Seagrasses of the World,* North Holland, Amsterdam, 1970.
25. **Fry, B. and Parker, P. L.,** Animal diet in Texas seagrass meadows: ^{13}C evidence for the importance of benthic plants, *Estuarine Coastal Mar. Sci.,* 8, 499, 1979.
26. **Kikuchi, T.,** Faunal relationships in temperate seagrass beds, in *Handbook of Seagrass Biology: An Ecosystem Perspective,* Phillips, R. C. and McRoy, C. P., Eds., Garland STPM Press, New York, 1980.
27. **Dawes, C. J.,** *Marine Botany,* John Wiley & Sons, New York, 1981.
28. **Lin, P.,** *Mangrove Vegetation,* Springer-Verlag, New York, 1988.
29. **Rützler, K. and Feller, C.,** Mangrove swamp communities, *Oceanus,* 30, 16, 1987.
30. **Lugo, A. E. and Snedaker, S. C.,** The ecology of mangroves, *Annu. Rev. Ecol. Syst.,* 5, 39, 1974.
31. **Tomlinson, P. B.,** *The Botany of Mangroves,* Cambridge University Press, Cambridge, England, 1986.
32. **Craighead, F. C.,** *The Trees of South Florida,* University of Miami Press, Coral Gables, FL, 1971.
33. **Mare, M. F.,** A study of a marine benthic community with special reference to the microorganisms, *J. Mar. Biol. Assoc. U.K.,* 25, 517, 1942.
34. **Fenchel, T.,** The ecology of marine microbenthos. IV. Structure and function of the benthic ecosystem, its chemical and physical factors and the microfauna communities with special reference to the ciliated protozoa, *Ophelia,* 6, 1, 1969.
35. **Fenchel, T. M.,** The ecology of micro- and meiobenthos, *Annu. Rev. Ecol. Syst.,* 9, 99, 1978.
36. **Daiber, F. C.,** *Animals of the Tidal Marsh,* Van Nostrand Reinhold, New York, 1982.
37. **Loveland, R. E. and Vouglitois, J. J.,** Benthic fauna, in *Ecology of Barnegat Bay, New Jersey,* Kennish, M. J. and Lutz, R. A., Eds., Springer-Verlag, New York, 1984, 135.
38. **Lippson, A. J., Haire, M. S., Holland, A. F., Jacobs, F., Jensen, J., Moran-Johnson, R. L., Polgar, T. T., and Richkus, W. A.,** *Environmental Atlas of the Potomac Estuary,* Johns Hopkins University Press, Baltimore, 1981.
39. **Sullivan, M. J.,** Diatom communities from a Delaware salt marsh, *J. Phycol.,* 11, 384, 1975.
40. **Whitney, D. E., Woodwell, G. M., and Howarth, R. W.,** Nitrogen fixation in Flax Pond: a Long Island salt marsh, *Limnol. Oceanogr.,* 20, 640, 1975.

41. **Van Raalte, C. D., Valiela, I., and Teal, J. M.,** Production of epibenthic salt marsh algae: light and nutrient limitation, *Limnol. Oceanogr.,* 21, 862, 1976.

42. **Reise, K.,** *Tidal Flat Ecology: An Experimental Approach to Species Interactions, Ecological Studies,* Vol. 54, Springer-Verlag, Berlin, 1985.

43. **Grant, J., Bathmann, U. V., and Mills, E. L.,** The interaction between benthic diatom films and sediment transport, *Estuarine Coastal Shelf Sci.,* 23, 225, 1986.

44. **Cammen, L. M. and Walker, J. A.,** The relationship between bacteria and microalgae in the sediment of a Bay of Fundy mudflat, *Estuarine Coastal Shelf Sci.,* 22, 91, 1986.

45. **Admiraal, W., Peletier, H., and Zomer, H.,** Observations and experiments on the population dynamics of epipelic diatoms from an estuarine mudflat, *Estuarine Coastal Shelf Sci.,* 14, 471, 1982.

46. **Baillie, P. W.,** Diatom size distributions and community stratification in estuarine intertidal sediments, *Estuarine Coastal Shelf Sci.,* 25, 193, 1987.

47. **Round, F. E.,** Benthic marine diatoms, *Oceanogr. Mar. Biol. Annu. Rev.,* 9, 83, 1982.

48. **Werner, D.,** *The Biology of Diatoms,* Blackwell Scientific, Oxford, 1977.

49. **McLusky, D. S.,** *The Estuarine Ecosystem,* Halsted Press, New York, 1981.

50. **Breitburg, D. L.,** Development of a subtidal epibenthic community: factors affecting species composition and the mechanisms of succession, *Oecologia,* 65, 173, 1985.

51. **Valiela, I.,** *Marine Ecological Processes,* Springer-Verlag, New York, 1984.

52. **Wetzel, R. L.,** An Experimental Study of Detrital Carbon Utilization in a Georgia Salt Marsh, Ph.D. thesis, University of Georgia, Athens, 1975.

53. **Pace, M. L., Shimmel, S., and Darley, W. M.,** The effect of grazing by a gastropod, *Nassarius obsoletus,* on the benthic microbial community of a salt marsh mud flat, *Estuarine Coastal Mar. Sci.,* 9, 121, 1979.

54. **Shanholtzer, S. F.,** Energy Flow, Food Habits, and Population Dynamics of *Uca pugnax* in a Salt Marsh System, Ph.D. thesis, University of Georgia, Athens, 1973.

55. **Haines, E. B. and Montague, C. L.,** Food sources of estuarine invertebrates analyzed using $^{13}C/^{12}C$ ratios, *Ecology,* 60, 48, 1979.

56. **Hughes, E. H.,** Estuarine Subtidal Food Webs Analyzed with Stable Carbon Isotope Ratios, M.S. thesis, University of Georgia, Athens, 1980.

57. **Kitting, C. L., Fry, B., and Morgan, M. D.,** Detection of inconspicuous epiphytic algae supporting food webs in seagrass meadows, *Oecologia,* 62, 145, 1984.

58. **Morgan, M. D.,** Grazing and predation of the grass shrimp *Palaemonetes pugio, Limnol. Oceanogr.,* 25, 896, 1980.

59. **Van Montfrans, J., Orth, R. J., and Hay, S. A.,** Preliminary studies of grazing by *Bittium varium* on eelgrass periphyton, *Aquat. Bot.,* 14, 75, 1982.

60. **McRoy, C. P. and McMillan, C.,** Production ecology and physiology of seagrasses, in *Seagrass Ecosystems: A Scientific Perspective,* McRoy, C. P. and Helfferich, C., Eds., Marcel Dekker, New York, 1977, 53.

61. **Twilley, R. R., Kemp, W. M., Staver, K. W., Stevenson, J. C., and Boynton, W. R.,** Nutrient enrichment of estuarine submersed vascular plant communities. I. Algal growth and effects on production of plants and associated communities, *Mar. Ecol. Prog. Ser.,* 23, 179, 1985.

62. **Marshall, N.,** Food transfer through the lower trophic levels of the benthic environment, in *Marine Food Chains,* Steele, J. H., Ed., Oliver and Boyd, Edinburgh, 1970, 52.

63. **Jones, J. A.,** Primary Productivity by the Tropical Marine Turtle Grass, *Thalassia testudinum* König, and Its Epiphytes, Ph.D. thesis, University of Miami, Coral Gables, FL, 1968.

64. **Penhale, P. A.,** Macrophyte-epiphyte biomass and productivity in an eelgrass (*Zostera marina* L.) community, *J. Exp. Mar. Biol. Ecol.,* 26, 211, 1977.

65. **Borum, J. and Wium-Andersen, S.,** Biomass and production of epiphytes on eelgrass (*Zostera marina* L.) in the Øresund, Denmark, *Ophelia,* 1(Suppl.), 57, 1980.

66. **Jones, J. A.,** Primary Productivity by the Tropical Marine Turtle Grass, *Thalassia testudinum* König, and Its Epiphytes, Abstract, Ph.D. dissertation, University of Miami, Coral Gables, FL, 1968, *Diss. Abstr. B,* 29(10-Botany), 3637, 1969.

67. **Orth, R. J. and Van Montfrans, J.,** Epiphyte-seagrass relationships with an emphasis on the role of micrograzing: a review, *Aquat. Bot.,* 18, 43, 1984.

68. **Harlin, M. M.,** Seagrass epiphytes, in *Handbook of Seagrass Biology: An Ecosystem Perspective,* Phillips, R. C. and McRoy, C. P., Eds., Garland STPM, New York, 1980, 117.

69. **Harrison, P. G. and Chan, A. T.,** Inhibition of the growth of microalgae and bacteria by extracts of eelgrass (*Zostera marina*) leaves, *Mar. Biol.,* 61, 21, 1980.

70. **Ramm, G.,** Structure of epiphytic diatom populations of the phytal of the Kiel Bight (western Baltic), in 4th Symp. on Recent and Fossil Marine Diatoms, Vol. 54, Simonsen, R., Ed., Proc. Beih. Nova Hedwigia, 1977, 379.

71. **Sullivan, M. J.,** Structural characteristics of a diatom community epiphytic on *Ruppia maritima, Hydrobiologia,* 53, 81, 1977.

72. **Sieburth, J. M. and Thomas, C. D.,** Fouling on eelgrass (*Zostera marina* L.), *J. Phycol.,* 9, 46, 1973.

73. **DeFelice, D. R. and Lynts, G. W.,** Benthic marine diatom associations: Upper Florida Bay (Florida) and associated sounds, *J. Phycol.,* 14, 25, 1978.
74. **Jacobs, R. P. W. M. and Noten, T. M. P. A.,** The annual pattern of the diatoms in the epiphyton of eelgrass (*Zostera marina* L.) at Roscoff, France, *Aquat. Bot.,* 8, 355, 1980.
75. **Novak, R.,** A study in ultra-ecology: microorganisms on the seagrass *Posidonia oceanica* (L.) Delile, *P.S.Z.N.I. Mar. Ecol.,* 5, 143, 1984.
76. **Lipps, J. H. and Valentine, J. W.,** The role of foraminifera in the trophic structure of marine communities, *Lethaia,* 3, 279, 1970.
77. **Lee, J. J., McEnery, M. E., Kennedy, E. M., and Rubin, H.,** A nutritional analysis of a sublittoral diatom assemblage epiphytic on *Enteromorpha* from a Long Island salt marsh, *J. Phycol.,* 11, 14, 1975.
78. **Lee, J. J., McEnery, M., Pierce, S., Freudenthal, H. D., and Muller, W. A.,** Tracer experiments in feeding littoral foraminifera, *J. Protozool.,* 13, 659, 1966.
79. **Lee, J. J., Tietjen, J. H., and Garrison, J. R.,** Seasonal switching in the nutritional requirements of *Nitocra typica*, a harpacticoid copepod from salt marsh aufwuchs communities, *Trans. Am. Microsc. Soc.,* 95, 628, 1976.
80. **Kikuchi, T. and Peres, J. M.,** Consumer ecology of seagrass beds, in *Seagrass Ecosystems: A Scientific Perspective,* McRoy, C. P. and Helfferich, C., Eds., Marcel Dekker, New York, 1977, 147.
81. **Wolff, W. J.,** Biotic aspects of the chemistry of estuaries, in *Chemistry and Biogeochemistry of Estuaries,* Olausson, E. and Cato, I., Eds., John Wiley & Sons, Chichester, 1980, 263.
82. **Colijn, F. and de Jonge, V. N.,** Primary production of microphytobenthos in the Ems-Dollard estuary, *Mar. Ecol. Prog. Ser.,* 14, 185, 1984.
83. **Terry, L. A. and Edyvean, R. G. J.,** Influences of microalgae on the corrosion of structural steel, in *Corrosion and Marine Growth on Offshore Structures,* Lewis, J. R. and Mercer, A. D., Eds., Ellis Horwood, Chichester, 1984, 38.
84. **Edyvean, R. G. J. and Moss, B. L.,** Microalgal communities on protected steel substrata in seawater, *Estuarine Coastal Shelf Sci.,* 22, 509, 1986.
85. **Baier, R. E.,** Initial events in microbial film formation, in *Marine Biodeterioration: An Interdisciplinary Study,* Costlow, J. D. and Tipper, R. C., Eds., Naval Institute Press, Annapolis, MD, 1984, 87.
86. **Little, B. J.,** Succession in microfouling, in *Marine Biodeterioration: An Interdisciplinary Study,* Costlow, J. D. and Tipper, R. C., Eds., Naval Institute Press, Annapolis, MD, 1984, 63.
87. **Cooksey, B., Cooksey, K. E., Miller, C. A., Paul, J. H., Rubin, R. W., and Webster, D.,** The attachment of microfouling diatoms, in *Marine Biodeterioration: An Interdisciplinary Study,* Costlow, J. D. and Tipper, R. C., Eds., Naval Institute Press, Annapolis, MD, 1984, 167.
88. **Cundell, A. and Mitchell, R.,** Microbial succession on a wooden surface exposed to the sea, *Int. Biodeterior. Bull.,* 13, 67, 1977.
89. **Mitchell, R., Ed.,** *Water Pollution Microbiology,* Vol. 2, John Wiley & Sons, New York, 1978.
90. **Mitchell, R. and Kirchman, D.,** The microbial ecology of marine surfaces, in *Marine Biodeterioration: An Interdisciplinary Study,* Costlow, J. D. and Tipper, R. C., Eds., Naval Institute Press, Annapolis, MD, 1984, 49.
91. **Mitchell, R.,** Negative Chemotaxis: A New Approach to Marine Fouling Control, Office of Naval Research Technical Report No. 1 — Biofouling, Office of Naval Research, Harvard University, Cambridge, MA, 1975.
92. **Mann, K. H.,** *Ecology of Coastal Waters: A Systems Approach,* University of California Press, Berkeley, 1982.
93. **Correll, D. L.,** Estuarine productivity, *BioScience,* 28, 646, 1978.
94. **Abbott, I. A. and Hollenberg, G. J.,** *Marine Algae of California,* Stanford University Press, Stanford, CA, 1976.
95. **Thom, R. M.,** Composition, habitats, seasonal changes and productivity of macroalgae in Grays Harbor estuary, Washington, *Estuaries,* 7, 51, 1984.
96. **Josselyn, M. N. and West, J. A.,** The distribution and temporal dynamics of the estuarine macroalgal community of San Francisco Bay, *Hydrobiologia,* 129, 139, 1985.
97. **Wilkinson, M.,** Estuarine benthic algae and their environment: a review, in *The Shore Environment,* Vol. 2, Price, J. H., Irvine, D. E. G., and Farnham, W. F., Eds., Systematics Assoc. Spec. Vol. 17(b), Academic Press, London, 1980, 213.
98. **Zechman, F. W. and Mathieson, A. C.,** The distribution of seaweed propagules in estuarine, coastal, and offshore waters of New Hampshire, U.S.A., *Bot. Mar.,* 28, 283, 1985.
99. **Tseng, C. K.,** Oceanographic factors and seaweed distribution, *Oceanus,* 26, 48, 1984.
100. **Charters, A. C., Neushul, M., and Coon, D.,** Effects of water motion on algal spore attachment, *Proc. Int. Seaweed Symp.,* 7, 243, 1972.
101. **Coon, D., Neushul, M., and Charters, A. C.,** The settling behavior of marine algal spores, *Proc. Int. Seaweed Symp.,* 7, 237, 1972.
102. **Norton, T. A. and Fetter, R.,** The settlement of *Sargassum muticum* propagules in stationary and flowing water, *J. Mar. Biol. Assoc. U.K.,* 61, 929, 1981.

103. **Norton, T. A.,** The resistance to dislodgement of *Sargassum muticum* germlings under defined hydrodynamic conditions, *J. Mar. Biol. Assoc. U.K.,* 63, 181, 1983.

104. **Wilkinson, M. and Rendall, D. A.,** The role of benthic algae in estuarine pollution assessment, in *Estuarine Management and Quality Assessment,* Wilson, J. G. and Halcrow, W., Eds., Plenum Press, New York, 1985, 71.

105. **den Hartog, C.,** Brackish water as an environment for algae, *Blumea,* 15, 31, 1967.

106. **Tittley, I.,** Zonation and seasonality of estuarine benthic algae: artificial embankments in the River Thames, *Bot. Mar.,* 28, 1, 1985.

107. **Tittley, I. and Price, J. H.,** The marine algae of the tidal Thames, *London Nat.,* 56, 10, 1977.

108. **Price, J. H., Tittley, I., and Honey, S. I.,** The benthic marine algal flora of Lincolnshire and Cambridgeshire: a preliminary survey, *Naturalist (Hull),* 102, 3, 1977.

109. **Schonbeck, M. and Norton, T. A.,** Factors controlling the upper limits of fucoid algae on the shore, *J. Exp. Mar. Biol. Ecol.,* 31, 303, 1978.

110. **Southward, A. J. and Southward, E. C.,** Recolonization of rocky shores in Cornwall after use of toxic dispersants to clean up the Torrey Canyon spill, *J. Fish. Res. Board Can.,* 35, 682, 1978.

111. **Pomeroy, W. M. and Stockner, J. G.,** Effects of environmental disturbance on the distribution and primary production of benthic algae on a British Columbia estuary, *J. Fish. Res. Board Can.,* 33, 1175, 1976.

112. **Davis, M. W. and McIntire, C. D.,** Production dynamics of sediment-associated algae in two Oregon estuaries, *Estuaries,* 4, 301, 1981.

113. **Neushul, M.,** Studies of subtidal marine vegetation in western Washington, *Ecology,* 48, 83, 1967.

114. **Thom, R. M.,** The Composition, Growth, Seasonal Periodicity, and Habitats of Benthic Algae on the Eastern Shore of Central Puget Sound, with Special Reference to Sewage Pollution, Ph.D. thesis, University of Washington, Seattle, 1978.

115. **Thom, R. M.,** Seasonality in low intertidal marine algal communities in central Puget Sound, Washington, USA, *Bot. Mar.,* 23, 7, 1980.

116. **Thom, R. M., Armstrong, J. W., Staude, C. P., Chew, K. K., and Norris, R. E.,** A survey of the attached marine flora at five beaches in the Seattle, Washington area, *Syesis,* 9, 267, 1976.

117. **Hodgson, J. M. and Waaland, J. R.,** Seasonal variation in the subtidal macroalgae of Fox Island, Puget Sound, Washington, *Syesis,* 12, 107, 1979.

118. **Pomeroy, W. M.,** Benthic Algal Ecology and Primary Pathways of Energy Flow on the Squamish River Delta, British Columbia, Ph.D. thesis, University of British Columbia, Vancouver, 1977.

119. **Loveland, R. E., Brauner, J. F., Taylor, J. E., and Kennish, M. J.,** Macroflora, in *Ecology of Barnegat Bay, New Jersey,* Kennish, M. J. and Lutz, R. A., Eds., Springer-Verlag, New York, 1984, 78.

120. **Taylor, J. E.,** The Ecology and Seasonal Periodicity of Benthic Marine Algae from Barnegat Bay, New Jersey, Ph.D. thesis, Rutgers University, New Brunswick, NJ, 1970.

121. **Moeller, H. W.,** A standing crop estimate of some marine plants in Barnegat Bay, *Bull. N.J. Acad. Sci.,* 9, 27, 1964.

122. **Chock, J. S. and Mathieson, A. C.,** Variations in New England estuarine seaweed biomass, *Bot. Mar.,* 26, 87, 1983.

123. **Mathieson, A. C. and Penniman, C. A.,** Species composition and seasonality of New England seaweeds along an open coastal-estuarine gradient, *Bot. Mar.,* 29, 161, 1986.

124. **Malinowski, K. C. and Ramus, J.,** Growth of the green algae *Codium fragile* in a Connecticut estuary, *J. Phycol.,* 9, 102, 1973.

125. **Mann, K. H.,** Macrophyte production and detritus food chains in coastal waters, *Mem. Ist. Ital. Idrobiol., Dott Marco de Marchi Pallanza Italy,* 29(Suppl.), 353, 1972.

126. **Breen, P. A. and Mann, K. H.,** Destructive grazing of kelp by sea urchins in eastern Canada, *J. Fish. Res. Board Can.,* 33, 1278, 1976.

127. **Chapman, A. R. O.,** Stability of sea urchin dominated barren grounds following destructive grazing of kelp in St. Margaret's Bay, eastern Canada, *Mar. Biol.,* 62, 307, 1981.

128. **Miller, R. J.,** Succession in sea urchin and seaweed abundance in Nova Scotia, Canada, *Mar. Biol.,* 84, 275, 1985.

129. **Warwick, R. M., Davey, J. T., Cree, J. M., and George, C. L.,** Faunistic control of *Enteromorpha* blooms: a field experiment, *J. Exp. Mar. Biol. Ecol.,* 56, 23, 1982.

130. **Kennish, M. J.,** *Ecology of Estuaries,* Vol. 1, CRC Press, Boca Raton, FL, 1986.

131. **Josselyn, M. N. and Mathieson, A. C.,** Seasonal influx and decomposition of autochthonous macrophyte litter in a north temperate estuary, *Hydrobiologia,* 71, 197, 1980.

132. **McLachlan, J.,** Macroalgae (seaweeds): industrial resources and their utilization, *Plant Soil,* 89, 137, 1985.

133. **Bonotto, S.,** Cultivation of plants: multicellular plants, in *Marine Ecology,* Vol. 3 (Part 1), Kinne, O., Ed., John Wiley & Sons, New York, 1976.

134. **Kiran, E., Teksoy, I., Gaven, K. C., Guler, E., and Guner, H.,** Studies on seaweeds for paper production, *Bot. Mar.,* 23, 205, 1980.

135. **Beeftink, W. G.,** Salt-marshes, in *The Coastline,* Barnes, R. S. K., Ed., John Wiley & Sons, Chichester, 1977, 93.

136. **Frey, R. W. and Basan, P. B.,** Coastal salt marshes, in *Coastal Sedimentary Environments,* 2nd ed., Davis, R. A., Ed., Springer-Verlag, New York, 1985, 225.

137. **Duncan, W. H.,** Vascular halophytes of the Atlantic and Gulf Coasts of North America north of Mexico, in *Ecology of Halophytes,* Reimold, R. J. and Queen, W. H., Eds., Academic Press, New York, 1974, 23.

138. **Dame, R. F.,** The importance of *Spartina alterniflora* to Atlantic Coast estuaries, *Rev. Aquat. Sci.,* 1, 639, 1989.

139. **Macdonald, K. B.,** Plant and animal communities of Pacific North American salt marshes, in *Wet Coastal Ecosystems,* Chapman, V. J., Ed., Elsevier, Amsterdam, 1977, 167.

140. **Macdonald, K. B.,** Coastal salt marsh, in *Terrestrial Vegetation of California,* Barbour, M. G. and Major, J., Eds., John Wiley & Sons, New York, 1977, 263.

141. **Frey, R. W. and Howard, J. D.,** Physical and biogenic processes in Georgia estuaries. II. Intertidal facies, *Geol. Assoc. Can. Short Course Notes,* 1, 183, 1980.

142. **Redfield, A. C.,** Development of a New England salt marsh, *Ecol. Monogr.,* 42, 201, 1972.

143. **Bertness, M. D. and Ellison, A. M.,** Determinants of pattern in a New England salt marsh plant community, *Ecol. Monogr.,* 57, 129, 1987.

144. **Schubauer, J. P. and Hopkinson, C. S.,** Above- and belowground emergent macrophyte production and turnover in a coastal marsh ecosystem, Georgia, *Limnol. Oceanogr.,* 29, 1052, 1984.

145. **Good, R. E., Good, N. F., and Frasco, B. R.,** A review of primary production and decomposition dynamics of the belowground marsh component, in *Estuarine Comparisons,* Kennedy, V. S., Ed., Academic Press, New York, 1982, 139.

146. **Gallagher, J. L. and Plumley, F. G.,** Underground biomass profiles and productivity in Atlantic coastal marshes, *Am. J. Bot.,* 66, 156, 1979.

147. **Linthurst, R. A. and Reimhold, R. J.,** Estimated net aerial primary productivity for selected estuarine angiosperms in Maine, Delaware, and Georgia, *Ecology,* 59, 945, 1978.

148. **Dame, R. F. and Kenny, P. D.,** Variability of *Spartina alterniflora* primary production in the euhaline North Inlet estuary, *Mar. Ecol. Prog. Ser.,* 32, 71, 1986.

149. **Pomeroy, L. R.,** Algae productivity in salt marshes of Georgia, *Limnol. Oceanogr.,* 4, 386, 1959.

150. **Teal, J. M.,** Energy flow in the salt marsh ecosystem of Georgia, *Ecology,* 43, 614, 1962.

151. **Gallagher, J. L. and Daiber, F. C.,** Primary production of edaphic algae communities in a Delaware salt marsh, *Limnol. Oceanogr.,* 19, 390, 1974.

152. **Zedler, J. B.,** Algae mat productivity: comparisons in a salt marsh, *Estuaries,* 3, 122, 1980.

153. **Valiela, I., Howes, B., Howarth, R., Giblin, A., Foreman, K., Teal, J. M., and Hobbie, J. E.,** Regulation of primary production and decomposition in a salt marsh ecosystem, in *Wetlands Ecology and Management,* Gopal, B., Turner, R. E., Wetzel, R. G., and Wigham, D. F., Eds., National Institute of Ecology (India) and International Scientific Publications, New Delhi, 1982, 151.

154. **Valiela, I., Teal, J. M., Allen, S. D., Van Etten, R., Goehringer, D., and Volkmann, S.,** Decomposition in salt marsh ecosystems: the phases and major factors affecting disappearance of above-ground organic matter, *J. Exp. Mar. Biol. Ecol.,* 89, 29, 1985.

155. **Wilson, J. O., Buchsbaum, R., Valiela, I., and Swain, T.,** Decomposition in salt marsh ecosystems: phenolic dynamics during decay of litter of *Spartina alterniflora, Mar. Ecol. Prog. Ser.,* 29, 177, 1986.

156. **Buchsbaum, R., Valiela, I., and Swain, T.,** The effect of phenolic and other constituents on feeding by Canada geese in a coastal marsh, *Oecologia,* 63, 343, 1984.

157. **Valiela, I., Koumjian, L., Swain, T., Teal, J. M., and Hobbie, J. E.,** Cinnamic acid inhibition of detritus feeding, *Nature,* 280, 55, 1979.

158. **Hartley, R. D. and Jones, E. C.,** Phenolic components and degradability of cell walls of grass and legume species, *Phytochemistry,* 16, 1531, 1977.

159. **Marinucci, A. C.,** Carbon and nitrogen fluxes during decomposition of *Spartina alterniflora* in a flow-through percolator, *Biol. Bull.,* 162, 54, 1982.

160. **Rice, D. L.,** The detritus nitrogen problem. New observations and perspectives from organic geochemistry, *Mar. Ecol. Prog. Ser.,* 9, 153, 1982.

161. **Rice, D. L. and Hanson, R. B.,** A kinetic model for detritus nitrogen: role of the associated bacteria in nitrogen accumulation, *Bull. Mar. Sci.,* 35, 326, 1984.

162. **Odum, E. P.,** The role of tidal marshes in estuarine production, *N.Y. State Conserv.,* 16, 12, 1961.

163. **Daiber, F. C.,** Salt marsh plants and future coastal salt marshes in relation to animals, in *Ecology of Halophytes,* Reimold, R. J. and Queen, W. H., Eds., Academic Press, New York, 1974, 475.

164. **Gerlach, S. A.,** Food-chain relationships in subtidal silty sand marine sediments and the role of meiofauna in stimulating bacterial productivity, *Oecologia,* 33, 55, 1978.

165. **Kuenzler, E. J.,** Structure and energy flow of a mussel population in a Georgia salt marsh, *Limnol. Oceanogr.,* 6, 191, 1961.

166. **Holle, P. A.,** Life history of the salt marsh snail *Melampus bidentatus* Say, *Nautilus,* 70, 90, 1957.

167. **Gosner, K. L.,** *A Field Guide to the Atlantic Seashore,* Houghton Mifflin, Boston, 1978.
168. **Bertness, M. D. and Miller, T.,** The distribution and dynamics of *Uca pugnax* (Smith) burrows in a New England salt marsh, *J. Exp. Mar. Biol. Ecol.,* 83, 211, 1984.
169. **Montague, C. L.,** A natural history of temperate western Atlantic fiddler crabs with reference to their impact on salt marshes, *Contrib. Mar. Sci.,* 23, 25, 1980.
170. **Montague, C. L.,** The influence of fiddler crab burrows and burrowing on metabolic processes in salt marsh sediments, in *Estuarine Comparisons,* Kennedy, V. S., Ed., Academic Press, New York, 1982, 283.
171. **Chalmers, A. G.,** Soil dynamics and the productivity of *Spartina alterniflora,* in *Estuarine Comparisons,* Kennedy, V. S., Ed., Academic Press, New York, 1982, 231.
172. **Bertness, M. D.,** Fiddler crab regulation of *Spartina alterniflora* production on a New England salt marsh, *Ecology,* 66, 1042, 1985.
173. **Barnwell, F. H. and Thurman, C. L., II,** Taxonomy and biogeography of the fiddler crabs (Ocypodidae: Genus *Uca*) of the Atlantic and Gulf Coasts of eastern North America, *Zool. J. Linn. Soc.,* 81, 23, 1984.
174. **Teal, J. M.,** Respiration of crabs in Georgia salt marshes and its relation to their ecology, *Physiol. Zool.,* 32, 1, 1959.
175. **Gleason, D. F. and Zimmerman, R. J.,** Herbivory potential of postlarval brown shrimp associated with salt marshes, *J. Exp. Mar. Biol. Ecol.,* 84, 235, 1984.
176. **Gunter, G.,** Habitat of juvenile shrimp (Family Penaeidae), *Ecology,* 42, 598, 1961.
177. **Weinstein, M. P.,** Shallow marsh habitats as primary nurseries for fishes and shellfish, Cape Fear River, North Carolina, *Fish. Bull. (U.S.),* 77, 339, 1979.
178. **Able, K. W., Heck, K. L., Jr., Fahay, M. P., and Roman, C. T.,** Use of salt-marsh peat reefs by small juvenile lobsters on Cape Cod, Massachusetts, *Estuaries,* 11, 83, 1988.
179. **Shenker, J. and Dean, J. M.,** The utilization of an intertidal salt marsh creek by larval and juvenile fishes: abundance, diversity, and temporal variation, *Estuaries,* 2, 154, 1979.
180. **Bozeman, E. L., Jr. and Dean, J. M.,** The abundance of estuarine larval and juvenile fish in a South Carolina intertidal creek, *Estuaries,* 3, 89, 1980.
181. **Kneib, R. T.,** Patterns in the utilization of the intertidal salt marsh by larvae and juveniles of *Fundulus heteroclitus* (Linnaeus) and *Fundulus luciae* (Baird), *J. Exp. Mar. Biol. Ecol.,* 83, 41, 1984.
182. **Byrne, D. M.,** Life history of the spotfin killifish, *Fundulus luciae* (Pisces: Cyprinodontidae), in Fox Creek Marsh, Virginia, *Estuaries,* 1, 211, 1978.
183. **Kneib, R. T.,** Habitat, diet, reproduction and growth of the spotfin killifish, *Fundulus luciae,* from a North Carolina salt marsh, *Copeia,* 1978, 164, 1978.
184. **Kneib, R. T., Stiven, A. E., and Haines, E. B.,** Stable carbon isotope ratios in *Fundulus heteroclitus* (L.) muscle tissue and gut contents from a North Carolina *Spartina* marsh, *J. Exp. Mar. Biol. Ecol.,* 46, 89, 1980.
185. **Smalley, A. E.,** Energy flow of a salt marsh grasshopper population, *Ecology,* 41, 672, 1980.
186. **Pfeiffer, W. J. and Wiegert, R. G.,** Grazers on *Spartina* and their predators, in *The Ecology of a Salt Marsh,* Pomeroy, L. R. and Wiegert, R. G., Eds., Springer-Verlag, New York, 1981, 87.
187. **Barnes, R. D.,** The ecological distribution of spiders in non-forest maritime communities at Beaufort, North Carolina, *Ecol. Monogr.,* 23, 315, 1953.
188. **Dale, P. E. R. and Hulsman, K.,** A critical review of salt marsh management methods for mosquito control, *Rev. Aquat. Sci.,* in press.
189. **Connell, W. A.,** Tidal inundation as a factor limiting distribution of *Aedes* spp. on a Delaware salt marsh, *Proc. N.J. Mosq. Exterm. Assoc.,* 27, 166, 1940.
190. **Ferrigno, F.,** A two-year study of mosquito breeding in the natural and untouched salt marshes of Egg Island, *Proc. N.J. Mosq. Exterm. Assoc.,* 45, 132, 1958.
191. **La Salle, R. N. and Knight, K. L.,** Effects of Salt Marsh Impoundments on Mosquito Populations, Tech. Rep. 92, Water Resources Research Institute, University of North Carolina, Chapel Hill, 1974, 85.
192. **Dukes, J. C., Axtell, R. C., and Knight, K. L.,** Additional Studies of the Effects of Salt Marsh Impoundments on Mosquito Populations, Tech. Rep. 102, Water Resources Research Institute, University of North Carolina, Chapel Hill, 1974, 38 pp.
193. **Dukes, J. C., Edwards, T. D., and Axtell, R. C.,** Associations of Tabanidae (Diptera) larvae with plant species in salt marshes, Carteret County, North Carolina, *Environ. Entomol.,* 3, 280, 1974.
194. **Meany, R. A., Valiela, I., and Teal, J. M.,** Growth, abundance and distribution of larval tabanids in experimentally fertilized plots on a Massachusetts salt marsh, *J. Appl. Ecol.,* 13, 323, 1976.
195. **Evans, P. D., Ruscoe, E. N. E., and Treherne, J. E.,** Observations on the biology and submergence behavior of some littoral beetles, *J. Mar. Biol. Assoc. U.K.,* 51, 375, 1971.
196. **Davis, L. V. and Gray, I. E.,** Zonal and seasonal distribution of insects in North Carolina salt marshes, *Ecol. Monogr.,* 36, 275, 1966.
197. **Luxton, M.,** The ecology of salt marsh Acarina, *J. Anim. Ecol.,* 36, 257, 1967.
198. **Luxton, M.,** The zonation of salt marsh Acarina, *Pedobiologia,* 7, 55, 1967.

199. **Oney, J.,** Final Report: Clapper Rail Survey and Investigation Study, *Georgia Game and Fisheries Commission,* Atlanta, GA, 1954, 50.
200. **Burger, J., Olla, B. J., and Winn, H. E., Eds.,** *Marine Birds,* Plenum Press, New York, 1980.
201. **Burger, J. and Olla, B. L., Eds.,** *Shorebirds: Breeding Behavior and Populations,* Plenum Press, New York, 1984.
202. **Burger, J. and Olla, B. L., Eds.,** *Shorebirds: Migration and Foraging Behavior,* Plenum Press, New York, 1984.
203. **Stewart, R. E.,** Waterfowl populations in the upper Chesapeake region, *U.S. Fish Wildl. Serv. Spec. Sci. Rep. Wildl.,* 65, 208, 1962.
204. **Perry, M. C.,** Waterfowl of Chesapeake Bay, in *Contaminant Problems and Management of Living Chesapeake Bay Resources,* Majumdar, S. K., Hall, L. W., Jr., and Austin, H. M., Eds., Pennsylvania Academy of Science, Easton, 1987, 94.
205. **Smith, T. J., III,** Alteration of salt marsh plant community composition by grazing snow geese, *Holarct. Ecol.,* 6, 204, 1982.
206. **Paradino, J. L. and Handley, C. O., Jr.,** Checklist of mammals of Assateague Island, *Chesapeake Sci.,* 6, 167, 1965.
207. **Shure, D. J.,** Ecological relationships of small mammals in a New Jersey barrier marsh habitat, *J. Mammal.,* 51, 267, 1970.
208. **Rozas, L. P. and Odum, W. E.,** Use of tidal freshwater marshes by fishes and macrofaunal crustaceans along a marsh stream-order gradient, *Estuaries,* 10, 36, 1987.
209. **Simpson, R. L., Good, R. E., Leck, M. A., and Whigham, D. F.,** The ecology of freshwater tidal wetlands, *BioScience,* 33, 255, 1983.
210. **Phillips, R. C. and Menez, E. G.,** *Seagrasses,* Smithsonian Institute Press, Washington, D. C., 1988.
211. **Dennison, W. C. and Alberte, R. S.,** Role of daily light period in the depth distribution of *Zostera marina* (eelgrass), *Mar. Ecol. Prog. Ser.,* 25, 51, 1985.
212. **Ranwell, D. S.,** *Ecology of Salt Marshes and Sand Dunes,* Chapman & Hall, London, 1972.
213. **McRoy, C. P. and Barsdate, R. J.,** Phosphate absorption in eelgrass, *Limnol. Oceanogr.,* 14, 6, 1970.
214. **McRoy, C. P., Barsdate, R. J., and Nebert, M.,** Phosphorus cycling in an eelgrass (*Zostera marina* L.) ecosystem, *Limnol. Oceanogr.,* 17, 58, 1972.
215. **Orth, R. J.,** The importance of sediment stability in seagrass communities, in *Ecology of Marine Benthos,* Coull, B. C., Ed., University of South Carolina Press, Columbia, 1977, 281.
216. **Phillips, R. C. and McRoy, C. P., Eds.,** *Handbook of Seagrass Biology: An Ecosystem Perspective,* Garland STPM Press, New York, 1980.
217. **Short, F. T. and Short, C. A.,** The seagrass filter: purification of estuarine and coastal waters, in *The Estuary as a Filter,* Kennedy, V. S., Ed., Academic Press, New York, 1984, 395.
218. **Mazzella, L. and Alberte, R. S.,** Light adaptation and the role of autotrophic epiphytes in primary production of the temperate seagrass, *Zostera marina* L., *J. Exp. Mar. Biol. Ecol.,* 100, 165, 1986.
219. **Howard, R. K. and Short, F. T.,** Seagrass growth and survivorship under the influence of epiphytic grazers, *Aquat. Bot.,* 24, 287, 1986.
220. **Summerson, H. C. and Peterson, C. H.,** Role of predation in organizing benthic communities of a temperate-zone seagrass bed, *Mar. Ecol. Prog. Ser.,* 15, 63, 1984.
221. **Taylor, J. D. and Lewis, M. S.,** The flora, fauna, and sediments of the marine grass beds of Mabe, Seychelles, *J. Nat. Hist.,* 4, 199, 1970.
222. **Decho, A. W., Hummon, W. D., and Fleeger, J. W.,** Meiofauna-sediment interactions around subtropical seagrass sediments using factor analysis, *J. Mar. Res.,* 43, 237, 1985.
223. **Thayer, G. W., Adams, S. M., and LaCroix, M. W.,** Structural and functional aspects of a recently established *Zostera marina* community, in *Estuarine Research,* Vol. 1, Cronin, L. E., Ed., Academic Press, New York, 1975, 518.
224. **Warme, J. E.,** Paleoecological aspects of a modern coastal lagoon, University of California, *Publ. Geol. Sci.,* 87, 1, 1971.
225. **Reise, K.,** Predation pressure and community structure of an intertidal soft bottom fauna, in *Biology of Benthic Organisms,* Keegan, B. F., Ceidigh, P. O., and Boaden, P. J. S., Eds., Pergamon Press, Elmsford, NY, 1977, 513.
226. **Kikuchi, T.,** An ecological study on animal communities of the *Zostera marina* belt in Tomioka Bay, Amakusa, Kyushu, Publ. *Amakusa Mar. Biol. Lab. Kyushu Univ.,* 1, 1, 1966.
227. **Santos, S. L. and Simon, J. L.,** Distribution and abundance of the polychaetous annelids in a south Florida estuary, *Bull. Mar. Sci.,* 24, 669, 1974.
228. **Brook, I. M.,** Comparative macrofaunal abundance in turtlegrass *(Thalassia testudinum)* communities in south Florida characterized by high blade density, *Bull. Mar. Sci.,* 28, 212, 1978.
229. **Peterson, C. H.,** Clam predation by whelks (*Busycon* spp.): experimental tests of the importance of prey size, prey density, and seagrass cover, *Mar. Biol.,* 66, 159, 1982.
230. **Lewis, F. G., III,** Crustacean epifauna of seagrass and macroalgae in Apalachee Bay, Florida, USA, *Mar. Biol.,* 94, 219, 1987.

231. **Young, D. K., Buzas, M. A., and Young, M. W.,** Species densities of macrobenthos associated with seagrass: a field experimental study of predation, *J. Mar. Res.,* 34, 577, 1976.

232. **Virnstein, R. W.,** The importance of predation by crabs and fishes on benthic infauna in Chesapeake Bay, *Ecology,* 58, 1199, 1977.

233. **Virnstein, R. W.,** Predator caging experiments in soft sediments: caution advised, in *Estuarine Interactions,* Wiley, M. L., Ed., Academic Press, New York, 1978, 261.

234. **Reise, K.,** Experiments on epibenthic predation in the Wadden Sea, *Helgol. Wiss. Meeresunters.,* 31, 51, 1978.

235. **Young, D. K. and Young, M. W.,** Regulation of species densities of seagrass-associated macrobenthos: evidence from field experiments in the Indian River estuary, Florida, *J. Mar. Res.,* 36, 569, 1978.

236. **Holland, A. F., Mountford, N. K., Hiegel, M. H., Kaumeyer, K. R., and Mihursky, J. A.,** Influence of predation on infaunal abundance in upper Chesapeake Bay, U.S.A., *Mar. Biol.,* 57, 221, 1980.

237. **Bell, J. D. and Westoby, M.,** Abundance of macrofauna in dense seagrass is due to habitat preference, not predation, *Oecologia,* 68, 205, 1986.

238. **Zieman, J. C., Iverson, R. L., and Ogden, J. C.,** Herbivory effects on *Thalassia testudinum* leaf growth and nitrogen content, *Mar. Ecol. Prog. Ser.,* 15, 151, 1984.

239. **Stephenson, R. L., Tan, F. C., and Mann, K. H.,** Stable carbon isotope variability in marine macrophytes and its implications for food web studies, *Mar. Biol.,* 81, 223, 1984.

240. **McKenzie, J. A.,** Carbon isotopes and productivity in the lacustrine and marine environment, in *Chemical Processes in Lakes,* Stumm, W., Ed., Wiley-Interscience, New York, 1985, 99.

241. **Stephenson, R. L., Tan, F. C., and Mann, K. H.,** Use of stable carbon isotope ratios to compare plant material and potential consumers in a seagrass bed and a kelp bed in Nova Scotia, Canada, *Mar. Ecol. Prog. Ser.,* 30, 1, 1986.

242. **Nichols, P. D., Klumpp, D. W., and Johns, R. B.,** A study of food chains in seagrass communities. III. Stable carbon isotope ratios, *Aust. J. Mar. Freshwater Res.,* 36, 683, 1985.

243. **Bach, S. D., Thayer, G. W., and LaCroix, M. W.,** Export of detritus from eelgrass *(Zostera marina)* beds near Beaufort, North Carolina, U. S. A., *Mar. Ecol. Prog. Ser.,* 28, 265, 1986.

244. **Thayer, G. W., Kenworthy, W. J., and Fonseca, M. S.,** Ecology of Eelgrass Meadows of the Atlantic Coast: A Community Profile, Publ. No. FWS/OBS-84/02, U.S. Fish and Wildlife Service, Washington, D.C., 1984.

245. **Nienhuis, P. H. and Groenendijk, A. M.,** Consumption of eelgrass *(Zostera marina)* by birds and invertebrates: an annual budget, *Mar. Ecol. Prog. Ser.,* 29, 29, 1986.

246. **Harrison, P. G. and Mann, K. H.,** Detritus formation from eelgrass (*Zostera marina* L.): the relative effects of fragmentation, leaching and decay, *Limnol. Oceanogr.,* 20, 924, 1975.

247. **Newell, S. Y., Fell, J. W., and Miller, C.,** Deposition and decomposition of turtlegrass leaves, *Int. Rev. Gesamten Hydrobiol.,* 71, 363, 1986.

248. **Heck, K. L., Jr. and Thoman, T. A.,** The nursery role of seagrass meadows in the upper and lower reaches of the Chesapeake Bay, *Estuaries,* 7, 70, 1984.

249. **Orth, R. J. and Heck, K. L., Jr.,** Structural components of eelgrass *(Zostera marina)* meadows in the lower Chesapeake Bay — fishes, *Estuaries,* 3, 278, 1980.

250. **Peterson, C. H.,** Enhancement of *Mercenaria mercenaria* densities in seagrass beds: is pattern fixed during settlement season or altered by subsequent differential survival, *Limnol. Oceanogr.,* 31, 200, 1986.

251. **den Hartog, C.,** Structure, function, and classification in seagrass communities, in *Seagrass Ecosystems: A Scientific Perspective,* McRoy, C. P. and Helfferich, C., Eds., Marcel Dekker, New York, 1977, 89.

252. **Fry, B.,** $^{13}C/^{12}C$ ratios and the trophic importance of algae in Florida *Syringodium filiforme* seagrass meadows, *Mar. Biol.,* 79, 11, 1984.

253. **Hootsmans, M. J. M. and Vermaat, J. E.,** The effect of periphyton-grazing by three epifaunal species on the growth of *Zostera marina* L. under experimental conditions, *Aquat. Bot.,* 22, 83, 1985.

254. **Zimmerman, R., Gibson, R., and Harrington, J.,** Herbivory and detritivory among gammaridean amphipods from a Florida seagrass community, *Mar. Biol.,* 54, 41, 1979.

255. **Van Montfrans, J., Wetzel, R. L., and Orth, R. J.,** Epiphyte-grazer relationships in seagrass meadows: consequences for seagrass growth and production, *Estuaries,* 7, 289, 1984.

256. **Borum, J.,** Development of epiphytic communities on eelgrass *(Zostera marina)* along a nutrient gradient in a Danish estuary, *Mar. Biol.,* 87, 211, 1985.

257. **Hellwig-Armonies, M.,** Mobile epifauna on *Zostera marina,* and infauna of its inflorescences, *Helgol. Wiss. Meeresunters.,* 42, 329, 1988.

258. **Hall, M. O. and Bell, S. S.,** Response of small motile epifauna to complexity of epiphytic algae on seagrass blades, *J. Mar. Res.,* 46, 613, 1988.

259. **Thayer, G. W., Stuart, H. H., Kenworthy, W. J., Ustach, J. F., and Hall, A. B.,** Habitat values of salt marshes, mangroves, and seagrasses for aquatic organisms, in *Wetland Functions and Values,* Greeson, P., Clark, J. R., and Clark, J. E., Eds., Proc. Natl. Symp. Wetlands, American Water Resources Association, Minneapolis, 1979, 235.

260. **Patriquin, D. G.,** The origin of nitrogen and phosphorus for growth of the marine angiosperm *Thalassia testudinum, Mar. Biol.,* 15, 35, 1972.

261. **Kirkman, H.,** Community structure in seagrasses in southwestern Australia, *Aquat. Bot.,* 21, 363, 1985.

262. **Thayer, G. W., Wolfe, D. A., and Williams, R. B.,** The impact of man on seagrass systems, *Sci. Am.,* 63, 288, 1975.

263. **Adams, S. M.,** The ecology of eelgrass, *Zostera marina* (L.), fish communities. I. Structural analysis, *J. Exp. Mar. Biol. Ecol.,* 22, 269, 1976.

264. **Hershner, C. and Wetzel, R. L.,** Submerged and emergent aquatic vegetation of the Chesapeake Bay, in *Contaminant Problems and Management of Living Chesapeake Bay Resources,* Majumdar, S. K., Hall, L. W., Jr., and Austin, H. M., Eds., Pennsylvania Academy of Science, Easton, 1987, 116.

265. **Ward, L. G., Kemp, W. M., and Boynton, W. R.,** The influence of waves and seagrass communities on suspended sediment dynamics in an estuarine embayment, *Mar. Geol.,* 59, 85, 1984.

266. **Heck, K. L., Jr. and Orth, R. J.,** Seagrass habitats: the roles of habitat complexity, competition, and predation in structuring associated fish and motile macroinvertebrate assemblages, in *Estuarine Perspectives,* Kennedy, V. S., Ed., Academic Press, New York, 1980, 449.

267. **Macnae, W.,** A general account of the fauna and flora of mangrove swamps and forests in the Indo-West Pacific region, *Adv. Mar. Biol.,* 6, 73, 1985.

268. **Chapman, V. J.,** Mangrove biogeography, in *Proc. Int. Biological Symp. on the Management of Mangroves,* Walsh, G., Snedaker, S., and Teas, H. L., Eds., University of Florida Press, Gainesville, 1975.

269. **Reimold, R. J. and Queen, W. H., Eds.,** *Ecology of Halophytes,* Academic Press, New York, 1974.

270. **Kuenzler, E. J.,** Mangrove swamp systems, in *Coastal Ecological Systems of the United States,* Vol. 1, Odum, H. T., Copeland, B. J., and McMahan, E. A., Eds., Conservation Foundation, Washington, D. C., 1974, 346.

271. **Sherrod, C. L. and McMillan, C.,** Black mangroves, *Avicennia germinans,* in Texas: past and present distribution, *Contrib. Mar. Sci.,* 24, 115, 1981.

272. **Bunt, J. S., Williams, W. T., and Clay, H. T.,** River water salinity and the distribution of mangrove species along several rivers in North Queensland, *Aust. J. Bot.,* 30, 401, 1982.

273. **Walsh, G. E.,** Mangroves, a review, in *Ecology of Halophytes,* Reimold, R. J. and Queen, W. H., Eds., Academic Press, New York, 1974, 51.

274. **Biebl, R.** Protoplasmatisch — okologische untersuchungen an mangrovealgen von Puerto Rico, *Protoplasma,* 55, 572, 1962.

275. **Bunt, J. S., Williams, W. T., and Bunt, E. D.,** Mangrove distribution in relation to tide at the seafront and up rivers, *Aust. J. Mar. Freshwater Res.,* 36, 481, 1985.

276. **Wolanski, E.,** An evaporation-driven salinity maximum zone in Australian tropical estuaries, *Estuarine Coastal Shelf Sci.,* 22, 415, 1986.

277. **Wolanski, E. and Ridd, P.,** Tidal mixing and trapping in mangrove swamps, *Estuarine Coastal Shelf Sci.,* 23, 759, 1986.

278. **Watson, J. G.,** Mangrove forests of the Malay Peninsula, *Malay. For. Rec.,* 6, 1, 1928.

279. **Davis, J. H., Jr.,** The ecology and geological role of mangroves in Florida, *Carnegie Inst. Washington Publ.,* 517 (Tortugas Lab. Pap. 32), 305, 1940.

280. **Chapman, V. J.,** Cambridge University expedition to Jamaica. I. A study of the environment of *Avicennia nitida* Jacq. in Jamaica, *J. Linn. Soc. London Bot.,* 52, 448, 1944.

281. **Almodovar, L. R. and Biebl, R.,** Osmotic resistence of mangrove algae around La Parguera, Puerto Rico, *Rev. Algol. (N.S.),* 6, 203, 1962.

282. **Taylor, W. R.,** Sketch of the character of the marine algal vegetation of the shores of the Gulf of Mexico, *Fish Bull. U.S.,* 55, 177, 1954.

283. **Taylor, W. R.,** Marine algae of the eastern tropical and subtropical coasts of the Americas, Sci. Ser. No. 21, University of Michigan Studies, Ann Arbor, 1960.

284. **Marath, K. V.,** A study of the subterranean algae flora of some mangrove swamps, *J. Indian Soc. Soil Sci.,* 13, 81, 1965.

285. **Tabb, D. C., Dubrow, D. L., and Manning, R. B.,** The Ecology of Northern Florida Bay and Adjacent Estuaries, Tech. Ser. No. 39, Florida Board Conservatory, Tallahassee, 1962.

286. **Golley, F., Odum, H. T., and Wilson, R. F.,** The structure and metabolism of a Puerto Rican red mangrove forest in May, *Ecology,* 43, 9, 1962.

287. **McCall, P. L. and Tevesz, M. J. S., Eds.,** *Animal-Sediment Relations: The Biogenic Alteration of Sediments,* Plenum Press, New York, 1982.

288. **Mattox, N. T.,** Studies on the biology of the edible oyster, *Ostrea rhizophorae* Guilding in Puerto Rico, *Ecol. Monogr.,* 19, 339, 1949.

289. **McConnaughey, B. H. and Zottoli, R.,** *Introduction to Marine Biology,* 4th ed., C. V. Mosby, St. Louis, 1983.

290. **Clough, B. F. and Attiwill, P. M.,** Primary productivity of mangroves, in *Mangrove Ecosystems in Australia: Structure, Function and Management,* Clough, B. F., Ed., Australian National University Press, Canberra, 1982, 213.

291. **Bunt, J. S., Boto, K. G., and Boto, G.,** A survey method for estimating potential levels of mangrove forest primary production, *Mar. Biol.,* 52, 123, 1979.

292. **Boto, K. G. and Bunt, J. S.,** Tidal export of particulate organic matter from a northern Australian mangrove system, *Estuarine Coastal Shelf Sci.,* 13, 247, 1981.

293. **Boto, K. G. and Wellington, J. T.,** Soil characteristics and nutrient status in a northern Australian mangrove forest, *Estuaries,* 7, 61, 1984.

294. **Boto, K. G., Bunt, J. S., and Wellington, J. T.,** Variations in mangrove forest productivity in northern Australia and Papua New Guinea, *Estuarine Coastal Shelf Sci.,* 19, 321, 1984.

295. **Hutchings, P. A. and Saenger, P.,** *The Ecology of Mangroves,* University of Queensland Press, Brisbane, Queensland, Australia, 1987.

296. **Christensen, B.,** Biomass and primary production of *Rhizophora apiculata* B_l in a mangrove in southern Thailand, *Aquat. Bot.,* 4, 43, 1978.

297. **Odum, W. E. and Heald, E. J.,** Trophic analysis of an estuarine mangrove community, *Bull. Mar. Sci.,* 22, 671, 1972.

298. **Lugo, A. E. and Snedaker, S. C.,** The ecology of mangroves, *Annu. Rev. Ecol. Syst.,* 5, 39, 1974.

299. **Day, J. W., Jr., Conner, W. H., Ley-Lou, F., Day, R. H., and Navarro, A. M.,** The productivity and composition of mangrove forests, Laguna de Terminos, Mexico, *Aquat. Bot.,* 27, 267, 1987.

300. **Woodroffe, C. D.,** Studies of a mangrove basin, Tuff Crater, New Zealand. I. Mangrove biomass and the production of detritus, *Estuarine Coastal Shelf Sci.,* 20, 265, 1985.

301. **Woodroffe, C. D.,** Studies of a mangrove basin, Tuff Crater, New Zealand. III. The flux of organic and inorganic particulate matter, *Estuarine Coastal Shelf Sci.,* 20, 447, 1985.

302. **Twilley, R. R.,** The exchange of organic carbon in basin mangrove forests in a southwest Florida estuary, *Estuarine Coastal Shelf Sci.,* 20, 543, 1985.

303. **Heald, E. J.,** The Production of Organic Detritus in a South Florida Estuary, Ph.D. thesis, University of Miami, Miami, FL, 1969.

304. **Twilley, R. R.,** Coupling of mangroves to the productivity of estuarine and coastal waters, in *Coastal-Offshore Ecosystem Interactions,* Jansson, B.-O., Ed., Springer-Verlag, Berlin 1988, 155.

305. **Vernberg, W. B. and Coull, B. C.,** Meiofauna, in *Functional Adaptations of Marine Organisms,* Vernberg, F. J. and Vernberg, W. B., Eds., Academic Press, New York, 1981, 147.

306. **Muus, B.,** The fauna of Danish estuaries and lagoons, *Medd. Dan. Fisk. Havunders.,* 5, 1, 1967.

307. **Fenchel, T.,** The ecology of marine microbenthos. I. The quantitative importance of ciliates as compared with metazoans in various types of sediments, *Ophelia,* 4, 121, 1967.

308. **Fenchel, T.,** The ecology of marine microbenthos. II. The food of marine benthic ciliates, *Ophelia,* 5, 73, 1968.

309. **Fenchel, T.,** The ecology of marine microbenthos. III. The reproductive potential of ciliates, *Ophelia,* 5, 123, 1968.

310. **Hartwig, E.,** Die ciliaten des gezeiten — sandstrandes der Nordseeinsel sylt. I. Systematik, *Mikrofauna Meeresbodens,* 18, 387, 1973.

311. **Hartwig, E.,** Die ciliaten des gezeiten-sandstrandes. II. Okologie, *Mikrofauna Meeresbodens,* 21, 3, 1973.

312. **Dragesco, J.,** Les ciliés mésopsammiques littoraux (systématique, morphologie, écologie), *Trav. Sta. Biol. (Roscoff, N.S.),* 12, 1, 1960.

313. **Fenchel, T.,** Aspects of the decomposition of seagrasses, in *Seagrass Ecosystems: A Scientific Perspective,* McRoy, C. P. and Helfferich, C., Eds., Marcel Dekker, New York, 1977, 123.

314. **Fenchel, T.,** Studies on the decomposition of organic detritus derived from the turtle grass *Thalassia testudinum,* *Limnol. Oceanogr.,* 15, 14, 1970.

315. **Fenchel, T.,** Aspects of decomposer food chains in marine benthos, *Verh. Dtsch. Zool. Ges. 65 Jahresversamml.,* 14, 14, 1972.

316. **Colwell, R. R.,** Bacteria, yeasts, viruses, and related microorganisms of the Chesapeake Bay, *Chesapeake Sci.,* 13(Suppl.), S67, 1972.

317. **Jørgensen, C. B. and Fenchel, T.,** The sulfur cycle of a marine sediment model system, *Mar. Biol.,* 24, 189, 1974.

318. **Rhoads, D. C.,** Organism-sediment relations on the muddy sea floor, *Oceanogr. Mar. Biol. Annu. Rev.,* 12, 263, 1974.

319. **Coull, B. C. and Bell, S. S.,** Perspectives of marine meiofaunal ecology, in *Ecological Processes in Coastal Marine Systems,* Livingston, R. J., Ed., Plenum Press, New York, 1979, 189.

320. **Tietjen, J. H.,** The ecology of shallow water meiofauna in two New England estuaries, *Oecologia,* 2, 251, 1969.

321. **Tietjen, J. H.,** Potential roles of nematodes in polluted ecosystems and the impact of pollution on meiofauna, in *Ecological Stress and the New York Bight: Science and Management,* Mayer, G. F., Ed., Estuarine Research Federation, Columbia, SC, 1982, 225.

322. **Hoffman, J. A., Katz, J., and Bertness, M. D.,** Fiddler crab deposit-feeding and meiofaunal abundance in salt marsh habitats, *J. Exp. Mar. Biol. Ecol.,* 82, 161, 1984.

323. **Bell, S. S.,** Short- and long-term variation in a high marsh meiofauna community, *Estuarine Coastal Mar. Sci.,* 9, 331, 1979.

324. **Dye, A. H. and Furstenburg, J. P.,** Estuarine meiofauna, in *Estuarine Ecology: With Particular Reference to Southern Africa,* Day, J. H., Ed., A. A. Balkema, Rotterdam, 1981, 179.
325. **Platt, H. M., Shaw, K. M., and Lambshead, P. J. D.,** Nematode species abundance patterns and their use in the detection of environmental perturbations, *Hydrobiologia,* 118, 59, 1984.
326. **Heip, C., Vincx, M., and Vranken, G.,** The ecology of marine nematodes, *Oceanogr. Mar. Biol. Annu. Rev.,* 23, 399, 1985.
327. **McIntyre, A. D.,** Ecology of marine meiobenthos, *Biol. Rev.,* 44, 245, 1969.
328. **Kerby, C.,** Life in the bottom-meiobenthos, in *Coastal Ecosystem Management,* Clark, J. R., Ed., John Wiley & Sons, New York, 1977, 656.
329. **Alongi, D. M.,** Intertidal zonation and seasonality of meiobenthos in tropical mangrove estuaries, *Mar. Biol.,* 95, 447, 1987.
330. **Phillips, F. E. and Fleeger, J. W.,** Meiofauna meso-scale variability in two estuarine habitats, *Estuarine Coastal Shelf Sci.,* 21, 745, 1985.
331. **Bell, S. S.,** Meiofauna-macrofauna interactions in a high marsh habitat, *Ecol. Monogr.,* 50, 487, 1980.
332. **Wieser, W.,** The meiofauna as a tool in the study of habitat heterogeneity: ecophysiological aspects, a review, *Cah. Biol. Mar.,* 16, 647, 1975.
333. **Jansson, B. O.,** Microdistribution of factors and fauna in marine sandy beaches, *Veroeff. Inst. Meeresforsch. Bremerhaven,* 2, 77, 1966.
334. **Reid, J. W.,** The Summer Meiobenthos of the Pamlico River Estuary, North Carolina, with Particular Reference to the Harpacticoid Copepods, M.S. thesis, North Carolina State University, Raleigh, 1970.
335. **Brickman, L. M.,** Base Food Chain Relationships in Coastal Salt Marsh Ecosystems, Ph.D. thesis, Lehigh University, Bethlehem, PA, 1972.
336. **Warwick, R. M. and Gee, J. M.,** Community structure of estuarine meiobenthos, *Mar. Ecol. Prog. Ser.,* 18, 97, 1984.
337. **Palmer, M. A.,** Dispersal of marine meiofauna: a review and conceptual model explaining passive transport and active emergence with implications for recruitment, *Mar. Ecol. Prog. Ser.,* 48, 81, 1988.
338. **Boaden, P. J. S. and Erwin, D. G.,** *Turbanella hyalina* versus *Protodriloides symbioticus:* a study in interstitial ecology, *Vie Milieu,* Suppl. 22, 479, 1971.
339. **Perkins, E. J.,** The food relationships of the microbenthos, with particular reference to that found at Whitstable, Kent, *Ann. Mag. Nat. Hist.,* 1, 64, 1958.
340. **Jansson, B. O.,** Quantitative and experimental studies of the interstitial fauna of four Swedish sandy beaches, *Ophelia,* 5, 1, 1968.
341. **Coull, B. C.,** Estuarine meiofauna: a review, trophic relationships and microbial interactions, in *Estuarine Microbiology and Ecology,* Stevenson, L. H. and Colwell, R. R., Eds., University of South Carolina Press, Columbia, 1973, 499.
342. **Dye, A. H. and Furstenburg, J. P.,** An ecophysiological study of the benthic meiofauna of the Swartkops estuary. II. The meiofauna: composition, distribution, seasonal fluctuations, and biomass, *Zool. Afr.,* 13, 19, 1977.
343. **Fenchel, T. and Riedl, R. J.,** The sulfide system: a new biotic community underneath the oxidized layer of marine sand bottoms, *Mar. Biol.,* 7, 255, 1970.
344. **Otl, J. and Schiemer, F.,** Respiration and anaerobiosis of free-living nematodes from marine and limnic sediments, *Neth. J. Sea Res.,* 7, 233, 1973.
345. **Coull, B. C.,** Hydrographic control of meiobenthos in Bermuda, *Limnol. Oceanogr.,* 14, 953, 1969.
346. **Jansson, B. O.,** The importance of tolerance and preference experiments for interpretation of mesopsmmon field experiments, *Helgol. Wiss. Meeresunters.,* 15, 41, 1967.
347. **Gerlach, S. A.,** Attraction to decaying organisms as a possible cause for patchy distribution of nematodes in a Bermuda beach, *Ophelia,* 16, 151, 1977.
348. **Hummon, W. D., Fleeger, J. W., and Hummon, M. R.,** Meiofauna-macrofauna interactions. I. Sand beach meiofauna affected by maturing *Limulus* eggs, *Chesapeake Sci.,* 17, 297, 1976.
349. **Lee, J. J., Tietjen, J. H., Mastropaolo, C., and Rubin, H.,** Food quality and the heterogeneous spatial distribution of meiofauna, *Helgol. Wiss. Meeresunters.,* 30, 272, 1977.
350. **Levin, S. A. and Paine, R. T.,** Disturbance, patch formation, and community structure, *Proc. Natl. Acad. Sci. U.S.A.,* 71, 2744, 1974.
351. **Heip, C. and Engels, P.,** Spatial segregation of copepod species from a brackish water habitat, *J. Exp. Mar. Biol. Ecol.,* 26, 77, 1977.
352. **Warwick, R. M.,** Population dynamics and secondary production of benthos, in *Marine Benthic Dynamics,* Tenore, K. R. and Coull, B. C., Eds., University of South Carolina Press, Columbia, 1980, 1.
353. **Woombs, M. and Laybourn-Parry, J.,** Growth, reproduction and longevity in nematodes from sewage treatment plants, *Oecologia,* 64, 168, 1984.
354. **Coull, B. C. and Dudley, B. W.,** Delayed naupliar development of meiobenthic copepods, *Biol. Bull.,* 150, 38, 1976.

355. **Heip, C. and Smol, N.,** On the importance of *Protohydra leuckarti* as a predator of meiobenthic populations, in *Proc. 10th European Symp. on Marine Biology,* Vol. 2, Persoone, G. and Jaspers, E., Eds., Universa Press, Wetteren, Belgium, 1976, 285.

356. **Gerlach, S. A.,** On the importance of marine meiofauna for benthos communities, *Oecologia,* 6, 176, 1971.

357. **Tietjen, J. H. and Lee, J. J.,** Feeding behavior of marine meiofauna, in *Ecology of Marine Benthos,* Coull, B. C., Ed., University of South Carolina Press, Columbia, 1977, 21.

358. **Heip, C.,** On the evolution of reproductive potentials in a brackish water meiobenthic community, *Microfauna Meeresboden,* 61, 105, 1977.

359. **Vranken, G., Herman, P. M. J., and Heip, C.,** Studies of the life history and energetics of marine and brackish-water nematodes. I. Demography of *Monhystera disjuncta* at different temperature and feeding conditions, *Oecologia,* 77, 296, 1988.

360. **Vranken, G. and Heip, C.,** The productivity of marine nematodes, *Ophelia,* 26, 429, 1986.

361. **Warwick, R. M., Joint, I. R., and Radford, P. J.,** Secondary production of the benthos in an estuarine environment, in *Ecological Processes in Coastal Environments,* Jefferies, R. L. and Davy, A. J., Eds., Blackwell Scientific, Oxford, 1979, 429.

362. **Goodman, K.,** The Ecology of Meiofauna Populations within the Ythan Estuary, Ph.D. thesis, University of Aberdeen, Scotland, 1980.

363. **Feller, R. J.,** Life History and Production of Meiobenthic Harpacticoid Copepods in Puget Sound, Ph.D. thesis, University of Washington, Seattle, 1977.

364. **Heip, C.,** The influence of competition and predation on production of meiobenthic copepods, in *Marine Benthic Dynamics,* Tenore, K. R. and Coull, B. C., Eds., University of South Carolina Press, Columbia, 1980, 167.

365. **Flint, R. W. and Younk, J. A.,** Estuarine benthos: long-term community structure variations, Corpus Christi Bay, Texas, *Estuaries,* 6, 126, 1983.

366. **Holland, A. F.,** Long-term variations of macrobenthos in a mesohaline region of Chesapeake Bay, *Estuaries,* 8, 93, 1985.

367. **Holland, A. F., Shaughnessy, A. T., and Hiegel, M. H.,** Long-term variation in mesohaline Chesapeake Bay macrobenthos: spatial and temporal patterns, *Estuaries,* 10, 227, 1987.

368. **Nichols, F. H.,** Abundance fluctuations among benthic invertebrates in two Pacific estuaries, *Estuaries,* 8, 136, 1985.

369. **Nichols, F. H. and Thompson, J. K.,** Time scales of change in the San Francisco Bay benthos, *Hydrobiologia,* 129, 121, 1985.

370. **Beukema, J. J. and Essink, K.,** Common patterns in the fluctuations of macrozoobenthic species living at different places on tidal flats in the Wadden Sea, *Hydrobiologia,* 142, 199, 1986.

371. **Coull, B. C.,** Long-term variability of meiobenthos: value, synopsis hypothesis generation and predictive modelling, in *Long-Term Changes in Coastal Benthic Communities,* Heip, C., Keegan, B. F., and Lewis, J. R., Eds., Dr. W. Junk, Dordrecht, 1987, 271.

372. **Franz, D. R. and Harris, W. H.,** Seasonal and spatial variability in macrobenthos communities in Jamaica Bay, New York — an urban estuary, *Estuaries,* 11, 15, 1988.

373. **Pearson, T. H. and Rosenberg, R.,** Macrobenthic succession in relation to organic enrichment and pollution of the marine environment, *Oceanogr. Mar. Biol. Annu. Rev.,* 16, 229, 1978.

374. **Gray, J. S.,** Pollution-induced changes in populations, *Philos. Trans. R. Soc. London Ser. B,* 286, 545, 1979.

375. **Grizzle, R. E.,** Pollution indicator species of macrobenthos in a coastal lagoon, *Mar. Ecol. Prog. Ser.,* 18, 191, 1984.

376. **Henderson, A. R.,** Long-term monitoring of the macrobenthos of the upper Clyde estuary, *Water Sci. Technol.,* 16, 359, 1984.

377. **Essink, K. and Beukema, J. J.,** Long-term changes in intertidal flat macrozoobenthos as an indicator of stress by organic pollution, *Hydrobiologia,* 142, 209, 1986.

378. **Luoma, S. N. and Cloern, J. E.,** The impact of waste-water discharge on biological communities in San Francisco Bay, in *San Francisco Bay: Use and Protection,* Kockelman, W. J., Conomos, T. J., and Leviton, A. E., Eds., American Association for the Advancement of Science, Pacific Division, San Francisco, CA, 1982, 137.

379. **Poff, N. L. and Matthews, R. A.,** Benthic macroinvertebrate community structural and functional group response to thermal enhancement in the Savannah River and a coastal plain tributary, *Arch. Hydrobiol.,* 106, 119, 1986.

380. **Rygg, B.,** Effect of sediment copper on benthic fauna, *Mar. Ecol. Prog. Ser.,* 25, 83, 1985.

381. **Jacobs, R. P. W. M.,** Effects of the "Amoco Cadiz" oil spill on the seagrass community at Roscoff with special reference to the benthic infauna, *Mar. Ecol. Prog. Ser.,* 2, 207, 1980.

382. **Sanders, H. L., Grassle, J. F., Hampson, G. R., Morse, L. S., Garner-Price, S., and Jones, C. C.,** Anatomy of an oil spill: long-term effects from the grounding of the barge Florida off West Falmouth, Massachusetts, *J. Mar. Res.,* 38, 265, 1980.

383. **Hellawell, J. M.,** *Biological Surveillance of Rivers,* Water Research Centre, Stevenage, England, 1978.
384. **Slack, K. V., Ferreira, R. F., and Averett, R. C.,** Comparison of four artificial substrates and the Ponar grab for benthic invertebrate collection, *Water Resour. Bull.,* 22, 237, 1986.
385. **Gaston, G. R. and Nasci, J. C.,** Trophic structure of macrobenthic communities in the Calcasieu estuary, Louisiana, *Estuaries,* 11, 201, 1988.
386. **Watling, L.,** Analysis of structural variations in a shallow estuarine deposit-feeding community, *J. Exp. Mar. Biol. Ecol.,* 19, 275, 1975.
387. **Boesch, D. F., Waas, M. L., and Virnstein, R. W.,** The dynamics of estuarine benthic communities, in *Estuarine Processes,* Vol. 1, Wiley, M., Ed., Academic Press, New York, 1976, 176.
388. **Buchanan, J. B., Sheader, M., and Kingston, P. F.,** Sources of variability in the benthic macrofauna off the south Northumberland Coast, 1971—1976, *J. Mar. Biol. Assoc. U.K.,* 58, 191, 1978.
389. **Price, R. and Warwick, R. M.,** Temporal variations in annual production and biomass in estuarine populations of two polychaetes, *Nephtys hombergi* and *Ampharete acutifrons, J. Mar. Biol. Assoc. U.K.,* 60, 481, 1980.
390. **Arntz, W. E. and Rumohr, H.,** Fluctuations of benthic macrofauna during succession and in an established community, *Sonderdruck Aus Bd.,* 31, 97, 1986.
391. **Wolff, W. J.,** Estuarine benthos, in *Estuaries and Enclosed Seas,* Ketchum, B. H., Ed., Elsevier, Amsterdam, 1983, 151.
392. **Stephenson, T. A. and Stephenson, A.,** *Life between Tidemarks on Rocky Shores,* W. H. Freeman, San Francisco, 1972.
393. **Denny, M. W.,** Wave forces on intertidal organisms: a case study, *Limnol. Oceanogr.,* 30, 1171, 1985.
394. **McQuaid, C. D. and Branch, G. M.,** Trophic structure of rocky intertidal communities: response to wave action and implications for energy flow, *Mar. Ecol. Prog. Ser.,* 22, 153, 1985.
395. **Jumars, P. A. and Nowell, A. R. M.,** Fluid and sediment dynamic effects on marine benthic community structure, *Am. Zool.,* 24, 45, 1984.
396. **Carriker, M. R.,** Ecology of estuarine benthic invertebrates: a perspective, in *Estuaries,* Lauff, G. H., Ed., Publ. 83, American Association for the Advancement of Science, Washington, D. C., 1967, 442.
397. **Gambi, M. C. and Giangrande, A.,** Distribution of soft-bottom polychaetes in two coastal areas of the Tyrrhenian Sea (Italy): structural analysis, *Estuarine Coastal Shelf Sci.,* 23, 847, 1986.
398. **Thrush, S. F. and Townsend, C. R.,** The sublittoral macrobenthic community of Lough Hyne, Ireland, *Estuarine Coastal Shelf Sci.,* 23, 551, 1986.
399. **Longbottom, M. R.,** The distribution of *Arenicola marina* L., with particular reference to the effects of particle size and organic matter in sediments, *J. Exp. Mar. Biol. Ecol.,* 5, 138, 1970.
400. **Newell, R.,** The role of detritus in the nutrition of two marine deposit-feeders, the prosobranch *Hydrobia ulvae* and the bivalve *Macoma balthica, Proc. Zool. Soc. London,* 1, 25, 1965.
401. **Fenchel, T. and Jansson, B. O.,** On the vertical distribution of the microfauna in the sediments of a brackish-water beach, *Ophelia,* 3, 161, 1966.
402. **Boaden, P. J. S.,** Anaerobiosis, meiofauna, and early metazoan evolution, *Zool. Scr.,* 4, 21, 1975.
403. **Fox, C. A. and Powell, E. N.,** Meiofauna and the sulfide system: the effects of oxygen and sulfide on the adenylate pool of three turbellarians and a gastrotrich, *Comp. Biochem. Physiol. A,* 85, 37, 1986.
404. **Brenchley, B. A.,** Disturbance and community structure: an experimental study of bioturbation in marine soft-bottom environments, *J. Mar. Res.,* 39, 767, 1981.
405. **Rhoads, D. C. and Boyer, L. F.,** The effects of marine benthos on physical properties of sediments: a successional perspective, in *Animal-Sediment Relations: The Biogenic Alteration of Sediments,* McCall, P. L. and Tevesz, M. J. S., Eds., Plenum Press, New York, 1982, 3.
406. **Rhoads, D. C. and Germano, J. D.,** Interpreting long-term changes in benthic community structure: a new protocol, *Hydrobiologia,* 142, 291, 1986.
407. **Hines, A. H. and Comtois, K. L.,** Vertical distribution of infauna in sediments of a subestuary of central Chesapeake Bay, *Estuaries,* 8, 296, 1985.
408. **Virnstein, R. W.,** Predation on estuarine infauna: response patterns of component species, *Estuaries,* 2, 69, 1979.
409. **Commito, J. A.,** The importance of predation by infaunal polychaetes in controlling the structure of a soft-bottom community in Maine, USA, *Mar. Biol.,* 68, 77, 1982.
410. **Ambrose, W. G., Jr.,** Role of predatory infauna in structuring marine soft-bottom communities, *Mar. Ecol. Prog. Ser.,* 17, 109, 1984.
411. **Dayton, P. K.,** Competition, disturbance, and community organization: the provision and subsequent utilization of space in a rocky intertidal community, *Ecol. Monogr.,* 41, 351, 1971.
412. **Rhoads, D. C., McCall, P. L., and Yingst, J. Y.,** Disturbance and production on the estuarine seafloor, *Am. Sci.,* 66, 577, 1978.
413. **Germano, J. D.,** Infaunal Succession in Long Island Sound: Animal-Sediment Interactions and the Effects of Predation, Ph.D. thesis, Yale University, New Haven, CT, 1983.

414. **Whitman, J. D.,** Refuges, biological disturbance, and rocky subtidal community structure in New England, *Ecol. Monogr.,* 55, 421, 1985.
415. **Johnson, R. G.,** Conceptual models of benthic marine communities, in *Models in Paleobiology,* Schopf, T. J. M., Ed., Freeman, Cooper, San Francisco, 1972, 148.
416. **Gallagher, E. D., Jumars, P. A., and Trueblood, D. D.,** Facilitation of soft-bottom benthic succession by tube builders, *Ecology,* 64, 1200, 1983.
417. **Sousa, W. P.,** The role of disturbance in natural communities, *Annu. Rev. Ecol. Syst.,* 15, 353, 1984.
418. **Sousa, W. P.,** Disturbance and patch dynamics on rocky intertidal shores, in *The Ecology of Natural Disturbance and Patch Dynamics,* Pickett, S. T. A. and White, P. S., Eds., Academic Press, Orlando, FL, 1985, 101.
419. **Connell, J. H. and Keough, M. J.,** Disturbance and patch dynamics of subtidal marine animals on hard substrata, in *The Ecology of Natural Disturbance and Patch Dynamics,* Pickett, S. T. A. and White, P. S., Eds., Academic Press, Orlando, FL, 1985, 125.
420. **Eckman, J. E.,** Small-scale patterns and processes in a soft substratum intertidal community, *J. Mar. Res.,* 37, 437, 1979.
421. **Rogal, U., Anger, K., Schriver, G., and Valentin, C.,** *In situ* investigations on small-scale local and short-term changes in sublittoral macrobenthos in Lubeck Bay (Western Baltic Sea), *Helgol. Wiss. Meeresunters.,* 31, 300, 1978.
422. **Rhoads, D. C. and Germano, J. D.,** Characterization of organism-sediment relations using sediment profile imaging: an efficient method of remote ecological monitoring of the seafloor (REMOTS® System), *Mar. Ecol. Prog. Ser.,* 8, 115, 1982.
423. **Osman, R. W. and Whitlatch, R. B.,** Patterns of species diversity: fact or artifact?, *Paleobiology,* 4, 41, 1978.
424. **McCall, P. L.,** Community patterns and adaptive strategies of the infaunal benthos of Long Island Sound, *J. Mar. Res.,* 35, 221, 1977.
425. **Moore, H. B.,** The muds of the Clyde Sea area. III. Chemical and physical conditions; rate and nature of sedimentation; and fauna, *J. Mar. Biol. Assoc. U.K.,* 17, 325, 1931.
426. **Sanders, H. L.,** Benthic studies in Buzzards Bay. I. Animal-sediment relationships, *Limnol. Oceanogr.,* 3, 245, 1958.
427. **Rhoads, D. C. and Young, D. K.,** The influence of deposit-feeding organisms on sediment stability and community trophic structure, *J. Mar. Res.,* 28, 150, 1970.
428. **Rhoads, D. C. and Young, D. K.,** Animal-sediment relations in Cape Cod Bay, Massachusetts. II. Reworking by *Molpadia oolitica* (Halothuroidea), *Mar. Biol.,* 11, 255, 1971.
429. **Young, D. K. and Rhoads, D. C.,** Animal-sediment relations in Cape Cod Bay, Massachusetts. I. A transect study, *Mar. Biol.,* 11, 242, 1971.
430. **Tenore, K. R., Boyer, L. F., Cal, R. M., Corral, J., Garcia-Fernandez, C., Gonzalez, N., Gonzalez-Gurriaran, E., Hanson, R. B., Iglesias, J., Krom, M., Lopez-Jamar, E., McClain, J., Pamatmat, M. M., Perez, A., Rhoads, D. C., de Santiago, G., Tietjen, J., Westrich, J., and Windom, H. L.,** Coastal upwelling in the Rias Bajas, N. W. Spain: contrasting the benthic regimes of the Rias de Arosa and de Muros, *J. Mar. Res.,* 40, 701, 1982.
431. **Wade, B.,** A description of a highly diverse soft-bottom community in Kingston Harbour, Jamaica, *Mar. Biol.,* 13, 57, 1972.
432. **Peterson, C. H.,** Predation, competitive exclusion and diversity in the soft-bottom communities of estuaries and lagoons, in *Ecological Processes in Coastal and Marine Systems,* Livingston, R. J., Ed., Plenum Press, New York, 1979, 233.
433. **Woodin, S. A.,** Biotic interactions in recent marine sedimentary environments, in *Biotic Interactions in Recent and Fossil Benthic Communities,* Tevesz, M. J. S. and McCall, P. L., Eds., Plenum Press, New York, 1983, 3.
434. **Reise, K.,** Predator control in marine tidal sediments, in *Proc. 19th European Marine Biology Symp.,* Gibbs, P. E., Ed., Cambridge University Press, Cambridge, 1985, 311.
435. **Schubert, A. and Reise, K.,** Predatory effects of *Nephtys hombergi* on other polychaetes in tidal flat sediments, *Mar. Ecol. Prog. Ser.,* 34, 117, 1986.
436. **Zajac, R. N. and Whitlatch, R. B.,** Population ecology of the polychaete *Nephtys incisa* in Long Island Sound and the effects of disturbance, *Estuaries,* 11, 117, 1988.
437. **Woodin, S. A.,** Refuges, disturbance, and community structure: a marine soft-bottom example, *Ecology,* 59, 274, 1978.
438. **Woodin, S. A.,** Disturbance and community structure in a shallow water sand flat, *Ecology,* 62, 1052, 1981.
439. **Jackson, J. B. C.,** Competition on marine hard substrata: the adaptative significance of solitary and colonial strategies, *Am. Nat.,* 111, 743, 1977.
440. **Diamond, J. M.,** Niche shifts and the rediscovery of interspecific competition, *Am. Sci.,* 66, 322, 1978.
441. **Wilson, W. H., Jr.,** The role of density dependence in a marine infaunal community, *Ecology,* 64, 295, 1983.
442. **Fletcher, W. T. and Creese, R. G.,** Competitive interactions between co-occurring herbivorous gastropods, *Mar. Biol.,* 86, 183, 1985.

443. **Lopez Gappa, J. J.,** Overgrowth competition in an assemblage of encrusting bryozoans settled on artificial substrata, *Mar. Ecol. Prog. Ser.,* 51, 121, 1989.

444. **Connell, J. H.,** Some mechanisms producing structure in natural communities: a model and evidence from field experiments, in *Ecology and Evolution of Communities,* Cody, M. L. and Diamond, J., Eds., Belknap Press, Cambridge, 1975, 460.

445. **Paine, R. T.,** Food webs: linkage, interaction strength, and community infrastructure, *J. Anim. Ecol.,* 49, 667, 1980.

446. **Mook, D.,** Responses of common fouling organisms in the Indian River, Florida, to various predation and disturbance intensities, *Estuaries,* 6, 372, 1983.

447. **Gee, J. M., Warwick, R. M., Davey, J. T., and George, C. L.,** Field experiments on the role of epibenthic predators in determining prey densities in an estuarine mudflat, *Estuarine Coastal Shelf Sci.,* 21, 429, 1985.

448. **Jensen, K. T. and Jensen, J. N.,** The importance of some epibenthic predators on the density of juvenile benthic macrofauna in the Danish Wadden Sea, *J. Exp. Mar. Biol. Ecol.,* 89, 157, 1985.

449. **Wilson, W. H., Jr.,** Importance of predatory infauna in marine soft-sediment communities, *Mar. Ecol. Prog. Ser.,* 32, 35, 1986.

450. **Lewin, R.,** Predators and hurricanes change ecology, *Science,* 221, 737, 1983.

451. **Flint, R. W. and Kalke, R. D.,** Biological enhancement of estuarine benthic community structure, *Mar. Ecol. Prog. Ser.,* 31, 23, 1986.

452. **Brault, S. and Bourget, E.,** Structural changes in an estuarine subtidal epibenthic community: biotic and physical causes, *Mar. Ecol. Prog. Ser.,* 21, 63, 1985.

453. **Reise, K.,** Predator exclusion experiments in an intertidal mud flat, *Helgol. Wiss. Meeresunters.,* 30, 263, 1977.

454. **Virnstein, R. W.,** Predation on estuarine infauna: patterns of component species, *Estuaries,* 2, 69, 1979.

455. **Reise, K.,** Moderate predation on meiofauna by the macrobenthos of the Wadden Sea, *Helgol. Wiss. Meeresunters.,* 32, 453, 1979.

456. **Wolff, W. J., Ed.,** *Ecology of the Wadden Sea,* Vol. 1, A. A. Balkema, Rotterdam, 1983.

457. **Botton, M. L.,** The importance of predation by horseshoe crabs, *Limulus polyphemus,* to an intertidal sand flat community, *J. Mar. Res.,* 42, 139, 1984.

458. **Commito, J. A. and Ambrose, W. G., Jr.,** Multiple trophic levels in soft-bottom communities, *Mar. Ecol. Prog. Ser.,* 26, 289, 1985.

459. **Whittaker, R. H.,** Gradient analysis of vegetation, *Biol. Rev.,* 42, 207, 1967.

460. **Boesch, D. F.,** A new look at the zonation of benthos along the estuarine gradient, in *Ecology of Marine Benthos,* Coull, B. C., Ed., University of South Carolina Press, Columbia, 1977, 245.

461. **Wolff, W. J.,** The estuary as a habitat: an analysis of data on the soft-bottom macrofauna of the estuarine area of the Rivers Rhine, Meuse, and Scheldt, *Zool. Verh.,* p. 126, 1973.

462. **McPherson, B. F., Sonntag, W. H., and Sabanskas, M.,** Fouling community of the Loxahatchee River estuary, Florida, 1980—81, *Estuaries,* 7, 149, 1984.

463. **Vernberg, F. J.,** Benthic macrofauna, in *Functional Adaptations of Marine Organisms,* Vernberg, F. J. and Vernberg, W. B., Eds., Academic Press, New York, 1981, 179.

464. **Wells, H. W.,** The fauna of oyster beds, with special reference to the salinity factor, *Ecol. Monogr.,* 31, 239, 1961.

465. **Gilles, R. and Jeuniaux, C.,** Osmoregulation and ecology in media of fluctuating salinity, in *Mechanisms of Osmoregulation in Animals,* Gilles, R., Ed., John Wiley & Sons, New York, 1979, 581.

466. **Pequeux, A., Gilles, R., and Bolis, L., Eds.,** *Osmoregulation in Estuarine and Marine Animals,* Springer-Verlag, New York, 1984.

467. **Drouin, G., Himmelman, J. H., and Béland, P.,** Impact of tidal salinity fluctuations on echinoderm and mollusc populations, *Can. J. Zool.,* 63, 1377, 1985.

468. **Schoffeniels, E. and Gilles, R.,** Ionregulation and osmoregulation in mollusca, in *Chemical Zoology,* Florkin, M., Ed., Academic Press, London, 1972, 393.

469. **Binyon, J.,** Salinity tolerance and ionic regulation, in *Physiology of Echinodermata,* Boolotian, R. A., Ed., John Wiley & Sons, New York, 1966, 359.

470. **Flint, R. W. and Kalbe, R. D.,** Benthos structure and function in a south Texas estuary, *Contrib. Mar. Sci.,* 28, 33, 1985.

471. **Richter, W.,** Distribution of the soft-bottom macroinfauna in an estuary of southern Chile, *Mar. Biol.,* 86, 93, 1985.

472. **Thorson, G.,** Reproduction and larval development of Danish marine bottom invertebrates with special reference to the plankton larvae in the sound (Øresund), *Medd. Komm. Dan. Fisk. Havunders. Ser. Plankton,* 4, 1, 1946.

473. **Thorson, G.,** Reproductive and larval ecology of marine bottom invertebrates, *Biol. Rev.,* 25, 1, 1950.

474. **Grahame, J. and Branch, G. M.,** Reproductive patterns of marine invertebrates, *Oceanogr. Mar. Biol. Annu. Rev.,* 23, 373, 1985.

475. **Day, R. and McEdward, L.,** Aspects of the physiology and ecology of pelagic larvae of marine benthic invertebrates, in *Marine Plankton Life Cycle Strategies,* Steidinger, K. A. and Walker, L. M., Eds., CRC Press, Boca Raton, FL, 1984, 93.

476. **Scheltema, R. S.,** The dispersal of the larvae of shoal-water benthic invertebrate species over long distances by ocean currents, in *4th European Marine Biology Symp.,* Crisp, D. J., Ed., Cambridge University Press, Cambridge, 1971, 7.

477. **Hines, A. H.,** Larval problems and perspectives in life histories of marine invertebrates, *Bull. Mar. Sci.,* 39, 506, 1986.

478. **Mileikovsky, S. A.,** Types of larval development in marine bottom invertebrates, their distribution and ecological significance: a reevaluation, *Mar. Biol.,* 10, 193, 1971.

479. **Mileikovsky, S. A.,** Types of larval development in marine bottom invertebrates: an integrated ecological scheme, *Thalassia Jugosl.,* 10, 171, 1974.

480. **Chia, F. S.,** Classification and adaptive significance of developmental patterns in marine invertebrates, *Thalassia Jugosl.,* 10, 121, 1974.

481. **Crisp, D. J.,** Energy relations of marine invertebrate larvae, *Thalassia Jugosl.,* 10, 103, 1974.

482. **Crisp, D. J.,** The role of the pelagic larva, in *Perspectives in Experimental Zoology,* Spencer-Davies, T., Ed., Pergamon Press, Oxford, 1976, 145.

483. **Jablonski, D.,** Larval ecology and macroevolution in marine invertebrates, *Bull. Mar. Sci.,* 39, 565, 1986.

484. **Jablonski, D. and Lutz, R. A.,** Larval ecology of marine benthic invertebrates: paleobiological implications, *Biol. Rev.,* 58, 21, 1983.

485. **Scheltema, R. S.,** On dispersal and planktonic larvae of benthic invertebrates: an eclectic overview and summary of problems, *Bull. Mar. Sci.,* 39, 290, 1986.

486. **Thorson, G.,** Length of pelagic larval life in the marine bottom invertebrates as related to larval transport by ocean currents, in *Oceanography,* Sears, M., Ed., American Association for the Advancement of Science, Washington, D. C., 1961, 455.

487. **Scheltema, R.,** Dispersal of marine invertebrate organisms: paleobiogeographic and biostratigraphic implications, in *Concepts and Methods of Biostratigraphy,* Kaufman, E. G. and Hazel, J. E., Eds., Dowden, Hutchinson, and Ross, Stroudsburg, PA, 1977, 73.

488. **Scheltema, R.,** On the relationship between dispersal of pelagic veliger larvae and the evolution of marine prosobranch gastropods, in *Marine Organisms Genetics, Ecology and Evolution,* Battaglia, B. and Beardmore, J., Eds., Plenum Press, New York, 1978, 303.

489. **Crisp, D. J.,** Genetic consequence of different reproductive strategies in marine invertebrates, in *Marine Organisms Genetics, Ecology and Evolution,* Battaglia, B. and Beardmore, J., Eds., Plenum Press, New York, 1978, 257.

490. **Bousfield, E. L.,** Ecological control of occurrence of barnacles in the Miramichi estuary, *Bull. Natl. Mus. Can.,* 137, 69, 1955.

491. **Sandifer, P. A.,** Distribution and abundance of decapod larvae in the York River estuary and adjacent lower Chesapeake Bay, Virginia, *Chesapeake Sci.,* 14, 235, 1973.

492. **Sandifer, P. A.,** The role of the pelagic larvae in recruitment to populations of adult crustaceans in the York River estuary and adjacent lower Chesapeake Bay, Virginia, *Estuarine Coastal Mar. Sci.,* 3, 269, 1975.

493. **Sulkin, S. D. and Van Heukelem, W. F.,** Ecological and evolutionary significance of nutritional flexibility in planktotrophic larvae of the deep sea red crab *Geryon quinquedens* and the stone crab *Menippe Mercenaria, Mar. Ecol. Prog. Ser.,* 2, 91, 1980.

494. **Norcross, B. L. and Shaw, R. F.,** Oceanic and estuarine transport of fish eggs and larvae: a review, *Trans. Am. Fish. Soc.,* 113, 153, 1984.

495. **Stancyk, S. E. and Feller, R. J.,** Transport of non-decapod invertebrate larvae in estuaries: an overview, *Bull. Mar. Sci.,* 39, 257, 1986.

496. **Roberts, M. H.,** Larval development of *Pagurus longicarpus* Say reared in the laboratory. III. Behavioral responses to salinity discontinuities, *Biol. Bull.,* 140, 489, 1971.

497. **Sulkin, S. D.,** Depth regulation of crab larvae in the absence of light, *J. Exp. Mar. Biol. Ecol.,* 13, 73, 1973.

498. **Sulkin, S. D., Van Heukelem, W., Kelly, P., and Van Heukelem, L.,** The behavioral basis of larval recruitment in the crab *Callinectes sapidus* Rathbun: a laboratory investigation of ontogenetic changes in geotaxis and barokinesis, *Biol. Bull.,* 159, 402, 1980.

499. **Todd, C. D. and Doyle, D. W.,** Reproductive strategies of marine benthic invertebrates: a settlement-timing hypothesis, *Mar. Ecol. Prog. Ser.,* 4, 75, 1981.

500. **Meadows, P. S. and Campbell, J. I.,** Habitat selection by aquatic invertebrates, *Adv. Mar. Biol.,* 10, 271, 1972.

501. **Crisp, D. J.,** Overview of research on marine invertebrate larvae, 1940—1980, in *Marine Biodeterioration: An Interdisciplinary Study,* Costlow, J. D. and Tipper, R. C., Eds., Naval Institute Press, Annapolis, MD, 1984, 103.

502. **Chia, F. S. and Rice, M. E., Eds.,** *Settlement and Metamorphosis of Marine Invertebrates,* Elsevier/North Holland, New York, 1978.

503. **Morse, D. E.,** Biochemical control of larval recruitment and marine fouling, in *Marine Biodeterioration: An Interdisciplinary Study*, Costlow, J. D. and Tipper, R. C., Eds., Naval Institute Press, Annapolis, MD, 1984, 134.

504. **Thorson, G.,** Light as an ecological factor in the dispersal and settlement of larvae of marine bottom invertebrates, *Ophelia*, 1, 167, 1964.

505. **Forward, R. B.,** The occurrence of a shadow response among brachyuran larvae, *Mar. Biol.*, 39, 331, 1977.

506. **Wheeler, D. E. and Epifanio, C. E.,** Behavioral responses to hydrostatic pressure in larvae of two species of xanthid crabs, *Mar. Biol.*, 46, 167, 1978.

507. **Crisp, D. J. and Meadows, P. S.,** The chemical basis of gregariousness in Cirripedes, *Proc. R. Soc. London Ser. B*, 156, 500, 1962.

508. **Maki, J. S. and Mitchell, R.,** Involvement of lectins in the settlement and metamorphosis of marine invertebrate larvae, *Bull. Mar. Sci.*, 37, 675, 1985.

509. **Morse, D. E.,** Neurotransmitter-mimetic inducers of larval settlement and metamorphosis, *Bull. Mar. Sci.*, 37, 697, 1985.

510. **Scheltema, R. S.,** Metamorphosis of the veliger larvae of *Nassarius obsoletus* (Gastropoda) in response to bottom sediment, *Biol. Bull.*, 120, 92, 1961.

511. **Hadfield, M. G. and Scheurer, D.,** Evidence for a soluble metamorphic inducer in *Phestilla:* ecological, chemical and biological data, *Bull. Mar. Sci.*, 37, 556, 1985.

512. **Hidu, H., Valleau, W. G., and Veitch, F. P.,** Gregarious setting in European oysters — a response to surface chemistry vs. waterborne pheromones, *Proc. Natl. Shellfish. Assoc.*, 68, 111, 1978.

513. **Coon, S. L. and Bonar, D. B.,** Induction of settlement and metamorphosis of the Pacific oyster *Crassostrea gigas* (Thunberg), by L-dopa and catecholamines, *J. Exp. Mar. Biol. Ecol.*, 94, 211, 1985.

514. **Cuomo, M. C.,** Sulphide as a larval settlement cue for *Capitella* sp. I, *Biogeochemistry*, 1, 169, 1985.

515. **Scheltema, R. S.,** Biological interactions determining larval settlement of marine invertebrates, *Thalassia Jugosl.*, 10, 263, 1974.

516. **Burke, R. D.,** The induction of metamorphosis of marine invertebrate larvae: stimulus and response, *Can. J. Zool.*, 61, 1701, 1983.

517. **Karlson, P. and Luscher, M.,** 'Pheromones': a new term for a class of biologically active substances, *Nature*, 183, 55, 1959.

518. **Burke, R. D.,** Pheromones and the gregarious settlement of marine invertebrate larvae, *Bull. Mar. Sci.*, 39, 323, 1986.

519. **Scheltema, R. S., Williams, I. P., Shaw, M. A., and Loudon, C.,** Gregarious settlement by the larvae of *Hydroides dianthus* (Polychaeta: Serpulidae), *Mar. Ecol. Prog. Ser.*, 5, 69, 1981.

520. **Butman, C. A.,** Larval settlement of soft-sediment invertebrates: the spatial scales of pattern explained by active habitat selection and the emerging role of hydrodynamical processes, *Oceanogr. Mar. Biol. Annu. Rev.*, 25, 113, 1987.

521. **Pechenik, J. A.,** The relationship between temperature, growth rate, and duration of planktonic life for larvae of the gastropod *Crepidula fornicata* (L.), *J. Exp. Mar. Biol. Ecol.*, 74, 241, 1984.

522. **Pechenik, J. A.,** Influence of temperature and temperature shifts on the development of chiton larvae, *Mopalia mucosa*, *Int. J. Invert. Reprod. Dev.*, 7, 3, 1984.

523. **Olson, R. R.,** The consequence of short-distance larval dispersal in a sessile marine invertebrate, *Ecology*, 66, 30, 1985.

524. **Scheltema, R. S. and Carlton, J. T.,** Method of dispersal among fouling organisms and possible consequences for range extension and geographical variation, in *Marine Biodeterioration: An Interdisciplinary Study*, Costlow, J. D. and Tipper, R. C., Eds., Naval Institute Press, Annapolis, MD, 1984, 127.

525. **De Lattin, G.,** *Grundriss der Zoogeographie*, Gustav Fisher, Jena, 1967.

526. **Maurer, D. and Watling, L.,** Studies on the oyster community in Delaware: the effects of the estuarine environment on the associated fauna, *Int. Rev. Gesamten Hydrobiol.*, 58, 161, 1973.

527. **Tenore, K. R.,** Macrobenthos of Pamlico River estuary, North Carolina, *Ecol. Monogr.*, 42, 51, 1972.

528. **Wass, M. L.,** A Check List of the Biota of Lower Chesapeake Bay with Inclusions from the Upper Bay and the Virginia Sea, Spec. Sci. Rep. No. 65, Virginia Institute of Marine Science, Gloucester Point, 1972.

529. **Day, J. H.,** The origin and distribution of estuarine animals in South Africa, *Monogr. Biol.*, 14, 159, 1964.

530. **Day, J. H.,** The ecology of Morrumbene estuary, Mocambique, *Trans. R. Soc. S. Afr.*, 41, 43, 1974.

531. **Carlton, J. T.,** Introduced invertebrates of San Francisco Bay, in *San Francisco Bay: The Urbanized Estuary*, Conomos, T. J., Ed., Pacific Division, American Association for the Advancement of Science, San Francisco, 1979, 427.

532. **Carlton, J. T.,** Transoceanic and interoceanic dispersal of coastal marine organisms: the biology of ballast water, *Oceanogr. Mar. Biol. Annu. Rev.*, 23, 313, 1985.

533. **Wolfe, D. A., Champ, M. A., Flemer, D. A., and Mearns, A. J.,** Long-term biological data sets: their role in research, monitoring, and management of estuarine and coastal marine systems, *Estuaries*, 10, 181, 1987.

534. **Holland, A. F. and Kjerfve, B.,** Preface: long-term estuarine and coastal data sets, *Estuaries*, 8(2A), 81, 1985.

535. **Phillips, F. X.,** The Ecology of the Benthic Macroinvertebrates of Barnegat Bay, New Jersey, Ph.D. thesis, Rutgers University, New Brunswick, NJ, 1972.

536. **Loveland, R. E., Moul, E. T., Busch, D. A., Sandine, P. H., Shafto, S. S., and McCarty, J.,** The Qualitative and Quantitative Analysis of the Benthic Flora and Fauna of Barnegat Bay before and after the Onset of Thermal Addition, Unpubl. Tech. Rep., Rutgers University, New Brunswick, NJ, 1972, 81 pp.

537. **Loveland, R. E., Edwards, P., Vouglitois, J. J., and Palumbo, D.,** The Qualitative and Quantitative Analysis of the Benthic Flora and Fauna of Barnegat Bay before and after the Onset of Thermal Addition, Unpubl. Tech. Rep., Rutgers University, New Brunswick, NJ, 1974, 78 pp.

538. **Loveland, R. E., Moul, E. T., Mountford, K., Sandine, P., Busch, D., Cohen, E., Kirk, N., Moskowitz, M., and Messing, C.,** The Qualitative and Quantitative Analysis of the Benthic Flora and Fauna of Barnegat Bay before and after the Onset of Thermal Addition, Unpubl. Tech. Rep., Rutgers University, New Brunswick, NJ, 1970, 30 pp.

539. **Boesch, D. F., Wass, M. L., and Virnstein, R. W.,** The Dynamics of Estuarine Benthic Communities, Unpubl. Tech. Rep., Virginia Institute of Marine Science, Gloucester Point, 1975.

540. **Sanders, H. L.,** Benthic studies in Buzzards Bay. III. The structure of the soft bottom community, *Limnol. Oceanogr.,* 5, 138, 1960.

541. **Sanders, H. L., Goudsmit, E. M., Mills, E. L., and Hampson, G. R.,** A study of the intertidal fauna of Barnstable Harbor, Massachusetts, *Limnol. Oceanogr.,* 7, 63, 1962.

542. **Boesch, D. F.,** Classification and community structure of macrobenthos in the Hampton Roads area, Virginia, *Mar. Biol.,* 21, 226, 1973.

543. **Mountford, N. K., Holland, A. F., and Mihursky, J. A.,** Identification and description of macrobenthic communities in the Calvert Cliffs region of the Chesapeake Bay, *Chesapeake Sci.,* 18, 360, 1977.

544. **Loi, T. N. and Wilson, B. J.,** Macroinfaunal structure and effects of thermal discharges in a mesohaline habitat of Chesapeake Bay, near a nuclear power plant, *Mar. Biol.,* 55, 3, 1979.

545. **Heck, K. L., Jr.,** Benthos, in *Ecological Studies in the Middle Reach of Chesapeake Bay: Calvert Cliffs,* Heck, K. L., Ed., Springer-Verlag, Berlin, 1987, 97.

546. **Diaz, R. J.,** Benthic resources of the Chesapeake Bay estuarine system, in *Contaminant Problems and Management of Living Chesapeake Bay Resources,* Majumdar, S. K., Hall, L. W., Jr., and Austin, H. M., Eds., Pennsylvania Academy of Science, Easton, 1987, 158.

547. **Holland, A. F., Mountford, N. K., and Mihursky, J. A.,** Temporal variation in upper bay mesohaline benthic communities. I. The 9-m mud habitat, *Chesapeake Sci.,* 18, 370, 1977.

548. **Grassle, J. F. and Grassle, J. P.,** Opportunistic life histories and genetic systems in marine benthic polychaetes, *J. Mar. Res.,* 32, 253, 1974.

549. **Tyler, M. A.,** Flow-induced variation in transport and deposition pathways in the Chesapeake Bay: the effect on phytoplankton dominance and anoxia, in *Estuarine Variability,* Wolfe, D. A., Ed., Academic Press, New York, 1986, 161.

550. **Officer, C. B., Biggs, R. B., Taft, J. L., Cronin, L. E., Tyler, M. A., and Boynton, W. R.,** Chesapeake Bay anoxia: origin, development, and significance, *Science,* 223, 22, 1984.

551. **Beukema, J. J.,** Quantitative data on the benthos of the Wadden Sea proper, in *Ecology of the Wadden Sea,* Vol. 1, Wolff, W. J., Ed., A. A. Balkema, Rotterdam, 1983, 4/135.

552. **Beukema, J. J., de Bruin, W., and Jansen, J. J. M.,** Biomass and species richness of the macro-benthic animals living on the tidal flats of the Dutch Wadden Sea: long term changes during a period with mild winters, *Neth. J. Sea Res.,* 12, 58, 1978.

553. **Remane, A.,** Die brackwasserfauna, *Verh. Dtsch. Zool. Ges.,* 36, 34, 1934.

554. **Remane, A.,** Ökologie des brackwassers, in *Die Biologie des Brachwassers,* Remane, A. and Schlieper, C., Eds., E. Schweizerbart´sche Verlagsbuchhandlung, Stuttgart, 1958, 1.

555. **Alexander, W. B., Southgate, B. A., and Basindale, R.,** Survey of the River Tees. II. The Estuary: Chemical and Biological, D. S. I. R. Water Pollut. Res., Tech. Paper No. 5, 1935.

556. **den Hartog, C.,** Die faunistische Gliederung im sudwestniederlandischen Delta-gebiet, *Int. Rev. Gesamten Hydrobiol.,* 46, 407, 1961.

557. **Heerebout, G. R.,** A classification system for isolated brackish inland waters, based on median chlorinity and chlorinity fluctuations, *Neth. J. Sea Res.,* 4, 494, 1970.

558. **Khlebovich, V. V.,** Some peculiar features of the hydrochemical regime and the fauna of mesohaline waters, *Mar. Biol.,* 2, 47, 1968.

559. **Kinne, O.,** Salinity. Animal-invertebrates, in *Marine Ecology,* Vol. 1 (Part 2), Kinne, O., Ed., Wiley-Interscience, London, 1971, 821.

560. **Sanders, H. L.,** Marine benthic diversity: a comparative study, *Am. Nat.,* 102, 243, 1968.

561. **Valentine, J. W.,** Resource supply and species diversity patterns, *Lethaia,* 4, 51, 1971.

562. **Valentine, J. W.,** *Evolutionary Ecology of the Marine Biosphere,* Prentice-Hall, Englewood Cliffs, NJ, 1973.

563. **Dayton, P. K. and Hessler, R. R.,** Role of biological disturbance in maintaining diversity in the deep sea, *Deep Sea Res.,* 19, 199, 1972.

564. **Walne, P. R.,** The importance of estuaries to commercial fisheries, in *The Estuarine Environment,* Barnes, R. S. K. and Green, J., Eds., Applied Science Publishers, London, 1972, 107.

565. **Hibbert, C. J.,** Biomass and production of a bivalve community on intertidal mudflats, *J. Exp. Mar. Biol. Ecol.,* 25, 249, 1976.

566. **Burke, M. V. and Mann, K. H.,** Productivity and production: biomass ratios of bivalve and gastropod populations in an eastern Canadian estuary, *J. Fish. Res. Board Can.,* 31, 167, 1974.

567. **Chambers, M. R. and Milne, H.,** Life cycle and production of *Nereis diversicolor* in the Ythan estuary, Scotland, *Estuarine Coastal Mar. Sci.,* 3, 133, 1975.

568. **Wolff, W. J. and de Wolf, L.,** Biomass and production of zoobenthos in the Grevelingen estuary, the Netherlands, *Estuarine Coastal Mar. Sci.,* 5, 1, 1977.

569. **Bahr, L. M.,** Energetic aspects of the intertidal oyster reef community at Sapelo Island, Georgia (U.S.A.), *Ecology,* 57, 121, 1976.

570. **Howe, S., Maurer, D., and Leathem, W.,** Secondary production of benthic molluscs from the Delaware Bay and coastal areas, *Estuarine Coastal Shelf Sci.,* 26, 81, 1988.

571. **Warwick, R. M. and Price, R.,** Macrofauna production in an estuarine mudflat, *J. Mar. Biol. Assoc. U.K.,* 55, 1, 1975.

572. **McLusky, D. S., Elliott, M., and Warnes, J.,** The impact of pollution on the distribution and abundance of aquatic oligochaetes in the Forth estuary, *Helgol. Wiss. Meeresunters.,* 33, 384, 1980.

573. **Sanders, H. L.,** Benthic studies in Buzzards Bay. I. Animal-sediment relationships, *Limnol. Oceanogr.,* 3, 245, 1958.

574. **Carey, D. A.,** Feeding strategies of high-volume surface deposit feeders, *Estuaries,* 8, 62A, 1985.

575. **May, E. B.,** The distribution of mud crabs (Xanthidae) in Alabama estuaries, *Proc. Natl. Shellfish. Assoc.,* 64, 33, 1974.

576. **Lubchenco, J. and Menge, B. A.,** Community development and persistence in a low rocky intertidal zone, *Ecol. Monogr.,* 48, 67, 1978.

577. **Paine, R. T.,** Food web complexity and species diversity, *Am. Nat.,* 100, 65, 1966.

578. **Connell, J. H.,** Community interactions on marine rocky intertidal shores, *Annu. Rev. Ecol. Syst.,* 3, 169, 1972.

579. **Arntz, W. E.,** Entwicklung von marinen Bodentiergemein-schaften bei Ausschluss von Raubern: nur artefakte?, *Meeresforschung,* 28, 189, 1981.

580. **Van Blaricom, G. R.,** Experimental analysis of structural regulation in a marine sand community exposed to oceanic swell, *Ecol. Monogr.,* 52, 283, 1982.

581. **Posey, M. H.,** Functional approaches to soft-substrate communities: how useful are they?, *Rev. Aquat. Sci.,* in press.

582. **Connell, J. H.,** Effects of competition, predation by *Thais lapillus,* and other factors on natural populations of the barnacle *Balanus balanoides, Ecol. Monogr.,* 31, 61, 1961.

583. **Menge, B. A. and Sutherland, J. P.,** Species diversity gradients: synthesis of the roles of predation, competition and temporal homogeneity, *Am. Nat.,* 110, 351, 1976.

584. **Mook, D.,** Effects of disturbance and initial settlement on fouling community structure, *Ecology,* 62, 522, 1981.

585. **Young, C. M. and Gotelli, N. J.,** Larval predation by barnacles: effects on patch colonization in a shallow subtidal community, *Ecology,* 69, 624, 1988.

586. **Aller, R. C.,** The effects of animal-sediment interactions on geochemical processes near the sediment-water interface, in *Estuarine Interactions,* Wiley, M. L., Ed., Academic Press, New York, 1978, 157.

587. **Aller, R. C.,** The effects of macrobenthos on chemical properties of marine sediment and overlying water, in *Animal-Sediment Relations: The Biogenic Alteration of Sediments,* McCall, P. L. and Tevesz, M. J. S., Eds., Plenum Press, New York, 1982, 53.

588. **Yingst, J. Y. and Rhoads, D. C.,** The role of bioturbation in the enhancement of bacterial growth rates in marine sediments, in *Marine Benthic Dynamics,* Tenore, K. R. and Coull, B. C., Eds., University of South Carolina Press, Columbia, 1980, 407.

589. **Taghon, G. L.,** The benefits and costs of deposit feeding in the polychaete *Abarenicola pacifica, Limnol. Oceanogr.,* 33, 1166, 1988.

590. **Eckman, J. E., Nowell, A. R. M., and Jumars, P. A.,** Sediment destabilization by animal tubes, *J. Mar. Res.,* 39, 361, 1981.

591. **Yingst, J. Y.,** Factors influencing rates of sediment ingestion by *Parastichopus parvimensis* (Clark), an epibenthic deposit-feeding holothurian, *Estuarine Coastal Shelf Sci.,* 14, 119, 1982.

592. **Grant, J.,** The relative magnitude of biological and physical sediment reworking in an intertidal community, *J. Mar. Res.,* 41, 673, 1983.

593. **Alongi, D. M.,** Effect of physical disturbance on population dynamics and trophic interactions among microbes and meiofauna, *J. Mar. Res.,* 43, 351, 1985.

594. **Luckenbach, M. W.,** Sediment stability around animal tubes: the roles of hydrodynamic processes and biotic activity, *Limnol. Oceanogr.,* 31, 779, 1986.

595. **Smith, J. N. and Schafer, C. T.,** Bioturbation processes in continental slope and rise sediments delineated by Pb-210, microfossil and textural indicators, *J. Mar. Res.,* 42, 1117, 1984.

596. **Probert, P. K.,** Disturbance, sediment stability, and trophic structure of soft-bottom communities, *J. Mar. Res.,* 42, 893, 1984.

597. **Wright, L. D.,** Benthic boundary layers of estuarine and coastal environments, *Rev. Aquat. Sci.,* 1, 75, 1989.

598. **Dobbs, F. C. and Whitlatch, R. B.,** Aspects of deposit-feeding by the polychaete *Clymenella torquata, Ophelia,* 21, 159, 1982.

599. **Rice, D. L., Bianchi, T. S., and Roper, E. H.,** Experimental studies of sediment reworking and growth of *Scoloplos* spp. (Orbiniidae: Polychaeta), *Mar. Ecol. Prog. Ser.,* 30, 9, 1986.

600. **Rice, D. L. and Whitlow, S. I.,** Early diagenesis of transition metals: a study of metal partitioning between macrofaunal populations and shallow sediments, in *The Fate and Effects of Pollutants, Proc. Maryland Sea Grant Symp.,* Maryland Sea Grant Program, University of Maryland Press, College Park, 1985, 21.

601. **Rice, D. L. and Whitlow, S. I.,** Diagenesis of transition metals in bioadvective marine sediments, in *Heavy Metals in the Environment,* Vol. 2, Lekkas, T. D., Ed., CEP Consultants, Edinburgh, 1985, 353.

602. **Officer, C. B. and Lynch, D. R.,** Bioturbation, sedimentation and sediment-water exchanges, *Estuarine Coastal Shelf Sci.,* 28, 1, 1989.

603. **Riemann, F. and Schrage, M.,** The mucus-trap hypothesis on feeding of aquatic nematodes and implications for biodegradation and sediment texture, *Oecologia,* 34, 75, 1978.

604. **Grenon, J.-F. and Walker, G.,** Biochemical and rheological properties of pedal mucus of the limpet, *Patella vulgata* L., *Comp. Biochem. Physiol. B,* 66, 451, 1980.

605. **Boer, P. L. De.,** Mechanical effects of microorganisms on intertidal bedform migration, *Sedimentology,* 28, 129, 1981.

606. **Rhoads, D. C. and Germano, J. D.,** Characterization of organism-sediment relations using sediment profile imaging: an efficient method of remote ecological monitoring of the seafloor (REMOTS™ system), *Mar. Ecol.,* 8, 115, 1983.

607. **Holme, N. A. and McIntyre, A. D., Eds.,** *Methods for the Study of Marine Benthos,* 2nd ed., Blackwell Scientific, Palo Alto, CA, 1984.

608. **Haderlie, E. C.,** A brief overview of the effects of macrofouling, in *Marine Biodeterioration: An Interdisciplinary Study,* Costlow, J. D. and Tipper, R. C., Eds., Naval Institute Press, Annapolis, MD, 1984, 163.

609. **Lindner, E.,** The attachment of macrofouling invertebrates, in *Marine Biodeterioration: An Interdisciplinary Study,* Costlow, J. D. and Tipper, R. C., Eds., Naval Institute Press, Annapolis, MD, 1984, 183.

610. Woods Hole Oceanographic Institution, *Marine Fouling and Its Prevention,* U.S. Naval Institute, Annapolis, MD, 1952.

611. **Costlow, J. D. and Tipper, R. C., Eds.,** *Marine Biodeterioration: An Interdisciplinary Study,* Naval Institute Press, Annapolis, MD, 1984.

612. **Benson, P. H., Brining, D. L., and Perrin, D. W.,** Marine fouling and its prevention, *Mar. Technol.,* 10, 30, 1973.

613. **Hillman, R. E.,** Environmental monitoring through the use of exposure panels, in *Fisheries and Energy Production: A Symposium,* Saila, S. B., Ed., D. C. Heath, Toronto, 1975, 55.

614. **Burkhart, M. D. and DePalma, J. R.,** Fouling, *Mariners Weather Log,* 21, 1, 1977.

615. **Sutherland, J. P. and Karlson, K. O.,** Development and stability of the fouling community at Beaufort, North Carolina, *Ecol. Monogr.,* 47, 425, 1977.

616. **Thorhaug, A. and Marcus, J.,** Macrofouling problems associated with ocean thermal energy conversion (OTEC) units, in *Proc. 5th Int. Congr. on Marine Corrosion and Fouling,* Editorial Garsi, Madrid, 1980, 225.

617. **Field, B.,** Marine biofouling and its control: history and state-of-the-art review, *Proc. Oceans 81,* 2, 542, 1981.

618. **Sutherland, J. P.,** The fouling community at Beaufort, North Carolina: a study in stability, *Am. Nat.,* 118, 499, 1981.

619. **Loveland, R. E. and Shafto, S. S.,** Fouling organisms, in *Ecology of Barnegat Bay, New Jersey,* Kennish, M. J. and Lutz, R. A., Eds., Springer-Verlag, New York, 1984, 226.

620. **Okamura, B.,** Formation and disruption of aggregations of *Mytilus edulis* in the fouling community of San Francisco Bay, California, *Mar. Ecol. Prog. Ser.,* 30, 275, 1986.

621. **Smedes, G. W.,** Seasonal changes and fouling community interactions, in *Marine Biodeterioration: An Interdisciplinary Study,* Costlow, J. D. and Tipper, R. C., Eds., Naval Institute Press, Annapolis, MD, 1984, 155.

622. **Osman, R. W.,** The establishment and development of a marine epifaunal community, *Ecol. Monogr.,* 47, 37, 1977.

623. **Hanson, C. H. and Bell, J.,** Subtidal and intertidal marine fouling on artificial substrata in northern Puget Sound, Washington, *Fish. Bull. (U.S.),* 74, 377, 1976.

624. **Shafto, S. S.,** The Marine Boring and Fouling Invertebrate Community of Barnegat Bay, New Jersey, M.S. thesis, Rutgers University, New Brunswick, NJ, 1974.

625. **Sutherland, J. P.,** The structure and stability of marine macrofouling communities, in *Marine Biodeterioration: An Interdisciplinary Study,* Costlow, J. D. and Tipper, R. C., Eds., Naval Institute Press, Annapolis, MD, 1984, 202.

626. **Hadfield, M. G.,** Settlement and recruitment of marine invertebrates: a perspective and some proposals, *Bull. Mar. Sci.,* 30, 418, 1986.

627. **Buss, L. W. and Jackson, J. B. C.,** Competitive networks: nontransitive competitive relationships in cryptic coral reef environments, *Am. Nat.,* 113, 223, 1979.

628. **Dean, T. A. and Hurd, L. E.,** Development in an estuarine fouling community: the influence of early colonists on later arrivals, *Oecologia,* 46, 295, 1980.

629. **Lane, C. E.,** The nutrition of *Teredo, Ann. N.Y. Acad. Sci.,* 77, 246, 1959.

630. **Turner, R. D.,** A Survey and Illustrated Catalogue of the Teredinidae, Museum of Comparative Zoology, Harvard University, Cambridge, MA, 1966.

631. **Richards, B. R., Hillman, R. E., and Maciolek, N. J.,** Shipworms, in *Ecology of Barnegat Bay, New Jersey,* Kennish, M. J. and Lutz, R. A., Eds., Springer-Verlag, New York, 1984, 201.

632. **Lane, C. E.,** Recent biological studies on *Teredo* — a marine wood-boring mollusc, *Sci. Monthly,* 10, 286, 1955.

633. **Lindgren, E. W.,** Treated piling systems, in *Coastal Ecological Systems of the United States,* Vol. 3, Odum, H. T., Copeland, B. J., and McMahan, E. A., Eds., Conservation Foundation, Washington, D. C., 1974, 301.

634. **Beckman, C. R., Menzies, J., and Wakeman, C. M.,** The biological aspects of attack on creosoted wood by *Limnoria, Corrosion,* 13, 162, 1957.

635. **Petersen, C. G. J.,** Valuation of the sea. I. Animal life of the sea bottom, its food and quantity, *Rep. Dan. Biol. Stn.,* 20, 3, 1911.

636. **Petersen, C. G. J.,** Valuation of the sea. II. The animal communities of the sea bottom and their importance for marine zoogeography, *Rep. Dan. Biol. Stn.,* 21, 1, 1913.

637. **Petersen, C. G. J.,** Appendix to report 21 (1911); on the distribution of animal communities on the sea bottom, *Rep. Dan. Biol. Stn.,* 22, 1, 1914.

638. **Petersen, C. G. J.,** On animal communities of the sea bottom in the Skagerrack, the Christiana Fjord and the Danish Waters, *Rep. Dan. Biol. Stn.,* 23, 3, 1915.

639. **Petersen, C. G. J.,** The sea bottom and its production of fish-food. A survey of the work done in connection with the valuation of the Danish waters from 1883—1917, *Rep. Dan. Biol. Stn.,* 25, 1, 1918.

640. **Thorson, G.,** Bottom communities, in *Treatise on Marine Ecology and Paleoecology,* Vol. 1, Hedgpeth, J. W., Ed., Mem. Geol. Soc. Am., Boulder, CO, 67, 461, 1957.

641. **Sanders, H. L.,** Oceanography of Long Island Sound, 1952—1954. X. The biology of marine bottom communities, *Bull. Bingham Oceanogr. Coll.,* 15, 345, 1956.

642. **Durand, J. B. and Nadeau, R. J.,** Water Resources Development in the Mullica River Basin. I. Biological Evaluation of the Mullica River — Great Bay Estuary, New Jersey, Unpubl. Tech. Rep., Water Resources Research Institute, Rutgers University, New Brunswick, NJ, 1972.

643. **Connell, J. H.,** A predator-prey system in the marine intertidal region. I. *Balanus glandula* and several predatory species of *Thais, Ecol. Monogr.,* 40, 49, 1970.

644. **Fonseca, M.S.,** Sediment stabilization by *Halophilia decipiens* in comparison to other seagrasses, *Estuarine Coastal Shelf Sci.,* 29, 501, 1989.

645. **Lie, U. and Evans, R.A.,** Long-term variability in the structure of subtidal benthic communities in Puget Sound, Washington, U.S.A., *Mar. Biol.,* 21, 122, 1973.

Chapter 5

FISHES

I. INTRODUCTION

Estuarine nekton consists of actively swimming pelagic organisms within the size range of 20 mm to 20 m which have the ability to move independently of water currents.[1] On the basis of body size, three nektonic categories have been established, namely, centimeter nekton, decimeter nekton, and meter nekton. A diversity of nektonic organisms occurs in estuaries from cuttle fish, swimming crabs, penaeid prawns, fishes, and crocodiles to occasional dolphins, dugongs, and turtles.[2] Because of their exceptional productivity, estuaries support high abundances and biomasses of fishes which, in some cases, rival those of oceanic upwelling regions.[3] Table 1 compares the biomass of fishes in selected estuaries with other ecosystems on earth. Estuaries serve as a unique environment for many marine species of teleosts.[4] Reduced salinity and lower incidence of piscivorous predators, together with a rich food supply, provide benefits to these finfishes.[5] While a few teleosts spend their entire life cycle in estuaries, most are seasonal members of estuarine communities, or utilize this habitat strictly as a migratory pathway between feeding and spawning areas (e.g., catadromous and anadromous species).[6] However, a contingent of marine juveniles has been classified as estuarine dependent.[7,8] Not to be forgotten are salt-tolerant freshwater populations that frequent the upper reaches of estuaries.[6]

As a component of the estuarine nekton, fish generally comprise the largest and most motile fraction. Their great mobility enables them to avoid unfavorable environmental conditions. Nonetheless, they exhibit some degree of temperature tolerance and osmoregulatory ability that help them adapt to potentially stressful conditions. The winter flounder (*Pseudopleuronectes americanus*), for example, can withstand near-freezing waters, altering its blood serum chemistry to depress the freezing point. Killifishes and mummichogs rapidly adjust to salinity changes, achieving salt balance via low permeability of the body surfaces and marked activity of the kidneys and salt glands in the gills.[3] The skin, scales, and mucous coatings minimize osmotic changes associated with salinity variations.[9]

Estuarine fish faunas are typified by the numerical dominance of a few species. The number of estuarine families with basic marine affinities also is small. The dominant species tend to be widespread, reflecting the broad tolerances and range of adaptations of these organisms.[10] Tropical estuaries usually have a greater diversity of fishes than temperate systems.[11] By far the most abundant forms are juveniles that use the estuarine environment as a nursery area. Many of these individuals belong to coastal populations with short-lived and euryhaline characteristics.[12] Although a few permanent residents can be found in estuaries at any time, most of the fish populations are seasonal migrants, moving into and out of these shallow ecosystems from the nearshore ocean.

Today, estuaries remain critical habitats for recreational and commercial fisheries. Estuarine nursery areas play a key role in maintaining commercial offshore stocks of fishes.[13] In the U.S., this relationship holds particularly strong for estuaries along the Atlantic and Gulf Coasts, but less so for those along the Pacific Coast.[9] McHugh[14] reported that more than 60% of the U.S. domestic commercial fish landings in 1980 by weight and more than 45% by dollar value occurred within 4.8 km of the coast. More recent statistics corroborate his findings (Table 2). He also specified that species which spend at least part of their lives within estuaries compose nearly two thirds of the U.S. commercial catch by dollar value and much of the marine recreational catch as well.[15,16] Seven of the ten species groups most important to commercial fisheries in the U.S. are typically marine, constituting 58% of the commercial catch.[17,18] According to McHugh,[16] estuarine-dependent fisheries in the U.S. exceeded $350 million in

TABLE 1
Biomass in Certain Ecosystems

Group	Area	g/m²
Birds	New Hampshire forest	0.04
Moose	Isle Royale, MI	0.7
Humans	U.S.	0.9
Fishes	Unpolluted rivers	1—5
	Georges Bank	1.6—7.4
	Atlantic salmon river, Matamek, Quebec	2.1—17.8
	Narragansett Bay	3.2
Large mammals	Central and East African grasslands	3.5—23.6
Fishes	Gulf of Mexico	5.6—31.6
	Flax Pond (Long Island) Estuary (annual average)	24.0
	California kelp bed	33.2—37.6
	Bermuda coral reef in summer	59.3
	Narrangansett Bay salt marsh embayment	69.2
Anchovy	Peruvian upwelling in autumn	216.7

From Haedrich, R. L. and Hall, C. A. S., *Oceanus*, 19, 55, 1976. With permission.

TABLE 2
U.S. Domestic Commercial Fish Landings in 1987, by Distance from Shore or off Foreign Coasts[a]

Distance from shore	Fishes	
	Weight	Dollar value
0 to 4.8 km	3431	897
4.8 to 320 km	5748	897
International waters (including foreign coasts)	604	276
Totals	9773	2070

[a] In millions of pounds and millions of dollars.

From Fisheries of the United States, 1987, Current Fishery Statistics 8700, U.S. Department of Commerce, NOAA, NMFS, May 1988.

1970. Louisiana was the leading fishing state in the U.S. at that time; the 502,250 MT (metric tons) of fish and shellfish landed commercially there consisted of about 99% estuarine species. In the Gulf of Mexico, estuarine species comprise almost the entire catch, accounting for more than 97% of the total landings.[13] Fisheries in the mid-Atlantic states likewise depend heavily on estuarine species.[16] The comprehensive work of Smith et al.[19] illustrates that five of the six most important commercial fishery species in the U.S. rely in some way on estuaries. The economic value of estuarine-dependent fisheries in the U.S. in more recent years has not diminished.[10,14]

The proximity of estuaries to landmasses renders them highly susceptible to pollution from human activities; this pollution threatens fish communities in many regions. Anthropogenic stresses on fish populations can be intense. Whereas much attention has been focused on the acute exposure of these populations to pollutants, sublethal and chronic exposures also debilitate resident and seasonal species. The mobility and migratory habits of fishes, however, make observations on anthropogenic effects more difficult to assess, and much evidence on pollution-induced changes in fish populations has been derived from laboratory experiments. Effects of man-made stresses on fishes in estuaries are often obscured by naturally occurring and poorly understood, long-term variations.[10,20]

This chapter deals with various aspects of estuarine fishes, their classification, community structure, population dynamics, and variations in space and time. An overview will be presented on the environmental adaptations, reproductive characteristics, feeding strategies, and migration patterns of estuarine fishes. Because this treatment will not be exhaustive on each of the subject areas, the reader will be alluded to pertinent literature citations where appropriate. The burgeoning research on the ichthyofauna of estuaries places limitations on a single, all-encompassing review. It is necessary, therefore, for the reader to examine the literature cited herein to obtain more comprehensive details of the complex topics addressed in this volume.

II. CLASSIFICATION

A. CLASSIFICATION OF PEARCY AND RICHARDS

Working on the Mystic River estuary, Connecticut, Pearcy and Richards[21] devised a classification scheme for estuarine ichthyofauna, categorizing species into one of five general assemblages based on the abundance of juveniles and adults, their seasonal occurrence, and the distribution of eggs and larvae (Table 3). A description of the five categories of classification follows: (1) resident species that spawn in the estuary and inhabit these waters for a substantial part of their lives; (2) summer migrants, usually southern species; (3) local marine forms, commonly found near the estuarine mouth, consisting of species indigenous to the local neritic waters; (4) diadromous species that undergo spawning migrations through the estuary to either freshwater or oceanic environments; and (5) freshwater fishes that normally occupy inland waters. Incomplete data on the distributions and life histories of various species and the lack of distinct natural divisions between some categories are drawbacks of this classification.

B. CLASSIFICATION OF PERKINS

Perkins[22] subdivided estuarine fishes into two broad groups, that is, diadromous and nondiadromous types. Diadromous species or transitory migrants, as noted above, are anadromous and catadromous populations that migrate between freshwater and seawater. The nondiadromous species conform to the other categories of Pearcy and Richards,[21] specifically residents, summer migrants, local marine forms, and freshwater fishes.

C. CLASSIFICATION OF DAY

The classification of Day[2,23] groups estuarine fishes into five classes according to their breeding and migratory habits. The largest group, marine migrants, contains juveniles of species that breed in the ocean and enter estuarine waters for purposes of feeding and protection. Two other classes of fishes are anadromous and catadromous species that migrate to and from their breeding grounds in freshwater and seawater, respectively. Estuarine residents breed and complete their life cycle within the estuary; only a few species represent this group. An anomalous group of fishes — not fitting into any well-defined class — makes up the fifth category.

D. CLASSIFICATION OF MOYLE AND CECH

Similar to Pearcy and Richards[21] and Day et al.,[2] Moyle and Cech[24] listed estuarine fishes under five broad categories. They include freshwater, diadromous, true estuarine, nondependent marine, and dependent marine fishes. Representatives of all five groups have been sampled in estuaries, although their relative abundance varies both temporally and spatially.

TABLE 3
Classification of the Fish of the Mystic River Estuary, Connecticut, According to Pearcy and Richards[21]

(I) Residents	(II) Summer migrants	(III) Local marine	(IV) Diadromous	(V) Freshwater
Cyprinodon variegatus	Synodus foetens	Raja erinacea	Alosa pseudoharengus	Esox niger
Fundulus heteroclitus	Strongylura marina	Anchoa mitchilli (I?)	Osmerus mordax	
F. majalis	Pomatomus saltatrix	Urophycis chuss	Anguilla rostrata	
Microgadus tomcod	Alectis crinitus	U. regius		
Apeltes quadracus	Seriola zonata	Tautoga onitis (I?)		
Gasterosteus aculeatus	Bairdiella chrysoura	Prionotus carolinus		
Pungitius pungitius	Leiostomus xanthurus	Myoxocephalus octodecemspino-		
Syngnathus fuscus	Menticirrhus saxatilis	sus		
Tautogolabrus adspersus	Mullus auratus	Cyclopterus lumpus		
Gobiosoma bosci	Stenotomus chrysops	Liparis atlanticus		
G. ginsburgi (II?)	Sphyraena borealis	Paralichthys oblongus		
Myoxocephalus aeneus	Mugil cephalus	Scophthalmus aquosus		
Pholis gunnellus	Eutropus microstomus			
Menidia beryllina	Paralichthys dentatus			
M. menidia	Echeneis naucrates			
Pseudopleuronectes americanus	Alutera schoepfi			
Opsanus tau	Balistes capriscus			
	Sphaeroides maculatus			
	Chilomycterus schoepfi			

From Pearcy, W. G. and Richards, S. W., *Ecology*, 43, 248, 1962. With permission.

E. CLASSIFICATION OF McHUGH

Using breeding, migratory, and ecological criteria, McHugh[9] formulated six classes of estuarine fishes. These are

1. Freshwater fishes that occasionally enter brackish waters
2. Truly estuarine species which spend their entire lives in the estuary
3. Anadromous and catadromous species
4. Marine species which pay regular seasonal visits to the estuary, usually as adults
5. Marine species which use the estuary primarily as a nursery ground, usually spawning and spending much of their adult life at sea but often returning seasonally to the estuary
6. Adventitious visitors which appear irregularly and have no apparent estuarine requirements

This classification scheme has gained wide acceptance among fisheries biologists. It is applicable to most estuarine systems.

1. Freshwater Fishes

This group of fishes, which has the capability of infiltrating brackish water zones, displays limited tolerance to seawater. However, they cannot be termed estuarine species. Although mainly inhabiting freshwater, they may stray into estuarine areas where salinity exceeds 10‰. In the Aransas River estuary, Texas, for example, five freshwater fishes (i.e., *Chaenobryttus gulosus, Cichlasoma cyanoguttatum, Gambusia affinis, Lepomis macrochirus,* and *Micropterus salmoides*) have been collected in waters with salinities as high as 17.4‰.[22] Other freshwater species gathered in this estuary are *Dorosoma petenense, Ictalurus furcatus, L. megalotis,* and *Notropis lutrensis*. McHugh[25] recovered the bluegill (*L. macrochirus*) as far downestuary as the mouth of the York River. Species of the genus *Esox* have been captured in a number of estuaries. Pearcy and Richards[21] listed the chain pickerel (*E. niger*) among the catch of the Mystic River estuary. In addition to this species, *E. americanus* has been taken in the finfish catch of the Slocum River estuary, a tributary of Buzzards Bay.[26] Other freshwater species identified in this estuary are *L. macrochirus* and the golden shiner *Notemigonus crysoleucas*. Tatham et al.[27] chronicled the succeeding freshwater strays in Barnegat Bay: *Aphredoderus sayanus, D. cepedianum, E. niger, Fundulus diaphanus, I. natalis, L. gibbosus,* and *N. crysoleucas*. These fishes show at least temporary tolerance to saline waters in estuaries. Day et al.[2] convey that the African species, *Sarotherodon (olim Tilapia) mosambicus,* is a rather unusually abundant freshwater fish with the ability to breed even in hypersaline water. Additional examples of freshwater fishes penetrating into the upper reaches of estuaries are the channel catfish (*I. punctatus*), white catfish (*I. catus*), mosquitofish (*Gambusia affinis*), and Sacramento splittail (*Pogonichthys macrolepidotus*).[24]

2. Truly Estuarine Species

Small species (e.g., ambassids, atherinids, some clupeids, gobies, stolephorids, and syngnathids) comprise the major fraction of fishes in this category.[2] As a group, the truly estuarine species are outnumbered by other populations, particularly marine forms. Thus, of a total of 70 species registered in the Laguna Madre, TX, Hedgpeth[28] uncovered nine estuarine residents. In St. Lucia, Wallace[29] considered 10 of 86 species to be estuarine residents. Tatham et al.[27] disclosed that 20 of 107 species were residents of Barnegat Bay, NJ.

Some of the more common members of the truly estuarine species group found in mid-Atlantic estuaries of the U.S. are the striped killifish (*Fundulus majalis*), Atlantic silverside (*Menidia menidia*), fourspine stickleback (*Apeltes quadracus*), and white perch (*Morone americanus*). Others documented in these estuaries include the naked goby (*Gobiosoma bosci*), northern pipefish (*Syngnathus fuscus*), and cunner (*Tautogolabrus adspersus*). Pearcy and Richards,[21] Moyle and Cech,[24] and Tatham et al.[27] quote additional species of this category.

3. Anadromous and Catadromous Species

a. Anadromous Species

Fishes which migrate from seawater to freshwater to spawn—anadromous types—have representatives in a number of families. Various species of basses (Serranidae), sea lampreys (Petromyzontidae), smelts (Clupeidae, Osmeridae, Salmonidae), and sturgeons (Acipenseridae) are anadromous. This group of fishes dominates migrations in cold northern estuaries, being nearly absent in tropical latitudes. The anadromous habit in estuaries declines with decreasing latitude. Only a few members of the Clupeidae (e.g., *Ethmalosa dorsalis* and *Hilsa ilisha*) purportedly migrate into the freshwaters of tropical regions. In boreal estuaries, species of Osmeridae and Salmonidae are the most important anadromous forms. The Southern Hemisphere lacks endemic anadromous fish.

The estuary is a multiple-use habitat for anadromous species. Adults utilize these coastal systems principally as a resting stop during their spawning migrations. The young individuals of many anadromous populations (e.g., bass, shad, and smelt), however, move into the estuary subsequent to hatching, where they feed and grow.[10] The striped bass, *Morone saxatilis*, provides as an excellent example of anadromy. The reproduction and early life history of this recreationally and commercially important fish are closely coupled to large estuaries. In the past, Chesapeake Bay, Delaware Bay, and the Hudson River have been essential spawning grounds of this species. In the Hudson River, striped bass spawn in late May or early June in freshwater, or almost in freshwater, with eggs and larvae drifting downstream for days or weeks. Young individuals, approximately 2 weeks old, also drift downestuary and gradually migrate vertically. They undergo a daily migration, moving toward surface waters at night and into deeper, saline waters during the day. This vertical translocation causes the young organisms to be swept seaward at night in the upper, fresher surface waters, and landward during the day in the deeper, more saline waters flowing upestuary. The diel vertical migration persists until the striped bass are about 6 weeks old, when they become stronger swimmers able to maintain their position against a current. This dynamic process enables the young striped bass to remain in an area with a rich food supply. As they grow, the juveniles spread into coastal waters, but are rarely caught 10 km beyond the shoreline.[3,30] Figure 1 depicts the life-history movements of the striped bass.

Other notable anadromous species include American shad (*Alosa sapidissima*), gizzard shad (*Dorosoma cepedianum*), blueback herring (*A. aestivalis*), alewife (*A. pseudoharengus*), and rainbow smelt (*Osmerus mordax*). Additionally, the sea lamprey (*Petromyzon marinus*) and the Atlantic sturgeon (*Acipenser oxyrhynchus*) are well-known anadromous fishes. Perhaps most familiar to some workers are the conspicuous migration runs of salmon (e.g., the Atlantic salmon, *Salmo salar*, and the pink salmon, *Oncorhynchus gorbuscha*).[9,10]

b. Catadromous Species

Fishes which migrate from freshwater to seawater to spawn—catadromous species—are few in number, freshwater eels (*Anguilla* spp.) probably being the most conspicuous. Mature eels travel from freshwater and brackish water habitats to spawn in the depths of the central ocean gyres. Subsequent to breeding far at sea, larval eels migrate to coastal waters and metamorphose into elvers. These forms, attaining immense numbers at times, pass through estuaries on their journey to freshwater areas.

Only eels and mullets are catadromous in the Northern Hemisphere; catadromy seems to be more common in the Southern Hemisphere, especially in cold-temperate waters. In estuaries of New Zealand, the flounder, *Rhombosolea retiaria*, appears to be catadromous.[31] Similarly, *Galaxias maculatus* is a catadromous species in the high latitudes of the Southern Hemisphere.[10,32]

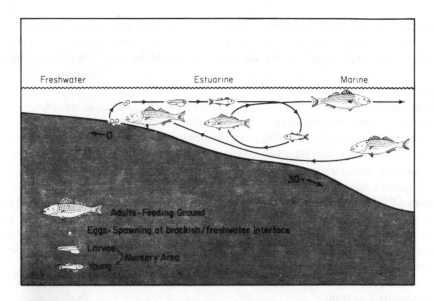

FIGURE 1. Anadromy of the striped bass, *Morone saxatilis*. Adults of the species migrate from coastal ocean water to, or almost to, freshwater to spawn, usually spawning near the interface of fresh and low salinity water. Young of the species feed in the estuary until mature and then repeat the cycle. (From Haedrich, R. L. and Hall, C. A. S., *Oceanus*, 19, 55, 1976. With permission.)

4. Seasonal Marine Migrants

Marine species that seasonally migrate into the estuary represent the majority of estuarine fishes. Tatham et al.,[27] for example, collected 34 species of warm-water migrants and 12 species of cool-water migrants in Barnegat Bay. The warm-water migrants constituted 65% of all fish taken in the estuary and the cool-water migrants, 3%

Species of marine migrants observed in estuaries belong to many families. The Clupeidae, Engraulidae, Sciaenidae, Sparidae, Gadidae, Chanidae, Synodontidae, and Belonidae are well represented. The flounder families Bothidae, Cynoglossidae, Pleuronectidae, and Soleidae also are examples. In numerous species of this group, spawning occurs offshore, even as distant as the outer margin of the continental shelf. After the eggs hatch, the larvae gradually move inshore to inhabit estuaries.[10] At a later stage, individuals emigrate from the estuary to coastal waters, but return annually to the estuary at regular seasonal periods. Species such as the horse-eye jack (*Caranx latus*) and the bluntnose stingray (*Dasyatis sayi*), which are warm-water migrants, as well as the Atlantic herring (*Clupea harengus harengus*) and the red hake (*Urophycis chuss*), which are cool-water migrants, exemplify seasonal marine migrants.

5. Marine Species Which Use The Estuary Primarily as a Nursery Ground

The estuary is a nursery area for a multitude of marine fishes, a few notable ones being the Atlantic menhaden (*Brevoortia tyrannus*), weakfish (*Cynoscion regalis*), spot (*Leiostomus xanthurus*), and Atlantic croaker (*Micropogonias undulatus*). The Atlantic menhaden, supporting a major commercial fishery along the Atlantic Coast of the U.S.,[33,34] forms a single coastal population. Found in Atlantic coastal waters from northern Florida to Nova Scotia,[35] this planktivorous fish is forage for predators (e.g., bluefish and weakfish). Segments of the coastal population invade estuaries, utilizing it as a nursery and feeding area. Many of these fish are immature.

The weakfish is a highly prized sport fish having greatest abundance centered around Chesapeake Bay, and ranging along the U.S. Atlantic Coast from Massachusetts Bay to southern Florida. Occasionally it strays as far north as Nova Scotia.[36] In the mid-Atlantic, weakfish enter estuaries from large coastal populations in the spring and emigrate in the fall. More than 95%

of all weakfish in some estuaries consist of young individuals that subsist on a variety of food items, but mainly mysid shrimp and sand shrimp as well as anchovies and silversides.[37,38] Smaller individuals ingest a higher percentage of invertebrates than larger forms which consume more fish. Weakfish spawn in nearshore oceanic and estuarine zones from late spring to fall.

Ranging from Massachusetts to Mexico, spot occupy estuaries in the summer and emigrate to the ocean in the fall. The fish overwinters and spawns mostly in offshore areas south to Cape Hatteras, but migrates northward into estuaries in the spring. Recreational and commercial fishing for spot intensifies during spring and fall migrations. Maximum abundance of the species along the Atlantic Coast takes place between Maryland and the Carolinas.[36] As a benthic feeder, spot consumes annelids, crustaceans, mollusks, small fish, and some plant matter.[39]

The Atlantic croaker, like spot, attains peak numbers along the U.S. Atlantic Coast from Chesapeake Bay to the Carolinas. Seasonal movements of this species parallel those of spot.[36] Inhabiting estuaries in late spring and summer, the Atlantic croaker emigrates to winter spawning grounds offshore, predominantly south of Chesapeake Bay.[40] It feeds on benthic organisms, principally annelids, crustaceans, mollusks, and fish.[36] The species is valuable to both recreational and commercial fisheries.[41]

6. Adventitious Visitors

Some fishes appear erratically in estuaries, occasionally entering the coastal systems when conditions are favorable. Along the mid-Atlantic Coast of the U.S., the blue runner (*Caranx crysos*) and bigeye scad (*Selar crumenophthalmus*) fall into this category. Along the Pacific Coast, the albacore (*Thunnus alalunga*) is an unpredictable visitor to estuaries. Pelagic fishes of the open ocean may stray here as well. For example, the Atlantic bonito (*Sarda sarda*), gag (*Mycteroperca microlepis*), king mackerel (*Scomberomorus cavalla*), and little tunny (*Euthynnus alletteratus*) sporadically move into bays and sounds.[9] Many other marine species are adventitious visitors, but generally remain unimportant members of the nektonic community of estuaries.

III. ADAPTATIONS TO ENVIRONMENTAL CONDITIONS

In comparison to other organisms, fish are better equipped to deal with the vagaries encountered in estuaries. When environmental conditions deteriorate, fish can easily swim away from affected areas. Even though fish populations can adjust their local environment by swimming away from unfavorable regions, their osmoregulatory ability and temperature tolerance allow them to cope more successfully with fluctuating habitats. Hence, salmon, which characteristically migrate from seawater to freshwater to spawn, acclimate quickly to rapid salinity reductions. Whereas much has been written about the role of salinity and temperature in governing the abundance and distribution of estuarine fishes, other factors may also be significant. For instance, oxygen depletions, elevated concentrations of suspended sediment, and red tides periodically inflict heavy mortalities on fish populations.[18] The structure of the estuarine fish community is sensitive to biological functions as well (e.g., reproduction, food habits, and interspecific interactions).[12] Both abiotic and biotic factors, therefore, strongly influence the species composition, abundance, and distribution of fish populations in estuaries.

A. PHYSICAL-CHEMICAL FACTORS
1. Salinity

Saila,[11] citing the work of Parry,[42] examined the osmoregulatory abilities of fishes. He sorted fish populations into five general groups based on their responses to salinity changes. These are (1) species which survive only in conditions of isosmotic constancy; (2) species which tolerate some degree of environmental change by changes in the ionic concentration of their fluids; (3) species or particular life-history stages of a species which may be semipermeable, allowing free water movement in or out in response to environmental changes; (4) species which reduce the permeability of their surfaces; and (5) species able to compensate actively for movements of water or solutes caused by external changes.

Estuarine fish must have the ability to regulate their internal environment to successfully inhabit the fluctuating environment of estuaries, where sudden floods and tidal changes abruptly alter salinities. Marine teleosts maintain the concentration of salts in their blood between 340 and 500 m-Osmol/l, which is well below that of seawater (approximately 1000 m-Osmol/l). Reductions in salt concentrations in the external environment can be tolerated by many marine teleosts to levels approaching the salt concentration in their blood. Green[43] delineated four mechanisms by which marine teleosts regulate the concentration of salts in their blood: (1) the body surface is relatively impermeable due to scales, thickening of the dermis, and production of mucus; (2) monovalent ions are actively secreted by the gills; (3) seawater is swallowed and water absorbed from the gut together with monovalent ions; and (4) only small quantities of urine production occur, thereby conserving water as much as possible.

The dominant estuarine fishes have broad salinity tolerances. Marine fishes are preadapted to reduced salinities, so that when salinity gradually decreases in an estuary, many of them will have little difficulty adapting to it.[2] Few species tolerate hypersaline conditions, though. Of the 70 species recorded by Hedgpeth[28] in the Laguna Madre of Texas, 18 tolerated salinities as high as 60‰. Wallace[29] registered 10 species at salinities greater than 70‰ in Lake St. Lucia.

The species richness of fish in estuaries commonly is greatest at the mouth. Few osmotic problems exist for marine teleosts in this area. When the salinity drops to approximately one third that of seawater, major osmotic problems may arise.

Truly estuarine fishes tend to adapt quickly to changes in salinity. Such is the case for the European flounder, *Platichthys flesus*, and the hogchoker, *Trinectes maculatus*. The European flounder adapts particularly well to sudden salinity changes. Juveniles of this species can live in freshwater. Adult *Gilchristella aestuarius*, *Hepsetia breviceps*, *Monodactylus falciformis*, and *Mugil cephalus* inhabiting South African estuaries survive direct transfer from seawater to freshwater.[2]

Some marine fishes display remarkable adaptation to lowered salinities. For example, the distribution of the herring, *Clupea harengus*, overlaps that of the northern pike, *Esox lucius*. Larvae of *C. harengus* actually have a greater salinity tolerance than adults. Other marine species show similar tolerances to lowered salinities.[9,10]

2. Temperature

A key physical factor, temperature greatly influences the biochemistry, physiology, and behavior of fishes.[44] As obligate poikilotherms, estuarine fishes lack the capability to regulate body temperature and must thermoregulate behaviorally, that is, by avoiding or selecting environmental temperatures to maximize survivorship.[45] The species distribution within an estuary will depend, in part, on thermal tolerance limits, and in terms of thermal distribution, other factors must be considered, including food availability, nutritional state, competition and/or predation, age, pathological condition, habitat requirements, social factors, time of day, and season.[44]

Temperature gradients clearly affect the species composition, abundance, and distribution of estuarine fishes. In the upper reaches of estuaries, water temperature is higher in summer and lower in winter than nearby coastal oceanic waters. In summer, surface water is warmer than bottom water, but this condition can reverse in winter. Seasonal differences such as these have a marked impact on the community structure, especially in high-latitude systems. Changing thermal gradients may act as barriers to some species, altering their migratory behavior. The bluefish, *Pomatomus saltatrix*, is an example.[44] Vertical as well as horizontal temperature gradients in estuaries elicit responses from fishes manifested in their distinctive distribution patterns. However, physiological or behavioral thermal limits make accurate predictions of their occurrence in nature extremely difficult. Within the thermal limits of a species, ecological factors other than temperature can control spatial distribution. Food availability, for instance, is a crucial factor that can modulate the position of fish in estuaries. Temperature, in turn, affects the organism's metabolism which can impact not only its feeding behavior, but also its reproduction and growth.[46-48]

In high latitude systems, where the importance of temperature exceeds that of salinity, the abundance of fish populations declines substantially in winter when much of the estuarine surface freezes. Most of the juveniles and resident fishes remain inactive in deeper waters at this time. Subsequent to the breakup of ice in spring, as light and temperature levels rise, the fishes migrate from the sea. Activity increases in summer only to decline once again by late fall. Similar, but less acute, seasonal changes are apparent in cold-temperate estuaries. Many juveniles arrive in the estuary from the ocean during the period from early spring to summer; predators follow these forage species. By midsummer the estuary contains an abundance of fishes which grow rapidly. The adults and most of the juveniles return to the ocean in the fall as temperatures diminish. However, a number of juveniles do not emigrate, but remain in the estuary for a couple of years. While temperature is of paramount importance in triggering fish migration in temperate and boreal waters, salinity effects are more significant than temperature in subtropical and tropical estuaries. The immigration of juveniles into the lower-latitude systems usually takes place subsequent to an increase in temperature. Moreover, seasonal temperature changes typically are smaller in tropical estuaries relative to salinity changes which have a greater overall effect on the fish populations.[2,49]

Brett[50] reviews the general concepts and theories of temperature regulation on fish. Effects of temperature on fish behavior are treated by Pitcher.[51] Pitcher and Hart[52] investigate the relationship of temperature to energy allocation, growth dynamics, and growth efficiency in fish.

3. Dissolved Oxygen

The concentration of dissolved oxygen in estuarine waters is a function of temperature, salinity, and pressure.[53] In addition to these physical factors, biological and pollutant controls on oxygen can alter dissolved oxygen levels.[54] Figure 2 plots oxygen solubility in water within the range of temperature and chlorosity observed in estuaries. Oxygen values are given in units of ml/l and mmol/l, but in the literature may also be presented as percent saturation. In an ecosystem where temperature and salinity undergo considerable spatial and temporal variation and where biological productivity is likewise markedly variable, oxygen readings will be constantly changing. Many estuaries, particularly those nearby metropolitan centers, are burdened with organic loading which in many cases has a dramatic impact on the amount of dissolved oxygen, especially in summer. Aston (p. 255)[54] recounts the effects of these biological factors:

> The surface sea waters and river waters mixing together in estuaries will normally be completely or nearly saturated with oxygen dissolved from the atmosphere. However, oxygen is consumed by natural biological consumption and by the oxidation of pollutant organic matter, e.g. sewage, if present. At present many estuaries receive both organic and nutrient pollutants from domestic and industrial sources, and the combination of these two contaminants can lead to major changes in the biological productivity and its associated oxygen demand. The high productivity in estuaries may, under certain circumstances of restricted circulation and flushing, lead to eutrophication.

Hydrographic conditions can contribute to acute oxygen deficiency. Broad expanses of Chesapeake Bay experience anoxia in late spring and summer.[55,56] The intensity of anoxia has been related to the degree of stratification of the water column and to the freshwater flow into the bay.[57] Episodes of anoxia seem to be most acute in years of high spring streamflow which creates a strongly stratified system that arrests mixing between surface and bottom waters. Such

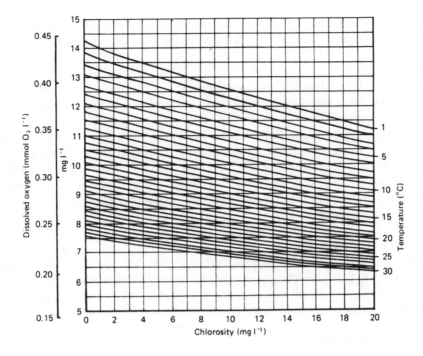

FIGURE 2. Solubility of oxygen in water within the range of temperature and chlorosity in estuaries. (From Head, P. C., *Practical Estuarine Chemistry: A Handbook,* Head, P. C., Ed., Cambridge University Press, Cambridge, 1985, 278. With permission)

conditions existed in Chesapeake Bay during the high streamflow year of 1983. A spring flow "pulse" governed the onset of stratification and the development of anoxia at this time. Oxygen concentrations below the pycnocline declined with a diminution in vertical exchange processes. As a consequence, vertical oxygen transfer was sharply curtailed.

In contrast to the moderately high flow year of 1983 in Chesapeake Bay, 1981 ushered in drought conditions, leading to lowered freshwater input and a weakly stratified system with considerable mixing of surface and bottom waters.[57] Hence, anoxic conditions during 1981 were less pervasive and persistent than during 1983. The baywide spread of anoxia in mid-June 1983 did not materialize in 1981. The anomolously high streamflow in 1984 also resulted in a severe case of oxygen depletion in deeper waters of the estuary.[56] The high streamflow effectively placed a "lid" on the system, mitigating the mixing of low-oxygen bottom waters and fostering anoxia. Low streamflow, although responsible for reducing the degree and duration of anoxia, does not totally preclude the depletion of oxygen in bottom waters.

The formation of an anoxic state in estuarine waters renders large areas uninhabitable for fish. As oxygen concentrations decrease from 4 to 0 ml/l, nektonic forms will be driven from affected areas. Migration routes can be effectively blocked, and fishes occasionally trapped in restricted areas.[9] Behavioral modification arises with decreasing concentrations of dissolved oxygen.[58] Mass mortality of fishes is likely when populations get trapped in waters devoid of oxygen. While extensive mortalities of fishes attributable to anoxia periodically occur, reliable estimates of fish mortalities often are not available.

Systems subjected to large variations in dissolved oxygen (e.g., some southern estuaries) form harsh environments for many fish populations. Although it is possible for mass mortalities of fishes to develop in these systems, altered distribution patterns are more common. For instance, fishes may not enter estuaries where oxygen concentrations have been substantially depleted because of low flow and excessive oxygen demand related either to very high primary production or to the discharge of sewage effluent which raises the volume of organic matter. The finfish species of the highly polluted Thames estuary in England swim up and down the estuary as oxygen values change. They avoid waters that are less than 10% saturated with oxygen.[24,59]

4. Other Factors

Physical and chemical factors, such as temperature, salinity, and dissolved oxygen, largely control the abundance and distribution of fishes in estuaries.[24] Biological factors, including interspecific competition and predation, are of secondary importance. Most fishes inhabit estuaries only part of the time and do so when conditions are most favorable. The abundance of food and shelter enhances the development of early life-history stages. When physical and chemical conditions deteriorate, however, they migrate to less stressful locales. Interspecific competition and predation may play a role in the seasonality of estuarine fish populations, but the principal movements of the ichthyofauna appear to be primarily in response to changes in temperature and salinity.[24]

B. BIOTIC FACTORS

Interspecific and intraspecific interactions affect the abundance and distribution of estuarine fishes albeit less so than major physical and chemical factors as related above. Moyle and Cech[24] subdivide interspecific interactions into predator-prey relationships, competition, and symbiosis. Symbiotic interactions include commensalism, mutualism, and parasitism.

1. Interspecific Interactions

a. Predator-Prey Relationships

Estuarine fishes are largely carnivores, only a few species being herbivores or detritivores. When present in the estuary, most fish have a dual role as predator and prey. The bay anchovy (*Anchoa mitchilli*), for example, consumes other small fishes as well as zooplankton, amphipods, decapods, and mysids, and it is a favorite food item of larger species (e.g., bluefish, *Pomatomus saltatrix*). Subtle or complex changes in feeding habits of numerous forms occur during their ontogeny which further muddles predator-prey relationships. Thus, gray mullet (family Mugilidae), when young, usually feed on plankton. Individuals less than about 20 mm in length mainly ingest algae; however, as they grow to between 20 and 40 mm in length, the mullet eat more animal food, amounting to approximately 22% of their diet. The fraction of organic detritus consumed by mullet increases significantly as they grow to lengths in excess of 40 mm.[43]

Predation by carnivorous fish can be so intense that prey populations, both invertebrates and fishes, can be locally depleted.[24] Virnstein[60] remarked how the density of benthic invertebrates in estuaries along the Atlantic Coast of the U.S. may be severely limited by the predaceous activity of spot (*Leiostomus xanthurus*). Rapid growth and short generation times of the benthic populations, however, tend to compensate for the high rates of predation. The bluefish, together with the weakfish, ingests large quantities of anchovies, menhaden, and silversides, but the mortality of these prey populations does not translate into catastrophic effects.

Because estuaries are excellent feeding grounds, young fish are subject to intense predation pressure. Rapid growth rates of the young fish provide a means whereby they can escape overexploitation. As growth accelerates, the prey organism quickly attains a size beyond the killing ability of many of the predators.[61] A rich food supply also allows the juveniles to reach adult (reproductive) status earlier than they could in environments where food is less plentiful.[24]

b. Competition

As Levinton (p. 70)[61] states, "Interspecific competition occurs when two or more species inhabit or interfere with one another as a result of the common use of resources." Competition between two species intensifies as ecological requirements become similar, that is, as niche requirements merge. Competitive displacement arises when extensive niche overlap develops between two species. A resource is generally divided between two species having similar ecological requirements because their differences become accentuated.[24] Theoretically, species with the same ecological requirements cannot coexist. Coexistence is fostered by species that evolve to minimize the overlap of resources.[61]

Competition between fishes in estuaries is sometimes difficult to detect due partially to the substantial variability of the environment. Niche dimensions must be accurately defined, and competitive effects demonstrated by establishing good criteria.[61] According to Levinton,[61] evidence for competition is derived from: (1) experimental manipulations, (2) laboratory experimental demonstrations of interference or niche overlap, (3) displacements in nature, and (4) continuity of niche space of coexisting field populations.

Fishes avoid competition by segregating ecologically, utilizing mechanisms such as differential exploitation, aggressive behavior, predation, habitat interference, habitat imprinting, and fugitive species.[24] Differential exploitation, while being the most important mechanism in many aquatic environments, perhaps is of less significance in estuaries where considerable overlap in food habits exists among species. For example, the sand shrimp, *Neomysis americana,* constitutes a major component of the diet of a variety of fish populations in Barnegat Bay, NJ.[27] Similarly, the opossum shrimp, *N. mercedis,* is a dietary component of numerous fishes in the Sacramento-San Joaquin estuary.[24]

c. Symbiosis

Thurman and Webber (p. 132)[62] have defined symbiotic relationships in marine ecosystems as "those in which two or more organisms are closely associated in a way that benefits at least one of the participants". A more generalized definition of symbiosis, "an interrelationship between two different species", has been proffered by Nybakken (p. 356).[63] Odum (p. 213)[64] endorses an even broader meaning of the word, simply describing it as "living together" without regard to the nature of the relationship between the organisms. Many scientists use the term to cover three specialized types of interactions (i.e., commensalism, mutualism, and parasitism). Other specialists employ the word symbiosis in the same sense as mutualism.

1. Commensalism

In general terms, commensalism refers to a relationship between organisms in which one participant benefits and the other remains unaffected. Among fish faunas, the remoras (Echeneidae) live in a commensal relationship with larger fishes, especially sharks, attaching to their host by means of a large sucker on the top of their head. The host fish transports the remoras to new food sources, but is otherwise unaffected by the smaller fish. The species, *Echeneis remora,* is always found in association with larger fishes. Another commensal relationship exists among certain species of gobies which inhabit the burrows of invertebrates. The burrows serve as a place of shelter for the gobies. *Clevelandia ios,* a small gobiid fish, takes refuge in tubes or burrows of other animals, emerging from them to feed.[63] Meanwhile the host animal does not gain anything from the relationship.[24] The commensal, *C. ios,* lives at times in association with three or more other commensals in the permanent U-shaped burrow of the echiurid worm, *Urechis caupo.* While *C. ios* prefers the upper end of the burrow, the scale worm, *Hesperonöe adventor* (which also snatches food from other tube-dwelling hosts), and the pinnotherid crab, *Scleroplax granulata,* extend farther into the tube in close proximity to *U. caupo.* Other commensals occasionally associated with the burrow of *U. caupo* are the clam, *Cryptomya californica,* and the shrimp, *Betaeus longidactylus.* Nybakken[63] invokes the term inquilism (under the heading of commensalism) to denote animals living in the home of another, or in its digestive tract, without being parasitic.

2. Mutualism

Several examples of mutualism have been detected among fishes in natural environments.

One of the most notable examples of mutualism is cleaning behavior which conspicuously occurs in reefs. The cleaning urasses (Labridae) of the Indo-Pacific region set up cleaning stations on reefs, attracting larger fish with their brilliant display of colors. When a larger fish arrives at a cleaning station, it remains motionless while the cleaning urasses remove ecoparasites from it. The mutualistic relationship here is apparent. The cleaning urasses obtain nourishment from the large fish, and potentially disease-causing, parasitic infestations are removed by them. Similarly, fish cleaning behavior has been established for several species of shrimp.[65]

Certain species of small gobies and burrow-dwelling shrimp exhibit mutualism. In this case, the shrimp constructs a burrow that also becomes the home of the goby. The goby, of course, gains by adapting to a burrow that provides shelter. Because the goby has superior vision than the shrimp, it warns the shrimp of predators which approach the burrow. In fact, signals may be communicated between the two organisms, accentuating a relationship that is mutually beneficial.

3. Parasitism

Organisms which derive nourishment and shelter from other individuals by living in or on them are known as parasites. Most marine organisms, including fish, contain parasites. However, few fishes themselves are parasitic, an exception being the pearlfishes (Carapidae) which live in association with sea cucumbers (Holothuria). Lampreys attach to the body of fishes, extracting blood and other fluids. Some workers categorize these organisms as an intermediary between parasites and predators.[24]

2. Intraspecific Interactions

Interactions between individuals of the same species parallel interspecific interactions. Intense intraspecific competition and predation have been well documented in fishes. The population size of some pelagic fishes (e.g., herrings and sardines) is affected by intraspecific competition. Cannibalism takes place in certain fish populations; for instance, adult striped bass (*Morone saxatilis*) consume large numbers of young-of-the-year forms in the Sacramento-San Joaquin estuary of California.[24] Nevertheless, the overall success of each year class of bass does not appear to be dependent upon this relationship. The young and adults of certain species segregate spatially, thereby minimizing intraspecific competition and predation among these life-history stages. Adult feeding grounds also do not overlap their breeding grounds. In addition, the ichthyoplankton often inhabit areas remote from those of the adults.

IV. FOOD HABITS AND DIET

Stomach content analyses of estuarine fishes reveal a wide variety of food items, reflecting the highly diverse feeding behavior of the ichthyofauna. The feeding ecology of these fishes is dynamic, frequently changing in time (with ontogenetic development) and in space. Their diet likewise changes in response to the availability of typical and atypical foods. Moreover, competition by other species for the foods available influences their diet. Information collected on the feeding habits of estuarine fishes is valuable in the study of trophodynamics.[66] It can be applied to investigations of conversion efficiencies or to the total energy demand of fish in an area.[67] Finally, an understanding of the feeding habits as well as the foods ingested by these organisms enables the ecologist to construct food chains and food webs necessary to ascertain the role that the species play in the ecosystem. Two comprehensive volumes that have broken new ground in fish nutrition are those of Cowey et al.[68] and Halver.[69]

As expressed by Stickney,[18] the feeding habits of fish relate to the manner in which they obtain their food, whereas the food habits strictly deal with the organisms' diet. Some fisheries biologists consider the quantity and quality of food as the two primary determinants of growth

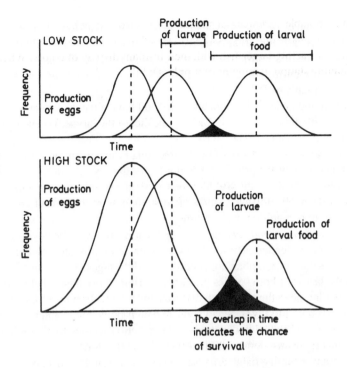

FIGURE 3.　The match of larval production to larval food. A greater chance
of match occurs at high stock (i.e., when recruitment becomes less variable).
(From Cushing, D. H., *Science and the Fisheries,* Edward Arnold, London,
1977. With permission.)

and survival of marine ichthyoplankton.[70-73] While this relationship is supported by both
laboratory and field inquiries, many questions remain unanswered and additional data must be
collected to verify its validity in the field.[74] In temperate waters, the food supply of marine
ichthyoplankton varies seasonally, being at lowest levels during the winter. Most fish in these
waters spawn in the spring, summer, or fall when the food supply is more plentiful for the larvae.
Spawning of the fish takes place during a fixed season, but the timing of the onset of spawning
and of peak production varies.[75] Larval production, therefore, is matched or mismatched to that
of the food available (Figure 3). The year-class strength of a species tends to be greater as the
chance of the match of larval production to that of its food rises. This phenomenon is neither
expected nor observed in tropical and subtropical waters where production continues year-round
and no need exists for a fixed spawning season.

Most juvenile fish become plankton feeders after the yolksac has been absorbed, and they
remain plankton feeders until adult organs evolve and somewhat specialized feeding ensues.[2]
Estuarine species of the Clupeidae, Engraulidae, and Stolephoridae develop a filtering mechanism of elongate gill rakers efficient in plankton feeding. Estuarine forms of Ambassids,
Monodactylids, and Syngnathids do not have a filtering mechanism, but snap up or suck in
individual amphipods, copepods, and mysids.

Microzooplankton in the size range 30 to 200 μm constitute ration for first-feeding fish
larvae.[76] Specifically, protozoans (Foraminifera, Radiolaria, Tintinnidae, and Ciliata other than
tintinnids) and metazoans (naupliar and postnaupliar stages of Copepoda, Heteropoda, and
Pteropoda) generally comprise this food source.[77] Evidence indicates that the size structure of
the planktonic community greatly affects the growth and survival of larval fish.[78] Indeed, the size
spectrum of the planktonic community may override the importance of the total plankton
abundance as a determinant of growth and survival of the larvae.

Hunter[76] and Govoni et al.[79] describe feeding strikes of the planktonic larvae of marine fishes on prey organisms. Initially the larvae perceive and recognize their prey, and subsequently, they flex their body axis and drive forth toward the food organism. Feeding strikes, largely visually elicited, occur mainly during the day. In a 12-h day, larval fishes may search 12 to 120 l of seawater for prey organisms.[76]

Both larval fish and their microzooplanktonic food have a patchy spatial distribution. The availability of prey organisms clearly influences the diet of the larvae. When presented with an assortment of prey organisms, the larval fish will be selective. Feeding selectivity is a function not only of food availability, but also of the interaction of the perception, recognition, capture, and successful ingestion of the prey.[79-81] A major factor in prey selection of larval fish is size; among the various dimensions, the width of the food appears to be most important because of its effect on ingestion.[79,82] However, other factors, such as mobility and color of the prey, influence perception, recognition, and capture.[83]

Larval spat (*Leiostomus xanthurus*) and larval Atlantic croaker (*Micropogonias undulatus*) are examples of selective feeders. These two sciaenids eat tintinnids, pteropods, pelecypods, and ostracods as well as the egg, naupliar, copepodid, and adult stages of copepods.[84] Govoni et al.[79] discerned that the larvae of both species primarily consumed organisms between 100 and 260 μm in width. Spot larvae in the size range of 1 to 5 and 5.01 to 10 mm selected copepodid and adult copepods (Stages CI to CVI), pelecypods, and pteropods. Small and large croaker larvae also opted for copepodid and adult copepods. Prey selection seemed to be contingent upon the width of the food organisms along with their behavior and color.

Juveniles of many estuarine fishes are voracious predators, ingesting copepods, amphipods, isopods, mysids, and ostracods in addition to other organisms. Meiofauna and small macrofauna serve as food sources, and predation impacts on them by juvenile benthic fish can be significant, particularly when actively selected by these predatory forms.[85,86] Juvenile salmonids in Alaska feed heavily on benthic and pelagic invertebrates as they descend streams and rivers to enter estuaries in spring and early summer.[87,88] At one time thought to be strictly planktivorous, Pacific salmon likewise incorporate benthic foods into their diet.

The food habits of large-sized estuarine species have frequently been based on investigations of juvenile forms. The most numerous species in estuaries rarely exceed a few grams in weight; thus, a large adult fish in these coastal systems may only weigh slightly more than 1 kg. As noted above, the food habits of estuarine fishes commonly change — in some instances dramatically — as the individuals age and grow. The larger fishes tend to become increasingly piscivorous as they grow.[18]

During the past two decades, research has been conducted on trophic interactions of estuarine fishes. Trophic resource partitioning has been suggested as a viable mechanism for reducing finfish competition. Oviatt and Nixon,[89] for example, reported resource partitioning in the finfish community of Narragansett Bay, RI, as did Haedrich and Haedrich[90] in the finfish community of the Mystic River estuary, MA.

Table 4 lists the food habits of selected fishes inhabiting estuaries along the Atlantic and Gulf Coasts of the U.S. A wide range of food items is evident in the diet of these organisms. While detritus and plant material compose a small fraction of the foods ingested, copepods, amphipods, mysids, and other crustaceans are a main staple. Fish constitute all or a significant portion of the diet of the inshore lizardfish (*Synodus foetens*), striped bass (*Morone saxatilis*), and gray snapper (*Lutjanus griseus*).

General feeding patterns of finfish communities have been determined in some estuaries, with the diet and trophic position of the predominant species being established. Darnell[91] constructed two food chains for Lake Pontchartrain, an estuarine system in Louisiana (U.S.). One food chain proceeds from small to large benthic invertebrates and small benthic fishes (e.g., young sciaenids). Large piscivorous fishes occupy the top of the chain. The second food chain consists of copepods (e.g., *Acartia*) near the base, passing upward through small fishes (e.g.,

TABLE 4
Primary Foods of Some Estuarine Fishes Along the U.S. Atlantic and Gulf Coasts

Scientific name	Common name	Food habits	Ref. (sources from Ref. 18)
Megalops atlanticus	Tarpon	Juveniles consume copepods, ostracods, grass shrimp, and fish	Harrington and Harrington (1960), Rickards (1968)
Brevoortia tyrannus	Atlantic menhaden	Algae, planktonic crustacea	Bigelow and Schroeder (1953), June and Carlson (1971), Kjelson et al. (1975)
Harengula jaguana	Scaled sardine	Harpacticoid copepods, amphipods, mysids, isopods, and chironomid larvae	Odum and Heald (1972)
Anchoa hepsetus	Striped anchovy	Copepods, mysids, isopods, mollusks, fish, zooplankton	Hildebrand and Schroeder (1928), Springer and Woodburn (1960), Carr and Adams (1973)
A. mitchilli	Bay anchovy	Zooplankton, fish, decapods, amphipods, mysids, detritus	Hildebrand and Schroeder (1928), Reid (1954), McLane (1955), Darnell (1958), Springer and Woodburn (1960), Odum (1971), Odum and Heald (1972), Carr and Adams (1973)
Synodus foetens	Inshore lizardfish	Fish	Linton (1905), Hildebrand and Schroeder (1928), Reid (1954), Reid (1955), Springer and Woodburn (1960), Carr and Adams (1973)
Arius felis	Sea catfish	Amphipods, decapods, insects, mollusks, copepods, schizopods, isopods, hydroids	Darnell (1958), Harris and Rose (1968), Odum and Heald (1972)
Opsanus tau	Oyster toadfish	Crustaceans, mollusks, polychaetes	R. R. Stickney, unpublished data
Fundulus majalis	Striped killifish	Mollusks, crustaceans, insects, fish	Bigelow and Schroeder (1953)
F. pulvereus	Bayou killifish	Insects, isopods	Simpson and Gunter (1956)
F. similis	Longnose killifish	Harpacticoid copepods, ostracods, barnacle larvae, insects, isopods, amphipods	Stickney and McGeachin (1978)
Cyprinodon variegatus	Sheepshead minnow	Plant detritus, small crustaceans, nematodes, diatoms, blue-green algae, filamentous algae, formas, insects	Simpson and Gunter (1956), Odum and Heald (1972), Stickney and McGeachin (1978)
Adinia xenica	Diamond killifish	Plant detritus, filamentous algae, amphipods, insects, small copepods, diatoms	Odum and Heald (1972)

TABLE 4 (continued)
Primary Foods of Some Estuarine Fishes Along the U.S. Atlantic and Gulf Coasts

Scientific name	Common name	Food habits	Ref. (sources from Ref. 18)
Lucania parva	Rainwater killifish	Insects, crustacean lar-vae, annelids, mysids, amphipods, cumaceans, copepods, plant detri-tus, small mollusks	Simpson and Gunter (1956), Odum and Heald (1972)
Gambusia affinis	Mosquito fish	Amphipods, chironom-ids, insects, algae	Odum and Heald (1972)
Poecilia latipinna	Sailfin molly	Algae, diatoms, vascular plant detritus, inorganic matter	Odum and Heald (1972)
Menidia beryllina	Tidewater silverside	Isopods, amphipods, co-pepods, mysids, detri-tus, algae, insects, barnacle larvae	Hildebrand and Schroe-der (1928), Reid (1954), McLane (1955), Darnell (1958), Sprin-ger and Woodburn (1960), Odum and Heald (1972), Carr and Adams (1973), Marsh (1973), Stickney and McGeachin (1978)
Membras martinica	Rough silverside	Copepods, barnacle lar-vae, amphipods, in-sects, shrimp, fish	Dixon (1974), Stickney and McGeachin (1978)
Morone saxatilis	Striped bass	Fish, crustaceans	Hollis (1952)
Lutjanus griseus	Gray snapper	Fish, crustaceans	Croker (1962), Odum and Heald (1972)
Diapterus plumieri	Striped mojarra	Mysids, amphipods, mollusks, ostracods, detritus, copepods	Odum and Heald (1972)
Eucinostomus gula	Silver jenny	Copepods, amphipods, mollusks, detritus, mys-ids	Odum and Heald (1972)
E. argenteus	Spotfin mojarra	Amphipods, copepods, mysids, mollusks, detri-tus	Odum and Heald (1972)
Archosargus probatoce-phalus	Sheepshead	Shrimp, mollusks, small fish, crabs, other crus-taceans, algae, plant detritus	Reid et al. (1956), Odum and Heald (1972), Ste-venson and Confer (1978)
Lagodon rhomboides	Pinfish	Fish, crustaceans, vascu-lar plants, algae, detri-tus, copepods, mysids, mollusks	Linton (1905), Smith (1907), Hildebrand and Schroeder (1928), McLane (1955), Han-son (1969), Odum and Heald (1972), Carr and Adams (1973), Kjelson et al. (1975), Adams (1976a)
Bairdiella chrysoura	Silver perch	Decapods, schizopods, copepods, mysids, am-phipods, polychaetes, ectoprocts, fish, detritus	Linton (1905), Hilde-brand and Schroeder (1928), Hildebrand and Cable (1930), Reid et al. (1956), Darnell (1958), Thomas (1971),

TABLE 4 (continued)
Primary Foods of Some Estuarine Fishes Along the U.S. Atlantic and Gulf Coasts

Scientific name	Common name	Food habits	Ref. (sources from Ref. 18)
			Odum and Heald (1972), Carr and Adams (1973), Stickney et al. (1975)
Cynoscion nebulosus	Spotted seatrout	Copepods, decapods, mysids, carideans, fish, mollusks	Moody (1950), Reid et al. (1956), Darnell (1958), Springer and Woodburn (1960), Odum and Heald (1972)
C. regalis	Weakfish	Polychaetes, copepods, amphipods, mysids, stomatopods, decapods, fishes	Thomas (1971), Stickney et al. (1975)
Leiostomus xanthurus	Spot	Polychaetes, copepods, isopods, amphipods, mysids, cumacea, fishes	Linton (1905), Welsh and Breder (1923), Hildebrand and Cable (1930), Reid (1954), Roelofs (1954), Darnell (1958), Diener et al. (1974), Stickney et al. (1975), Stickney and McGeachin (1978)
Micropogonias undulatus	Atlantic croaker	Polychaetes, mollusks, amphipods, isopods, copepods, decapods, stomatopods, mysids, cumacea, ascidians, fish	Linton (1905), Welsh and Breder (1923), Gunter (1945), Roelofs (1954), Reid et al. (1956), Darnell (1958), Diener et al. (1974), Stickney et al. (1975), Roussel and Kilgen (1975), Stickney and McGeachin (1978)
Stellifer lanceolatus	Star drum	Amphipods, isopods, copepods, cumaceans, mysids, stomatopods, decapods, fish	Stickney et al. (1975)
Chaetodipterus faber	Atlantic spadefish	Small crustaceans, annelids, detritus, ctenophores	Smith (1907), Hildebrand and Schroeder (1928), Breder (1948)
Mugil cephalus	Striped mullet	Algae, detritus, vascular plants, crustaceans, bacteria, diatoms	Darnell (1958), Odum (1968), Odum and Heald (1972), Moriarity (1976)
Gobiosoma robustum	Code goby	Amphipods, mysids, insect larvae, cladocerans, algae, detritus, mollusks	Odum and Heald (1972)
Ancylopsetta quadrocellata	Ocellated flounder	Mysids, copepods, other crustaceans, polychaetes, fish	Stickney et al. (1974)
Citharichthys spilopterus	Bay whiff	Mysids, other crustaceans, fish	Stickney et al. (1974)
Etropus crossotus	Fringed flounder	Polychaetes, mollusks, copepods, isopods	Stickney et al. (1974)

TABLE 4 (continued)
Primary Foods of Some Estuarine Fishes Along the U.S. Atlantic and Gulf Coasts

Scientific name	Common name	Food habits	Ref. (sources from Ref. 18)
Scophthalmus aquosus	Windowpane	Mysids, other crustaceans, fish	Stickney et al. (1974)
Trinectes maculatus	Hogchoker	Annelids, algae, amphipods, detritus, foraminifera, plant seeds, copepods, insect larvae, mollusks, cumaceans	Hildebrand and Schroeder (1928), Reid (1954), McLane (1955), Darnell (1958), Odum and Heald (1972), R. R. Stickney (unpublished data)
Symphurus plagiusa	Blackcheek tonguefish	Mollusks and crustaceans	Stickney (1976)

From Stickney R. R., *Estuarine Ecology of the Southeastern United States and Gulf of Mexico*, Texas A & M University, College Station, 1984. With permission.

Anchoa, Brevoortia, and young sciaenids), and ending with large piscivores.[91] Darnell[91] surmised that detritus was a potentially valuable food for the estuarine fauna. Juvenile and small adult fishes in Lake Pontchartrain are intermediaries in the food chain, providing a means by which energy of microcrustaceans, clams, penaeid and palaemonid shrimp, and blue crabs transfers to the piscivores.[92]

Diener et al.,[93] dissecting the stomach contents of 40 species of fishes in Clear Lake, TX, a small bay at the northwestern perimeter of the Galveston Bay system, could not identify the two primary food chains advocated by Darnell[91] for Lake Pontchartrain. Principal items in the diet of the 40 species consisted of copepods, grass shrimp, river shrimp, mysids, penaeids, polychaetes, and detritus (Table 4). Shallow-water, nearshore species were not adequately represented in the study; however, Alexander[94] later assessed the diet and trophic position of finfishes from nearshore habitats of West Bay, located on the southwestern margin of the Galveston Bay system. Juveniles of the sheepshead minnow (*Cyprinodon variegatus*), tidewater silverside (*Menidia peninsulae*), pinfish (*Lagodon rhomboides*), spot (*Leiostomus xanthurus*), and striped mullet (*Mugil cephalus*) numerically dominated the community. The sheepshead minnow and striped mullet, both bottom-feeding herbivores, primarily consumed microalgae and vascular plant material. Small fish and crustaceans comprised much of the diet of the gulf killifish, with microalgae, vascular plant material, and small benthic animals being a minor component. With both a planktonic diet of copepods and meroplankton and a benthic diet largely of microalgae, the tidewater silverside showed the most varied feeding habit. The pinfish and spot had an omnivorous diet, ingesting microalgae, vascular plant matter, and meiobenthos from the bottom.

The food web of Barnegat Bay, NJ, incorporates two interlocking components: (1) a detritus-generated component; and (2) a phytoplankton (grazing)-generated component (see Chapter 6, Section II.A.2.c and Section II.B.2.c).[95] In the detritus-generated food web, detritus mostly composed of dead eelgrass (*Zostera marina*), benthic algae, and salt marsh grasses (*Spartina alterniflora*) is the major energy source. Micro-, meio-, and macrofauna ingest the microbes attached to the detritus and are, in turn, preyed on by carnivorous invertebrates and fishes. The piscivorous bluefish and weakfish act as top predators in the system. In the grazing-generated food web, live phytoplankton, and, to a lesser degree, live microalgae, macroalgae, and vascular plants are the principal energy sources. Herbivores (e.g., epifaunal benthos, zooplankton, and

TABLE 5
Feeding Strategies of Representative Fishes of Barnegat Bay, NJ

Species	Feeding strategy
Alosa aestivalis (blue herring)	Primarily planktivorous; consumes copepods, mysids, ostracods, small fishes
Ammodytes americanus (sand lance)	Principally feeds on small crustaceans
Anchoa mitchilli (bay anchovy)	Primarily planktivorous; consumes mysids, copepods, organic detritus, small fishes
Anguilla rostrata (American eel)	Omnivorous; consumes annelids, crustaceans, echinoderms, mollusks, eelgrass, small fishes
Bairdiella chrysoura (silver perch)	Omnivorous; consumes calanoid copepods, annelids, fishes, organic detritus
Brevoortia tyrannus (Atlantic menhaden)	Planktivorous; detritivorous; principally filter-feeds on phytoplankton; also feeds on zooplankton and organic detritus
Caranx hippos (crevalle jack)	Fish predator
Centropristis striata (black sea bass)	Omnivorous; benthophagous; feeds on crustaceans, mollusks, fishes, plants
Cynoscion regalis (weakfish)	Primarily a fish predator; mostly consumes anchovies and silversides; also feeds on small crustaceans
Fundulus heteroclitus (mummichog)	Omnivorous; consumes small animals and plants
Gasterosteus aculeatus (threespine stickleback)	Omnivorous; feeds on algae, copepods, fish eggs
Gobiosoma spp. (gobies)	Consume small invertebrates and fishes
Leiostomus xanthurus (spot)	Detritivore; benthophagous; consumes organic detritus and microbenthos
Menidia menidia (Atlantic silverside)	Omnivorous; feeds on small crustaceans, annelids, algae, insects
Morone americana (White perch)	Omnivorous; consumes annelids, crustaceans, fishes, organic detritus
M. saxatilis (striped bass)	Primarily a fish predator; also feeds on annelids, crustaceans, mollusks
Opsanus tau (oyster toadfish)	Omnivorous; principally consumes annelids, crustaceans, fishes
Paralichthys dentatus (summer flounder)	Predator; feeds on fishes, crustaceans, mollusks
Pomatomus saltatrix (bluefish)	Fish predator; consumes large numbers of Atlantic menhaden and bay anchovy
Pseudopleuronectes americanus (winter flounder)	Benthophagous; feeds on annelids, crustaceans, mollusks
Sphoeroides maculatus (northern puffer)	Primarily consumes small crustaceans
Stenotomus chrysops (scup)	Benthophagous; feeds on annelids, crustaceans, small fishes
Syngnathus sp. (pipefish)	Primarily feeds on amphipods and copepods
Tautog onitis (tautog)	Consumes crustaceans, mollusks, and other invertebrates

From Kennish, M. J. and Loveland, R. E., *Ecology of Barnegat Bay, New Jersey,* Kennish, M. J. and Lutz, R. A., Eds., Springer-Verlag, New York, 1984, 302. With permission.

nekton) consume this live plant matter. Primary, secondary, and tertiary carnivores occupy succeeding trophic levels. Once again, top predators include piscivorous fishes.

Table 5 depicts the feeding strategies of representative finfish taxa in the estuary. Only a few species eat detritus, the most notable being spot (*L. xanthurus*). Many of the populations are omnivorous; the American eel (*Anguilla rostrata*), black sea bass (*Centropristis striata*), mummichog (*Fundulus heteroclitus*), Atlantic silverside (*Menidia menidia*), and oyster toadfish (*Opsanus tau*) fall into this category, subsisting on a mixture of planktonic and benthonic organisms as well as detritus. Piscivorous species of repute include the crevalle jack (*Caranx hippos*), weakfish (*Cynoscion regalis*), striped bass (*Morone saxatilis*), summer flounder (*Paralichthys dentatus*), and bluefish (*Pomatomus saltatrix*).

Estuarine fishes encompass a wide variety of feeding types. Some herbivorous forms ingest benthic and epiphytic diatoms, filamentous algae (e.g., *Enteromorpha*), and seagrasses (e.g., *Ruppia, Posidonia,* and *Zostera*). However, the existence of specialized herbivores among fish communities in estuaries is highly unlikely.[2] Decaying plant matter, detritus, and attached epifloral and epifaunal populations directly support a few fish populations. A number of species of the Bothidae, Blennidae, Lethrinidae, Lutjanidae, Gobiidae, Platycephalidae, Pleuronecti-dae, Sciaenidae, and Sillaginidae are benthic feeders that remove significant quantities of benthic invertebrates. Whitfield and Blaber[96] have sorted predatory fishes into two groups. The first group consists of high speed swimmers with lunate or forked tails (e.g., *Caranx ignobilis* and *Lichia amia*). The second, slower group of piscivores has truncate or rounded tails (e.g., *Elops machnata*). The volume of food eaten by estuarine fishes is a function of their age and physiological activity, as well as the water temperature and caloric value of the food.[2]

V. POPULATION DYNAMICS

Topics related to the population dynamics of fish center around growth, mortality, and recruitment. Knowledge of the age of a fish is essential to investigations of its population dynamics and resource management. The age composition of a population is useful in the assessment of survival and mortality rates. Research on age and growth rates of fish populations continue to be critical areas of study by fisheries biologists.

Methods used in estimating growth rates of fish were determined early, having been evolved by 1910. Procedures on survival rates progressed more slowly, with many new methods proposed in the literature during the last three decades.[97] By the 1930s, the Graham-Schaefer model and the Russell equation supplied the basic principles of fish population dynamics. Development of the yield per recruit model by Beverton and Holt[98] and formulation of the concept of dependence of recruitment on parent stock led to great advances in fisheries science.[75] The overall production of a fish stock and work on stock/recruitment problems are subjects of intense interest in the field that have resulted in a burgeoning volume of information over the years.

A. PARAMETERS
1. Age

Data on the age of fishes are principally acquired by employing three methods, namely, mark and recapture of individuals, length-frequency distributions, and interpretation of layers recorded in the hard parts of specimens.[99] The recovery of marked fish of known age, while costly and time consuming, provides an accurate data base. It is most effectively utilized, however, in conjunction with other aging methods.

Length-frequency distributions of fish of one age tend to follow a normal distribution pattern. By plotting the length of the fish on the X-axis and frequency on the Y-axis, a succession of peaks appears on the graph. A count of the peaks yields an approximation of age, which in many cases suffices only for the first 2 to 4 years because of the coalescing of the peaks due to increased dispersion of the data. Although the overlapping in length distribution reduces the reliability of the method for older age groups, it often is used as an effective check on the scale method of age determination. The length-frequency method suffers from several other disadvantages that affect its usage. First, a particular year class of fish may be absent or poorly represented in a sample. Second, size groupings of fish sometimes are not indicative of year classes due to irregular episodes of hatching. Third, a portion of a year class of fish may be equal in age but not in size owing to differential development under variable conditions. Fourth, fish of a certain size tend to school together.[99]

Age determinations of fish generally are accomplished by interpretations made on the selective hard parts of the organism. Scales and otoliths serve as the most popular structures for aging fish and calculating their growth. Spines and rays, in addition to bony structures (e.g., vertebrae, dentary bone, and cross sections of other bony structures), serve as alternate structures for estimating age. Both ctenoid and cycloid scales are commonly utilized in aging research, the ctenoid scales usually removed from spiny fishes and the cycloid scales, from fishes with soft-rayed fins. Prerequisites have been established for using scales in age and growth studies of fish. Conditions that must be met include: (1) yearly formation of an annulus at approximately the same time each year, (2) proportional growth of the scale to the growth of the fish, and (3) constancy in the number and identity of scales throughout ontogeny.[99] Fisheries biologists must be able to differentiate the true annulus from false annuli to obtain precise data on age and growth.

For scaleless fish or fish with irregularities in sculpturing that prevent accurate interpretations, otoliths represent a favorable alternative in age determinations. Otoliths are calcareous concretions within the labyrinth system of fishes;[100] three otoliths exist on each side of the head cavities at the base of the semicircular canals and are named astricus, sagitta, and lapillus. The largest of the three mineralized structures, the sagitta, is the most intensely studied otolith. Subdaily, daily, and multidaily growth patterns have been delineated in fish sagittae.[101] Fish otoliths are not only valuable as indicators of age, but also as continuous recorders of somatic growth, temperature regime, and resource availability.[102-105] As such, they monitor conditions of the environment throughout the ontogeny of the animal.

When otoliths cannot be studied, annular rings in spines and rays can be counted to ascertain the organism's age. Preparations of the spines and rays for age identification involve cutting transverse sections near the base. This procedure does not require sacrificing the fish as in the case of obtaining data from otoliths.

2. Growth

Most fish grow throughout ontogeny, increasing in length or mass with age albeit at a gradually decreasing rate with age. Growth is not continuous, however, and somatic growth checks can be produced from changes in physical-chemical conditions (e.g., sudden fluctuations in temperature, salinity, and light) and biological processes (e.g., spawning events). Moyle and Cech (p. 100)[24] explain growth simply as "the change in size (length, weight) over time or, energetically, as the change in calories stored as somatic and reproductive tissue". The rate at which a fish grows is mainly a function of the amount of food energy ingested (I), measured in calories, which must emerge as energy expended for metabolism (M) or growth (G), or as energy excreted (E).[106,107] This relationship may be written as

$$I = M + G + E \tag{1}$$

In this energy equation, M incorporates the calories expended for digesting food, for body maintenance, and for movement, and E includes excreted energy as feces, urea, and ammonia. The remaining energy accounts for growth (G) of the organism.

The growth rate of fish is generally determined by observing changes in body weight (mass) or length per unit of time. Most investigators employ one of five methods to establish growth rates. They are (1) mark-and-recapture experiments, (2) length-frequency distributions, (3) back calculations from rings on scales and otoliths, (4) growth measurements of fish reared experimentally in controlled environments, and (5) radiocarbon uptake.[24]

Several models have been devised in fisheries research to evaluate growth. The von Bertalanffy, logistic, and Gompertz equations comprise three commonly used models applied to the study of finfish growth. The Walford plot also is a valuable technique. Pitcher and Hart[52] examine these models, and much of the material presented here is derived from their work. They express the von Bertalanffy equation as

$$l_t = L\infty\left(1 - \exp\left(-k\left(t - t_0\right)\right)\right) \qquad \text{for length} \qquad (2)$$

and

$$w_t = W\infty\left(1 - \exp\left(-k\left(t - t_0\right)\right)\right)^3 \qquad \text{for weight} \qquad (3)$$

where l_t is the length and w_t is the weight of the organism, respectively, at time t (which usually represents the age of the fish), $L\infty$ is the length and $W\infty$ is the weight at infinity, and age k is a growth parameter measuring how quickly the length approaches $L\infty$ or $W\infty$. Larger values of k cause $L\infty$ or $W\infty$ to be approached more rapidly. The derivative for Equation 1 has the form

$$l' = k\left(L\infty - l\right) \qquad (4)$$

which can be arranged to

$$l' + kl = kL\infty \qquad (5)$$

with the initial condition $l_0 = l\ (t_0)$.[108]

Based on the von Bertalanffy equation, fish grow toward a theoretical length or weight, with the rate of change of size declining as the length approaches the maximum value. It can be fitted by a least squares method,[109] or alternatively via the Walford plot.[97] The model fits most observed fish growth, particularly beyond the point of inflection in the absolute growth curve.[99]

The logistic equation for weight is

$$W_t = \frac{W\infty}{1 + \exp\left(-g\left(t - t_0\right)\right)} \qquad (6)$$

where g refers to the instantaneous rate of growth as w_t tends to zero, and t_0 signifies the inflection point of the curve. All other terms in the equation are the same as defined above. The rate of change of weight equals zero at the inflection point. As in the case of the von Bertalanffy model, the logistic reflects size tending to an asymptote.

Another curve in which weight at age tends to an asymptote is the Gompertz. It differs from the von Bertalanffy and logistic curves, however, in that the curve shows an asymmetrically placed inflection point, at an age less than half way up the curve. This equation reads

$$W_t = W\infty \exp\left(-g\left(t - t_0\right)\right) \qquad (7)$$

where g equals the instantaneous rate of growth when $t = t_0$, and t_0 is the inflection point of the curve at $W\infty/\exp(1)$. Ricker[97,110] explains methods for fitting the Gompertz model.

The Walford plot is an effective method for estimating growth parameters. This technique, although not as accurate as current computer programs that calculate parameters (e.g., the parameters $L\infty$, K, t_0, or l_0 in the von Bertalanffy equation), nonetheless yields a useful graphic solution for estimating the parameters. When plotting the length of a fish at age one against the length of the next younger age, a straight line results for many populations.[111] The von Bertalanffy growth curve will fit the data if the resulting Walford plot follows a straight line having a slope of less than one. The growth coefficient K can be estimated by taking the natural logarithm of the Walford slope, which equals e^{-K}, with sign changed. The point at which the growth line intersects a 45 line drawn through the origin gives an estimate of the ultimate length $L\infty$ (Figure 4).

The Walford plot is also worthwhile in tagging experiments. By tagging fish and measuring

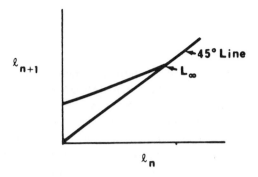

FIGURE 4.   The Walford growth transformation of plotting length at age n + 1 against length at age n. (From Everhart, W. H., Eipper A. W., and Youngs, W. D., *Principles of Fishery Science*, Cornell University Press, Ithaca, 1975. With permission.)

their size at recapture, growth curves can be drawn from data on individuals not aged directly. Moreover, growth estimates on ages not readily available from scales or otoliths can be generated via the Walford line.[99]

3. Mortality

Fisheries biologists focus on two sources of mortality when dealing with fish populations: natural and fishing. The death rates of fishes in the absence of fishing are poorly known for many populations. The death caused by fishing, often exceeding the natural death rate, tends to obscure mortality investigations.[75] In spite of this complication, it is unequivocal that the specific death rate of fishes drops with age, thus paralleling the specific growth rate. The decrease of the specific death rate with age, therefore, indicates that the death of fishes from metamorphosis onward follows an exponential function that is modified toward an asymptote at infinite age. The effects of fishing and natural mortality may be expressed as

$$N_t = N_0 e^{-(F + M)t} = N_0 e^{-Ft} e^{-Mt} \tag{8}$$

where N_t is the number of fish at time t, N_0, the number of fish at time 0, F, the fishing mortality, and M, the natural mortality.

The estimation of survival and expectation of fishing are two important parameters considered in many fisheries management programs. As Everhart et al. (p. 106)[99] expound on Equation 8 "... survival is equal to the probability of living for a specified time period which is equal to the probability of not being caught by fishing, e^{-Ft}, multiplied by the probability of not dying from natural causes, e^{-Mt}. Mortality is the complement of survival; if A is mortality, then A = 1 – S ... If death from fishing and natural death are the two risks, then the expectation of death from fishing, E, is the chance or probability that a fish dies from fishing during the time period being considered — when both fishing and natural risks are present. The expectation of death from natural causes, D, is the probability that a fish dies of natural causes during the time period — when both natural and fishing risks are present."

In evaluating the expectation of death associated with two risks — fishing and natural death — the expectation of death from fishing, E, represents the probability that a fish dies from fishing during an interval when both fishing and natural risks exist. The expectation of death from natural causes, D, is the probability that a fish dies from natural causes during an interval when both natural and fishing risks exist. The expectation of death from fishing may be computed as

$$\frac{F}{Z} = \frac{E}{(1 - S)}$$

(9)

where Z refers to the force of total mortality and the other symbols, as previously specified. This relationship holds only when F/Z remains constant. It has been conventionally calculated for an annual time period, although other time periods also can be used.[99]

Fish typically are characterized by smaller numerical populations, greater longevity, and lower mortality rates than invertebrates. In addition, they have more restricted spawning periods. Reproductive curves of fish usually incorporate an underlying assumption that density-dependent mortality operates in a population. Disease, predation, cannibalism, and depleted ration are examples of factors responsible for density-dependent types of mortality. They arise from actions of the fish population in contrast to density-independent types of mortality that are unrelated to any population action. Examples of density-independent types of mortality include those incurred from storms, floods, temperature and salinity extremes, and anthropogenic impacts on the environment. In essence, environmental variation can account for a substantial fraction of the total mortality due to density-independent effects.

Density-dependent factors normally elicit compensatory responses in mortality of fishes during the critical early life-history stages. Consequently, a dense adult stock would favor higher mortality among the larvae, and a sparse adult population, lower mortality among the larvae. The long-term effect of compensation is to move a population toward a numerical average.

Survivorship rates are inversely related to fecundity; hence, high fecundities in fishes correlate with high death rates, especially during the embryonic and larval stages.[24] For pelagic larvae, even small fluctuations in survival rates translate into significant changes in the adult population size. Fecundity increases with female size according to the equation

$$F = aL^b$$

(10)

where F equals fecundity, L, fish length, and a and b, constants derived from the data.[112] The following section deals with the survivorship of viviparous, ovoviparous, and oviparous fish larvae.

4. Reproduction and Recruitment

Fish exhibit three reproductive conditions (i.e., bisexual, hermaphroditic, and parthenogenetic). The most prevalent type of condition is bisexual in which sperm and egg development occurs in separate male and female individuals. Fewer hermaphrodites — those species bearing both sexes in one individual — inhabit estuarine and marine waters. The development of young without fertilization (parthenogenesis) is rare, having been reported in the tropical Amazon molly, *Poecilia formosa*. In this fish, the sperm incites the egg to develop, but does not take part in heredity.[113]

Early developmental strategies of fish fall into three categories, namely, viviparous, ovoviparous, and oviparous strategies. In viviparous fish, eggs hatch inside of the female, and young meet their nutritional, excretory, and respiratory requirements via a connection with maternal tissues. In ovoviviparity, eggs hatch within the female fish and developing young obtain nutrition from yolk present in the egg. Oviparity differs from the other strategies in that eggs hatch outside of the body of the female and may or may not be protected by the parents. Among the various fish orders and families, viviparity takes place in the elasmobranchs and in the teleost orders Atheriniformes, Ophidiiformes, Gadiformes, and Perciformes. Ovoviviparity is found in the elasmobranchs, Coelacanthiformes, and the teleost orders Atheriniformes, Ophidiiformes, Gadiformes, Scorpaeniformes, and Perciformes. The most common strategy, oviparity, has been observed or is suspected in all orders of gnathostomous fishes, in the elasmobranchs, and in the Coelacanthiformes.[114-116]

The reproductive potential of estuarine and marine fishes varies among viviparous, ovoviparous, and oviparous forms. The viviparous species have low fecundity. However, the young generally develop to a larger size prior to hatching, and, on the average, a high percentage of them survive. Because of their large size, the young of viviparous fishes are less susceptible to predation and typically exert greater control in obtaining food. The population sizes tend to be more constant through time.

In contrast, oviparous fishes produce many more eggs, which are unprotected and with limited yolk reserves. Survival is closely linked to the duration of the planktonic egg and/or larval stage. The dispersal ranges of these species greatly surpass those of the viviparous forms as do tneir rates of mortality, especially through the free embryo and larval stages. The number of planktonic eggs and early stage larvae lost as currents transport them to unfavorable environments is high. Factors limiting survival of ichthyoplankton are less than optimal salinity and temperature levels, predation, inadequate ration, feeding effects, and anthropogenic impacts. Regardless of the sensitivities of the ichthyofauna, a planktonic existence early in ontogeny confers certain advantages for a species. Oftentimes, for example, the ichthyoplankton are removed from areas of intense predation and spread over wide areas. This dispersal reduces competition for available food resources since the larvae are not confined to a restricted area. In addition, a planktonic habit decreases the probability that a localized biological perturbation will devastate an entire year class through direct effects on the species or its food. Larval dispersal is of major significance to benthic-dwelling species whose adults do not migrate long distances.[115]

Some investigators link the reproductive success of fish populations to fecundity. Estimating fecundity may be a relatively simple process for some fishes, when their eggs mature synchronously and are shed in a batch over a brief time interval each year.[117] In these populations, fecundity can be computed directly by enumerating the ripening eggs generated per female. In other species — termed multiple spawners, batch spawners, serial spawners, or functional spawners — fish shed multiple batches of eggs successively within a single spawning season.[118,119] The assessment of fecundity of multiple spawners has not been highly successful due to problems in deciphering annual egg production or the number of batches of eggs spawned annually.[119]

Fertility, rather than fecundity, should be a better measure of reproductive success. Whereas fecundity signifies the number of eggs generated per female, fertility denotes the actual number of young produced. Because of the dispersal capability of young fishes, fertility cannot be measured as easily as fecundity. Nevertheless, fecundity approximates fertility in many species, notably viviparous fishes.[24]

Recruitment variability has been related to the variability in larval survival, which is partially dependent on the feeding success of the larval fish.[120] Some larval fish are more advanced ontogenetically than others and, accordingly, have greater capacities to deal with vagaries in food supply and other environmental conditions. Rothschild and DiNardo,[120] for instance, scrutinized recruitment variability of marine and anadromous fishes and the relationship of recruitment variability to life history and morphological characteristics (i.e., relative fecundity, egg size, incubation time, size of hatching, duration of the yolk-sac larval stage, eye development, and mouth development). They disclosed that recruitment variability of marine fish seems to be less than that of anadromous fish. Having become more specialized than marine fish, the anadromous forms require more suitable environmental conditions per egg and, as a consequence of being more highly tuned to a particular set of environmental conditions, are susceptible to population stress incurred from environmental perturbations.

Major sources of recruitment losses arise from predation as well as food and feeding effects. Predation by schooling juvenile and adult fish clearly imposes significant impacts on fish eggs and larvae. The feeding of predators on patches of eggs and larvae can profoundly increase mortality rates. Indeed, the relative year-class strength of adult fish stocks may be coupled to the spatial patchiness of eggs and larvae. Early life-history survival of fishes, therefore, would appear to be regulated by the physical and biological mechanisms controlling the spatial patchiness of eggs and larvae.[121] A lack of adequate food availability and/or an inability of larval fish to obtain prey are factors that can affect the year-class strength of a species.[115,122,123]

Recruitment may be defined simply as the addition of new members to a fish population. However, recruitment to a fishery occurs at a stage in ontogeny when the fish initially is vulnerable to capture by gear used in the fishery.[99] Quantitative treatment of recruitment has met with limited success, particularly in providing predictive estimates of recruitment.

General stock-recruitment models have been introduced in an effort to properly manage fisheries. Two models gaining widespread acceptance over the years are those of Beverton and Holt[98] and Ricker.[124] Pitcher and Hart (p. 182)[52] differentiate these two models in the following way

> The Ricker recruitment equation describes a family of humped curves with low recruitment at high stock levels, whereas the Beverton and Holt model covers a family of asymptotic curves exhibiting constant recruitment beyond a certain stock density. The Ricker curve implies strong density dependence, increasing geometrically over a certain range of stock densities. The Beverton and Holt curve implies an arithmetically progressive reduction in the recruitment rate as stock density increases.

Ricker's recruitment curve can be defined by the equation

$$R = \alpha S \exp(-\beta S) \tag{11}$$

where R specifies the recruits, S is the adult spawning stock, and α and β represent the parameters of the curve.

The asymptotic recruitment curve of Beverton and Holt is delineated by

$$R = \frac{1}{\alpha' + \beta' / S} \tag{12}$$

where α' and β' are new parameters and R and S, the same as above. The asymptote increases while the curvature is reduced as α decreases. As β declines, the asymptote is approached more rapidly.[52]

Larval survival and recruitment modeling has been attempted, but only a few models have proven to be useful in fisheries research. The gadoid recruitment model of Jones and Hall[125] and the bioenergetic model for winter flounder by Laurence[126] are both valuable. Information on these two models is presented by Pitcher and Hart.[52]

Many workers have related the seasonal variation in the species composition of estuarine fish faunas to their reproductive habits and migratory behavior. Haedrich and Haedrich,[90] for example, concentrate on the seasonal dominance patterns of fishes in Mystic River, MA. In the spring, the anadromous species, *Alosa pseudoharengus,* is dominant, being supplanted by metamorphosing juveniles and adults of the winter flounder, *Pseudopleuronectes americanus,* in the summer and fall, respectively. The anadromous fish, *Osmerus mordax,* predominates during its winter spawning migration.

Healthy temperate estuaries display marked seasonal changes in their local fish communities.[3,127] Seasonal alterations in the fish faunas of temperate systems provide an important mechanism for resource sharing.[10] Seasonality promotes a turnover of fish populations. This enables multiple use of the estuary by the fish community; those systems harboring the most species exist in regions of high seasonal turnover.[3] In less equitable environments with static conditions, fewer species are evident, and usually only the most hardy species will inhabit systems heavily stressed by pollution. Where seasonal changes are acute, fish abundance and/or diversity may have multimodal peaks over the year. Several conditions create this multimodality, including spawning migrations, visits by predatory fishes (e.g., bluefish, *Pomatomus saltatrix*, and jacks, *Carangidae*), and expanding populations of young fish spawned within the estuary.[10]

Long-term studies of fish communities corroborate the findings of short-term investigations over the span of a single year — that the overall abundance and diversity of fishes in estuaries are reasonably predictable as a function of season. Hence, during a 4-year period, McErlean et al.[128] monitored a cyclical pattern in abundance and diversity of fishes in the Patuxent River estuary in Maryland (U.S.). Similar patterns were connoted by Livingston et al.[129] in Apalachicola Bay, FL and by Hillman et al.[20] in Long Island Sound, NY. However, these two studies also exposed rather significant temporal changes in species composition from one year to the next.

VI. SPECIES ABUNDANCE AND DIVERSITY

Estuaries are physically unstable areas characterized by large spatial and temporal variations in temperature, salinity, oxygen concentration, turbidity, and other factors.[130] Temporally, such variations take place in the short term (e.g., over a tidal cycle and season) and long term (e.g., from year to year). Yet, despite these variations, the basic structure of estuarine fish communities is reasonably stable,[131] and the fishes often have more or less predictable patterns of abundance and distribution.[24] Not uncommonly, however, estuarine fish populations change dramatically in response to environmental perturbations; these population changes can be permanent even though the predominantly estuarine species have broad temperature tolerances and strong osmoregulatory abilities.[3] The species composition of estuarine communities changes constantly, attesting to the variable environmental conditions and the limitations of the tolerances of the fish populations to alterations in the habitat. The findings of Richardson and Pearcy[132] and Priede et al.[133] demonstrate how variations in salinity, temperature, dissolved oxygen, substrate, detritus, and river discharge contribute to annual variations in the fish populations.

Moyle and Cech[24] attribute the stability of the basic structure of estuarine fish communities to four principal conditions. These are (1) the regular distribution of fish populations along environmental gradients of temperature, salinity, and other variables; (2) seasonal migration and emigration movements of fishes into and out of the estuaries; (3) the dominance of estuaries by relatively few species; and (4) the robust food webs. The authors derive examples of these effects from estuaries along the Gulf of Mexico.

An overview of estuarine fish faunas can be drawn from investigations conducted worldwide. Estuaries support large stocks of fish, primarily juvenile forms. Temperate estuaries serve as major nursery grounds for fishes.[10] While the abundance and biomass of fishes in estuaries are large, the species diversity is generally low. Less than 100 species of fishes are frequently recorded in estuaries. Bechtel and Copeland,[134] for example, registered 55 species in Galveston Bay, TX, and Richards and Castagna,[135] 70 species in Chesapeake Bay. Dahlberg and Odum[136] collected 70 species in a Georgia salt marsh-estuary, which compares favorably with the 73 species recovered by Livingston[137] in the Econfina-Fenholloway system of Florida. In a survey of California bays and estuaries, Horn and Allen[138] calculated an average of 66 species of fish for an area of approximately 5 ha.

Only a few species generally dominate estuarine fish faunas as mentioned above, and they usually are widespread forms having broad tolerances and a wide range of adaptations.[10] Thus, Webb[139] determined that 9 of 28 species of fish in the Avon-Heathcote estuary, New Zealand, accounted for 90% of the total number of individuals. Oviatt and Nixon[89] likewise documented an overriding abundance of a few fish species in Narragansett Bay, with the 10 most abundant forms (of a total of 99 sampled) comprising 90% of all individuals. In spite of the dominance by relatively few populations, the estuarine fish community experiences temporal changes in species composition because of the basic mobility of the fishes and their seasonal migrations coupled to reproductive and feeding cycles. Although at first glance estuarine fish faunas seem to be simple in many respects, seasonal and annual variations result in considerable complexity within the communities.[140]

VII. FISH COMMUNITIES

A. TYPE EXAMPLES
1. Barnegat Bay

Extensive studies of the ichthyofauna of Barnegat Bay were conducted from September 1975 through August 1981 using trawls, seines, gill nets, and bongo nets.[141] Most samples were taken along the western perimeter of the bay, with fewer collections made near Barnegat Inlet and extensive shoals in the eastern segment (Figure 5). Tatham et al.[27] and Vouglitois et al.[141] explain the sampling programs and life-history studies performed during the 6-year period.

According to Tatham et al.,[27] the fish community of Barnegat Bay is characteristic of mid-Atlantic estuaries and embayments, in general, and representative of the fish communities of New Jersey coastal bays. As elsewhere, only a few species numerically dominate the fish community of this estuary. For example, the bay anchovy (*Anchoa mitchilli*), Atlantic silverside (*Menidia menidia*), fourspine stickleback (*Apeltes quadracus*), spot (*Leiostomus xanthurus*), and winter flounder (*Pseudopleuronectes americanus*) constituted more than 90% of the total number of fish sampled in bay collections from September 1975 through August 1978, being responsible for 57.9, 22.1, 4.2, 3.8, and 1.9% of all individuals, respectively. During this period, 107 species belonging to 57 families were recovered in the bay (Table 6).

Based on their spatial and temporal occurrence and their relative abundance within or outside the bay, the fishes were classified into five general assemblages: (1) residents (20 species) occupying the estuary year-round, (2) warm-water migrants (34 species) abundant primarily from April through November, (3) cool-water migrants (12 species) present from November through April, (4) marine strays (42 species), and (5) freshwater strays (7 species). Resident species comprised 31% of all fish collected, warm-water   migrants 65%, cool-water migrants 3%, and marine and freshwater strays 1%. Most individuals were either small forage fishes, principally resident in the estuary, or young and juveniles of marine species present only seasonally. The diversity of fishes peaked from late summer through midfall (41 to 47 species per month). Warm-water migrants accounted for an increase in diversity from spring through fall. The number of species dropped sharply in the winter when only a few residents and cool-water migrants (13 species) inhabited the bay.

Both resident and migratory fishes used the bay as a spawning area, with most reproduction triggered in the spring, summer, and winter. *Anchoa mitchilli, Gobiosoma* spp. (gobies), *M. menidia*, and *Syngnathus fuscus* (northern pipefish) were the main spawners during the spring and summer months, and *Ammodytes* sp. (sand lance) and *P. americanus,* during the summer months. From May through October, the young of most resident species (19 of 20 species) and the young of many warm-water migrants (21 of 34 species) utilized the estuary as a nursery area.

Larval and young stages of *Ammodytes* sp., *Micropogonias undulatus*, and *P. americanus* use the estuary as a nursery in the winter and early spring, as do young *Apeltes quadracus*. Immature *P. americanus* live in the bay year-round. Table 7 specifies the usage of Barnegat Bay by residents, warm-water migrants, and cool-water migrants.

The absolute abundance of fishes in Barnegat Bay is highest from May through November due to the arrival of warm-water migrants and recruitment from spawning populations in the estuary. Far fewer individuals are present during the winter, although an increase in abundance becomes evident as early as March or April (Figure 6). Larvae and juveniles attain maximum numbers in the spring and summer months (Figure 7). Annual variations in absolute abundance of 50 to 100% are not unusual. Fluctuations in environmental conditions that influence reproductive success may precipitate such large variations in abundance.

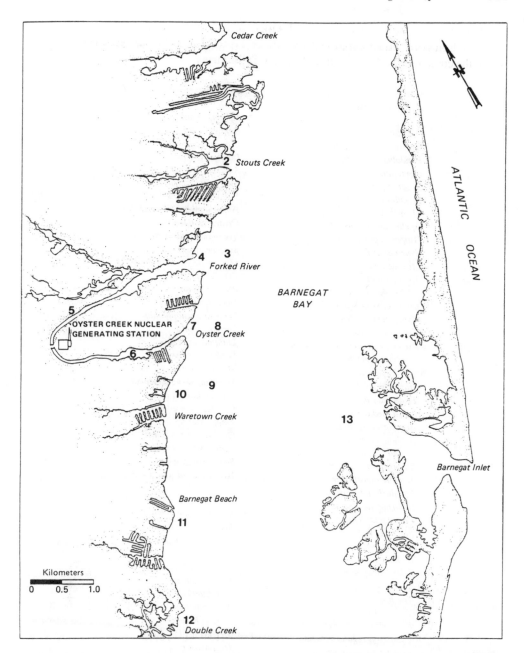

FIGURE 5. Fish sampling locations in Barnegat Bay from September 1975 through August 1978. (From Tatham, T. R., Thomas, D. L., and Danila, D. J., *Ecology of Barnegat Bay, New Jersey,* Kennish, M. J. and Lutz, R. A., Eds., Springer-Verlag, New York, 1984, 241. With permission.)

Menidia menidia and *A. mitchilli* are the two most abundant fishes in the bay. Vouglitois et al.[141] found that *M. menidia* comprised 31% and *A. mitchilli* 27% of the total finfish catch from 1975 to 1981. Tatham et al.[27] listed *A. mitchilli* and *M. menidia* as the first and second most abundant fishes, respectively, each year during their 3-year survey. Congregating in the shore zone from spring through fall, *M. menidia* migrates to deeper waters of the estuary or coastal ocean in the winter. Similarly, the maximum abundance of *A. mitchilli* arises between May and October each year, but an offshore and southerly migration from the estuary to continental-shelf wintering grounds takes place in the fall. A bimodal peak in the abundance of trawl catches of

TABLE 6
Temporal and Spatial Occurrence of Fishes Taken in Barnegat Bay and Western Tributaries

Residents
  Anguilla rostrata[a]
  Opsanus tau
  Ophidion marginatum
  Cyprinodon variegatus
  Fundulus heteroclitus
  F. majalis
  Lucania parva
  Menidia beryllina
  M. menidia
  Apeltes quadracus
  Hippocampus erectus?
  Syngnathus fuscus
  Morone americana
  Tautoga onitis
  Tautogolabrus adspersus
  Chasmodes bosquianus
  Hypsoblennius hentzi
  Gobiosoma bosci
  Pseudopleuronectes americanus[b]
  Trinectes maculatus
Warm-water migrants
  Dasyatis sayi
  Elops saurus
  Alosa mediocris
  Brevoortia tyrannus
  Anchoa mitchilli
  Synodus foetens
  Strongylura marina
  Membras martinica
  Fistularia tabacaria
  Centropristis striata
  Pomatomus saltatrix
  Alectis ciliaris?
  Caranx cryosos
  C. hippos
  Selene setapinnis?
  S. vomer
  Trachinotus carolinus
  T. falcatus
  Bairdiella chrysoura
  Cynoscion regalis
  Leiostomus xanthurus
  Menticirrhus saxatilis
  Pogonias cromis
  Mugil cephalus
  M. curema
  Sphyraena borealis
  Astroscopus guttatus
  Prionotus carolinus
  P. evolans
  Aralichthys dentatus
  Aluterus schoepfi
  Monacanthus hispidus
  Sphoeroides maculatus
  Chilomycterus schoepfi?

Local marine strays
  Mustelus canis
  Squalus acanthias
  Raja eglantaria?
  Conger oceanicus
  Etrumeus teres
  Anchoa hepsetus?
  Engraulis eurystole
  Pollachius virens
  Hyporhamphus unifasciatus
  Oostethus lineatus?
  Sygnathus louisianae?
  Morone saxatilis
  Mycteroperca microlepis
  Pristigenys alta
  Rachycentron canadum
  Decapterus punctatus
  Selar crumenophthalmus
  Seriola zonata
  Lutjanus griseus
  Stenotomus chrysops
  Mullus auratus
  Pseudupeneus maculatus
  Chaetodipterus faber
  Chaetodon ocellatus?
  Ammodytes sp.?
  Gobiosoma ginsburgi?
  Peprilus triacanthus
  Scorpaena plumieri
  Hemitripterus americanus
  Myoxocephalus octodecemspinosus
  Dactylopterus volitans
  Scophthalmus aquosus
  Symphurus plagiusa
  Lactophrys quadricornis?
  L. triquerter?
Cool-water migrants
  Alosa aestivalis
  A. pseudoharengus
  A. sapidissima
  Clupea harengus
  Merluccius bilinearis
  Urophycis chuss
  U. regia
  Gasterosteus aculeatus
  Micropogonias undulatus
  Myoxocephalus aenaeus
  Etropus microstomus
  Pseudopleuronectes americanus[b]
Freshwater strays
  Dorosoma cepedianum
  Esox niger
  Notemigonus crysoleucas
  Ictalurus natalis
  Aphredoderus sayanus
  Fundulus diaphanus
  Lepomis gibbosus

TABLE 6 (continued)
Temporal and Spatial Occurrence of Fishes Taken in Barnegat Bay and Western Tributaries

Note: ? indicates some uncertainty.

[a]   Catadromous.
[b]   Immature individuals are resident year-round, adults are cool-water migrants.

From Tatham, T. R., Thomas, D. L., and Danila, D. J., *Ecology of Barnegat Bay, New Jersey*, Kennish, M. J. and Lutz, R. A., Eds., Springer-Verlag, New York, 1984, 241. With permission.

TABLE 7
Usage of Barnegat Bay as a Spawning and Nursery Area by Resident Species, Warm-Water Migrants, and Cool-Water Migrants

Species	Spawning area	Nursery Area	
		Significant usage[a]	Minor usage[b]
Resident species			
Anguilla rostra	X[c]	X	
Opsanus tau	X	X	
Ophidion marginatum			X
Cyrpinodon variegatus	X	X	
Fundulus heteroclitus	X	X	
F. majalis	X	X	
Lucania parva	X	X	
Menidia beryllina	X	X	
M. menidia	X	X	
Apeltes quadracus	X	X	
Hippocampus erectus	X	X	
Syngnathus fuscus	X	X	
Morone americana	X	X	
Tautoga onitis	X	X	
Tautogolabrus adspersus	X	X	
Chasmodes bosquianus	X	X	
Hypsoblennius hentzi	X	X	
Gobiosoma bosci	X	X	
Pseudopleuronectes americanus[d]	X	X	
Trinectes maculatus	X	X	
Cool-water migrants			
Alosa aestivalis	X[e]		X
A. pseudoharengus	X[e]		X
A. sapidissima			
Clupea harengus			X
Merluccius bilinearis			X
Urophycia chuss			X
U. regia			X
Gasterosteus aculeatus	X		X
Micropogonias undulatus		X[f]	
Etropus microstomus			X
Pseudopleuronectes americanus[g]	X	X	
Warm-water migrants			
Dasyatis sayi			
Elops saurus			
Brevoortia tyrannus		X	

TABLE 7 (continued)
Usage of Barnegat Bay as a Spawning and Nursery Area by Resident Species,
Warm-Water Migrants, and Cool-Water Migrants

Species	Spawning area	Nursery Area Significant usage[a]	Minor usage[b]
Warm-water migrants (continued)			
Alosa mediocris			
Anchoa mitchilli	X	X	
Synodus foetens			X
Strongylura marina		X	
Membras martinica		X	
Fistularia tabacaria			
Centropristis striata			X
Pomatomus saltatrix		X	
Alectis ciliaris			
Caranx crysos			X
C. hippos		X	
Selene setapinnis			
S. vomer		X	
Trachinotus carolinus			
T. falcatus			
Bairdiella chrysoura		X	
Cynoscion regalis		X	
Leiostomus xanthurus		X	
Menticirrhus saxatilis			
Pogonias cromis			
Mugil cephalus			X
M. curema		X	
Sphyraena borealis			
Astroscopus guttatus			X
Prionotus carolinus			X
P. evolans			X
Paralichthys dentatus			X
Aluterus schoepfi			
Monacanthus hispidus			
Sphoeroides maculatus	X	X[f]	
Chilomycterus schoepfi			

[a]   Significant usage denotes that larvae and young were common to abundant.
[b]   Minor usage denotes that larvae and young were occasional or uncommon.
[c]   Catadromous species.
[d]   Immature.
[e]   Zich (1977) reported spawning migrations in some bay tributaries.
[f]   Probably significant usage when species is abundant.
[g]   Adult.

From Tatham, T. R., Thomas, D. L., and Danila, D. J., *Ecology of Barnegat Bay, New Jersey*, Kennish, M. J. and Lutz, R. A., Eds., Springer-Verlag, New York, 1984, 241. With permission.

A. mitchilli has been elucidated; an initial peak corresponds to May or June and a second, often larger, one to September or October.[140] Bay anchovy eggs and larvae numerically dominate ichthyoplankton samples in the estuary, composing up to 98 and 56% of the annual egg and larval catches, respectively.

In sum, the community structure, seasonal patterns, and population trends of the fish

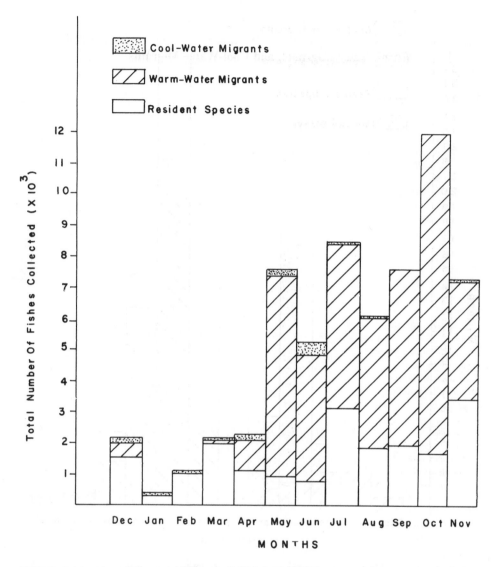

FIGURE 6. Monthly abundance of resident fishes, warm-water migrants, and cool-water migrants in Barnegat Bay. (From Tatham, T. R., Thomas, D. L., and Danila, D. J., *Ecology of Barnegat Bay, New Jersey,* Kennish, M. J. and Lutz, R. A., Eds., Springer-Verlag, New York, 1984, 241. With permission.)

community of Barnegat Bay are similar to those of the larger New Jersey estuaries of Delaware Bay and Raritan Bay, as well as other coastal bays from Sandy Hook to Cape May. Forage fishes and juveniles numerically dominate the communities, utilizing the system primarily as a nursery area. Adult marine forms spawn or feed in the bay, but typically inhabit oceanic waters. Warm-water and cool-water migrants appear seasonally, occasionally being present in greater numbers than resident species. Warm-water migrants are more abundant than cool-water migrants, and account for large numbers of fish in the bay from July through November. At this time, young of resident and warm-water migrants coexisting in the estuary reach maximum population sizes. The fish community, therefore, is characterized by: (1) numerical dominance of a few species, (2) forage fishes and juveniles, (3) seasonal occurrence of warm-water and cool-water migrants, and (4) large fluctuations in the size of populations.[27,141]

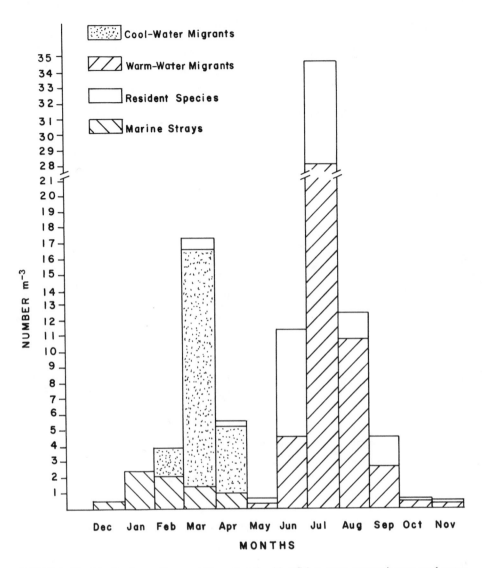

FIGURE 7.   Monthly abundance of larvae and juveniles of resident fishes, warm-water migrants, cool-water migrants, and marine strays in Barnegat Bay. (From Tatham, T. R., Thomas, D. L., and Danila, D. J., *Ecology of Barnegat Bay, New Jersey*, Kennish, M. J. and Lutz, R. A., Eds., Springer-Verlag, New York, 1984, 241. With permission.)

2. Chesapeake Bay

Orth and Heck[142] have described the structure of the fish community in the lower Chesapeake Bay, comparing fishes in vegetated and unvegetated areas of the system. They concentrated on three study areas dominated by eelgrass, *Zostera marina*, sampling at a station in the mouth of the York River, a station in Mobjack Bay near Browns Bay, and a station on the western side of the Delmarva Peninsula off Church Neck at the mouth of Hungass Creek (Figure 8). Monthly samples were collected from September 1976 to December 1977 using a 4.9-m otter trawl with 1.9-cm mesh wings and a 0.6-cm mesh cod end liner.

During the 14-month survey program, 24,182 fish belonging to 48 species were sampled in the field (Table 8). *Leiostomus xanthurus* comprised 63% of all fish collected, *Syngnathus fuscus* 14%, *Anchoa mitchilli* 9%, and *Bairdiella chrysoura* 5%. The remaining 44 species constituted less than 10% of the samples. The number of species and their abundances differed

FIGURE 8. Sampling sites of fish collected in the lower Chesapeake Bay. (From Orth, R. J. and Heck, K. L., Jr., *Estuaries*, 3, 278, 1980. With permission.)

between the vegetated and unvegetated areas. Despite a spatial separation of less than 500 m between the two habitats, vegetated areas showed a higher density and diversity of ichthyofauna than unvegetated areas. Results of other work substantiate these findings.[143] Benthic macrovegetation increases the complexity of the habitat and associated faunal communities. *Zostera* beds are excellent feeding habitats for fish,[144] and the grass meadows also may be refuges for fish from predators.

More fish were captured at night than during the day due to increased feeding activity at night, trawl avoidance during the day, or lower nighttime temperatures in the grass area. Seasonally, the density and diversity of the ichthyofauna increased in the spring and early summer concomitant with the rise in biomass of the eelgrass. As water temperature and eelgrass biomass declined in the fall and winter, so did fish abundance and species richness. Both the decreasing water temperature and biomass of eelgrass triggered the movement of fish, with warm-water forms migrating to deeper waters of the estuary or to coastal oceanic waters between November and March. The warm-water migrants reentered the bay in the spring and summer.

Horwitz[145] also discerned acute seasonal migrations of fishes in Chesapeake Bay. Taking monthly 7.6-m semiballoon trawl samples in the middle reach of the bay in the vicinity of Calvert Cliffs, MD from 1968 to 1981, Horwitz[145] discovered great variations in abundance of fish populations related to natural environmental variability, man-induced environmental changes, changes in fishing intensity, and/or biological interactions. Trends in abundance followed a rather uniform decline for some forms (e.g., *Apeltes quadracus, Bairdiella chrysoura, Clupea*

TABLE 8
Abundances of Fishes Collected at Three Stations in the Lower Chesapeake Bay[a]

Species	Station 1	Station 2	Station 3	Total
Leiostomus xanthurus	3,925	3,666	7,683	15,274
Syngnathus fuscus	1,489	1,024	795	3,308
Anchoa mitchilli	493	571	949	2,013
Bairdiella chrysoura	175	393	721	1,289
Menidia menidia	95	317	65	475
Syngnathus floridae	121	93	79	293
Apeltes quadracus		8	236	244
Hypsoblennius hentzi	30	51	105	186
Fundulus heteroclitus	8	113	5	126
Orthropristis chrysoptera	12	7	104	123
Tautoga onitis	29	19	51	99
Lagodon rhomboides	34	44	18	96
Paralichthys dentatus	35	20	33	88
Centropristis striata	11	15	57	83
Chasmodes bosquianus	19	32	23	74
Gobiosoma bosci	18	16	11	45
Cyprinodon variegatus		44		44
Morone saxatilis	13	8	21	42
Pseudopleuronectes americanus	27	7	2	36
Micropogonias undulatus	20	8	5	33
Opsanus tau	1	15	15	31
Cynoscion nebulosus	6	9	10	25
Anguilla rostrata	6	15		21
Sphoeroides maculatus	1	2	15	18
Mycteroperca microlepis		2	13	15
Lucania parva		14		14
Gobiesox strumosus	3	6	3	12
Urophycis regius	3	7		10
Sciaenops ocellata		8	1	9
Trinectes maculatus	5		3	8
Chilomycterus schoepfi	1	2	4	7
Menidia beryllina		6		6
Synodus foetens		4	2	6
Hippocampus erectus	2	3		5
Brevoortia tyrannus	2		1	3
Pomatomus saltatrix		3		3
Prionotus carolinus	3			3
Monacanthus hispidus			3	3
Alosa pseudoharengus		2		2
Chaetodon ocellatus	1		1	2
Aluterus schoepfi			1	1
Fistularia tabacaria			1	1
Fundulus majalis		1		1
Peprilus paru	1			1
P. triacanthus			1	1
Scophthalmus aquosus			1	1
Stenotomus chrysops			1	1
Rissola marginata			1	1
Total number of species	31	36	37	48
Total number of individuals	6,589	6,555	11,040	24,182

[a] Data collected from unvegetated areas and night trawls not included.

From Orth, R. J. and Heck, K. L., Jr., *Estuaries*, 3, 278, 1980. With permission.

harengus, Menidia menidia, Menticirrhus americanus, and *Sphoeroides maculatus*), a rather uniform increase for others (e.g., *Brevoortia tyrannus, Dorosoma cepedianum, Gasterosteus aculeatus, Hypsoblennius hentzi, Menidia beryllina,* and *Synodus foetens*), highest abundances during the middle years of sampling (e.g., *Centropristis striata, Prionotus carolinae,* and *Symphurus plagiusa*), and irregular patterns of high and low abundance. Table 9 is a taxonomic list of fishes captured in trawl samples over the long-term sampling period.

Three of the 58 species collected (i.e., *Anchoa mitchilli, Leiostomus xanthurus,* and *Micropogonias undulatus*) either consistently or frequently dominated the fish communities. *A. mitchilli* and *L. xanthurus* were most abundant each year, and common species included *Brevoortia tyrannus, M. undulatus, Morone americana, Pseudopleuronectes americanus,* and *Trinectes maculatus.* In each year of sampling, the three most abundant species made up 88 to 99% of the total catch (Table 10). Factors implicated in declining abundances of fishes over the years of study were changes in climate, large temperature changes, reductions in dissolved oxygen, decreases in submerged macrophytes, fluctuations in river inflow, and direct or indirect anthropogenic impacts.

Setzler-Hamilton[146] examined the early life-history stages of fishes in Chesapeake Bay. Dovel[147] dealt with the ichthyoplankton of the mesohaline and oligohaline waters of the upper bay. Olney[148] investigated the eggs and larvae of the polyhaline lower bay.

Fish spawn throughout the year in the estuary. Based on the work of Setzler-Hamilton,[146] eggs of *A. mitchilli* and *T. maculatus* are most abundant in the upper bay, while the larvae of *A. mitchilli, Gobiosoma bosci,* and *M. americana* dominate larval catches there. *A. mitchilli* predominates in ichthyoplankton collections of the lower bay, being responsible for up to 96% of all eggs and 88% of all larvae sampled. Ichthyoplankton of *Cynoscion regalis* rank as high as second in abundance in the lower bay; other important taxa include *T. maculatus, G. ginsburgi, Hypsoblennius hentzi, Paralichthys dentatus,* and *Symphurus plagiusa.* In mesohaline waters of the estuary, eggs of *A. mitchilli* and *T. maculatus* may comprise 99% or more of all fish eggs collected.[147]

The polyhaline lower bay has a peak diversity of fish eggs and larvae in July and August. The maximum diversity of ichthyoplankton in the upper bay appears somewhat earlier in June. In general, most larvae and juveniles occur in Chesapeake Bay during the summer months.

3. Terminos Lagoon

Yanez-Arancibia et al.[149] studied the species composition and ecology of the fish community of the Terminos Lagoon, a tropical estuary in the southern Gulf of Mexico at Campeche, Mexico (Figure 9). The Terminos Lagoon, the largest coastal lagoon in the southern Gulf of Mexico covering an area of approximately 2500 km², undergoes dramatic shifts in climate during the year. Water temperature ranges from 25 to 36°C and precipitation from 1200 to 2000 mm/year. The dry season extends from January through May, but from June through September, afternoon and evening showers commonly erupt on a daily basis. Winter storms or nortes strike from October to March.

From July 1976 to March 1979, 173 trawl collections (9-m trawl; mouth size 5 m; mesh size 1.9 cm) were made at nine localities in the estuary (Figure 9). The sampling yielded 21,734 fish belonging to 121 species. In earlier surveys, Toral and Resendez[150] and Castro[151] retrieved an additional 46 species from the lagoon (Table 11). Only a few species numerically dominated the community, with 17 species comprising 82% of the catch (Table 12). These 17 species had broad distributions in the estuary, being captured at most of the sampling locations. Only 15 of the 121 species sampled by trawl were taken at all stations. Most of the fishes consisted of seasonal or cyclical community members (54 species; 45% of the total number of species) and occasional visitors (55 species; 45%) to the estuary. A small number of fishes (12 species; 10%) were permanent residents. Nonresident species probably used the estuary largely for spawning or as a nursery area.

TABLE 9
Scientific and Common Names of Species Collected in Trawl Samples in the Vicinity of Calvert Cliffs, MD

Scientific name	Common name
Dasyatis sayi	Bluntnose stingray
Rhinoptera bonasus	Cownose ray
Conger oceanicus	Conger eel
Anguilla rostrata	American eel
Acipenser brevirostrum	Shortnose sturgeon
Alosa aestivalis	Blueback herring
A. pseudoharengus	Alewife
A. sapidissima	American shad
Brevoortia tyrannus	Menhaden
Clupea harengus	Atlantic herring
Dorosoma cepedianum	Gizzard shad
Anchoa hepsetus	Striped anchovy
A. mitchilli	Bay anchovy
Umbra pygmaea	Eastern mudminnow
Ictalurus catus	White catfish
Synodus foetens	Inshore lizardfish
Opsanus tau	Oyster toadfish
Gobiesox strumosus	Skilletfish
Urophycis rigia	Spotted hake
Microgadus tomcod	Atlantic tomcod
Merluccius bilinearus	Silver hake
Strongylura marina	Atlantic needlefish
Menidia beryllina	Tidewater silverside
M. menidia	Atlantic silverside
Apeltes quadracus	Fourspine stickleback
Gasterosteus aculeatus	Threespine stickleback
Hippocampus erectus	Lined seahorse
Syngnathus fuscus	Northern pipefish
Prionotus carolinus	Northern searobin
P. evolans	Striped searobin
Morone americana	White perch
M. saxatilis	Striped bass
Centropristis striata	Black sea bass
Pomatomus saltatrix	Bluefish
Bairdiella chrysoura	Silver perch
Cynoscion nebulosus	Spotted seatrout
C. regalis	Weakfish
Leiostomus xanthurus	Spot
Menticirrhus americanus	Southern kingfish
Micropogonias undulatus	Croaker
Pogonias cromis	Black drum
Sciaenops ocellata	Red drum
Chaetodipterus faber	Atlantic spadefish
Astroscopus guttatus	Northern stargazer
Hypsoblennius hentzi	Feather blenny
Chasmodes bosquianus	Striped blenny
Ammodytes americanus	American sand lance
Gobiosoma bosci	Naked goby
Mugil curema	White mullet
Peprilus triacanthus	Butterfish
P. alepidotus	Harvestfish
Paralichthys dentatus	Summer flounder
Scophthalmus aquosus	Windowpane
Pseudopleuronectes americanus	Winter flounder
Trinectes maculatus	Hogchoker

TABLE 9 (continued)
Scientific and Common Names of Species Collected in Trawl Samples in the Vicinity of Calvert Cliffs, MD

Scientific name	Common name
Symphurus plagiusa	Blackcheek tonguefish
Aluterus schoepfi	Orange filefish
Sphoeroides maculatus	Northern puffer

From Horwitz, R. J., *Ecological Studies in the Middle Reach of Cheapeake Bay: Calvert Cliffs*, Heck, K. L., Jr., Ed., Springer-Verlag, Berlin, 1987, 167. With permission.

A distinctive trophic structure of fishes is manifested in the estuary. The majority of species (51%) consisted of second order consumers dominated by demersal forms feeding principally on microbenthic fauna (i.e., crustaceans, polychaetes, and mollusks). Fewer populations (27%) were higher order consumers that ingested first- and second-order consumer fish and macrobenthos. First-order consumer fish — herbivores, detritivores, or consumers — which utilize detritus as a major dietary component constituted only 22% of all the species.

An east to west migration pattern of fish in the lagoon paralleled prevailing currents. Juvenile forms concentrated in Puerto Real Inlet and in the easternmost regions of the lagoon. Through time, their migratory route tracked south and westward with individuals emigrating from the estuary to the Gulf of Mexico via Carmen Inlet (Figure 10). Juveniles attained maximum abundance in the estuary during August and September at the time of high river flow and peak productivity.

The Terminos Lagoon contains many more fish species than most temperate estuarine systems.[152] Despite the higher diversity of ichthyofauna in tropical systems, biomass and fish production may not be greater.[3] The information function H^1 of Shannon and Weaver[153] (which increases as species richness and evenness increase), calculated for the ichthyofauna of the Terminos Lagoon, ranged from 0.53 to 2.50. Biomass values, in turn, varied between 0.95 and 11.3 g wet weight per square meter. The estuary serves as an important habitat for many populations of fish, supplying food, shelter, spawning areas, nursery grounds, and migratory pathways.

4. San Francisco Bay

One of the largest estuarine basins of the world is that formed by San Francisco Bay and the delta of the Sacramento and San Joaquin Rivers (Figure 11).[154,155] The fishes of this system have recently been the target of elaborate investigations.[156,157] Historically, the fisheries of San Francisco Bay have been important as a food source for the region and as a recreational pursuit for sport fishermen. A gradual shift in the character of the fisheries has occurred over the years, from primarily commercial to principally recreational. Insidious pollution of the environment accelerated the rapid decline of certain fishery resources. The diversion of riverine discharges for agricultural and domestic purposes has been detrimental to commercially and recreationally important species. Legislation restricting the harvest of anadromous fishes (e.g., salmon, striped bass, and sturgeon) significantly reduced commercial catches.[158] At one time, the American shad (*A. sapidissima*), chinook salmon (*Oncorhynchus tshawytscha*), striped bass (*Morone saxatilis*), green sturgeon (*Acipenser medirostris*), and white sturgeon (*A. transmontanus*) supported commercial fisheries in the bay, but these fisheries were terminated in 1956, 1956, 1935, 1917, and 1917, respectively.[156] Clam and oyster industries, likewise once bountiful, have disappeared from the bay.[158] The Pacific herring (*Clupea harengus pallasi*) remains the only major commercial fishery in the estuary; the American shad (*A. sapidissima*), chinook salmon (*O. tshawytscha*), California halibut (*Paralichthys californicus*), striped bass (*M. saxatilis*), starry

TABLE 10
Rank Abundance of the Ten Most Common Species in Trawl Samples in the Vicinity of Calvert Cliffs, MD[a,b]

Species	1969	1970	1971	1972	1973	1974	1975	1976	1977	1978	1979	1980	1981	Total
Anchoa mitchilli	1	1	1	1	1	1	2	2	2	1	1	1	1	16
Leiostomus xanthurus	2	2	2	2	2	2	1	1	1	2	2	2	2	23
Micropogonias undulatus	3	8	4	7	5	3	3	3	8	3	8	3	5	63
Morone americanus	—	6	3	4	3	5	7	5	7	—	6	7	—	86
Pseudopleuronectes americanus	10	—	10	5	—	9	6	4	4	5	3	6	3	87
Trinectes maculatus	4	5	6	3	4	7	5	—	9	10	4	—	8	87
Brevoortia tyrannus	—	—	—	—	8	9	4	7	5	4	5	3	4	93
Cynoscion regalis	6	9	7	6	—	6	—	9	3	7	7	—	6	99
Menidia menidia	5	7	5	10	7	—	—	6	—	—	—	5	7	117
Alosa aestivalis	7	10	—	—	6	—	8	6	—	6	—	—	—	122
A. pseudoharengus	—	4	—	—	—	8	—	10	10	8	9	8	10	124
Paralichthys dentatus	—	—	8	9	10	—	9	—	—	9	—	—	—	126
Morone saxatilis	—	3	—	—	9	10	—	8	—	—	10	10	—	128
Urophycis regius	—	—	9	—	—	—	—	—	—	—	—	—	—	135
Peprilus alepidotus	—	—	—	8	—	—	—	—	—	—	—	—	—	139
Bairdiella chrysoura	8	—	—	—	—	—	—	—	—	—	—	—	—	140
Anguilla rostrata	9	—	—	—	—	—	—	—	—	—	—	—	—	140
Apeltes quadracus	—	—	—	—	—	—	—	—	—	—	—	—	9	141
Anchoa hepsetus	—	—	—	—	—	—	—	—	—	—	—	—	—	141
Proportion of total CPUE in three most abundant species	89	94	88	88	98	96	99	93	95	96	97	97	97	

[a] Ranks are based on the average monthly catch-per-unit-effort.

[b] — equals rank greater than 10. Total of ranks computed with rank greater than 10 counted as 11.

From Horwitz, R. J., *Ecological Studies in the Middle Reach of Chesapeake Bay: Calvert Cliffs*, Heck, K. L., Jr., Ed., Springer-Verlag, Berlin, 1987, 167. With permission.

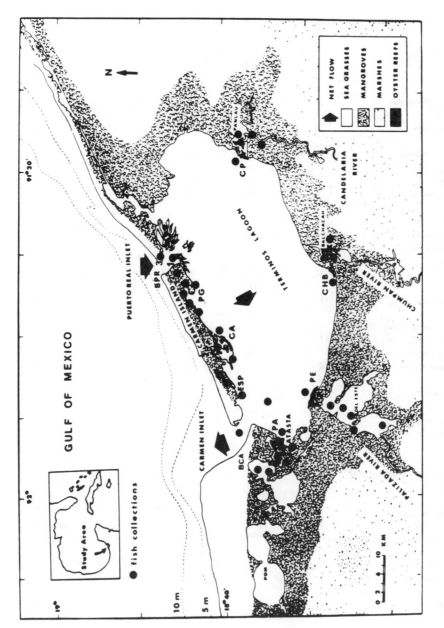

FIGURE 9. Terminos Lagoon in Campeche, Mexico. Localities of fish collections are BPR = Puerto Real Inlet; Carmen Island inner areas: PG = Punta Gorda, CA = Bajos del Cayo, ESP = Estero Pargo; BCA = Carmen Inlet; Fluvial-Lagoon systems: CP = Candelaria- Panlau, CHB = Chumpan-Balchacah, PC = Palizada-del Este, PA = Pom-Atasta. (From Yanez-Arancibia, A., Linares, F. A., and Day, J. W., Jr., *Estuarine Perspectives*, Kennedy, V. S., Ed., Academic Press, New York, 1980, 465. With permission.)

TABLE 11
Type of Inhabitant (T.I.), Trophic Category (T.C.), and Frequency of Occurrence of Fishes in the Terminos Lagoon[a-e]

Species	BPR-1	BPR-3	PG	CA	ESP	CP	CHB	PE	PA	T.I.	T.C.
Urolophus iamaicienses	2	1	5	3	7					Sed	3
Dasyatis sabina						10		1	7	Cv	3
Himatura schmardae			3	1						Occ	3
Dasyatis hastatus			1							Occ	3
Elops saurus					3					Occ	3
Albula vulpes			4							Occ	2
Sardinella macrophthalmus		1208	38	4	81	3	18		6	Cv	1
S. humeralis								1		Occ	1
Opisthonema oglinum		38	2			1		15	1	Cv	1
Anchovia sp.		2								Occ	1
Anchoa mitchilli mitchilli			52	9	3751	8	23	8	44	Cv	1
A. hepsetus hepsetus	1	7							4	Occ	1
A. lamprotaenia		49	770		20				18	Cv	1
Centengraulis edentulus		3	108	3	38	29	11	4	115	Cv	1
Anchoviella sp.		2								Occ	1
Synodus foetens		1	3	2	3				8	Cv	3
Cyprinodon variegatus		7	1	285						Occ	1
Ictalurus meridionalis									9	Occ	2
Bagre marinus						7	5	3	14	Cv	2—3
Arius felis	8	207	137	54	75	5	1	9	74	Sed	2—3
A. melanopus		2	600		91	257	418	225	141	Sed	2—3
Gobiosox strumosus	1									Occ	2
Opsanus beta		1	61	38	18	1	5			Cv	2
Nautopaedium porosissimun							4	1	5	Cv	2
Hemirhamphus brasiliensis	11		1							Occ	2
Hyporhamphus unifasciatus		26	16				1		1	Occ	2
Tylosurus raphidoma	4		1	1	1					Occ	3
T. acus	2				1					Occ	3
Strongylura marina			6							Occ	3
Poeciliopsis sp.					1					Occ	1
Chriodorus aterinoides			9	1						Cv	2
Syngnathus rosseau	1		5	1						Cv	3
S. machayi		1		2		2				Cv	3
S. scoveili				2	2					Cv	3
Hippocampus hudsonius punctulatus	1		2		5					Cv	3
Scorpaena plumieri	2									Occ	3
S. grandicornis		1								Occ	3
Prionotus carolinus			10	21	2				1	Occ	3
Dactylopterus volitans	6									Occ	3
Centropomus undecimalis			3		30	1	3		8	Cv	3
C. paralellus					2					Cv	3
Epinephellus guttatus	3		2	5	1	1	1			Cv	3
Diplectrum radiale		3								Occ	3
Echeneis neucrates			10	1						Occ	3
Caranx hippos					9	3			7	Cv	2
C. ruber[f]					1					Occ	2
C. latus					1					Occ	2
Chloroscombrus chrysurus		24	2	1	8	5		1	12	Cv	2
Selene vomer			41		4				3	Cv	2
Oligoplites saurus		1	8	1	2	1			2	Cv	2
Trachinotus falcatus[f]		2	9		1					Occ	2
Lutianus griseus	3		13	16	39	2				Cv	3
L. synagris	3	5	1							Cv	3
Lutjanus analis	1				2					Cv	3

TABLE 11 (continued)

Type of Inhabitant (T.I.), Trophic Category (T.C.), and Frequency of Occurrence of Fishes in the Terminos Lagoon[a-e]

Species	BPR-1	BPR-3	PG	CA	ESP	CP	CHB	PE	PA	T.I.	T.C.
Lobotes surinamensis					1					Occ	3
Eucinostomus gula	36	5	1434	903	621	393	5	1	275	Sed	1
E. argenteus		54	60	86	43	22			14	Cv	1
E. melanopterus			1	3	3					Occ	1
Gerres cinereus				2						Cv	1
Eugerres plumieri			25		12	1	20	13	1	Cv	1
Diapterus rhombeus			6	1	171	343	1	1	11	Cv	1
D. evermani					24					Occ	1
Orthopristis chrysopterus	84	4	154	27	94					Cv	2
O. poeyi	24		92	17						Cv	2
Anisotremus virginicus	6			3						Cv	2
A. spleniatus	2		15							Cv	2
Bathystoma rimator	25		1							Cv	2
Haemulon plumieri	68		145	171	104					Cv	2
H. bonariense			35	28	8					Cv	2
Calamus penna		1	1							Occ	2
Archosargus probatocephalus	2		10	1	9	3	1		1	Cv	2
A. unimaculatus	9	300	793	128	426					Cv	2
Calamus calamus	1									Occ	2
Menticirrus martinicences		1	1							Cv	2
M. saxantilis		24								Cv	2
Micropogon furnieri				4	35	214	2	36	274	Cv	2
Bairdiella chrysoura	41	271	179	179	168	401	13	4	51	Cv	2
B. rhonchus				10	47	117	35	9	12	Cv	2
Cynoscion nebulosus	3		19	17	7	61		2	6	Cv	3
C. nothus						9		3	18	Cv	3
C. arenarius									22	Cv	3
Equetus acuminatus	1			1						Occ	2
Odontoscion dentex	42	2	3							Occ	2
Corvula sancta-lucia	30									Occ	2
Chaetodipterus faber		9	35	17	12	12	2	2	32	Sed	1
Chaetodon ocellatus	1									Occ	1
Pomacanthus arcuatus			1							Occ	1
Cichlasoma urophthalmus			221	363	3	22				Sed	1—2
C. fenestratum								9		Occ	1—2
Mugil curema					32				4	Cv	1
Polynemus octonemus						42		5	17	Cv	2
Novaculichthys infirmus	1									Occ	1
Scarus noyesi	1				4					Occ	2
Nicholsina ustus	20		3		19					Cv	2
Hypsoblennius hentzi			1							Occ	2
Gobionellus oceanicus						1			1	Occ	1
Gobiosoma bosci									1	Sed	1
Trichiurus lepturus		1			12			2	25	Cv	2
Scomberomorus maculatus									1	Occ	3
Cytharichthys spilopterus			1	4	26	2	5	3	122	Sed	3
Ancyclopsetta quadrocellata			2							Occ	3
Etropus crossotus			1						2	Cv	3
Achirus lineatus		1	15	2	1	4		5	15	Sed	2
Gymnachirus melas			1					5		Occ	2
Trinectes maculatus								5		Occ	2
Symphurus plagiusa			1						1	Occ	2
Balistes capriscus	1									Occ	2

TABLE 11 (continued)
Type of Inhabitant (T.I.), Trophic Category (T.C.), and Frequency of Occurrence of Fishes in the Terminos Lagoon[a-e]

Species	BPR-1	BPR-3	PG	CA	ESP	CP	CHB	PE	PA	T.I.	T.C.
Monacanthus hispidus	3	5	24	1	1					Cv	2
M. ciliatus	1									Occ	2
Alutera schoefi		5	2		3					Occ	2
Lactoprhys tricornis	5		40	11	6					Sed	2
L. bicaudalis	1									Sed	2
Sphoeroides marmoratus	1		8	122	3					Cv	2—3
S. testudineus	19	45	148	14	148	20	8	24	172	Sed	2—3
S. nephelus			6	5	7					Occ	2
S. spengleri	1									Occ	2—3
S. sp.				1	5			2	3	Occ	2
Lagocephalus lavigeatus		1	1							Occ	2—3
Diodon hystrix	45									Occ	2—3
Chilomycterus schoepfi	4			30	22	3				Occ	2—3
C. antennatus				9	31	12				Cv	2—3

Note: Castro (1978) reported the following additional species for Terminos Lagoon: *Carcharhinus limbatus, C. porosus, C. leucas, Sphyrna tiburo, Pristis pectinatus, P. perotteti, Rhinobatos lentiginosus, Narcine brasillensis, Dasyatis americana, Megalops atlanticus, Brevoortia guntheri, Strongylura timucu, Menidia beryllina, Syngnathus louisianae, S. floridae, Oostethus lineatus, Lutjanus cyanopterus, L. apodus, Ulaema lefroyi, Eucinostomus habana, Rypticus saponaceus, Promicrops itaiara, Epinephellus morio, Diplodus caudimacula, Lagodon rhomboides, Cynoscion jamaiciensis, Umbrina coroides, Stillifer lanceolatus, Mugil cephalus, M. tricodon, Sphyraena barracuda, Gobiomorus dormitator, Dormitator maculatus, Awaous tajasica, Gobionellus hastatus, Astroscopus ygraecum, Batrachoides surinamensis, Stephanolepis hispidus, Atherinomorus stipes.* Toral and Resendes (1974) also reported the following species: *Cichlosoma meeki, C. aureum, C. pearsi, C. sextafaciatum, C. friedrichsthali, C. heterospilum, Petenia splendida.*

[a]   Occ = occasional visitor.
[b]   Cv = cyclical or seasonal visitor.
[c]   Sed = resident (typically estuarine) species.
[d]   1, 2, and 3 refer to first-, second-, and third-order consumers, respectively.
[e]   See Figure 9 for sampling localities.
[f]   Species collected with different methods and not processed quantitatively.

From Yanez-Arancibia, A., Linares, F. A., and Day, J. W., Jr., *Estuarine Perspectives*, Kennedy, V. S., Ed., Academic Press, New York, 1980, 465. With permission.

flounder (*Platichthys stellatus*), white (*Acipenser transmontanus*) and green (*A. medirostris*) sturgeon, jacksmelt (*Atherinopsis californiensis*), brown rockfish (*Sebastes auriculatus*), staghorn sculpin (*Leptocottus armatus*), white catfish (*Ictalurus cactus*), brown smoothhound shark (*Mustelus henlei*), leopard shark (*Triakis semifasciata*), and various surf perch support large recreational fisheries.[156]

More than 100 species of fish have been recorded in the San Francisco Bay ecosystem.[156,159,160] In a survey of the fishery resources of the bay between January 1980 and December 1982, Armor and Herrgesell[156] identified 74 species of adult fish including 15 introduced forms (Table 13). Resident species, such as the starry flounder (*P. stellatus*), leopard shark (*T. semifasciata*), splittail (*Pogonichthys macrolepidotus*), and shiner perch (*Cymatogaster aggregata*), accounted for the bulk of the species. Other important forms were anadromous fishes (e.g., American shad, *A. sapidissima*; chinook salmon, *O. tshawytscha*; and striped bass, *M.*

TABLE 12
The 17 Most Abundant Species of
Fish Taken in Trawl Samples in
the Terminos Lagoon from July
1976 to March 1979

Species	Total number
Anchoa mitchilli	3895
Eucinostomus gula	3673
Arius melanopus	1734
Archosargus unimaculatus	1656
Sardinella macrophthalmus	1358
Bairdiella chrysoura	1307
Anchoa lamprotaenia	857
Cichlasoma urophthalmus	609
Sphoeroides testudineus	598
Arius felis	570
Diapterus rhombeus	534
Cetengraulis edentulus	311
Eucinostomus argenteus	279
Cytharichthys spilopterus	163
Opsanus beta	124
Chaetodipterus faber	121
Cynoscion nebulosus	115

From Yanez-Arancibia, A., Linares, F. A., and Day, J. W., Jr., *Estuarine Perspectives,* Kennedy, V. S., Ed., Academic Press, New York, 1980, 465. With permission.

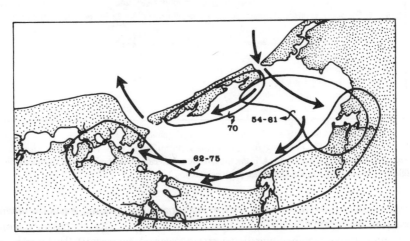

FIGURE 10.   Migratory pathways of juvenile fishes in the Terminos Lagoon. Arrows illustrate a general migration pathway of east to west through the system which follows prevailing currents. (From Yanez-Arancibia, A., Linares, F. A., and Day, J. W., Jr., *Estuarine Perspectives,* Kennedy, V. S., Ed., Academic Press, New York, 1980, 465. With permission.)

saxatilis) which spawned in freshwater streams and subsequently returned to the estuary as young-of-the-year forms and adults. The young-of-the-year fish, whether spawned offshore (e.g., English sole, *Parophys vetulus*) or in the bay (e.g., Pacific herring, *C. harengus pallasi*), generally remained in the estuary for 4 to 8 months prior to emigrating to the Pacific Ocean.

The following discussion on the distribution and abundances of fishes in the San Francisco

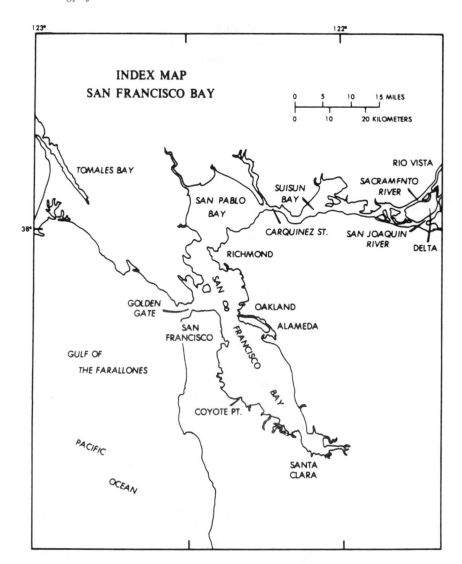

FIGURE 11.   Map of San Francisco Bay and environs. (From Krone, R. B., *San Francisco Bay: The Urbanized Estuary*, Conomos, T. J., Ed., Pacific Division, American Association for the Advancement of Science, San Francisco, CA, 1979, 85. With permission.)

Bay system is drawn heavily from the work of Armor and Herrgesell.[156] Fishes of the bay can be separated into two groups depending on their seasonal trends: (1) species having no consistent seasonal trend and whose catches over the 3-year study period (January 1980 to December 1982) appeared to be randomly distributed (e.g., white sturgeon, *A. transmontanus*; brown smooth-hound shark, *M. henlei*; and Pacific tomcod, *Microgadus proximus*); and (2) species having obvious seasonal cycles of occurrence each year (e.g., Pacific herring, *C. harengus pallasi*; striped bass, *M. saxatilis*; and northern anchovy *Engraulis mordax*). Fishes belonging to this second group experienced seasonal shifts in abundance. For example, their catches were two to tenfold greater in spring and summer than in winter and early spring. The summer peaks in abundance consisted of young-of-the-year fish which were spawned in the bay in the winter and early spring, adults and young-of-the-year fish which migrated into the bay and spawned there, and fish which moved into the estuary after reaching capturable size.

TABLE 13
Scientific and Common Names of Fishes Collected in the San Francisco Estuarine Ecosystem

Family name Scientific name	Common name	Family name Scientific name	Common name
Petromyzontidae		Bythitidae	
Lampetra tridentata	Pacific lamprey	*Brosmophycis margin-ata*	Red brotula
Carcharhinidae		Cyprinodotidae	
Mustelus henlei	Brown smoothhound	*Lucania parva*	Rainwater killifish
Triakis semifasciata	Leopard shark	Poeciliidae	
Squalidae		*Gambusia affinis*	Mosquitofish
Squalus acanthias	Spiny dogfish	Atherinidae	
Rajidae		*Atherinops affinis*	Topsmelt
Raja binoculata	Big skate	*Atherinopsis califor-niensis*	Jacksmelt
		Menidia audens	Mississippi silverside
Myliobatidae		Gastrosteidae	
Myliobatis californica	Bat ray	*Gasterosteus aculeatus*	Threespine stickleback
Acipenseridae		Syngnathidae	
Acipenser medirostris	Green sturgeon	*Syngnathus leptorhyn-chus*	Bay pipefish
A. transmontanus	White sturgeon		
		Percichthyidae	
Clupeidae		*Morone saxatilis*	Striped bass
Alosa sapidissima	American shad	Centrarchidae	
Clupea harengus pal-lasi	Pacific herring		
Dorosoma petenense	Threadfin shad	*Lepomis macrochirus*	Bluegill
Engraulidae		Percidae	
Engraulis mordax	Northern anchovy	*Percina macrolepida*	Bigscale logperch
Salmonidae		Sciaenidae	
Oncorhynchus kisutch	Silver salmon	*Genyonemus lineatus*	White croaker
O. tshawytscha	King salmon	Embiotocidae	
Salmo gairdnerii	Steelhead	*Amphistichus argen-teus*	Barred surfperch
		A. koelzi	Calico surfperch
Osmeridae		*A. rhodoterus*	Redtail surfperch
Allosmerus elongatus	Whitebait smelt	*Cymatogaster aggre-gata*	Shiner perch
Hypomesus nipponen-sis	Wakasagi		
		Embiotoca jacksoni	Black perch
H. pretiosus	Surf smelt	*Hyperprosopon argen-teum*	Walleye surfperch
H. transpacificus	Delta smelt		
		Hypsurus caryi	Rainbow seaperch
Spirinchus starksi	Night smelt	*Micrometrus minimus*	Dwarf perch
S. thaleichthys	Longfin smelt	*Phanerodon furcatus*	White seaperch
Bathylagidae		*Rhacochilus toxotes*	Rubberlip seaperch
Bathylagus pacificus	Pacific blacksmelt	*Hysterocarpus traski*	Tule perch
Myctophidae		*Rhacochilus vacca*	Pile perch
Stenobrachius leucop-sarus	Northern lampfish		
Tarletonbeania crenu-laris	Blue lanternfish	Labridae	
Cyprinidae		*Oxyjulis californica*	Senorita
Carassius auratus	Goldfish	Clinidae	
Cyprinus carpio	Carp	*Gibbonsia metzi*	Striped kelpfish
Pogonichthys macrole-pidotus	Splittail	*Neoclinus uninotatus*	Onespot fringehead
Ptychocheilus grandis	Sacramento squawfish	Pholidae	
Catastomidae		*Apodichthys flavidus*	Penpoint gunnel
Catastomus occiden-talis	*Sacramento sucker*	*Pholis ornata*	Saddleback gunnel

TABLE 13 (continued)
Scientific and Common Names of Fishes Collected in the San Francisco Estuarine Ecosystem

Family name / Scientific name	Common name	Family name / Scientific name	Common name
Ictaluridae		Ammodytidae	
Ictalurus cactus	White catfish	*Ammodytes hexapterus*	Pacific sandlance
I. melas	Black bullhead	Gobiidae	
I. nebulosus	Brown bullhead	*Lepidogobius lepidus*	Bay goby
I. punctatus	Channel catfish	*Acanthogobius flavimanus*	Yellowfin goby
Batrachoididae		*Ilypnus gilberti*	Cheekspot goby
Porichthys notatus	Plainfin midshipman	*Clevelandia ios*	Arrow goby
Gobiesocidae		*Tridentiger trigonocephalus*	Chameleon goby
Gobiesox maeandricus	Northern clingfish	*Coryphopterus nicholsii*	Blackeye goby
Gadidae		*Gillichthys mirabilis*	Longjaw mudsucker
Microgadus proximus	Pacific tomcod	Stromateidae	
		Peprilus simillimus	Pacific pompano
Scorpaenidae		Agonidae	
Sebastes auriculatus	Brown rockfish	*Odontopyxis trispinosa*	Pygmy poacher
S. melanops	Black rockfish	Cyclopteridae	
Hexagrammidae		*Liparis pulchellus*	Showy snailfish
Hexagrammos decagrammus	Kelp greenling	Bothidae	
Ophiodon elongatus	Lingcod	*Citharichthys sordidus*	Pacific sanddab
Oxylebius pictus	Painted greenling	*C. stigmaeus*	Speckled sanddab
Cottidae		*Paralichthys californicus*	California halibut
Hemilepidotus spinosus	Brown Irish lord	Pleuronectidae	
Leptocottus armatus	Staghorn sculpin	*Hypsopsetta guttulata*	Diamond turbot
Oligocottus synderi	Fluffy sculpin	*Parophrys vetulus*	English sole
Scorpaenichthys marmoratus	Cabezon	*Platichthys stellatus*	Starry flounder
Artedius harringtoni	Scalyhead sculpin	*Pleuronichthys decurrens*	Curlfin turbot
A. notospilotus	Bonehead sculpin	*P. verticalis*	Hornyhead turbot
Oligocottus maculosus	Tidepool sculpin	*Psettichthys melanostictus*	Sand sole
		Cynoglossidae	
Cottus asper	Prickly sculpin	*Symphurus atricauda*	California tonguefish

From Armor, C. and Herrgesell, P. L., *Hydrobiologia*, 129, 211, 1985. With permission.

Fish distributions and abundances may be regulated by freshwater flow into the estuary. During 1980 and 1982, high winter flows lowered salinities throughout the bay and enabled some freshwater species to spread into downstream regions. In wet years, the threadfin shad (*Dorosoma petenense*) extended into central and southern San Francisco Bay, while the splittail (*Pogonichthys macrolepidotus*) and the rainwater killifish (*Lucania parva*) became more abundant in San Pablo Bay. Several anadromous fishes were more numerous downstream in San Pablo Bay, the most prominent being the striped bass (*M. saxatilis*). During freshets, the northern anchovy (*E. mordax*) and Pacific herring (*C. harengus pallasi*) also translocated downstream.

Armor and Herrgesell[156] perceived at least three time scales of altered fish distributions and

TABLE 14
Species Response to Water Year Type in the San Francisco Bay Estuary. Number in Parenthesis is that Fish's Rank in the 15 Most Abundant Species

	Wet response	Dry response	Mixed response
Fresh	Threadfin shad	Sacramento squawfish	Inland silverside
	Carp	Tuleperch	Splittail
	Prickly sculpin		White catfish
Anadromous	White sturgeon	Pacific lamprey	American shad (14)
	Green sturgeon		King salmon
	Steelhead		Striped bass (5)
Estuarine	Threespine stickleback		Delta smelt
	Yellowfin goby (10)		Bay goby (11)
	Longfin smelt (2)		
	Staghorn sculpin (8)		
	Starry flounder (12)		
Marine-estuarine	Pacific herring (3)	Arrow goby	White croaker (9)
	Cheekspot goby	Walleye surfperch	Northern anchovy (1)
			Plainfin midshipman (15)
			Shiner perch (4)
Marine	Leopard shark	Bat ray	Night smelt
	Pile perch	White seaperch	Bay pipefish
	Speckled sanddab (7)	Jacksmelt (13)	Barred surfperch
	Diamond turbot	Black perch	Brown rockfish
	Sand sole	Rubberlip seaperch	Lingcod
	California tonguefish	Pacific butterfish	England sole (6)
	Brown smoothhound	Bonyhead sculpin	Dwarf perch
	Spiny dogfish		Big skate
	Pacific tomcod		Surf smelt
	Topsmelt		Curlfin turbot
	Snowy snailfish		

From Armor, C. and Herrgesell, P. L., *Hydrobiologia*, 129, 211, 1985. With permission.

catches — short-term (3 to 4 weeks), seasonal, and interannual — related to variations of freshwater inflow. They also differentiated fishes into wet-response, dry-response, and mixed-response species (Table 14), depending upon when they reached peak abundance. A population was considered to be a dry-response species if it achieved peak baywide abundance during 1981, a dry year characterized by low freshwater inflow into the estuary. Conversely, if peak abundance occurred during 1980 and 1982 when wet conditions and high freshwater inflow prevailed, the fish was categorized as a wet-response species. A mixed-response species was one reaching peak abundance during either 1980 or 1981 and 1982. Analysis of trawl catches showed that 24 of the species studied were wet-response species, 22 mixed-response species, and 12 dry-response species (Table 15). Wet years, therefore, seemed to enhance the catches of more species, particularly estuarine forms which were most abundant during 1980 and 1982. In terms of the relative importance of individual populations, 6 of the 15 most abundant species exhibited a wet-year response, 8 of the 15 most abundant forms showed mixed-year responses, and only 1 of 15 species displayed a dry-year response.

Some factors associated with increased freshwater inflow affect the distribution and abundance of fishes in San Francisco Bay. The magnitude of freshwater inflow, for instance, certainly affects other environmental conditions, such as salinity, temperature, currents, food supply, and turbidity which can modulate fish migrations and spawning success. Increased freshwater flow can foster higher food production in the bay either directly by the transport of more food into the system or indirectly by the nutrient enhancement of primary production.

TABLE 15
Number of Species with Highest Catches during
Various Year Types in the San Francisco Bay
Estuary

Salinity preference group	Wet response	Mixed response	Dry response
Freshwater	3	3	2
Anadromous	3	3	1
Estuarine	5	2	0
Marine-estuarine	2	4	2
Marine	11	10	7
Total	24	22	12

From Armor, C. and Herrgesell, P. L., *Hydrobiologia*, 129, 211, 1985. With permission.

Larval fish survival also may be altered by seasonal freshwater inflow. The downstream movement of freshwater and estuarine fishes from riverine systems into the bay and the recruitment of certain marine species from oceanic areas into the bay are coupled to the volume of freshwater inflow.[156]

VIII. SUMMARY AND CONCLUSIONS

Fish are a major component of the nektonic community of estuaries and represent the largest and most motile faunal elements of these coastal ecosystems. Their great mobility confers significant advantages on them over sessile organisms, enabling them to avoid unfavorable environmental conditions and predatory attacks. Estuarine populations display some degree of temperature tolerance and osmoregulatory ability.

The ichthyofauna of estuaries are characterized by the numerical dominance of a few species which tend to be widespread, reflecting the broad tolerances and range of adaptations of these fishes. Juveniles, which use the estuarine environment as a nursery area, constitute the most abundant forms. Many of these young fishes enter estuaries from coastal populations. The fish communities of estuaries harbor numerous migratory species that seasonally move into and out of these ecotones. A smaller number of populations are permanent residents.

Estuarine environments have long been known for their exceptional recreational and commercial fisheries. Much of the U.S. commercial catch and marine recreational catch embodies species which inhabit estuaries at least part of their lives. The bulk of the fisheries landings in the Gulf of Mexico (greater than 95%) has been attributed to estuarine species. Similar landings data exist for the mid-Atlantic states. The economic value of estuarine-dependent fisheries raises concern over the potential impact of anthropogenic stresses on natural populations of fish.

A number of classification schemes have been formulated for estuarine fishes. Pearcy and Richards[21] proposed five general assemblages of estuarine ichthyofauna based on the abundance of juveniles and adults, their seasonal occurrence, and the distribution of eggs and larvae. These encompass: (1) resident species that spawn in the estuary and inhabit estuarine waters for much of their lives; (2) summer migrants, typically southern species, occupying estuarine habitats during the warm months; (3) local marine forms commonly occurring near the mouth of estuaries and usually indigenous to local neritic waters; (4) diadromous species passing through estuaries on spawning migrations to either freshwater or oceanic environments; and (5) freshwater species normally found in inland waters.

Perkins[22] grouped estuarine fishes into two broad assemblages of diadromous and nondiadromous species. Anadromous and catadromous populations form the diadromous category. Residents, summer migrants, local marine species, and freshwater populations comprise the nondiadromous category.

Employing breeding and migratory habits, Day[2,23] separated estuarine fishes into five classes. Marine migrants are the largest class of fishes, containing juveniles of species that breed in the ocean and use the estuary for feeding and protection. Two other classes incorporate anadromous and catadromous species — those migrating to and from breeding grounds in freshwater and seawater — utilizing estuarine waters at certain times of the year. Species completing their life cycle in the estuary are certified as residents. Fishes not belonging to any of the four groups mentioned above are listed in an anomalous category.

The classification of Moyle and Cech[24] categorizes fishes into five broad classes: (1) freshwater, (2) diadromous, (3) true estuarine, (4) nondependent marine, and (5) dependent marine fishes. A similar classification, involving six rather than five categories, has been proffered by McHugh.[9] His scheme focuses on breeding, migratory, and ecological criteria as follows: (1) freshwater fishes that occasionally enter brackish waters; (2) truly estuarine species which spend their entire lives in the estuary; (3) anadromous and catadromous species; (4) marine species which pay regular visits to the estuary, usually as adults; (5) marine species which use the estuary primarily as a nursery ground, spawning and spending much of their adult life at sea, but often returning seasonally to the estuary; and (6) adventitious visitors which appear irregularly and have no apparent estuarine requirements.

The structure of an estuarine fish community depends on abiotic and biotic factors. Among abiotic factors, both chemical and physical conditions — especially salinity, temperature, and dissolved oxygen — strongly influence the species composition, abundance, and distribution of fishes in estuarine waters. Fishes that dominate the community in estuaries tend to have a broad salinity tolerance. The truly estuarine forms typically adapt readily to salinity changes. The adaptability of some marine fishes to lower salinities seems remarkable. However, fewer osmotic problems are encountered by marine teleosts at the mouth of estuaries; as a result, the species richness of the fish community generally peaks in this region. Progressing upestuary, osmotic problems arise for various populations, and at a salinity of approximately one third that of seawater, major osmotic difficulties may develop.

In the estuarine environment, fishes must deal with temperature as well as salinity gradients. In temperate and boreal regions, in particular, seasonal temperature changes have a marked effect on the community structure. The migratory patterns of many species, for example, are strongly coupled to seasonal temperature levels. Changing thermal gradients can act as barriers to certain species (e.g., bluefish), thereby affecting their migratory behavior.

In order to maximize survivorship in estuarine habitats, fishes thermoregulate behaviorally, avoiding or selecting environmental temperatures. However, the observed distribution of a species in an estuary reflects its response to other factors as well, such as food availability, nutritional state, competition, predation, habitat requirements, and so forth. Temperature responses probably are most acute in cold-temperate and high-latitude systems rather than subtropical and tropical waters where salinity plays a paramount role. As temperature increases in the spring in temperate and boreal regions, marine fishes (largely juvenile forms) migrate into estuaries in search of food and shelter. Activity in these highly seasonal environs peaks in the summer with the arrival of summer migrants, but declines by late fall as many populations emigrate to the nearshore ocean sparked by diminishing temperatures. The abundance of fishes decreases greatly in the winter; at this time, residents remain inactive in deeper waters while the estuarine surface often freezes.

Estuaries burdened by excessive oxygen consumption due to natural biological removal, together with the oxidation of pollutant organic matter, place additional demands on fish populations. The development of anoxia, as in Chesapeake Bay in summer, creates inhospitable conditions for fishes. When dissolved oxygen concentrations drop below 4 ml/l and approach 0 ml/l, ichthyofauna are impacted. In severe cases, migration routes can be effectively blocked by oxygen-depleted water masses that spread over broad areas, persisting for months. Fishes entering waters devoid of oxygen can be trapped, increasing their risk of death. Mass mortality of fishes ascribable to low dissolved oxygen levels has been periodically documented. Other factors shown to occasionally inflict heavy losses on fish populations are large quantities of suspended sediment and red tides.

Biotic factors — interspecific and intraspecific interactions — also can profoundly affect a fish community. Among interspecific interactions, predator-prey relationships, competition, and symbiosis (i.e., commensalism, mutualism, and parasitism) are important. Intraspecific interactions parallel interspecific interactions, with predation, competition, and even cannibalism between individuals of the same species potentially regulating the abundance of some populations.

Estuarine fishes exhibit dynamic feeding strategies that commonly change in time and space, and with ontogeny. As some fish age and grow, they become increasingly piscivorous, especially the larger species. Whereas a few fishes are basic herbivores, consuming benthic and epiphytic diatoms, filamentous algae, or seagrasses, the juveniles of numerous forms are voracious predators, ingesting copepods, amphipods, isopods, mysids, ostracods, and other invertebrates. Top predators, such as the bluefish, remove substantial biomasses of forage fishes (e.g., bay anchovy). Only a few estuarine species appear to be detritivores, the spot being a notable example. Many populations are omnivorous (e.g., Atlantic silverside and black sea bass) or carnivorous (e.g., striped bass and weakfish).

Research on the population dynamics of fishes involves collecting data on growth, mortality, and recruitment. Age determinations are essential to this work as well as to the proper management of fisheries resources. Estimates on age can be obtained by three methods: (1) mark-and-recapture experiments, (2) length-frequency distributions, and (3) interpretations of layers recorded in hard parts (i.e., scales and otoliths).

The von Bertalanffy, logistic, and Gompertz equations are three of the most frequently used relationships in fisheries research for assessing the growth of fish populations. In models applying the von Bertalanffy equation, fish grow toward a theoretical length or weight; the rate of change of size declines as the length approaches a maximum value. The logistic curve also reveals size tending to an asymptote. The Gompertz model, while transcribing a curve whereby length or weight at age tends to an asymptote, differs from the von Bertalanffy and logistic models in that the curve generated has an asymmetrically placed inflection point, at an age less than halfway up the curve. A valuable method of estimating growth parameters is the Walford plot.

Investigations of ichthyofaunal mortality consider both natural and fishing death. Fishery management programs evaluate both types of mortality. The death rates of fish populations in the absence of fishing is poorly known.

Density-dependent and density-independent mortality of fish populations have been ascertained in estuarine communities. Density-dependent mortality (e.g., disease and predation) arises from actions of the fish populations, whereas density-independent mortality is unrelated to any population action (e.g., temperature extremes and storms). Compensatory responses in mortality of fishes due to density-dependent factors normally originate during the critical early life-history stages.

Viviparous, ovoviparous, and oviparous estuarine fishes have different reproductive potentials. Viviparous forms are characterized by low fecundity, but the young develop to a large size prior to hatching. Oviparous fishes, in contrast, produce many more eggs, typified by limited yolk reserves. These unprotected eggs enter the plankton and are subject to a large wastage of numbers. Survivorship is much lower than for viviparous fishes, especially through the free embryo and larval stages, and it depends greatly on the length of the planktonic egg and/or larval stage. Survival of the ichthyoplankton hinges on multiple factors (e.g., temperature, salinity, predation, food availability, feeding effects, and pollution). Fishes with high fecundities also have high mortality rates.

Recruitment variability has been related to the variability in larval survival. Major sources of recruitment losses are predation, low food availability, and feeding effects. The spatial clustering of eggs and larvae and their subsequent removal by predators impose significant limitations on recruitment success.

The management of estuarine fisheries employs the application of general stock-recruitment models. The models of Beverton and Holt[98] and Ricker[124] are widely used in recruitment studies. The Ricker model delineates a series of curves with low recruitment at high stock levels. The Beverton and Holt recruitment model, in turn, entails a group of asymptotic curves that depict constant recruitment beyond a certain stock density.

In conclusion, the ichthyofauna of estuaries consist primarily of juveniles that use the protective habitat largely as a nursery area. The abundance and biomass of fishes typically are high, but species diversity generally is low. Only a few species dominate the community; they usually are widespread forms with broad tolerances and a wide range of adaptations. Fish communities of Barnegat Bay, Chesapeake Bay, the Terminos Lagoon, and San Francisco Bay provide examples of these characteristics.

REFERENCES

1. **Lincoln, R. J., Boxshall, G. A., and Clark, P. F.,** *A Dictionary of Ecology, Evolution and Systematics,* Cambridge University Press, Cambridge, 1982.
2. **Day, J. H., Blaber, S. J. M., and Wallace, J. H.,** Estuarine fishes, in *Estuarine Ecology: With Particular Reference to Southern Africa,* Day, J. H., Ed., A. A. Balkema, Rotterdam, 1981, 197.
3. **Haedrich, R. L. and Hall, C. A. S.,** Fishes and estuaries, *Oceanus,* 19, 55, 1976.
4. **Henderson, P. A.,** The structure of estuarine fish communities, *J. Fish Biol.,* 33 (Suppl. A), 223, 1988.
5. **Claridge, P. N. and Potter, I. C.,** Abundance, movements and size of gadoids (Teleostei) in the Severn estuary, *J. Mar. Biol. Assoc. U.K.,* 64, 771, 1984.
6. **Potter, I. C., Claridge, P. N., and Warwick, R. M.,** Consistency of seasonal changes in an estuarine fish assemblage, *Mar. Ecol. Prog. Ser.,* 32, 217, 1986.
7. **Fortier, L. and Leggett, W. C.,** Fickian transport and the dispersal of fish larvae in estuaries, *Can. J. Fish. Aquat. Sci.,* 39, 1150, 1982.
8. **Claridge, P. N., Potter, I. C., and Hardisty, M. W.,** Seasonal changes in movements, abundance, size composition, and diversity of the fish fauna of the Severn estuary, *J. Mar. Biol. Assoc. U.K.,* 66, 229, 1986.
9. **McHugh, J. L.,** Estuarine nekton, in *Estuaries,* Lauff, G. H., Ed., Publ. 83, American Associaton for the Advancement of Science, Washington, D. C., 1967, 581.
10. **Haedrich, R. L.,** Estuarine fishes, in *Estuaries and Enclosed Seas,* Ketchum, B. H., Ed., Elsevier Scientific, Amsterdam, 1983, 183.
11. **Saila, S. B.,** Some aspects of fish production and cropping in estuarine systems, in *Estuarine Research,* Vol. 1, Cronin, L. E., Ed., Academic Press, New York, 1975, 473.
12. **Sheridan, P. F. and Livingston, R. J.,** Cyclic trophic relationships of fishes in an unpolluted, river-dominated estuary in north Florida, in *Ecological Processes in Coastal and Marine Systems,* Livingston, R. J., Ed., Plenum Press, New York, 1979, 143.
13. **Gunter, G.,** Some relationships of estuaries to the fisheries of the Gulf of Mexico, in *Estuaries,* Lauff, G. H., Ed., Publ. 83, American Association for the Advancement of Science, Washington, D. C., 1967, 621.
14. **McHugh, J. L.,** *Fishery Management,* Springer-Verlag, New York, 1984.
15. **McHugh, J. L.,** Management of estuarine fisheries, in *A Symposium on Estuarine Fisheries,* American Fisheries Society, Spec. Publ. 3, Bethesda, MD, 1966, 133.
16. **McHugh, J. L.,** Estuarine fisheries: are they doomed?, in *Estuarine Processes,* Vol. 1, Wiley, M., Ed., Academic Press, New York, 1976, 15.
17. **McHugh, J. L.,** Are estuaries necessary?, *Commer. Fish. Rev.,* 30, 37, 1968.
18. **Stickney, R. R.,** *Estuarine Ecology of the Southeastern United States and Gulf of Mexico,* Texas A & M University, College Station, TX, 1984.

19. **Smith, R. F., Swartz, A. H., and Massman, W. H., Eds.,** A symposium on estuarine fisheries, *Trans. Am. Fish. Soc.,* Suppl. 95, 1, 1966.

20. **Hillman, R. E., Davis, N. W., and Wennemer, J.,** Abundance, diversity, and stability of shore-zone fish communities in an area of Long Island Sound affected by the thermal discharge of a nuclear power station, *Estuarine Coastal Mar. Sci.,* 5, 355, 1977.

21. **Pearcy, W. B. and Richards, S. W.,** Distribution and ecology of fishes of the Mystic River estuary, Connecticut, *Ecology,* 43, 248, 1962.

22. **Perkins, E. J.,** *The Biology of Estuaries and Coastal Waters,* Academic Press, London, 1974.

23. **Day, J. H.,** The ecology of South African estuaries. Part 1. A review of estuarine conditions in general, *Trans. R. Soc. S. Afr.,* 33, 53, 1951.

24. **Moyle, P. B. and Cech, J. J., Jr.,** *Fishes: An Introduction to Ichthyology,* Prentice-Hall, Englewood Cliffs, NJ, 1982.

25. **McHugh, J. L.,** The pound-net fishery in Virginia. II. Species composition of landings reported as menhaden, *Commer. Fish. Rev.,* 22, 1, 1960.

26. **Hoff, J. G. and Ibara, R. M.,** Factors affecting the seasonal abundance, composition and diversity of fishes in a southeastern New England estuary, *Estuarine Coastal Mar. Sci.,* 5, 665, 1977.

27. **Tatham, T. R., Thomas, D. L., and Danila, D. J.,** Fishes of Barnegat Bay, in *Ecology of Barnegat Bay, New Jersey,* Kennish, M. J. and Lutz, R. A., Eds., Springer-Verlag, New York, 1984, 241.

28. **Hedgpeth, J. W.,** Ecological aspects of the Laguna Madre, a hypersaline estuary, in *Estuaries,* Lauff, G. H., Ed., Publ. 83, American Association for the Advancement of Science, Washington, D. C., 1967, 408.

29. **Wallace, J. H.,** The estuarine fishes of the East Coast of Africa. I. Species composition and length distribution in the estuarine and marine environments. II. Seasonal abundance and migrations, *Invest. Rep. Oceanogr. Res. Inst. Durban.,* 40, 1, 1975.

30. **Talbot, G. B.,** Estuarine environmental requirements and limiting factors for striped bass, *Trans. Am. Fish. Soc.,* 95 (Suppl.), 37, 1966.

31. **McDowall, R. M.,** The role of estuaries in the life cycles of fishes in New Zealand, *Proc. N.Z. Ecol. Soc.,* 23, 27, 1976.

32. **McDowall, R. M.,** The galaxiid fishes of New Zealand, *Bull. Mus. Comp. Zool. Harv. Univ.,* 139, 341, 1970.

33. **Nicholson, W. R.,** Movements and population structure of Atlantic menhaden indicated by tag returns, *Estuaries,* 1, 141, 1978.

34. **Blomo, V. J.,** Distribution of economic impacts from proposed conservation measures in the U.S. Atlantic menhaden fishery, *Fish. Res.,* 5, 23, 1987.

35. **Bigelow, H. B. and Schroeder, W. C.,** Fishes of the Gulf of Maine, *U.S. Bur. Fish. Wildl. Serv. Fish. Bull.,* p. 53, 1953.

36. **Grosslein, M. D. and Azarovitz, T. R.,** Fish Distribution, MESA New York Bight Atlas Monogr. 15, New York Sea Grant Institute, Albany, 1982.

37. **Hildebrand, S. F. and Schroeder, W. C.,** Fishes of Chesapeake Bay, *Bull. U S. Bur. Fish.,* 1, 43, 1928.

38. **deSylva, D. P., Kalber, F. A., and Shuster, C. N., Jr.,** Fishes and ecological conditions in the shore zone of the Delaware River estuary, with notes on other species collected in deeper water, Mar. Lab. Inf. Ser. Publ. No. 5, University of Delaware, Lewes, 1962.

39. **Hildebrand, S. F. and Cable, L. E.,** Development and life history of fourteen teleostean fishes at Beaufort, North Carolina, *Bull. U.S. Bur. Fish.,* 46, 383, 1930.

40. **Bearden, C. M.,** Distribution and abundance of Atlantic croaker, *Micropogon undulatus,* S.C. Contrib. Bears Bluff Lab., 40, 1, 1964.

41. **McHugh, J. L. and Ginter, J. J. C.,** Fisheries, MESA New York Bight Atlas Monogr. 16, New York Sea Grant Institute, Albany, 1978.

42. **Parry, G.,** Osmotic adaptation in fishes, *Biol. Rev.,* 41, 392, 1966.

43. **Green, J.,** *The Biology of Estuarine Animals,* University of Washingon Press, Seattle, 1968.

44. **Olla, B. L., Studholme, A. L., and Bejda, A. J.,** Behavior of juvenile bluefish *Pomatomus saltatrix* in vertical thermal gradients: influence of season, temperature acclimation, and food, *Mar. Ecol. Prog. Ser.,* 23, 165, 1985.

45. **Neill, W. H., Magnuson, J. J., and Chipman, G.,** Behavioral thermoregulation by fishes: a new experimental approach, *Science,* 176, 1443, 1972.

46. **Elliott, J. M.,** Some aspects of thermal stress on freshwater teleosts, in *Stress and Fish,* Pickering, A. D., Ed., Academic Press, London, 1981, 209.

47. **FitzGerald, G. J. and Wootton, R. J.,** Behavioral ecology of sticklebacks, in *The Behavior of Teleost Fishes,* Pitcher, T. J., Ed., Johns Hopkins University Press, Baltimore, 1986, 409.

48. **Weatherley, A. H. and Gill, H. S.,** *The Biology of Fish Growth,* Academic Press, Orlando, FL, 1987.

49. **Longhurst, A. R. and Pauly, D.,** *Ecology of Tropical Oceans,* Academic Press, Orlando, FL., 1987.

50. **Brett, J. R.,** Fishes, functional responses, in *Marine Ecology, a Comprehensive, Integrated Treatise on Life in Oceans and Coastal Waters,* Vol. 1, Kinne, O., Ed., Wiley-Interscience, New York, 1970, 515.

51. **Pitcher, T. J., Ed.,** *The Behavior of Teleost Fishes,* Johns Hopkins University Press, Baltimore, 1986.

52. **Pitcher, T. J. and Hart, P. J. B.,** *Fisheries Ecology,* AVI Publishing, Westport, CT, 1982.
53. **Head, P. C.,** Data presentation and interpretation, in *Practical Estuarine Chemistry: A Handbook,* Head, P. C., Ed., Cambridge University Press, Cambridge, 1985, 278.
54. **Aston, S. R.,** Nutrients, dissolved gases, and general biogeochemistry in estuaries, in *Chemistry and Biogeochemistry of Estuaries,* Olausson, E. and Cato, I., Eds., John Wiley & Sons, Chichester, 1980, 233.
55. **Officer, C. B., Biggs, R. B., Taft, J. L., Cronin, L. E., Tyler, M. A., and Boynton, W.,** Chesapeake Bay anoxia: origin, development, and significance, *Science,* 223, 22, 1984.
56. **Seliger, H. H., Boggs, J. A., and Biggley, W. H.,** Catastrophic anoxia in the Chesapeake Bay in 1984, *Science,* 228, 70, 1985.
57. **Tyler, M. A.,** Flow-induced variation in transport and deposition pathways in the Chesapeake Bay: the effect on phytoplankton dominance and anoxia, in *Estuarine Variability,* Wolfe, D. A., Ed., Academic Press, Orlando, FL, 1986, 161.
58. **Bejda, A. J., Studholme, A. J., and Olla, B. L.,** Behavioral responses of red hake, *Urophycis-chuss,* to decreasing concentrations of dissolved oxygen, *Environ. Biol. Fishes,* 19, 261, 1987.
59. **Arthur, D. R.,** Constraints on the fauna in estuaries, in *River Ecology,* Whitton, B. A., Ed., University of California Press, Berkeley, 1975, 514.
60. **Virnstein, R. W.,** The importance of predation by crabs and fishes on benthic infauna in Chesapeake Bay, *Ecology,* 58, 1199, 1977.
61. **Levinton, J. S.,** *Marine Ecology,* Prentice-Hall, Englewood Cliffs, NJ, 1982.
62. **Thurman, H. V. and Webber, H. H.,** *Marine Biology,* Charles E. Merrill Publishing, Columbus, OH, 1984.
63. **Nybakken, J. W.,** *Marine Ecology: An Ecological Approach,* Harper & Row, New York, 1982.
64. **Odum, E. P.,** *Fundamentals of Ecology,* 3rd ed., W. B. Saunders, Philadelphia, 1971.
65. **Jonasson, M.,** Fish cleaning behaviour of shrimp, *J. Zool.,* 213, 117, 1987.
66. **Simenstad, C. A. and Cailliet, G. M., Eds.,** *Contemporary Studies on Fish Feeding,* Develop. Environ. Biol. Fishes Ser., Dr. W. Junk Publishers, Dordrecht, The Netherlands, 1986.
67. **Worobec, M. N.,** Field estimates of the daily ration of winter flounder *Pseudopleuronectes americanus* (Walbaum) in a southern New England salt pond, *J. Exp. Mar. Biol. Ecol.,* 77, 183, 1984.
68. **Cowey, C. B., Mackie, A. M., and Bell, J. G., Eds.,** *Nutrition and Feeding in Fish,* Academic Press, Orlando, FL, 1986.
69. **Halver, J. E., Ed.,** *Fish Nutrition,* 2nd ed., Academic Press, Orlando, FL, 1988.
70. **Dekhnik, T. V. and Sinyvkova, V. I.,** Survival of marine fish larvae in relation to food availability, *J. Ichthyol.,* 16, 294, 1976.
71. **Werner, R. G. and Blaxter, J. H. S.,** The effect of prey density on mortality, growth, and food consumption in larval herring (*Clupea harengus* L.), *Rapp. P.V. Reun. Cons. Int. Explor. Mer,* 178, 405, 1981.
72. **Fortier, L. and Leggett, W. C.,** Small-scale covariability in the abundance of fish larvae and their prey, *Can. J. Fish. Aquat. Sci.,* 41, 502, 1984.
73. **Frank, K. T.,** Independent distributions of fish larvae and their prey: natural paradox or sampling artifact, *Can. J. Fish. Aquat. Sci.,* 45, 48, 1988.
74. **Leggett, W. C.,** The dependence of fish larval survival on food and predator densities, in *The Role of Freshwater Outflow on Coastal Marine Ecosystems,* Vol. G7, Skreslet, S., Ed., NATO ASI Series, Springer-Verlag, Berlin, 1986, 117.
75. **Cushing, D. H.,** *Science and the Fisheries,* Edward Arnold, London, 1977.
76. **Hunter, J. R.,** Feeding ecology and predation of marine fish larvae, in *Marine Fish Larvae: Morphology, Ecology, and Relation to Fisheries,* Lasker, R., Ed., Washington Sea Grant Program, Seattle, 1981, 34.
77. **Beers, J. R. and Stewart, G. L.,** Micro-zooplankton in the euphotic zone at five locations across the California current, *J. Fish. Res. Board Can.,* 24, 2053, 1967.
78. **Frank, K. T. and Leggett, W. C.,** Effect of prey abundance and size on the growth and survival of larval fish: an experimental study employing large volume enclosures, *Mar. Ecol. Prog. Ser.,* 34, 11, 1986.
79. **Govoni, J. J., Ortner, P. B., Al-Yamani, F., and Hill, L. C.,** Selective feeding of spot, *Leiostomus xanthurus,* and Atlantic croaker *Micropogonias undulatus,* larvae in the northern Gulf of Mexico, *Mar. Ecol. Prog. Ser.,* 28, 175, 1986.
80. **Eggers, D. M.,** The nature of prey selection by planktivorous fish, *Ecology,* 58, 46, 1977.
81. **Drenner, R. W., Strickler, J. R., and O'Brien, W. J.,** Capture probability: the role of zooplankton escape in the selective feeding of planktivorous fish, *J. Fish. Res. Board Can.,* 35, 1370, 1978.
82. **Dabrowski, K. and Bardega, R.,** Mouth size and predicted food size preferences of larvae of three cyprinid fish species, *Aquaculture,* 40, 41, 1984.
83. **Stoecker, D. K. and Govoni, J. J.,** Food selection by young larval gulf menhaden (*Brevoortia patronus*), *Mar. Biol.,* 80, 299, 1984.
84. **Govoni, J. J., Hoss, D. E., and Chester, A. J.,** Comparative feeding of three species of larval fishes in the northern Gulf of Mexico: *Brevoortia patronus, Leiostomus xanthurus,* and *Micropogonias undulatus, Mar. Ecol. Prog. Ser.,* 13, 189, 1983.

85. **Evans, S.,** Production, predation, and food niche segregation in a marine shallow soft-bottom community, *Mar. Ecol. Prog. Ser.,* 10, 147, 1983.

86. **Morais, L. de and Bodiou, J. Y.,** Predation on meiofauna by juvenile fish in a western Meditterranean flatfish nursery ground, *Mar. Biol.,* 82, 209, 1984.

87. **Healey, M. C.,** Utilization of the Nanaimo River estuary by juvenile chinook salmon, *Oncorhynchus tshawytacha, Fish. Bull. (U.S.),* 77, 654, 1980.

88. **Wolf, E. G., Morson, B., and Fucik, K. W.,** Preliminary studies of food habits of juvenile fish, China Poot Marsh and Potter Marsh, Alaska, 1978, *Estuaries,* 6, 102, 1983.

89. **Oviatt, C. A. and Nixon, S. W.,** The demersal fish of Narragansett Bay: an analysis of community structure, distribution and abundance, *Estuarine Coastal Mar. Sci.,* 1, 361, 1973.

90. **Haedrich, R. L. and Haedrich, S. O.,** A seasonal survey of the fishes in the Mystic River, a polluted estuary in downtown Boston, Massachusetts, *Estuarine Coastal Mar. Sci.,* 2, 59, 1974.

91. **Darnell, R. M.,** Food habits of fishes and larger macroinvertebrates of Lake Pontchartrain, Louisiana, an estuarine community, *Publ. Inst. Mar. Sci. Univ. Tex.,* 5, 353, 1958.

92. **Matlock, G. C. and Garcia, M. A.,** Stomach contents of selected fishes from Texas bays, *Contrib. Mar. Sci.,* 26, 95, 1983.

93. **Diener, R. A., Inglis, A., and Adams, G. B.,** Stomach contents of fishes from Clear Lake and tributary waters, a Texas estuarine area, *Contrib. Mar. Sci.,* 18, 7, 1974.

94. **Alexander, S. K.,** Summer diet of finfish from nearshore habitats of West Bay, Texas, *Tex. J. Sci.,* 35, 93, 1983.

95. **Kennish, M. J. and Loveland, R. E.,** Trophic relationships, in *Ecology of Barnegat Bay, New Jersey,* Kennish, M. J. and Lutz, R. A., Eds., Springer-Verlag, New York, 1984, 302.

96. **Whitfield, A. K. and Blaber, S. J. M.,** Food and feeding ecology of piscivorous fishes at Lake St. Lucia, Zululand, *J. Fish Biol.,* 13, 675, 1978.

97. **Ricker, W. E.,** Computation and Interpretation of Biological Statistics of Fish Populations, Bull. 191, Fisheries Research Board of Canada, Ottawa, 1975.

98. **Beverton, R. J. H. and Holt, S. J.,** On the dynamics of exploited fish populations, *Min. Agric. Fish Food (U.K.) Fish. Invest. London,* Ser. 2, 19, 1957.

99. **Everhart, W. H., Eipper, A. W., and Youngs, W. D.,** *Principles of Fishery Science,* Cornell University Press, Ithaca, New York, 1975.

100. **Reznick, D., Lindbeck, E., and Bryga, H.,** Slower growth results in larger otoliths: an experimental test with guppies (*Poecilia reticulata*), *Can. J. Fish. Aquat. Sci.,* 46, 108, 1989.

101. **Pannella, G.,** Growth patterns in fish sagittae, in *Skeletal Growth of Aquatic Organisms: Biological Records of Environmental Change,* Rhoads, D. C. and Lutz, R. A., Eds., Plenum Press, New York, 1980, 519.

102. **Campana, S. E. and Neilson, J. D.,** Microstructure of fish otoliths, *Can. J. Fish. Aquat. Sci.,* 42, 1014, 1985.

103. **Bradford, M. J. and Green, G. H.,** Size and growth of juvenile chinook salmon back-calculated from otolith growth increments, in *Age and Growth of Fish,* Summerfelt, R. C., Ed., Iowa State University Press, Ames, 1987, 453.

104. **Jones, C. and Brothers, E. B.,** Validation of the otolith increment technique for striped bass, *Morone saxatilis,* larvae reared under suboptimal feeding conditions, *Fish. Bull. (U.S.),* 85, 171, 1987.

105. **Secor, D. H. and Dean, J. M.,** Somatic growth effects on the otolith — fish size relationship in young pond-reared striped bass, *Morone saxatilis, Can. J. Fish. Aquat. Sci.,* 46, 113, 1989.

106. **Brett, J. R. and Groves, T. D.,** Physiological energetics, in *Fish Physiology,* Vol. 9, Hoar, W. S., Randall, D. J., and Brett, J. R., Eds., Academic Press, New York, 1979, 279.

107. **Hoar, W. S., Randall, D. J., and Donaldson, E. M., Eds.,** *Fish Physiology,* Vol. 11 (Part A), Academic Press, Orlando, FL, 1988.

108. **Gallucci, V. F. and Quinn, T. J.,** Reparameterizing, fitting, and testing a simple growth model, *Trans. Am. Fish. Soc.,* 108, 14, 1979.

109. **Allen, K. R.,** A method for fitting growth curves of the von Bertalanffy type to observed data, *J. Fish. Res. Board Can.,* 23, 163, 1966.

110. **Ricker, W. E.,** Growth rates and models, in *Fish Physiology,* Vol. 8, Hoar, W. S., Randall, D. J., and Brett, J. R., Eds., Academic Press, New York, 1979, 677.

111. **Walford, L. A.,** A new graphic method of describing the growth of animals, *Biol. Bull.,* 90, 141, 1946.

112. **Bagenal, T. B.,** Aspects of fish fecundity, in *Ecology of Freshwater Fish Production,* Gerking, S. D., Ed., John Wiley & Sons, New York, 1978, 75.

113. **Lagler, K. F., Bardach, J. E., and Miller, R. R.,** *Ichthyology,* John Wiley & Sons, New York, 1962.

114. **Breeder, C. M. and Rosen, D. E.,** *Modes of Reproduction in Fishes,* National History Press, Garden City, NY, 1966.

115. **Leiby, M. M.,** Life history and ecology of pelagic fish eggs and larvae, in *Marine Plankton Life Cycle Strategies,* Steidinger, K. A. and Walker, L. M., Eds., CRC Press, Boca Raton, FL, 1984, 121.

116. **Potts, G. W. and Wooton, R. J., Eds.,** *Fish Reproduction: Strategies and Tactics,* Academic Press, Orlando, FL, 1984.

117. **Bagenal, T. B.,** A short review of fish fecundity, in *The Biological Basis of Freshwater Fish Production,* Gerking, S. D., Ed., Blackwell Scientific, Oxford, 1967, 89.

118. **Snyder, D. E.,** Fish eggs and larvae, in *Fisheries Techniques,* Nielsen, L. A. and Johnson, D. L., Eds., American Fisheries Society, Bethesda, MD, 1983, 165.

119. **Conover, D. O.,** Field and laboratory assessment of patterns in fecundity of a multiple spawning fish: the Atlantic silverside *Menidia menidia, Fish. Bull. (U.S.),* 83, 331, 1985.

120. **Rothschild, B. J. and DiNardo, G. T.,** Comparison of recruitment variability and life history data among marine and anadromous fishes, *Am. Fish. Soc. Symp.,* 1, 531, 1987.

121. **McGurk, M. D.,** Natural mortality of marine pelagic fish eggs and larvae: role of spatial patchiness, *Mar. Ecol. Prog. Ser.,* 34, 227, 1986.

122. **Houde, E. D.,** Critical food concentrations for larvae of three species of subtropical marine fishes, *Bull. Mar. Sci.,* 28, 395, 1978.

123. **Houde, E. D. and Schekter, R. C.,** Feeding by marine fish larvae: developmental and functional responses, *Environ. Biol. Fishes,* 5, 315, 1980.

124. **Ricker, W. E.,** Stock and recruitment, *J. Fish. Res. Board Can.,* 11, 559, 1954.

125. **Jones, R. and Hall, W. B.,** A simulation model for studying the population dynamics of some fish species, in *The Mathematical Theory of the Dynamics of Biological Populations,* Bartlett, M. S. and Hiorns, R. W., Eds., Academic Press, London, 1973, 35.

126. **Laurence, G. C.,** A bioenergetic model for the analysis of feeding and survival potential of winter flounder, *Pseudopleuronectes americanus,* larvae during the period from hatching to metamorphosis, *Fish. Bull. (U.S.),* 75, 529, 1977.

127. **Haedrich, R. L.,** Diversity and overlap as measures of environmental quality, *Water Res.,* 9, 945, 1975.

128. **McErlean, A. J., O'Connor, S. F., Mihursky, J. A., and Gibson, C. I.,** Abundance, diversity, and seasonal patterns of estuarine fish populations, *Estuarine Coastal Mar. Sci.,* 1, 19, 1973.

129. **Livingston, R. J., Kobylinski, G. J., Lewis, F. G., III., and Sheridan, P. F.,** Long-term fluctuations of epibenthic fish and invertebrate populations in Apalachicola Bay, Florida, *Fish. Bull. (U.S.),* 74, 311, 1976.

130. **Day, J., Hall, C., Kemp, M., and Yanez-Arancibia, A.,** *Estuarine Ecology,* John Wiley & Sons, New York, 1989.

131. **Livingston, R. J.,** Diurnal and seasonal fluctuations of organisms in a north Florida estuary, *Estuarine Coastal Mar. Sci.,* 4, 373, 1976.

132. **Richardson, S. L. and Pearcy, W. G.,** Coastal and oceanic fish larvae in an area of upwelling off Yaquina Bay, Oregon, *Fish. Bull. (U.S.),* 75, 125, 1977.

133. **Priede, I. G., Solbé, J. F. de L. G., Nott, J. E., O'Grady, K. T., and Cragg-Hine, D.,** Behaviour of adult Atlantic salmon, *Salmo salar* L., in the estuary of the River Ribble in relation to variations in dissolved oxygen and tidal flow, *J. Fish Biol.,* 33(Suppl. A), 133, 1988.

134. **Bechtel, T. J. and Copeland, B. J.,** Fish species diversity indices as indicators of pollution in Galveston Bay, Texas, *Contrib. Mar. Sci.,* 15, 103, 1970.

135. **Richards, C. E. and Castagna, M.,** Marine fishes of Virginia's eastern shore (inlet and marsh, seaside waters), *Chesapeake Sci.,* 11, 235, 1970.

136. **Dahlberg, M. D. and Odum, E. P.,** Annual cycles of species occurrence, abundance, and diversity in Georgia estuarine fish populations, *Am. Midl. Nat.,* 83, 382, 1970.

137. **Livingston, R. J.,** Impact of kraft pulp-mill effluents on estuarine and coastal fishes in Apalachee Bay, Florida, U.S.A., *Mar. Biol.,* 32, 19, 1975.

138. **Horn, M. H. and Allen, L. G.,** Numbers of species and faunal resemblance of marine fishes in California bays and estuaries, *Bull. South Calif. Acad. Sci.,* 75, 159, 1976.

139. **Webb, B. F.,** Fish populations of the Avon-Heathcote estuary. I. General ecology, distribution, and length frequency, *N. Z. J. Mar. Freshw. Res.,* 6, 570, 1972.

140. **Wheeler, A.,** Fish in estuaries: a summing-up, *J. Fish Biol.,* 33 (Suppl. A), 251, 1988.

141. **Vouglitois, J. J., Able, K. W., Kurtz, R. J., and Tighe, K. A.,** Life history and population dynamics of the bay anchovy in New Jersey, *Trans. Am. Fish. Soc.,* 116, 141, 1987.

142. **Orth, R. J. and Heck, K. L., Jr.,** Structural components of eelgrass (*Zostera marina*) meadows in the lower Chesapeake Bay — fishes, *Estuaries,* 3, 278, 1980.

143. **Thayer, G. W., Adams, S. M., and LaCroix, M. W.,** Structural and functional aspects of a recently established *Zostera marina* community, in *Estuarine Research,* Vol. 1, Cronin, L. E., Ed., Academic Press, New York, 1975, 518.

144. **Ryer, C. H. and Orth, R. J.,** Feeding ecology of the northern pipefish, *Syngnathus fuscus,* in a seagrass community of the lower Chesapeake Bay, *Estuaries,* 10, 330, 1987.

145. **Horwitz, R. J.,** Fish, in *Ecological Studies in the Middle Reach of Chesapeake Bay Calvert Cliffs,* Heck, K. L., Jr., Ed., Springer-Verlag, Berlin, 1987, 167.

146. **Setzler-Hamilton, E. M.,** Utilization of Chesapeake Bay by early life history stages of fishes, in *Contaminant Problems and Management of Living Chesapeake Bay Resources,* Majumdar, S. K., Hall, L. W., Jr., and Austin, H. M., Eds., Pennsylvania Academy of Science, Easton, 1987, 63.

147. **Dovel, W. L.,** Fish Eggs and Larvae of the Upper Chesapeake Bay, Spec. Rep. No. 4, National Research Institute, University of Maryland, College Park, MD, 1971, 71 pp.

148. **Olney, J. E.,** Eggs and early larvae of the bay anchovy, *Anchoa mitchilli,* and the weakfish, *Cynoscion regalis,* in lower Chesapeake Bay with notes on associated ichthyoplankton, *Estuaries,* 6, 20, 1983.

149. **Yanez-Arancibia, A., Linares, F. A., and Day, J. W., Jr.,** Fish community structure and function in Terminos Lagoon, a tropical estuary in the southern Gulf of Mexico, in *Estuarine Perspectives,* Kennedy, V. S., Ed., Academic Press, New York, 1980, 465.

150. **Toral, S. and Resendez, A.,** Los ciclidos (Pisces: Perciformes) de la Laguna de Términos y sus afluentes, *Rev. Biol. Trop.,* 21, 254, 1974.

151. **Castro, J. L.,** Catalogo sistematico de los peces marinos que penetran a las aquas continentales de México con aspectos zoogeografico y ecologicos, *Dep. Pesca Méx. Ser. Cient.,* 19, 1, 1978.

152. **Moore, R.,** Variations in the diversity of summer estuarine fish populations in Aransas Bay, Texas, 1966—1973, *Estuarine Coastal Mar. Sci.,* 6, 495, 1978.

153. **Shannon, C. E. and Weaver, W.,** *The Mathematical Theory of Communication,* University of Illinois Press, Urbana, 1963.

154. **Conomos, T. J., Ed.,** *San Francisco Bay: The Urbanized Estuary,* Pacific Division, American Association for the Advancement of Science, San Francisco, CA, 1979.

155. **Kockelman, W. J., Conomos, T. J., and Leviton, A. E., Eds.,** *San Francisco Bay: Use and Protection,* Pacific Division, American Association for the Advancement of Science, San Francisco, CA, 1982.

156. **Armor, C. and Herrgesell, P. L.,** Distribution and abundance of fishes in the San Francisco Bay estuary between 1980 and 1982, *Hydrobiologia,* 129, 211, 1985.

157. **Spratt, J. D.,** Variation in the growth rate of Pacific herring from San Francisco Bay, California, *Calif. Fish. Game,* 73, 132, 1988.

158. **Smith, S. E. and Kato, S.,** The fisheries of San Francisco Bay: past, present, and future, in *San Francisco Bay: The Urbanized Estuary,* Conomos, T. J., Ed., Pacific Division, American Association for the Advancement of Science, San Francisco, CA, 1979, 445.

159. **Miller, D. J. and Lea, R. N.,** Guide to the coastal marine fishes of California, *Calif. Fish. Game Bull.,* p. 157, 1972.

160. **Moyle, P. B.,** *Inland Fishes of California,* University of California Press, Berkeley, 1976.

Chapter 6

TROPHIC RELATIONSHIPS

I. INTRODUCTION

The movement of nutrient materials through the biota of estuarine and marine communities occurs in a predictable sequence of trophic (eating) relationships.[1] The trophic structure of a community centers around the concept of the food chain, which groups organisms into categories or trophic levels (i.e., producers, consumers, and decomposers) for the delineation of food energy transfer within the system.[2] Hence, the base of the food chain consists of autotrophs that fix carbon via photosynthesis and provide energy for consumer organisms (i.e., heterotrophs). At successively higher trophic levels, primary consumers serve as a food source for secondary consumers which, in turn, are consumed by tertiary consumers (i.e., carnivores). Decomposers (i.e., saprophagic bacteria and fungi) assimilate dead plant and animal matter, transforming it into dissolved organic matter to meet their energy requirements, while releasing mineral nutrients useful for autotrophic growth. At each trophic level, approximately 80 to 90% of the potential energy is lost as heat, thereby limiting the length of food chains to three or four trophic levels.[3,4]

Two hypotheses have been postulated to explain the low natural variation existing in food chain length: (1) species atop long food chains cannot recover sufficiently quickly following disturbances; and (2) the number of trophic levels present in a community reflects the amount of energy entering at the base of the food chain.[3] The average or maximal food chain length does not appear to be constrained by environmental variability alone.[5] Several inherent assumptions operate in food chain analyses; for example, simplified food chains assume that a single species or several species at a single trophic level are consumed by one species or a group of species at the next higher trophic level. Additionally, food chain structure assumes that a species maintains the same trophic position in the food chain throughout ontogeny. Food chains typically consider only the adult stages of the dominant species.[6] In natural aquatic systems, more complex feeding relations exist, and it is nearly impossible to depict them accurately in a single diagram.[7] Even the dominant trophodynamic pathways can be circuitous and problematical.

Many consumer organisms in estuarine and marine ecosystems feed at several trophic levels. Multiple feeding strategies result in a network of relationships referred to as a food web. Whereas some consumer organisms have specific food habits, others exhibit broader requirements (e.g., omnivores), not being restricted to the preceding trophic level as an intermediate energy source and possibly deriving energy from more than one trophic level.[8] It follows from this description, therefore, that a food web involves an interlocking pattern of producer, consumer, and decomposer organisms, and, illustrated in a diagram, the feeding relationships by species transcribe a network of links or arrows demonstrating alternative and multiple feeding strategies (Figure 1). As food webs increase in complexity, ecosystems attain greater stability, becoming more resilient to environmental perturbations.[9]

Two basic types of food webs are recognized — the grazing and detrital food webs. Usually an estuary is dominated by one of these two food webs,[10] although in simplest form, both represent a major interlocking energy flow and not isolated entities.[11] Odum and Biever (p. 360)[11] state that ... "the separation between grazing and detrital pathways is distinct only at the primary producer-primary consumer transfer. At secondary consumer levels, energy flow breaks up into complex food web networks with many organisms deriving nourishment (energy and nutrients) from both primary sources."

Recently, the concept of a microbial loop has emerged for pelagic food webs in which bacteria process nonliving organic matter while being grazed by protozoans; the microbial food web is

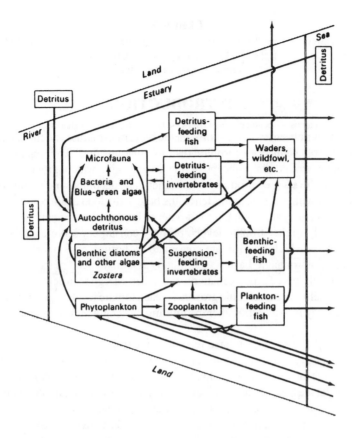

FIGURE 1.   A simplified estuarine food web. (From Barnes, R. S. K., *Estuarine Biology,* Edward Arnold, London, 1974. With permission.)

the ultimate food resource for metazooplankton.[12] The microbial loop has stirred a "sink or link" debate among marine scientists concerning the relative importance of microbes as a sink for fixed carbon via respiratory losses or as a source of food for the metazoan food web (see Chapter 1, Section I).[13,14] Sherr and Sherr (p. 1225)[15] consider this controversy to be a nonissue, however, viewing the microbial loop as "a component (and integral) part of a larger microbial food web, which includes all pro- and eucaryotic unicellular organisms, both autotrophic and heterotrophic. From this standpoint, the entire microbial food web, not simply phytoplankton, supports the metazoan food web."

II. ESTUARINE FOOD WEBS

A. DETRITUS FOOD WEB

1. Detritus

Phytoplankton, benthic algae, and detritus constitute the principal food sources in estuaries, with detritus often being most important.[9,16] The bulk of living plant material in estuarine ecosystems remains ungrazed, and a large fraction of it enters the detrital, or nonliving, organic carbon pool. For example, zooplankton graze inefficiently on phytoplankton in these coastal ecosystems; as a result, about 50% of the net phytoplankton production becomes available to benthic fauna as living cells or ultimately as detritus.[2] More than 90% of the primary production of benthic macrophytes passes to detritus food webs — approximately one third as dissolved organic matter (DOM) and two thirds as particulate organic matter (POM) — because of the low exploitation rates of grazing herbivores on the live plant matter.[17] Much of the decaying

TABLE 1
Detritus-Generated and Phytoplankton-Generated Food Webs in Estuaries

Detritus-Generated Food Webs

Detritus — benthos (epifauna) — benthophagous fishes
Detritus — benthos (infauna) — benthophagous fishes (rays)
Detritus — benthos — benthophagous fishes — large fish predators (sharks)
Detritus — small benthos — larger invertebrates and small benthic fishes — large fishes
Detritus — large detritivorous fishes (mullet): "telescoping" of food chain
Detritus — benthos — large predators (rays)
Detritus — micronekton — intermediate predators (snappers, croaker)
Detritus — zooplankton — small fishes and invertebrates
Detritus — zooplankton — small fishes and invertebrates — larger fishes

Phytoplankton-Generated Food Webs

Phytoplankton — zooplankton — planktivorous, pelagic, and benthopelagic fishes
Phytoplankton — zooplankton — planktivorous fishes — large fish predators
Phytoplankton — phytoplanktonic fishes (menhaden): summer
Phytoplankton — zooplankton — menhaden: winter
Phytoplankton — zooplankton — large carnivores
Phytoplankton — (dinoflagellates) — mullet: alteration of usual feeding habits

Modified from de Sylva, D. P., *Estuarine Research,* Vol. 1, Cronin, L. E., Ed., Academic Press, New York, 1975, 420. With permission.

macrophytes accumulates from fringing salt marsh grasses, mangrove swamps, and seagrasses which have high rates of productivity (see Chapter 4, Section III.B). The amount of detritus derived from these autochthonous sources far exceeds that of allochthonous sources from river discharges and tidal waters.

Detritus may be defined in general terms as both freshly dead organisms and all biogenic material in various stages of microbial decomposition,[10,18] encompassing particulate as well as dissolved nonliving organic matter.[19,20] The concentration of detritus in estuaries generally ranges from 0.1 to greater than 125 mg/l.[21,22] While decaying vascular plants account for much of this concentration, biodeposits (feces and pseudofeces of animals) can be substantial. Of the total macrophyte biomass that converts to detritus, an unknown quantity is recycled through fecal pellet and pseudofeces generation and coprophagy among benthic organisms and zooplankton.[23] Systems characterized by high secondary benthic and fish production typically have high detrital inputs.[24-27]

Questions abound in regard to the nutritive value of detritus for the benthos and the complex feeding relationships regulating energy flow in detritus food webs.[26,28,29] Potential foods for detritivores are morphous material largely composed of plant remains (e.g., vascular plant debris and diatom frustules),[29] amorphous material (e.g., humic geopolymers, microbial exudates, and absorbed molecules),[30,31] and dissolved molecules.[32,33] The relative importance of these foods is poorly understood and a matter of controversy in the literature.[29]

Although detritus food webs can be of several types (Table 1), the base of each one is structured by similar components, notably, detritus and an associated microbial community.[34] Indeed, the structure and complexity of the detrital community control the amount of energy available to upper-trophic-level consumers.[35] A succession of organisms colonizes detrital particles beginning with bacteria and followed by fungi, algae, ciliates, flagellates, and larger grazers.[36,37] The fungi effectively decompose detrital cellulose and lignins.[38] Bacterial populations associated with the detritus support a diverse assemblage of micrometazoans. Together with fungi, the bacteria degrade the detrital substrate, assimilating nutrients and transforming the POM into DOM to meet their energy demands. Higher organisms (e.g., ciliates, nematodes, ostracods, and harpacticoids) consume the bacteria, whereas larger heterotrophic consumers, (e.g., ciliates, nematodes, rhabdocoels, and halacarids) ingest the micrometazoans and each other. Many of the microfauna are omnivorous. Detritus-feeding macrofauna, in turn, do not effectively assimilate the organic detritus; although numerous species possess the capability to digest cellulose of plant detritus,[39,40] the energetic significance of this process has yet to be put into perspective relative to energy requirements.

Fenchel[41] discriminated between two broad groups of animals associated with detritus: (1) microfauna which actively select their food particles (e.g., protozoans, rotifers, turbellarians, gastrotrichs, ostracods, nematodes, and harpacticoids); and (2) meiofaunal and macrofaunal detritus feeders which browse on detrital surfaces or unselectively ingest the substrate (e.g., amphipods, gastropods, bivalves, polychaetes, and oligochaetes). Among the microfauna, rotifers and turbellarians are principally carnivorous, and the ciliates, omnivorous. Zooflagellates have a steady diet of bacteria. The macrofaunal detritus feeders are mainly deposit-feeding benthic invertebrates. The deposit feeders (e.g., *Capitella capitata* and *Nereis succinea*) swallow large volumes of sediment, satisfying their nutritional requirements from the organic fraction of the ingested sediment, namely, the organic debris and sediment-associated microbes.[29] Both microbial and detrital foods appear to be part of the diet of many deposit feeders. This benthic group attains greatest abundance in fine estuarine sediments, being spatially separated from suspension feeders whose survival decreases in muddy habitats because fine sediments interfere with their delicate filtering apparatus. Suspension feeders obtain their nutrition by filtering the water of food particles and detritus. The resuspension of sedimentary particles by bioturbating infauna negatively affects the survival or growth of the suspension feeders via burial or clogging of filtering organs.[42,43] This organismal interaction, referred to as trophic group amensalism,[42] influences the local distribution of the suspension feeders. Larger predators, such as birds, crabs, and fish, prey on the macrofaunal detritus and suspension feeders and, by doing so, may impact the benthic community structure.[44-47] Figure 2 illustrates terrestrial, riverine, estuarine, and marine detrital sources that yield nutrition for the micro-, meio-, and macrofauna of estuaries.

Barnes[16] assessed the detritus-feeding activity of mudflat macrofauna. He discerned five different modes by which the macrofauna obtain detritus. First, a number of sedentary polychaetes, mainly belonging to the families Spionidae, Terebellidae, and Ampharetidae, extend tentacles across or through the sediment, collecting food particles and transporting them to their mouth by means of ciliary tracts or peristaltic muscle contractions. Second, various bivalves utilize their inhalent siphon to suck detritus from the sediment surface; their gill lamellae sieve and sort this material, with the pseudofeces being rejected and the remainder passed by cilia to the mouth. Third, suspension-feeding bivalves with short inflexible siphons (e.g., *Cardium*) process benthic algae and surface detritus set into suspension by tidal action. Fourth, arthropods (e.g., *Corophium* spp.) use their appendages to mechanically sort through the sediment, selectively removing detrital particles. Fifth, deposit-feeding polychaetes (e.g., *Arenicola, Capitella,* and *Nereis*) consume the sediment, organic detritus, and associated microbes together, digesting the mass for its nutritive content.

Orthochemical processes (i.e., leaching and sorption), microbial attack, and mechanical fragmentation cause the breakdown of detritus in estuarine environments.[48] Of the major vascular plant sources, salt marsh plants are most resistant to this degradation followed by mangroves and seagrasses.[28,49] The gradual breakdown of plant material in estuaries generates fine detrital particles. A recent interpretation of this process reveals that chemical and microbial precipitation of DOM produces amorphous organic matter, with plant exudation, animal excretion, and microbial breakdown contributing to the dissolved organic pool. In addition, amorphous organic matter is generated by humification reactions involving bacterial exoenzymes.[29-31,50]

Detritus feeders tend to have low assimilation efficiencies. They continually rework

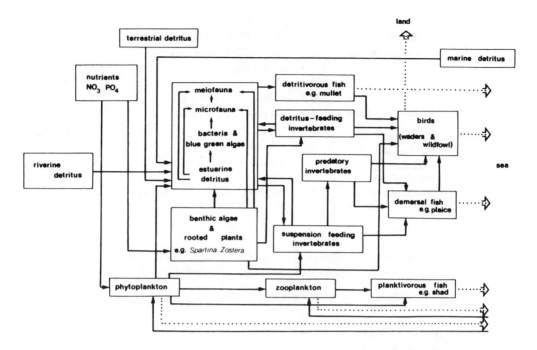

FIGURE 2.   Generalized estuarine food web illustrating the feeding relationships of detritus and grazing food chains. Note that detritus and phytoplankton occupy the base of the food web. Dotted lines signify losses from the estuarine system. (From Barnes, R. S. K., *Estuarine Biology,* Edward Arnold, London, 1984. With permission.)

sediments in search of food, actively burrowing through grains and modifying physical-chemical properties, especially in proximity to the sediment-water interface. The feeding, respiratory, and locomotor activities of these organisms create advection of water and sediments on the estuarine seafloor, thereby increasing erosion rates.[51] These activities also affect the sedimentary bacteria which regulate nutrient remineralization and sediment organic content.[29]

Deposit feeders constantly strip microbes from detrital particles during feeding and digestion as alluded to in Chapter 4, Section IV. The microbes recolonize the particles repeatedly and serve as a renewable food source. Microbial stripping alone, however, seems to be insufficient to meet the energy requirements of most of these benthic macrofauna, since microbial abundance in the sediments is too low in relation to measured ingestion rates and absorption efficiencies of the macrofauna.[29,52] According to Lopez and Levinton,[29] the only deposit feeders likely to meet their metabolic demands from microbial sources are surface deposit feeders inhabiting intertidal mudflats, where benthic microalgae grow abundantly and are an important food resource. Such is not the case in subtidal and subsurface sediments. While some workers emphasize the potential significance of organic detritus in the nutrition of detritus feeders, weaknesses in the microbial stripping theory have raised questions regarding its viability. One pressing problem of detrital utilization is that most detritus feeders lack the necessary enzymes to hydrolyze cellulose, xylan, and other structural components of detritus, which limits their ability to assimilate the material.[53] Furthermore, many of them do not have the gut retention for efficient digestion of these complex molecules.[29]

Most probably, both organic detritus and microbes are dietary components of many detritivores. Whereas sediment microbes obviously represent a protein source for detritus feeders, the bulk of the carbon requirements of these animals may be satisfied by absorption of amorphous detritus.[29] The radiolabeling of detritus and sedimentary organic matter promises to advance our knowledge of the feeding processes of detritus feeders, particularly in respect to selection and absorption of detritus.[54-58]

TABLE 2
Production for the Duplin River Marsh and Estuary as a
Whole, Prorated on the Basis of 21% Subtidal and 79%
Intertidal Area

Producer population	g C per square meter per year	Above–ground (%)	Total (%)
Spartina whole plants	1216		84
Spartina roots	608		42
Spartina shoots	608	73	42
Benthic algae	150	18	10
Phytoplankton	79	9	6
Total	1445		
Total above–ground	758		

From Pomeroy, L. R., Darley, W. M., Dunn, E. L., Gallagher, J. L., Haines, E. B., and Whitney, D. M., *The Ecology of a Salt Marsh, Ecological Studies,* Vol. 38, Pomeroy, L. R. and Wiegert, R. G., Eds., Springer-Verlag, New York, 1981, 39. With permission.

2. Type Detritus Food Webs
a. Georgia Salt Marsh Systems
1. Detritus Production

Trophic relationships in the salt marshes of Georgia have been investigated for more than 30 years.[59-62] Detritus production in coastal Georgia marshes is mainly attributed to two vascular plants, the cordgrass, *Spartina alterniflora* Loisel, and the black rush, *Juncus roemerianus* Scheele, which account for 70 to 90% of the primary production.[61] Although these plants are responsible for most of the primary production of the salt marshes, recent research has shown that algal populations supply a greater proportion of the carbon production than once believed possible. When prorated on the basis of area occupied in the marshes of Sapelo Island, for example, algal production amounted to one third of the above-ground production of *Spartina* (Table 2).[63] Nevertheless, nearly all of the production of *S. alterniflora* and *J. roemerianus* enters the detritus food web (about 90% of the above-ground primary production of the plants becomes detritus),[64] while much of the algal production, which is a high quality food source, goes directly to grazing consumers.

Above-ground production of salt marsh grasses at Sapelo Island has been measured on numerous occasions. Gallagher et al.[65] recorded annual production estimates of short-form *S. alterniflora* (1350 g dry weight per square meter per year) and tall-form *S. alterniflora* (3700 g dry weight per square meter per year), respectively (Table 3). Annual above-ground production of *J. roemerianus* equaled 2200 g dry weight per square meter per year; because this species covers only 6% of the marsh surface,[66] its contribution to the pool of primary production is rather slight. Gallagher and Plumley[67] found that the annual below-ground production of both short- and tall-form *S. alterniflora* surpassed 2000 g dry weight per square per year, which is substantially less than the annual below-ground production of *J. roemerianus* (3480 g dry weight per square meter per year). These values are among the most reliable generated on this salt marsh system.

Reimold et al.[61] proffered mean monthly and annual detritus production figures for *S. alterniflora* and *J. roemerianus* in Georgia salt marshes. For short- and tall-form *S. alterniflora*, monthly detritus production amounted to 113.6 and 197.9 g/m^2, respectively. Monthly detritus production of *J. roemerianus* paralleled that of tall-form *S. alterniflora*, equaling 188.4 g/m^2. For the aboveground portion of these plants, the mean annual detritus production, when weighted for the percentage of the watershed occupied by each stand, was 1845.8 g/m^2/year.

TABLE 3
Net Annual Aerial Primary Production Estimates for
Spartina alterniflora and *Juncus roemerianus*

	g dry weight per square meter		
Streamside	*S. alterniflora* high marsh	Whole marsh	*J. roemerianus*
3300[a]	2200[a]	2288[a]	1500[c]
3700[b]	1350[b]	1538[b]	2200[b]
2000[c]	400[c]	528[c]	

[a] From Odum and Fanning (1973).
[b] From Gallagher et al. (1980).
[c] From Gallagher et al. (1972).

From Pomeroy, L. R., Darley, W. M., Dunn, E. L., Gallagher, J. L., Haines, E. B., and Whitney, D. M., *The Ecology of a Salt Marsh, Ecological Studies,* Vol. 38, Pomeroy, L. R. and Wiegert, R. G., Eds., Springer-Verlag, New York, 1981, 39. With permission.

2. Microbes

Bacteria and fungi quickly colonize the leaves and stems of vascular plants after the plants senesce and die. Not all of the detritus from the salt marsh grasses undergoes microbial attack. As documented by Gallagher et al.,[68] leaching losses from *S. alterniflora* approach 6.1 g C per square meter per year. In addition to the organic matter leached by tidal water, grazing herbivores remove a small percentage of the aboveground primary production.

Fungi initially colonize decaying *Juncus* and *Spartina* and are quickly followed by bacteria and small metazoans.[69] While microbes are first to manipulate the vascular plant detritus, the ultimate fate of this material is somewhat enigmatic. Moran et al. (p. 1525)[70] recount the following information: "... At one extreme, it is possible that the fate of detrital carbon is largely that of respiratory losses from the microbial community, and thus none or very little of this material supports the production of higher trophic levels. At the other extreme, detritus may be the sole source of carbon and energy to higher order consumers in the food web, as well as to the microorganisms. In this latter scenario, production of metazoans would be based on either direct ingestion of intact or partially degraded detrital material or, more likely, on carbon derived from a microbial food web."

3. Macroconsumers

Macroconsumer grazing, notably by fiddler crabs (e.g., *Uca pugnax*),[71] snails (e.g., *Littorina irrorata*),[72,73] and herbivorous fish (e.g., *Mugil cephalus*),[74] is a major pathway for the loss of algal biomass. Because of the stressful conditions of the salt marsh environment, the species diversity of macroconsumers is low. However, the extremely productive plant populations foster high secondary production.[75,76] The biomass of macroconsumers may be greater than 15 g C per square meter in the intertidal zone.[77] The density of macroconsumers also can be high; over a square meter of marsh surface area, it is not unusual to observe 80 to 200 *U. pugnax*, 400 to 700 *L. irrorata* or *Ilyanassa obsoleta*, and 7 to 8 *Geukensia demissa*.[60,78-81] *I. obsoleta* occurs in densities of 500 to 1600 individuals per square meter on mudflats.[82] The salt marsh habitat supports recreationally and commercially important species as well, including brown shrimp (*Penaeus aztecus*), white shrimp (*P. setiferus*), blue crabs (*Callinectes sapidus*), hard clams (*Mercenaria mercenaria*), and oysters (*Crassostrea virginica*). Fishes (e.g., *Cyprinodon variegatus, Fundulus heteroclitus, F. luciae,* and *Mugil cepalus*), mammals (e.g., *Mustela vison, Oryzomys palustris,* and *Procyon lotor*), and birds (e.g., *Rallus longirostris* and *Telmatodytes*

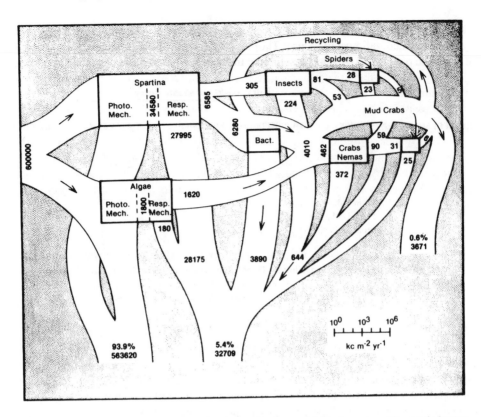

FIGURE 3.   Energy flow in Georgia salt marshes depicted in a box model. The two outward flows at the bottom delineate heat losses. The outward flow at the right shows net secondary production exported from the marsh. (From Montague, C. L., Bunker, S. M., Haines, E. B., Pace, M. L., and Wetzel, R. L., *The Ecology of a Salt Marsh, Ecological Studies,* Vol. 38, Pomeroy, L. R. and Wiegert, R. G., Eds., Springer-Verlag, New York, 1981, 69. With permission. [Adapted from Teal, J. M., *Ecology,* 43, 614, 1962.])

palustris) are vertebrate predators at or near the top of the food chain. The classic work of Teal[60] describes energy flow in Georgia salt marshes (Figure 3).

Stable carbon isotope ratio analysis of food sources and marsh fauna has greatly facilitated study of the structure of detritus food webs in Georgia salt marshes.[83] Five general macrofaunal feeding types have been characterized in these marshes. They include: (1) deposit feeders, (2) suspension feeders, (3) deposit-suspension feeders, (4) predator-scavengers, and (5) strict predators.

Deposit feeders — Salt marsh soils and unvegetated sediments in the intertidal and shallow subtidal zones contain a rich food supply comprised of vascular plant detritus and an associated microbial community (microfauna e.g., ciliates and foraminiferans, and meiofauna e.g., nematodes, gastrotrichs, rotifers, and turbellarians).[84] Fiddler crabs (*Uca pugilator, U. pugnax, and U. minax*), polychaete worms, and various species of gastropods are deposit feeders that peruse the marsh habitat in search of food. The three fiddler crabs display little overlap in their spatial distribution.[85,86] *U. pugilator* prefers sandy substrates, whereas *U. pugnax* lives in muddy sediments beneath *S. alterniflora. U. minax* has a low tolerance to higher salinities and congregates in marshes subjected to some freshwater inflow. Deposit feeders constantly reingest organic matter in their feces; the organically enriched excretion products serve as ideal substrates for decomposer colonization, yielding additional food for the system.[87]

Suspension feeders — Suspension-feeding invertebrates filter and assimilate organic matter from overlying waters. Typified by bivalves in the marsh proper (e.g., *G. demissa* and *Polynesoda caroliniana*) and in neighboring tidal saline creeks (e.g., *M. mercenaria* and *C.*

virginica), the suspension feeders consume detrital particles and attached microbes along with phytoplankton and sedimentary grains. They sort the organic and inorganic particles, ejecting the inorganic particles as pseudofeces. Similar to the deposit feeders, these organisms produce organically enriched feces that is recolonized by microbes, thereby adding to the total energy of the salt marsh.

Deposit-suspension feeders — Some estuarine fauna have the capability of feeding as deposit or suspension feeders. Examples of deposit-suspension feeding invertebrates inhabiting the Georgia salt marsh biotope are the grass shrimp (*Palaemonetes pugio*), brown shrimp (*Penaeus aztecus*), and white shrimp (*P. setiferus*) which subsist on bottom deposits as well as suspended matter.[88,89] The striped mullet (*Mugil cephalus*), a detritus-feeding finfish, sorts through sediments in search of detritus, but retains the ability to graze on phytoplankton patches.[90]

Predator-scavengers — Other estuarine animals behave as both scavengers and predators depending on the availability of food. Hence, they are scavenging opportunists that do well on a diet of carrion and live prey. The blue crab, *Callinectes sapidus*, falls into this category, ingesting carrion if present, but otherwise being a voracious predator, attacking bivalves, shrimp, snails, crabs, and other organisms.[91,92] The shark, juvenile Atlantic tarpon, and white catfish also have dual feeding strategies, scavenging only opportunely, but remaining primarily predators.[93-96]

Strict carnivores — At the top of the detritus food web are strict carnivores, relying nearly exclusively on live prey as an energy source. The striped bass (*Morone saxatilis*) and the inshore lizardfish (*Synodus foetens*) provide examples. Only a few species fit into this category in the salt marsh habitat; most of the macrofauna here exhibit omnivorous feeding habits. This environment favors food generalists, possibly because of spatial and temporal variability in the quality and quantity of food, nutrient imbalances, and periods of food scarcity.[97]

b. Florida Mangroves

1. Detritus Production

Odum and Heald[10] dealt with detritus production and trophic relationships in a mangrove community of the North River estuary in southern Florida. As in other mangrove systems, primary production in the North River estuary is high, and most of this production enters the detritus food web. For example, grazing insects strip only about 5% of the leaves on mangrove trees prior to their abscission.[98] Most of the leaves fall to the estuarine surface where they are dispersed by currents and wave action. Detritus production from the mangroves far exceeds 8 t dry weight per hectare per year. Once colonized by bacteria and fungi, it is transformed to an excellent, high protein food source for detritivores.

The red mangrove, *Rhizophora mangle*, which covers approximately 1050 ha, accounts for more than 85% of the total production of vascular plant detritus in the North River basin. *Juncus* and sawgrass from bordering marshes supply most of the remaining detritus; algal contributions to the detrital pool are small. Leaves, twigs, leaf scales, and flowers of the red mangrove drop to the estuarine surface at a rate of 12,400 t dry weight per year or 8.8 t dry per hectare per year (2.41 dry g/m^2/d). Most of this material (83%) consists solely of dead leaves. Bacteria and fungi rapidly invade the detrital mass, and protozoans subsequently graze on the bacteria. Fungi, including species of *Alternaria, Cladosporium, Dendryphaiella, Fusarium, Mucor, Nigrospora,* and *Phomopsis,* heavily coat mangrove detritus and probably play a significant role in its breakdown. The protozoan-bacteria-fungi-detritus complex is an essential energy source for consumer organisms of the estuarine mangrove community.

2. Consumer Organisms

The flow of energy through the heterotrophic community occurs by means of at least four pathways: (1) dissolved organic substances → microorganisms → higher consumers, (2) dissolved organic substances → sorption on sedimentary grains → higher consumers either directly or by way of microorganisms, (3) leaf material → higher consumers, and (4) leaf material → bacteria and fungi → higher consumers. The first three pathways are quantitatively less significant than the fourth pathway. In general terms, the primary route of energy flow in the system is mangrove leaf detritus → bacteria and fungi → detritus consumers → lower carnivores → higher carnivores.

Based on the digestive-tract contents of approximately 120 species, Odum and Heald[10] devised a sequence of heterotrophic levels for the North River ecosystem, assigning individual species to trophic positions corresponding to numerical indices: herbivores (1.0), carnivores feeding on herbivores (2.0), and carnivores feeding on carnivores (3.0 and 4.0). They calculated a numerical index for each species to determine its trophic position. Table 4 lists the trophic sequence of primary to top carnivores according to the index values. The major trophic categories had the following range of mean index values: (1) herbivores (1.1 to 1.3), (2) omnivores (1.4 to 1.8), (3) lower carnivores (2.0 to 2.8), (4) middle carnivores (3.1 to 3.3), and (5) top carnivores (3.5 to 3.7).

About one third of the species studied were categorized as detritus consumers. The digestive-tract contents of these organisms contained at least 20% vascular plant detritus by volume on an annual basis. Grouped among the detritus consumers were herbivorous and omnivorous crustaceans, nematodes, polychaetes, mollusks, insect larvae, and a few fishes. They had the capability of assimilating algae, portions of vascular plant detritus, microbes, and perhaps DOM sorbed upon inorganic particles.

Detritus cycling is a key process operating in the detritus food web of the North River estuary. It involves multiple reuse of detritus particles, especially fecal material, with microbial colonization of the particles and subsequent ingestion by detritus consumers being repeated continuously. Figure 4A delineates the cyclical nature of the process, emphasizing the utilization and reutilization of fecal detrital particles.

Clark[99] formulated a food chain for the black mangrove (*Avicennia*) located at higher tidal levels in the Florida mangrove swamp than *Rhizophora*.[2] In the black mangrove, mosquito and midge larvae are the main detritivores. They provide food for a variety of forage fishes which, in turn, are eaten by carnivorous fishes and birds (Figure 4B).

By analyzing the diets of aquatic organisms, Odum and Heald[10] developed a schematic representation of the North River estuarine food web (Figure 5). In this diagram, broad arrows signify the primary energy flow pathways. Narrow arrows depict less important food chains. The pathway of dissolved leaf material parallels the dotted lines.

c. Temperate Seagrasses
1. Detritus Production

Few organisms utilize live seagrass plants as a food source.[100] Some echinoderms, gastropods, fishes, green turtles, dugongs, manatees, and waterfowl (e.g., geese and swans) graze on seagrasses; however, the total amount of plant tissue consumed is small, oftentimes less than 5%. Kikuchi[101] remarked that the only significant seagrass grazing, by sea urchins and parrot fish, has been reported in the Caribbean Sea. The majority of grazing herbivores feeds on epiphytes and algal felt attached to the seagrass leaves. Most of the seagrass production (greater than 90%) ultimately enters the detritus food web.

2. Barnegat Bay

Zostera marina is the principal detrital plant source in Barnegat Bay. It forms thick beds along the margins of the estuary, mainly at depths of 1 m or less.[102] Polychaetes and amphipods are the dominant macrofaunal detritivores in the bay (Table 5). *Pectinaria gouldii*, occurring in densities up to 20,000 individuals per square meter, and *Ampelisca* spp. represent the predominant detritus feeders.

TABLE 4
Trophic Position of North River Herbivores, Omnivores, and Carnivores Based on
Function Trophic Indices[a]

Heterotrophic level	Mean value	Range of values	Principal components of diet
Herbivores			
Group A			
Heterotrophic bacteria, yeasts, and fungi	1.1	1.0—4.0	Plant material
Ciliates, *Frontonia marina* and *Strombidium* sp.	1.1	1.0—1.1	Microalgae
Copepods (at least five species)	1.1	1.0—2.0	Microalgae, detritus
False mussel, *Congeria leucophaeata*	1.1	1.0—1.1	Microalgae, detritus
Scorched mussel, *Brachidontes exustus*	1.1	1.0—1.1	Microalgae, detritus
Eastern oyster, *Crassostrea virginica*	1.1	1.0—1.1	Microalgae, detritus
Chironomid midge larvae (three species)	1.1	1.0—1.1	Detrius, microalgae
Nematodes (undetermined number of species)	1.1	1.0—1.1	Detritus, microalgae
Ostracods (undetermined number of species)	1.1	1.0—1.1	Microalgae, detritus
Cumaceans, *Cyclaspis varians* and *Oxyurostylis* sp.	1.1	1.0—1.1	Microalgae, detritus
Isopods (at least three species)	1.1	1.0—1.1	Microalgae, detritus
Group B			
Striped mullet, *Mugil cephalus*	1.2	1.0—2.0	Microalgae, detritus
Sailfin molly, *Poeciliia latipinna*	1.2	1.0—2.0	Detritus, microalgae
Group C			
Sheepshead minnow, *Cyprinodon variegatus*	1.3	1.0—2.0	Detritus, microalgae
Diamond killifish, *Adinia xenica*	1.3	1.0—2.0	Detritus, microalgae
Amphipods, *Melita nitida, Grandidierella bonnieri, Corophium lacustre*	1.3	1.0—2.0	Detritus
Omnivores			
Group A			
Polychaetes, *Neanthes succinea, Nereis pelagica*	1.4	1.0—2.0	Detritus, microalgae
Crab, *Rhithropanopeus harrisii*	1.4	1.0—3.0	Detritus, small animals
Snapping shrimp, *Alpheus heterochaelis*	1.4	1.0—2.0	Detritus, small animals
Goldspotted killifish, *Floridichthyes carpio*	1.5	1.0—2.0	Small animals, detritus
Caridean shrimp, *Palaemonetes intermedius, P. paludosus*	1.5	1.0—3.0	Small animals, detritus
Nematodes (undetermined number of species)	1.5	1.0—4.0	Plant and animal material
Group B			
Pink shrimp, *Penaeus duorarum*	1.8	1.0—3.0	Small animals, detritus
Mosquito fish, *Gambusia affinis*	1.8	1.0—2.0	Small animals, algae
Least killifish, *Heterandria formosa*	1.8	1.0—2.0	Small animals, algae
Crested goby, *Lophogobius cyprinoides*	1.8	1.0—3.0	Small animals, detritus
Lower Carnivores			
Most larval and postlarval fishes	2.0	1.0—3.0	Zooplankton eggs
Copepods (undetermined number of species)	2.0	1.0—3.0	Small zooplankton
Pinfish, *Lagodon rhomboides*	2.3	1.0—3.0	Crustaceans
Sheepshead, *Archosargus probatocephalus*	2.4	1.0—3.0	Crustaceans, mollusks
Juveniles of most middle and top carnivores	2.5	2.0—3.0	Crustaceans, mollusks
Blue crab, *Callinectes sapidus*	2.5	1.0—4.0	Mollusks, crustaceans
Scaled sardine, *Harengula pensacolae*	2.5	2.0—3.0	Crustaceans
Bay anchovy, *Anchoa mitchilli*	2.5	1.0—3.0	Crustaceans
Marsh killifish, *Fundulus confluentus*	2.5	1.0—3.0	Crustaceans, midge larvae
Gulf killifish, *F. grandis*	2.5	2.0—3.0	Crustaceans
Rainwater killifish, *Lucania parva*	2.5	1.0—3.0	Crustaceans, midge larvae
Bluefin killifish, *L. goodei*	2.5	2.0—3.0	Crustaceans, ostracods
Spotted sunfish, *Lepomis punctatus*	2.5	2.0—3.0	Assorted small animals

TABLE 4 (continued)
Trophic Position of North River Herbivores, Omnivores, and Carnivores Based on Function Trophic Indices[a]

Heterotrophic level	Mean value	Range of values	Principal components of diet
Lower Carnivores (continued)			
Silver jenny, *Eucinostomus gula*	2.5	2.0—3.0	Crustaceans
Spotfin mojarra, *E. argentius*	2.5	2.0—3.0	Crustaceans
Striped mojarra, *Diapterus plumieri*	2.5	2.0—3.0	Crustaceans
Code goby, *Gobiosoma robustum*	2.5	1.0—3.0	Crustaceans, midge larvae
Clown goby, *Microgobius gulosus*	2.5	1.0—3.0	Crustaceans
Frillfin goby, *Bathygobius soporator*	2.5	2.0—3.0	Crustaceans
Tidewater silversides, *Menidia beryllina*	2.5	2.0—3.0	Crustaceans, insects
Hogchoker, *Trinectes maculatus*	2.5	2.0—3.0	Assorted small animals
Lined sole, *Achirus lineatus*	2.5	2.0—3.0	Assorted small animals
Common eel, *Anguilla rostrata*	2.6	2.0—3.0	Assorted small animals
Skilletfish, *Gobiesox strumosus*	2.6	2.0—3.0	Crustaceans
Silver perch, *Bairdiella chrysoura*	2.8	2.0—3.5	Crustaceans, small fishes
Gulf toadfish, *Opsanus beta*	2.8	2.0—3.5	Crustaceans, mollusks
Sea catfish, *Arius felis*	2.8	1.0—4.0	Assorted animals
Middle Carnivores			
Wood stork, *Mycteria americana*	3.1	2.0—3.5	Fishes, crustaceans
White ibis, *Eudocinus albus*	3.1	2.0—3.5	Fishes, crustaceans
Great blue heron, *Ardea herodias*	3.1	2.0—3.5	Fishes, crustaceans
Little blue heron, *Florida caerulea*	3.1	2.0—3.5	Fishes, crustaceans
Louisiana heron, *Hydranassa tricolor*	3.1	2.0—3.5	Fishes, crustaceans
Green heron, *Butorides virescens*	3.1	2.0—3.5	Fishes, crustaceans
Great white heron, *Ardea occidentalis*	3.1	2.0—3.5	Fishes, crustaceans
Common egret, *Casmerodius albus*	3.1	2.0—3.5	Fishes, crustaceans
Inshore lizardfish, *Synodus foetus*	3.1	2.0—3.5	Fishes
Needlefishes, *Strongylura* spp.	3.1	2.0—3.5	Fishes, crustaceans
Leatherjacket, *Oligoplites saurus*	3.2	2.0—4.0	Crustaceans, fishes
Ladyfish, *Elops saurus*	3.2	2.0—4.0	Crustaceans, fishes
Gafftopsail catfish, *Bagre marinus*	3.2	2.0—4.0	Fishes, crustaceans
Snook, *Centropomus undecimalis* and *C. pectinatus*	3.2	2.0—4.0	Fishes, crustaceans
Spotted seatrout, *Cynoscion nebulosus*	3.2	2.0—4.0	Crustaceans, fishes
Red drum, *Sciaenops ocellata*	3.2	2.0—4.0	Crustaceans, mollusks
Juveniles of certain top carnivores	3.2	2.0—4.0	Fishes, crustaceans
Jewfish, *Epinephelus itajara*	3.2	2.0—4.0	Fishes, crustaceans
Crevalle jack, *Caranx hippos*	3.2	2.0—4.0	Crustaceans, fishes
Florida gar, *Lepisosteus platyrhincus*	3.3	2.0—4.0	Fishes, crustaceans
Top Carnivores			
Tarpon, *Megalops atlantica*	3.5	2.0—4.0	Fishes, crustaceans
Barracuda, *Sphyraena barracuda*	3.5	2.0—4.0	Fishes, crustaceans
American alligator, *Alligator mississipiensis*	3.6	2.0—4.0	Fishes, reptiles
Bull shark, *Carcharhinus leucas*	3.6	2.0—4.0	Fishes, crustaceans
Bald eagle, *Haliaeetus leucocephalus*	3.7	2.0—4.0	Fishes
Osprey, *Pandion haliaetus*	3.7	2.0—4.0	Fishes

[a]   1.0 = strict herbivore; 2.0 = carnivore feeding on herbivores; 3.0 = carnivore feeding on 2.0 carnivores; 4.0 = carnivore feeding on 3.0 carnivores.

From Odum, W. E. and Heald, E. J., *Estuarine Research*, Vol. 1, Cronin, L. E., Ed., Academic Press, New York, 1975, 265. With permission.

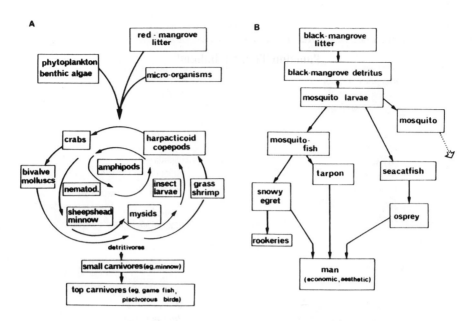

FIGURE 4. Detrital food chains in Florida mangrove systems. (A) The red mangrove (Odum and Heald, 1975); (B) the black mangrove (Clark, 1977). (From Boaden, P. J. S. and Seed, R., *An Introduction to Coastal Ecology,* Blackie & Son, Glasgow, 1985. With permission.)

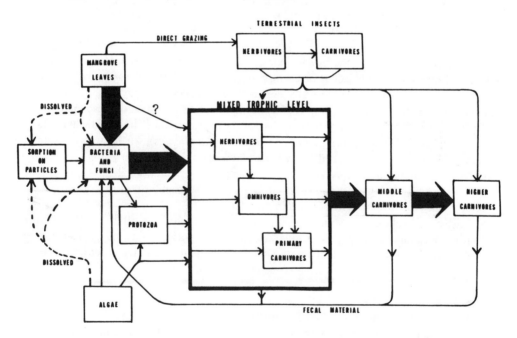

FIGURE 5. Conceptual model of the North River estuarine food web. (From Odum, W. E. and Heald, E. J., *Estuarine Research,* Vol. 1, Cronin, L. E., Ed., Academic Press, New York, 1975, 265. With permission.)

Few fishes in the estuary consume vascular plant detritus. Spot, *Leiostomus xanthurus*, ingests *Z. marina* detritus, but it subsists on other food items as well, such as polychaetes and harpacticoid copepods. Omnivorous fishes, in addition to spot, which incorporate detritus in their diet are the black sea bass (*Centropristis striata*), oyster toadfish (*Opsanus tau*), mummichog (*Fundulus heteroclitus*), Atlantic silverside (*Menidia menidia*), and American eel (*Anguilla rostrata*).

TABLE 5
Dominant Macrofaunal Detritus Feeders of Barnegat Bay

Polychaeta
Amphitrite ornata
Capitella capitata
Clymenella torquata
C. zonalis
Maldane sarsi
Maldanopsis elongata
Nephtys incisa
N. picta
Pectinaria gouldii
Polycirrus eximius
Polydora ligni
Scoloplos fragilis
S. robustus

Mysidacea
Neomysis americana

Amphipoda
Ampelisca abdita
A. macrocephala

Amphipoda (continued)
A. vadorum
A. verrilli
Gammarus lawrencianus
G. mucronatus
Corophium spp.

Decapoda
Palaemonetes pugio
P. vulgaris

Gastropoda
Bittium alternatum

Bivalvia
Solemya vellum
Yoldia limatula
Macoma balthica
M. tenta
Pitar morrhuana
Tellina agilis

From Kennish, M. J. and Loveland, R. E., *Ecology of Barnegat Bay, New Jersey*, Kennish, M. J. and Lutz, R. A., Eds., Springer-Verlag, New York, 1984, 302. With permission.

Major predatory fishes in the bay include the bluefish (*Pomatomus saltatrix*), weakfish (*Cynoscion regalis*), crevalle jack (*Caranx hippos*), and striped bass (*Morone saxatilis*). Over the years, the striped bass has gradually declined in numbers. The bay anchovy (*Anchoa mitchilli*) and Atlantic silverside (*M. menidia*) supply much of the ration of these predators.

A group of benthic macroinvertebrates in the estuary is carnivorous. The Atlantic horseshoe crab (*Limulus polyphemus*), lobed moon shell (*Polinices duplicatus*), whelks (*Busycon* spp.), oyster drills (*Eupleura caudata* and *Urosalpinx cinerea*), and the sea star (*Asterias forbesi*), which ingest various life-history stages of other macroinvertebrates, fall into this trophic category. The hard clam, *Mercenaria mercenaria*, serves as susceptible prey of these predators, especially when the bivalve is less than 20 mm in length.

Kennish and Loveland[103] constructed a simplified detritus food web for Barnegat Bay (Figure 6). Detritus accumulating largely from eelgrass, salt marsh grasses, and biodeposits fuels the base of the food web. This detritus supports a diverse community of primary (detritivores) and secondary (omnivorous and carnivorous) consumers.

B. GRAZING FOOD WEB
1. Phytoplankton and Benthic Algal Production
The grazing food chain results from a transfer of food energy from live plants to grazing herbivores and ultimately to carnivores. The interconnection of grazing food chains involving complex interlocking patterns of feeding relationships defines the grazing food web. Because of the generally higher vascular plant and detritus production in estuaries relative to phytoplankton and benthic algal production, grazing food webs are often less important than detritus food webs in these shallow coastal systems. Grazing food webs attain greatest prominance in larger, deeper estuaries where vascular plants are confined to the marginal zones.

Herbivorous grazers, unlike detritivores, have a direct effect on primary producers. By

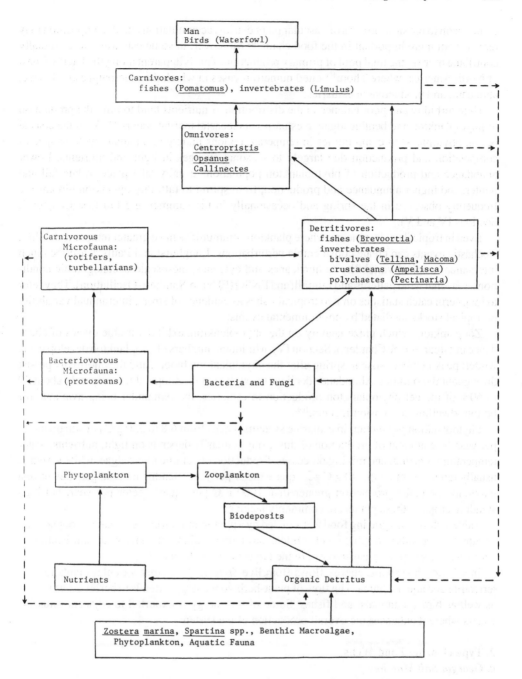

FIGURE 6. Schematic diagram of the detritus-generated food web of Barnegat Bay. (From Kennish, M. J. and Loveland, R. E., *Ecology of Barnegat Bay, New Jersey*, Kennish, M. J. and Lutz, R. A., Eds., Springer-Verlag, New York, 1984, 302. With permission.)

consuming the live plant, herbivores not only remove the producer organism but also the product. Thus, the grazing of large quantities of the plant photosynthetic apparatus can impact net production, especially if the plants are removed early in the growing season.[104]

While production of phytoplankton and benthic algae commonly is an order of magnitude (or more) less than that of vascular plants in estuaries, the more rapid utilization of algae in comparison to the slow process of vascular plant detritus accumulation and decomposition may render them more important in the food chain.[105] Moreover, in some estuaries, algae actually contribute more to the total pool of primary production. This is apparent along the Pacific Coast of North America, where Thom[106] cited numerous cases in which the macroalgae are diverse, abundant, and productive in estuaries.

High turbidity and poor balance in the distribution of nutrients tend to limit the production of phytoplankton and benthic algae in estuaries and coastal marine waters.[107] As in the coastal ocean, phytoplankton communities in temperate regions undergo an annual cycle in species composition and production due largely to seasonal changes in light and nutrients. Lower abundance and production of phytoplankton populations usually take place in late fall and winter, and higher abundance and production, from spring to fall. Phytoplankton blooms are frequently observed in the spring and occasionally in the summer and fall (see Chapter 2, Sections IV and V).

Even in tropical environments, where plankton communities have greater temporal stability, stochastic events can create considerable perturbations. Longhurst and Pauly[108] ascribe much interannual tropical variability to hurricanes and cyclones, mesoscale oceanographic conditions, seasonal wind patterns and rainfall, and ENSO (El Nino/Southern Oscillation). They refer to long-term catch statistics on two tropical fishes as evidence of strong interannual variability in tropical stocks mediated by environmental events.

Zooplankton, which graze heavily on the phytoplankton, exhibit variable cycles of abundance in estuaries (see Chapter 3, Section IV). Minimum numbers of zooplankton develop in the winter; peak numbers arise in spring after the phytoplankton bloom, and they normally persist throughout the summer. Abundance declines through the fall. Zooplankton crop only about 50 to 60% of the net phytoplankton production in estuaries, the remainder being available for bottom-dwelling suspension feeders.[109]

Phytoplankton production in estuarine systems ranges from 6.8 to 530 g C per square meter per year. The amount of production of this group primarily depends on light, nutrients, water temperature, grazing, and mixing processes. Production of benthic microalgae, while variable, usually ranges from 25 to 240 g C per square meter per year. Benthic macroalgae may have a maximum potential production greater than 1000 g C per square meter per year, but high standing crops of these plants are commonly lacking.[9]

Table 1 chronicles grazing food webs fueled by a phytoplankton energy source. Zooplankton occupy the secondary trophic level. Middle carnivores include planktivorous and benthophagous fishes, and the top carnivores are the largest fish predators.[110]

In salt marsh systems, energy flows from live *Spartina* and other vascular plants to both terrestrial and aquatic fauna. The flow of plant-herbivore energy extends to herbivorous insects as well as birds, mammals, and fishes. Some of the energy is exported to adjacent estuarine waters where it enhances the overall production of the system.

2. Type Grazing Food Webs
a. Georgia Salt Marshes

Estuaries bordered by extensive salt marshes, such as those along the southeast coast of the U.S., have classically been viewed as detritus-based systems. Most of the primary production in the form of *Spartina* roots, rhizomes, shoots, stems, and leaves pass to the detritus food web. Only a few insects, which are consumed by a variety of aquatic and terrestrial carnivores, and a limited number of vertebrate grazers feed directly on live marsh plants.[109,111] Whereas some ecologically interesting forms structure the grazing food web of the salt marsh system,[112] they account for only a small portion of the total energy flow as grazers of *Spartina* tissue. Teal[60] estimates that only 31 g C per square meter per year is lost to herbivores; terrestrial grazers do not appear to be a significant factor in the overall carbon flux of the salt marsh.[111]

The largest fraction of energy flowing from *Spartina* to the grazing community is apportioned to the grasshopper, *Orchelimum fidicinium*, and the plant hopper, *Prokelisia marginata*.[104] These two insects assimilate an estimated 304 k cal/m²/year which equals about 4.6% of the net *Spartina* production.[7,60] *O. fidicinium* belongs to a leaf-chewing guild that embodies two other important members, the squareback crab (*Sesarma reticulatum*) and adults of the weevil (*Lissorhoptrus chapini insularis*). *Sesarma* and *Lissorhoptrus* are mainly restricted to the low marsh, whereas *Orchelimum* has a broader distribution over the marsh. As much as 5% of the net primary production of *Spartina* may be lost to *O. fidicinium* during its active growth period.[104] *O. fidicinium* is prey of lycosid and araneid spiders, long-billed marsh wrens (*Telmatodytes palustris griseus*), and possibly cattle egrets (*Bubulcus ibis*).

A more diverse and productive sap-sucking guild of insects feeds on the cordgrass, the most important species being the plant hopper (*Delphacodes detecta*), the mirid (leaf) bug (*Trigonotylus whleri*), the lygaeid bug (*Ischnodemus badius*), and the armored scale (*Haliaspis spartinae*). Plant hoppers are the main aboveground herbivorous consumers of the cordgrass in Georgia salt marshes.[104]

Predatory arachnids consume most of the marsh insects. Acarina are dominant predators of the microarthropod community. Mesostigmated mites, larvae of trombidiid (chigger) mites, and ectoparasites of spiders numerically predominate among the predatory Acarina.[104] Marsh wrens may regulate population sizes of arthropod predators.[113] Another salt marsh bird consuming insects and spiders is the clapper rail, *Rallus longirostris*, which also ingests large quantities of *Sesarma reticulatum*. The grapsid crab constitutes a favorite staple of other birds as well, namely, herons, ibises, and egrets. Killifish feed on insects and spiders that collect on the surface of tidal streams. Rice rats (*Oryzomys palustris*), raccoons (*Procyon lotor*), and minks (*Mustela vison*) increase egg and nestling mortality of the long-billed marsh wren.[104]

A somewhat different grazing scheme can be drafted for the aquatic habitats of the marsh system, that is, areas where water covers the marsh surface at high tide as well as in tidal creeks. In this area of the marsh, two types of animal populations are encountered: (1) those residing permanently in the water (e.g., plankton, crabs, and fishes) and (2) those residing permanently in the marsh (e.g., xanthid crabs, fiddler crabs, and mussels). Major primary producers in the water are phytoplankton and benthic algae, with macroalgae being sparse throughout the system. The combined production of phytoplankton and benthic microflora is much less than Spartina production. Wiegert et al.[111] assigned 80% of the total net primary production integrated over the watershed (4.6 g C per square meter per day) to rooted plants, 10% to phytoplankton, and 10% to microalgae in sediments.

Epibenthic grazing by fiddler crabs, snails, and mullet appears to remove much algal biomass.[63] While the microalgal community, consisting mainly of diatoms and filamentous blue-green algae, occupies approximately the same parts of the marsh as *S. alterniflora*, the macroalgal community is poorly represented. Extremes of temperature, dessication, high turbidity, and rapid sedimentation in the lower areas of the marsh lead to a paucity of benthic macroalgae.[114]

Several investigators have addressed the issue of benthic algal production in or near the Sapelo Island salt marshes. Pomeroy[115] estimated that gross algal production in the marsh approached 200 g C per square meter per year, and net production was not less than 180 g C per square meter per year. Later surveys of epibenthic algal production conducted in the marshes near Sapelo Island yielded net production figures of about 190 g C per square meter per year.[63]

Although Ragotzkie[116] recorded negligible phytoplankton production in estuarine waters of Georgia due largely to high turbidity, more recent findings reveal that production may be much greater. According to Pomeroy et al.,[63] the net production of phytoplankton in salt marsh estuaries of the southeastern U.S. ranges from less than 100 to almost 400 g C per square meter per year. Phytoplankton production in the Duplin River and adjacent Doboy Sound amounts to 375 g C per square meter per year.[63] Primary consumers (i.e., zooplankton and benthic suspension feeders) assimilate most of this production. Secondary and tertiary consumers, including omnivorous and carnivorous macrobenthic invertebrates and fishes, feed on the primary consumers and on each other. Hence, the grazing food web in the aquatic regime of the salt marsh system is similar to that of other estuaries where benthic algae and phytoplankton form the base of the food web.

b. Chesapeake Bay

Chesapeake Bay, perhaps the most productive estuary on earth, is a plankton-based ecosystem in which phytoplankton and bacteria components support zooplankton intermediaries that provide ration for higher-trophic-level benthonic and nektonic consumers.[117] As is typical of temperate estuaries, nanoplankton (microflagellates and small diatoms) and larger phytoplankton (diatoms and dinoflagellates) taxa supply most of the primary production in the bay. Zooplankton composed of three broad groups (micro-, meso-, and macrozooplankton) depend on phytoplankton production ranging from 74 to 851 g C per square meter per year in the midbay.[118] The microzooplankters graze much of the phytoplankton in the water column; benthic suspension feeders strain substantial quantities of phytoplankton from late spring to early summer, and grazing pressure from macrozooplankton assemblages peaks from July to September. The spring recruitment of juvenile benthic macroinvertebrate populations creates a maximum benthic demand on phytoplankton in the water column during May and June.[118]

The primary groups of microzooplankton in the estuary are the protozoan phyla Sarcodina and Ciliophora, the Rotifera, and nauplii of the Copepoda. Of the mesozooplankton taxa, the Chaetognatha, Ctenophora, Schyphozoa, and juvenile and adult Copepoda and Cladocera predominate, although the larvae of benthic invertebrates can be seasonally abundant. The macrozooplankton encompass a diverse assemblage of organisms such as amphipods, isopods, mysid shrimp, true shrimp, hydromedusae, comb jellies, and true jellyfishes.[117]

Horwitz[119] separated the ichthyofauna of Chesapeake Bay into six feeding categories based on their food habits. These categories are (1) feeders on phytoplankton-detritus; (2) feeders on zooplankton; (3) feeders on a combination of zooplankton, benthos, and detritus; (4) feeders on benthic and epifaunal invertebrates; (5) feeders on fish; and (6) feeders on other substances. Only two species, *Brevoortia tyrannus* and *Mugil curema*, directly consume phytoplankton and are classified as phytoplankton-detritus feeders. The most common zooplanktivorous fishes include *Anchoa mitchilli*, *Menidia menidia*, *Alosa aestivalis*, *A. pseudoharengus*, and *A. sapidissima*. In addition, *Anchoa hepsetus*, *Clupea harengus*, *Dorosoma cepedianum*, and *Ammodytes americanus* are zooplanktivorous. *Leiostomus xanthurus*, *Micropogonias undulatus*, and *Bairdiella chrysoura* subsist on a diverse diet of zooplankton, benthos, and detritus. A large number of species prefer benthic and epifaunal invertebrate prey, examples being *Apeltes quadracus*, *Gasterosteus aculeatus*, *Gobiosoma bosci*, and *Syngnathus fuscus*. Type carnivorous fishes which prey on smaller ichthyofaunal populations are the conspicuous piscivores, *Cynoscion regalis*, *Morone saxatilis*, and *Pomatomus saltatrix*.

Analysis of the relative abundances of ichthyofaunal trophic groups disclosed that the three most common species in trawl samples (i.e., *A. mitchilli*, *L. xanthurus*, and *M. undulatus*) fed on a mixture of benthos, zooplankton, and detritus. For these fishes, both the detritus-based and grazing-based food webs are significant in their nutrition. They were the dominant group of fishes sampled throughout the study period, comprising 92 to 99% of all individuals caught from 1973 to 1981 (Table 6). Over the study period 1969 to 1981, the relative abundances of zooplanktivores, piscivores, and species feeding heavily on mysids decreased. Because of diminishing vegetation in the estuary,[120] small fishes which fed among vegetation and hard substrates declined during the 12-year sampling period. The shifts in the abundance of feeding groups suggested a diminution in trophic diversity in the bay.[119]

c. Barnegat Bay

Phytoplankton populations are the key components of the base of the grazing food web of

TABLE 6
Trophic Composition of Trawl Samples at Four Stations in the Mid-Chesapeake Bay, 1969—1981. Some Species May Comprise More Than One Group

Predominant food type(s)	Proportion of annual CPUEs		CPUE[a] (catch per 30 min)	
	Median	Range	Median	Range
Zooplankton-benthos-detritus	93.2	77.6—99.0	1121	469—3144
Zooplankton-benthos-detritus[b,c]	0.019	0.002—0.34	0.12	0.021—4.1
Zooplankton[b]	0.57	0.069—3.7	7.9	2.1—43.7
Benthos[c]	2.9	0.48—19.9	30.6	5.6—155.3
Polychaetes	2.4	0.36—18.5	25.5	4.2—144.1
Mysids	0.38	0.017—4.2	7.1	0.19—17.2
Mysids, mysids + fish	1.6	0.10—6.3	17.6	2.0—112.0
Variety of small benthos epifauna	0.30	0.026—3.2	3.6	0.30—38.8
Fish	0.60	0.95—4.2	11.4	2.1—106.3
Phytoplankton-detritus	0.28	0.002—5.4	4.5	0.028—63.6
Other	0.045	0.006—0.11	0.52	0.094—8.6

[a] Catch-per-unit effort.
[b] Excluding *Anchoa mitchilli*.
[c] Excluding *Leiostomus xanthurus* and *Micropogonias undulatus*.

From Horwitz, R. J., *Ecological Studies in the Middle Reach of Chesapeake Bay: Calvert Cliffs*, Heck, K. L., Jr., Springer-Verlag, Berlin, 1987, 167. With permission.

Barnegat Bay, although vascular plants and benthic macroalgae comprise a minor energy source, supplying food for grazing epifauna. Microflagellates and dinoflagellates numerically dominate the phytoplankton community in summer, and diatoms are most abundant in winter. Zooplankton, suspension-feeding benthos, and planktivorous fishes consume the phytoplankton. Much of the subsequent discussion on the grazing food web of the estuary is based on the work of Kennish and Loveland[103] and Kennish.[121]

Phytoplankton blooms in the late winter dominated by *Thalassiosira nordenskioldii* and *Detonula confervacea* yield forage for zooplankton populations which increase in abundance in the spring. *Acartia hudsonica, Pseudocalanus minutus, Eurytema* spp., and *Temora longicornis* reach high densities in spring and are eaten by mysids (e.g., *Neomysis americana*), ctenophores (e.g., *Mnemiopsis leidyi* and *Pleurobrachia* sp.), cnidarians (e.g., *Obelia* spp., *Rathkea octopunctata*, and *Sarsia* spp.), chaetognaths (e.g., *Sagitta* spp.), and fishes (e.g., *Pseudopleuronectes americanus* and *Ammodytes americanus*). Microzooplankton supply food for ichthyoplankton, *A. americanus*, and *P. americanus*, during the spring.

The main phytoplankton consumers in the estuary in the summer are copepods (especially *Acartia tonsa, Oithona colcarva*, and *Paracalanus* spp.) and meroplankton (e.g., larvae of barnacles, bivalves, ectoprocts, gastropods, and polychaetes). At this time, benthic invertebrates — filter and suspension feeders (e.g., *Mercenaria mercenaria, Mulinia lateralis*, and *Mytilus edulis*) as well as deposit feeders (e.g., *Macoma balthica* and *Tellina agilis*) — actively feed on phytoplankton that settle to the estuarine seafloor. The planktivorous-detritivorous fish, *Brevoortia tyrannus*, which inhabits the bay from spring to fall, removes large quantities of phytoplankton during the warm months.

Mnemiopsis leidyi is a voracious predator on zooplankton in summer and can completely decimate copepod populations in the estuary. Larvae of *Anchoa mitchilli* and *Gobiosoma* spp. subsist on microzooplankton during the summer. In addition, adult *A. mitchilli, B. tyrannus*, and *Menidia menidia* ingest holoplankton and meroplankton; they are, in turn, prey for larger predatory fishes (e.g., *Cynoscion regalis, Morone saxatilis*, and *Pomatomus saltatrix*). Predation pressure diminishes in the fall as fishes migrate to oceanic waters and phytoplankton and zooplankton biomasses decrease.

The grazing food web of the bay is, in part, rather speculative because of multiple feeding strategies of consumer organisms. Food energy does not transfer directly from phytoplankton through herbivorous copepods to bay anchovy, bluefish, and weakfish. Instead, omnivory plays a vital role, with zooplankton (e.g., copepods) and fishes (e.g., American eel and weakfish) feeding on more than one trophic level. Figure 7 presents a schematic diagram of the grazing food web of Barnegat Bay.

C. SECONDARY AND TERTIARY CONSUMERS IN ESTUARIES

Although the pathway of energy transfer differs among grazing and detritus food webs in estuaries, both food web components terminate similarly, with common secondary and tertiary consumers. Clearly, omnivory and carnivory influence the complexity of the food webs.[122] In the plankton communities they studied, Sprules and Bowerman[123] hypothesized that structural stability is enhanced by a relatively high incidence of cannibalism, reciprocal predation, life-history omnivory, and close linkage between omnivory and high trophic position.

Many of the populations on the higher trophic levels of estuarine food chains are predators, chiefly fish, birds, and some carnivorous invertebrates. The effects of these three groups of organisms are perhaps most evident on mudflats and in shallow subtidal waters easily accessible to field observations. Predation on lower trophic levels by these organisms can exacerbate variations in upper trophic levels.[110] Estuarine fish, mostly predators, are particularly noteworthy in this respect.

1. Invertebrate Predators

It may be convenient to subdivide this group into a migratory epifaunal component (e.g., crabs and shrimp) and an infaunal component (e.g., errant polychaetes). The mobile epifauna often serve a dual role as carnivores and scavengers (e.g., *Callinectes sapidus*). Mobile shrimp and prawns (e.g., Crangon crangon) and opossum shrimp (e.g., *Neomysis integer* and *Praunus flexuosus*) move from deeper water in winter to shallower water in summer where they feed on a variety of items, such as plant fragments, zooplankton, and benthic fauna. The seasonal movements of these omnivores, therefore, parallel those of certain species of crabs.[18] For example, the shore crab, *Carcinus maenas,* distributes throughout the estuary in winter, but migrates farther upestuary in summer. These seasonal movements of shore crabs have been perceived in the Yealm estuary in England[124] and the Isefjord in Denmark.[125] In these systems, the crabs have adapted to the high secondary benthic production in summer by migrating onshore.

Several polychaetes are well-known predators of the benthos, consuming large numbers of meiofauna and smaller polychaetes. Examples include *Nephtys incisa, Glycera* spp., and perhaps some *Nereis* spp. In addition to these predators, nemertines (e.g., *Lineus ruber*) prey heavily on small polychaetes.[16] The opisthobranch, *Retusa obtusa,* is a predatory estuarine species that eats *Hydrobia.*[16,18] Other predatory gastropods ingesting detritus feeders and suspension feeders belong to various genera (e.g., *Aglaja, Busycon, Chelidonura, Eupleura,* and *Polinices*). Additionally, carnivorous arthropods (e.g., *Limulus*) and echinoderms (e.g., *Asterias*) can completely ravage populations of primary consumers in estuaries.

2. Fishes

While carnivorous invertebrates affect the structure of bottom communities,[42-45] the dominant predators tend to be fishes and birds. Most estuarine fishes are predators, the detritus-feeding gray mullet (*Mugil cephalus*) being an exception. It not only ingests detritus, but at some ontogenetic stages, browses on algae and grazes on plankton.[16]

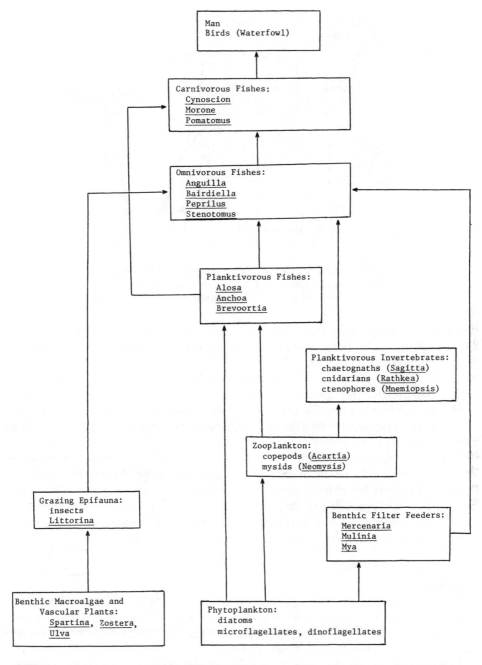

FIGURE 7. A schematic diagram of the grazing food web of Barnegat Bay. (From Kennish, M. J. and Loveland, R. E., *Ecology of Barnegat Bay, New Jersey,* Kennish, M. J. and Lutz, R. A., Eds., Springer-Verlag, New York, 1984, 302. With permission.)

In European estuaries, flatfishes subsist on a range of invertebrate forage. For instance, plaice (*Pleuronectes platessa*) feeds primarily on polychaetes, and the flounder (*Platichthys flesus*) consumes amphipods, mysids, and shrimp.[109] Brill (*Rhombus laevis*) attacks smaller fish and crustaceans; dab ingests the tentacular fan of *Sabella.*[16] Primary carnivores typically encountered in British estuaries are gobies (*Gobius*), herring (*Clupea*), and mackerel (*Scomber*) which constitute a portion of the diet of secondary carnivores, such as flatfish, dogfish, and blue sharks

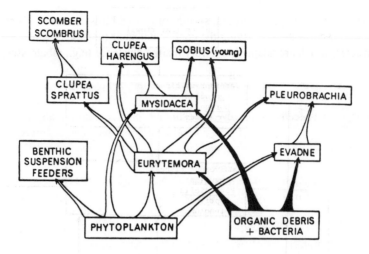

FIGURE 8.   Simplified food web of a British estuary revealing feeding relationships of secondary consumers. (From Green, J., *The Biology of Estuarine
Animals*, University of Washington Press, Seattle, 1968. With permission.)

(Figure 8).[7,110] Some of the estuarine fishes remove a wide variety of prey species. The goby
(*Pomatoschistus microps*), for example, eats chironomids, harpacticoids, oligochaetes, amphipods, gastropods, and barnacles.[7,16]

de Sylva[110] described the food web of the Delaware River estuary. Primary carnivores in the
estuary consist of silversides, anchovies, and several other small fishes that feed on invertebrates
at the bottom and in the water column (e.g., amphipods, decapods, isopods, and mysids). They
provide forage for secondary carnivores (e.g., barracuda, bluefish, and striped bass). Sharks are
tertiary carnivores in the system.

Changes in the feeding habits of some carnivorous fish during ontogeny obfuscate the
structure of estuarine food webs. Stomach content analysis of striped bass in Minas Basin and
Cobequid Bay, Nova Scotia, for instance, reveals that young-of-the-year fish consume invertebrates, but as the organisms age, they become increasingly piscivorous. Older individuals
subsist on invertebrates as well as other fishes.[126] A common dietary progression with increasing
age of estuarine fishes is zooplankton to detritus to macroinvertebrates, and (in some cases) to
fishes.[109]

3. Birds

A key predatory element of estuarine ecosystems, often overlooked in food web analyses, is
the avifauna. Most estuarine avifauna consume benthic infauna — crustaceans, mollusks, and
polychaetes — inhabiting tidal flats. Conspicuous estuarine avifauna are ducks and geese as
well as gulls, terns, shore birds, and wading birds.[109]

Both primary and secondary seabirds frequent estuarine habitats. The predominant primary
seabirds found here belong to the order Charadriiformes (i.e., gulls, terns, skuas, skimmers, and
auks). The secondary seabirds, not considered to be true seabirds, are mainly freshwater birds
that resort seasonally to coasts in pursuit of food. Divers, grebes, seaducks (e.g., eiders,
mergansers, and scoters) and phalaropes comprise this group.[2] However, the largest populations
of birds occurring in estuaries generally are waders (e.g., oystercatchers, plovers, and sandpipers) and wildfowl (e.g., ducks, geese, and swans) that commonly have inland breeding grounds,
similar to many secondary seabirds. Dense flocks of ducks, geese, and waders may be seen
overwintering on estuarine mudflats. Some classification schemes designate this latter group of
birds as littoral species, principally restricted to the intertidal zone or very shallow inshore

waters when not migrating. The littoral species of coastal birds contrast with neritic species, which prefer coastal waters overlying the continental shelf (e.g., auks, gulls, and terns), and oceanic species, which are adapted to a truly marine existence (e.g., albatrosses, petrels, and shearwaters).[2]

Several recent books give in depth coverage of the behavior, ecology, and evolution of seabirds.[47,127,128] These publications underscore the significance of seabirds in coastal food webs. While some species feed exclusively on estuarine phanerogams, many populations exert predation pressure on a variety of pelagic and benthic fauna. Hence, egrets and herons are predators of finfishes, and brent geese and widgeon consume plants. Most of the littoral species of birds observed in estuaries, however, feed on benthic invertebrates. On mudflats, coastal seabirds probe sediments at low tide in search of benthic infauna. The type of prey taken is linked to bill morphology. Birds with short bills, such as plovers and sandpipers, remove surface-dwelling invertebrates (e.g., *Hydrobia*), whereas those with medium-length bills, such as dunlins, redshanks, and other waders, search the upper layers of sediments for crustaceans, polychaetes, and mollusks. Birds with long bills (e.g., curlews, godwits, and willets) dig deep within mudflat sediments to remove invertebrate prey. Much of the prey may be partitioned due to the zonation of feeding birds with respect to the water's edge. Thus, godwits and redshanks capture food in shallow waters at the water's edge, dunlins haunt wet mud at the water's edge, and plovers seek prey from drier sediments on mudflats.

McLusky[18] reviewed the diet and methods of feeding of avifauna in European estuaries. The feeding behavior of predatory birds on mudflats is coupled to the tides and the activity of the prey in response to the tides. Along the intertidal zone at low water, the redshank (*Tringa totanus*) and shelduck (*Tadorna tadorna*) eat *Hydrobia* and *Macoma,* in addition to other infauna, near the water's edge. In shallow water at low tide, mussels *(Mytilus edulis)* are favorite prey of eider ducks (*Somateria m. mollissima*). Other birds dive for their prey at high water. Enlisted within this group of predatory birds are diving ducks (e.g., scaup, *Aythya marila,* and goldeneye, *Bucephala clangula*), cormorants (*Phalacrocorax carbo*), and mergansers (*Mergus serrator*). Predatory activity of avifauna is often manifested most clearly on rocky shorelines.[129]

Some birds pursue certain prey species. The godwit, *Limosa lapponica*, principally subsists on the lugworm, *Arenicola marina*. The grey plover, *Pluvialis squatarola,* searches for the ragworm, *Nereis diversicolor*. The knot (*Calidris canutus*), similar to *T. tadorna*, ingests *Hydrobia*. The curlew, *Numenius arquata,* stalks for *N. diversicolor* and *Carcinus maenas*. *Cardium edule, Macoma balthica,* and *Mytilus edulis* are the primary species of prey of the oystercatcher, *Haematopus ostralegus*. *Tringa totanus* sorts through sediments for the amphipod, *Corophium volutator,* but also picks up *C. maenas, Crangon* spp., *Hydrobia ulvae, Nereis* spp., and other invertebrates when available. Many benthic infaunal populations of prey migrate vertically in mudflat sediments in response to the presence of avifaunal predators and/or tidal inundations.

As conveyed above, one of the most important categories of estuarine birds in England incorporates waders, notably the oystercatcher (*H. ostralegus*), knot (*C. canutus*), dunlin (*C. alpina*), redshank (*T. totanus*), godwit (*L. lapponica*), turnstone (*A. interpres*), grey plover (*P. squatarola*), curlew (*N. arquata*), ringed plover (*Charadrius hiaticula*), and sanderline (*Crocethia alba*). Waders, which consume a diversity of prey species even though certain infauna may be preferred, remove substantial biomasses of benthic invertebrates. Prater,[130] for example, reported that a single oystercatcher can consume 315 *Cardium* per day; a single redshank, 40,000 *Corophium* per day; and a single knot, 730 *Macoma* per day. This predation crops an estimated 4 to 20% of the invertebrate prey populations annually.[16]

The importance of wading birds in estuaries is exemplified by the Severn estuary.[131] Principal feeding sites of wading birds along the fringes of this estuary exist on relatively stable fine sediments near the top of the shore. At peak abundance, the wading bird population in this system averages 88,500 individuals. Eight of the populations are largely intertidal foraging species present in very high numbers. The Welsh Grounds and Bridgewater Bay serve as the two predominant areas of occupation by the wading birds. The recorded density of wading birds in the Taff estuary in Cardiff's dockland equals 1900 waders per square kilometer.

The most common species of estuarine birds in the Severn estuary are the lapwing (*Vanellus vanellus*) and the dunlin (*Calidris alpina*); the highest average monthly counts of these two species in the estuary amount to 16,930 and 47,000 individuals, respectively.[132] While some of the birds disperse throughout the estuary, others have limited distributions, either in space or in time. Thus, the knot (*C. canutus*) attains maximum numbers in winter along the northern shore and is not abundant during any other season. The black-tailed godwit (*Limosa limosa*) is only common during the fall in Bridgewater Bay. Of the dunlin populations, the Siberian breeding, *Calidris alpina alpina,* dominates in winter, but in spring and fall, the Icelandic breeding, *C. a. shinzii,* and Greenland breeding, *C. a. arctica*, also invade the region. In addition to the regular seasonal changes in the distribution of waders in the estuary, nonperiodic deviations in distributions have been ascribed to the erosion of intertidal sediments and consequent losses of feeding grounds resulting from increased wave action during prolonged storms.[133]

Research on the predatory activity of avifauna in European and North American estuaries indicates the potential significance of this group in modulating energy flow through these ecosystems.[46,128,134,135] For example, of the total summer production by mussels (1300 kcal/m^2) in the Ythan estuary, England, oystercatchers consumed 93 kcal, eider ducks 275 kcal, and herring gulls 112 kcal.[136] In the Grevelingen estuary, the Netherlands, estuarine birds have been shown to remove 3.42 g/m^2/year of the total production of 50.34 to 57.43 g/m^2/year by benthic primary consumers.[137] Predatory birds have the greatest impact on intertidal macrofauna of mudflat environments. In many estuaries of England and the Netherlands, peak numerical abundances of birds occur during the winter as the waders and wildfowl migrate from eastern and northern Europe. However, they can be detected in large numbers during the other seasons of the year as well.

In recent years, avifauna have been exposed to decreasing areas of coastal wetlands and beaches due to natural processes and man-induced impacts.[138] Disturbances of habitat from demolition, beach cleanup, and construction for development have adversely affected waterfowl and migrating shorebirds. Burger[139,140] traced the effects of people on shorebirds of Delaware, Raritan, and Jamaica bays in the mid-Atlantic Bight (U.S.). She discovered that shorebirds shift sites continually when faced with human disturbances. Unequivocally, foraging is interrupted, and, in the case of shorebirds on migration stopover, individuals may be pressed for foraging time, causing them to search for food in suboptimal areas and at night.[141] In the case of laughing gulls, Burger[142] discerned depressed foraging efficiency for 60 to 90 min subsequent to the initiation of beach cleanup work. Given that birds play a potentially significant role in the energy flow and food web structure of estuaries, anthropogenic impacts on these organisms — and ultimately on all estuarine biota — can be profound.

III. SUMMARY AND CONCLUSIONS

Estuaries are nutrient-rich environments characterized by high primary and secondary production. A complex food web exists in estuarine ecosystems structured by two major interlocking components of energy flow, that is, detrital and grazing pathways. The distinction between detritus and grazing food webs occurs at the primary producer-primary consumer levels. At secondary consumer levels and above, a network of heterotrophs derives energy and nutrients from both primary pathways; consequently, the separation between grazing and detrital food webs becomes obscured among upper-trophic-level organisms.

The base of detrital food webs is detritus — a general term denoting both particulate and dissolved nonliving organic matter. The bulk of detritus results from the accumulation of plant material, primarily benthic macrophytes (i.e., seagrasses, salt marsh grasses, and mangroves), that largely remain ungrazed while alive and enter the detritus pool upon death. Because of low exploitation rates by grazing herbivores, more than 90% of the primary production by benthic macrophytes passes to detritus food webs in estuaries. Another potentially significant source of detritus consists of biodeposits, feces and pseudofeces, of animals.

Much estuarine detritus is consumed directly by detritivores living on or in bottom sediments. The detritus forms a substrate for the growth of bacteria, fungi, and microalgae which provide a rich food supply for primary consumers. The value of detritus as a food, therefore, depends both on its own chemical composition as well as its associated bacteria, microalgae, and protozoans. Many detritivores appear to require both organic detritus and the attached microbial community to meet their energy needs.

Bacteria and fungi are decomposers that attack the detrital substrate, regenerating nutrients and transforming the POM into DOM which they assimilate. Higher organisms, such as ciliates, nematodes, ostracods, and harpacticoids, feed on the bacteria; larger consumers, including rhabdocoels and halacarids, consume the micrometazoans and each other. Detritus-feeding macrofauna strip microbes from the detrital particles during feeding and digestion. The microbes recolonize the particles repeatedly, and, by doing so, become a renewable food source, albeit an insufficient one to meet the total energy demands of many detritivores. Organisms colonize detrital particles in a successional pattern, beginning with bacteria and fungi and followed by algae, ciliates, flagellates, and larger grazers.

The detritivores are essential foods for secondary consumers (larger invertebrates, fishes, and birds) in estuaries. A diversity of fishes and birds, many of which have omnivorous feeding habits, exhibit mutiple feeding strategies. Thus, they generally ingest organisms from more than one trophic level, accounting for the complex network of feeding relationships manifested as links or arrows on detrital food web diagrams. As food web complexity increases, so does the stability of the ecosystems.

Based principally upon vascular plant detritus, detritus food webs are common features of shallow estuaries, particularly those characterized by extensive communities of salt marsh grasses, mangroves, or seagrasses. Herbivorous and omnivorous crustaceans, mollusks, nematodes, polychaetes, waterfowl, and fishes constitute major groups of detritus feeders in these systems. Typically, shallow estuaries exhibit much lower primary production in the water column than in bottom floral communities. In addition, a few herbivorous grazers remove only a limited amount of live plant biomass. The salt marsh estuaries of Georgia and the shallow estuarine waters of southern Florida bordered by mangrove swamps provide excellent examples of systems dominated by the detritus food web. This type of food web also predominates in temperate estuaries supported by seagrass beds (e.g., Barnegat Bay, NJ).

Phytoplankton and other live plant communities form the base of grazing food webs. This type of food web is most evident in deeper, clearer systems. Zooplankton and zooplankton grazers (e.g., bay anchovies) are critical primary and secondary consumers of grazing food webs.

Although this food web is most conspicuous in larger estuaries (e.g., Chesapeake Bay and Long Island Sound), every estuary, regardless of size, possesses components of both types of food webs. However, one type of food web usually dominates over the other in a given estuary. Hence, despite the greater importance of detritus food webs in salt marsh- and seagrass-dominated systems, like those encountered near Sapelo Island, GA, and in Barnegat Bay, NJ, respectively, grazing food webs have also been ascertained in both areas.

Clearly, phytoplankton play an essential role in grazing food webs, supporting a multitude of zooplankton populations. Benthic micro- and macroalgae likewise supply food for a host of benthic invertebrate grazers (e.g., *Littorina* spp. and *Uca* spp.). Copepods, meroplankton of the benthos (e.g., larvae of barnacles, bivalves, ectoprocts, gastropods, and polychaetes), and suspension-feeding mollusks (e.g., *Mercenaria mercenaria, Mulinia lateralis,* and *Mytilus edulis*) graze heavily on phytoplankton. Zooplankton, in turn, are a dietary component of numerous organisms, especially other zooplankton (e.g., *Mnemiopsis* spp.), ichthyoplankton (e.g., larvae of *Anchoa mitchilli* and *Gobiosoma* spp.), and adult ichthyofauna (e.g., *A. mitchilli, Brevoortia tyrannus,* and *Menidia menidia*). Larger predatory fishes (e.g.. *Cynoscion regalis, Morone saxatilis* and *Pomatomus saltatrix*) feed on the smaller secondary consumers, *A. mitchilli* and *M. menidia.*

Similar secondary and tertiary consumers occupy the upper trophic levels of detritus and grazing food chains. Many of these consumers are predators, but many feed at more than one trophic level. The chief predators of estuaries are carnivorous invertebrates (e.g., *Asterias, Eupleura, Glycera, Limulus,* and *Acteocina*), fishes (e.g., *Pleuronectes* and *Pomatomus*), and birds (e.g., *Haematopus ostralegus, Limosa lapponica, Tadorna tadorna,* and *Tringa totanus*). Much of the research on the predatory activity of estuarine birds has been conducted in European systems, where various categories of birds, particularly waders, remove a substantial portion of the total secondary production of benthic invertebrate populations in the intertidal and shallow subtidal zones of estuaries.

REFERENCES

1. **Odum, H. T.,** *Systems Ecology,* John Wiley & Sons, New York, 1983.
2. **Boaden, P. J. S. and Seed, R.,** *An Introduction to Coastal Ecology,* Blackie & Son, London, 1985.
3. **Pimm, S. L. and Kitching, R. L.,** The determinants of food chain lengths, *Oikos,* 50, 302, 1987.
4. **Pimm, S. L.,** Energy flow and trophic structure, in *Concepts of Ecosystem Ecology: A Comparative View,* Pomeroy, L. R. and Alberts, J. J., Eds., Springer-Verlag, New York, 1988, 263.
5. **Briand, F. and Cohen, J. E.,** Environmental correlates of food chain length, *Science,* 238, 956, 1987.
6. **Levinton, J. S.,** *Marine Ecology,* Prentice-Hall, Englewood Cliffs, NJ, 1982.
7. **Green, J.,** *The Biology of Estuarine Animals,* University of Washington Press, Seattle, 1968.
8. **Whicker, F. W. and Schultz, V.,** *Radioecology: Nuclear Energy and the Environment,* Vol. 1, CRC Press, Boca Raton, FL, 1982.
9. **Kennish, M. J.,** *Ecology of Estuaries,* Vol. 1, CRC Press, Boca Raton, FL, 1986.
10. **Odum, W. E. and Heald, E. J.,** The detritus-based food web of an estuarine mangrove community, in *Estuarine Research,* Vol. 1, Cronin, L. E., Ed., Academic Press, New York, 1975, 265.
11. **Odum, E. P. and Biever, L. J.,** Resource quality, mutualism, and energy partitioning in food chains, *Am. Nat.,* 124, 360, 1984.
12. **Azam, F.,** The ecological role of water-column microbes in the sea, *Mar. Ecol. Prog. Ser.,* 10, 257, 1983.
13. **Ducklow, H. W., Purdie, D. A., Williams, P. J. le B., and Davis, J. M.,** Bacterioplankton: a sink for carbon in a coastal marine plankton community, *Science,* 232, 865, 1986.
14. **Sherr, E. B., Sherr, B. F., and Albright, L. J.,** Bacteria: link or sink, *Science,* 235, 88, 1987.
15. **Sherr, E. B. and Sherr, B. F.,** Role of microbes in pelagic food webs: a revised concept, *Limnol. Oceanogr.,* 33, 1225, 1988.
16. **Barnes, R. S. K.,** *Estuarine Biology,* Edward Arnold, London, 1974.
17. **Mann, K. H.,** *Ecology of Coastal Waters: A Systems Approach,* University of California Press, Berkeley, 1982.
18. **McLusky, D. S.,** *The Estuarine Ecosystem,* Halsted Press, New York, 1981.
19. **Rich, P. H. and Wetzel, R. G.,** Detritus in the lake ecosystem, *Am. Nat.,* 112, 57, 1978.
20. **Wetzel, R. G.,** *Limnology,* 2nd ed., W. B. Saunders, Philadelphia, 1983.
21. **Fenchel, T. M. and Jørgensen, B. B.,** Detritus food chains of aquatic ecosystems: the role of bacteria, *Adv. Microb. Ecol.,* 1, 1, 1977.
22. **Wolff, W. J.,** Biotic aspects of the chemistry of estuaries, in *Chemistry and Biogeochemistry of Estuaries,* Olausson, E. and Cato, I., Eds., John Wiley & Sons, Chichester, 1980, 263.
23. **Turner, J. T. and Ferrante, J. G.,** Zooplankton fecal pellets in aquatic ecosystems, *BioScience,* 29, 670, 1979.
24. **Jeffries, H. P.,** Fatty acid and ecology of a tidal marsh, *Limnol. Oceanogr.,* 17, 433, 1972.
25. **Jeffries, H. P.,** Diets of juvenile Atlantic menhaden (*Brevoortia tyrannus*) as determined from fatty acid composition of gut contents, *J. Fish. Res. Board Can.,* 32, 587, 1975.
26. **Tenore, K. R.,** Food chain pathways in detrital feeding benthic communities: a review, with observations on sediment resuspension and detrital recycling, in *Ecology of Marine Benthos,* Coull, B. C., Ed., University of South Carolina Press, Columbia, 1978, 37.

27. **White, D. C., Livingston, R. J., Bobbie, R. J., and Nickels, J. S.,** Effects of surface composition, water column chemistry, and time of exposure on the composition of the detrital microflora and associated macrofauna in Apalachicola Bay, Florida, in *Ecological Processes in Coastal and Marine Systems,* Livingston, R. J., Ed., Plenum Press, New York, 1979, 83.

28. **Zieman, J. C., Macko, S. A., and Mills, A. L.,** Role of seagrasses and mangroves in estuarine food webs: temporal and spatial changes in stable isotope composition and amino acid content during decomposition, *Bull. Mar. Sci.,* 35, 380, 1984.

29. **Lopez, G. R. and Levinton, J. S.,** Ecology of deposit-feeding animals in marine sediments, *Q. Rev. Biol.,* 62, 235, 1987.

30. **Bowen, S. H.,** Evidence of a detritus food chain based on consumption of organic precipitates, *Bull. Mar. Sci.,* 35, 440, 1984.

31. **Rice, D. L. and Hanson, R. B.,** A kinetic model for detritus nitrogen: role of the associated bacteria in nitrogen accumulation, *Bull. Mar. Sci.,* 35, 326, 1984.

32. **Stephens, G. C.,** Uptake of naturally occurring primary amines by marine annelids, *Biol. Bull.,* 149, 397, 1975.

33. **Feral, J.-P.,** Effect of short-term starvation on the biochemical composition of the apodous holothurian *Leptosynapta galliennei* (Echinodermata): possible role of dissolved organic material as an energy source, *Mar. Biol.,* 86, 297, 1985.

34. **Sibert, J. R. and Naiman, R. J.,** The role of detritus and the nature of estuarine ecosystems, in *Marine Benthic Dynamics,* Tenore, K. R. and Coull, B. C., Eds., University of South Carolina Press, Columbia, 1980, 311.

35. **Christian, R. R. and Wetzel, R. L.,** Interaction between substrate, microbes, and consumers of *Spartina* detritus in estuaries, in *Estuarine Interactions,* Wiley, M. L., Ed., Academic Press, New York, 1978, 93.

36. **Lopez, G. R., Levinton, J. S., and Slobodkin, L. B.,** The effect of grazing by the detritivore, *Orchestia grillus* on *Spartina* litter and its associated microbial community, *Oecologia,* 30, 111, 1977.

37. **Morrison, S. J., King, J. D., Bobbie, R. J., Bechtold, R. E., and White, D. C.,** Evidence for microfloral succession on allochthonous plant litter in Apalachicola Bay, Florida, U.S.A., *Mar. Biol.,* 41, 229, 1977.

38. **Parkinson, P.,** Terrestrial decomposition, in *Productivity of World Ecosystems,* National Academy of Science, Washington, D. C., 1975, 55.

39. **Yokoe, Y. and Yasumasu, I.,** The distribution of cellulase in invertebrates, *Comp. Biochem. Physiol.,* 13, 323, 1964.

40. **Crosby, N. D. and Reid, R. G. B.,** Relationships between food phylogeny and cellulose digestion in the Bivalvia, *Can. J. Zool.,* 49, 617, 1971.

41. **Fenchel, T.,** Aspects of the decomposition of seagrasses, in *Seagrass Ecosystems,* McRoy, C. P. and Helfferich, C., Eds., Marcel Dekker, New York, 1977, 123.

42. **Rhoads, D. C. and Young, D. K.,** The influence of deposit-feeding organisms on sediment stability and community trophic structure, *J. Mar. Res.,* 28, 150, 1970.

43. **Posey, M. H.,** Functional approaches to benthic communities: how useful are they?, *Rev. Aquat. Sci.,* in press.

44. **Virnstein, R. W.,** Predator caging experiments in soft sediments: caution advised, in *Estuarine Interactions,* Wiley, M. L., Ed., Academic Press, New York, 1978, 261.

45. **Valiela, I.,** *Marine Ecological Processes,* Springer-Verlag, New York, 1984.

46. **Reise, K.,** *Tidal Flat Ecology: An Experimental Approach to Species Interactions,* Springer-Verlag, New York, 1985.

47. **Croxall, J. P., Ed.,** *Seabirds: Feeding Ecology and Role in Marine Ecosystems,* Cambridge University Press, New York, 1987.

48. **Tenore, K. R. and Rice, D. L.,** A review of trophic factors affecting secondary production of deposit-feeders, in *Marine Benthic Dynamics,* Tenore, K. R. and Coull, B. C., Eds., University of South Carolina Press, Columbia, 1980, 325.

49. **Zieman, J. C.,** Quantitative and dynamic aspects of the ecology of turtle grass, *Thalassia testudinum,* in *Estuarine Research,* Vol. 1, Cronin, L. E., Ed., Academic Press, New York, 1975, 541.

50. **Rice, D. L.,** The detritus nitrogen problem: new observations and perspectives from organic geochemistry, *Mar. Ecol. Prog. Ser.,* 9, 153, 1982.

51. **Aller, R. C.,** The effects of macrobenthos on chemical properties of marine sediment and overlying water, in *Animal-Sediment Relations,* McCall, P. L. and Tevesz, M. J. S., Eds., Plenum Press, New York, 1982, 53.

52. **Findlay, S. and Meyer, J. L.,** Significance of bacterial biomass and production as an organic carbon source in lotic detrital systems, *Bull. Mar. Sci.,* 35, 318, 1984.

53. **Tenore, K. R., Cammen, L., Findlay, S. E. G., and Phillips, N.,** Perspectives of research on detritus: do factors controlling the availability of detritus to macroconsumers depend on its source?, *J. Mar. Res.,* 40, 473, 1982.

54. **Banks, C. W. and Wolfinbarger, L., Jr.,** A rapid and convenient method for radiolabelling detritus with [^{14}C] acetic anhydride, *J. Exp. Mar. Biol. Ecol.,* 53, 115, 1981.

55. **Lopez, G. R. and Crenshaw, M. A.,** Radiolabelling of sedimentary organic matter with ^{14}C-formaldehyde: preliminary evaluation of a new technique for use in deposit-feeding studies, *Mar. Ecol. Prog. Ser.,* 8, 283, 1982.

56. **Wolfinbarger, L., Jr. and Crosby, M. P.,** A convenient procedure for radiolabelling detritus with [^{14}C] dimethylsulfate, *J. Exp. Mar. Biol. Ecol.,* 67, 185, 1983.

57. **Stephenson, R. L., Tan, F. C., and Mann, K. H.,** Stable carbon isotope variability in marine macrophytes and its implications for food web studies, *Mar. Biol.,* 81, 223, 1984.
58. **Nichols, P. D., Klumpp, D. W., and Johns, R. B.,** A study of food chains in seagrass communities. III. Stable carbon isotope ratios, *Aust. J. Mar. Freshwater Res.,* 36, 683, 1985.
59. **Smalley, A. E.,** Energy flow of a salt marsh grasshopper population, *Ecology,* 41, 672, 1960.
60. **Teal, J. M.,** Energy flow in the salt marsh ecosystem of Georgia, *Ecology,* 43, 614, 1962.
61. **Reimold, R. J., Gallagher, J. L., Linthurst, R. A., and Pfeiffer, W. J.,** Detritus production in coastal Georgia salt marshes, in *Estuarine Research,* Vol. 1, Cronin, L. E., Ed., Academic Press, New York, 1975, 217.
62. **Pomeroy, L. R. and Wiegert, R. G., Eds.,** *The Ecology of a Salt Marsh, Ecological Studies,* Vol. 38, Springer-Verlag, New York, 1981.
63. **Pomeroy, L. R., Darley, W. M., Dunn, E. L., Gallagher, J. L., Haines, E. B., and Whitney, D. M.,** Primary production, in *The Ecology of a Salt Marsh, Ecological Studies,* Vol. 38, Pomeroy, L. R. and Wiegert, R. G., Eds., Springer-Verlag, New York, 1981, 39.
64. **Smalley, A. E.,** The Role of Two Invertebrate Populations, *Littorina irrorata* and *Orchelimum fidicinium* in the Energy Flow of a Salt Marsh Ecosystem, Ph.D. thesis, University of Georgia, Athens, 1959.
65. **Gallagher, J. L., Reimold, R. J., Linthurst, R. A., and Pfeiffer, W. J.,** Aerial production, mortality, and mineral accumulation-export dynamics in *Spartina alterniflora* and *Juncus roemerianus* plant stands, *Ecology,* 61, 303, 1980.
66. **Gallagher, J. L., Reimold, R. J., and Thompson, D. E.,** Remote sensing and salt marsh productivity, in Proc. 38th Annu. Meet., American Society of Photogrammetry, Washington, D. C., 1972, 338.
67. **Gallagher, J. L. and Plumley, F. G.,** Underground biomass profiles and productivity in Atlantic coastal marshes, *Am. J. Bot.,* 66, 156, 1979.
68. **Gallagher, J. L., Pfeiffer, W. J., and Pomeroy, L. R.,** Leaching and microbial utilization of dissolved organic matter from leaves of *Spartina alterniflora, Estuarine Coastal Mar. Sci.,* 4, 467, 1976.
69. **Gessner, R. V., Goos, R. D., and Sieburth, J. M.,** The fungal microcosm of the internodes of *Spartina alterniflora, Mar. Biol.,* 16, 269, 1979.
70. **Moran, M. A., Legovic, T., Benner, R., and Hodson, R. E.,** Carbon flow from lignocellulose: a simulation analysis of a detritus-based ecosystem, *Ecology,* 69, 1525, 1988.
71. **Haines, E. B. and Montague, C. L.,** Food sources of estuarine invertebrates analyzed using $^{13}C/^{12}C$ ratios, *Ecology,* 60, 48, 1979.
72. **Kraeuter, J. N. and Wolf, P. L.,** The relationship of marine macroinvertebrates to salt marsh plants, in *Ecology of Halophytes,* Reimold, R. J. and Queen, W. H., Eds., Academic Press, New York, 1974, 449.
73. **Wetzel, R. L.,** An Experimental Study of Detrital Carbon Utilization in a Georgia Salt Marsh, Ph.D. thesis, University of Georgia, Athens, 1975.
74. **Hughes, E. H.,** Estuarine Subtidal Food Webs Analyzed with Stable Carbon Isotope Ratios, M.S. thesis, University of Georgia, Athens, 1980.
75. **Odum, E. P.,** The role of tidal marshes in estuarine production, *NY State Conservationist,* 15, 12, 1961.
76. **Wiegert, R. G. and Evans, F. C.,** Investigations of secondary productivity in grasslands, in *Secondary Productivity of Terrestrial Ecosystems,* Petrusewicz, K., Ed., Polish Academy of Science, Warsaw, 1967, 499.
77. **Montague, C. L., Bunker, S. M., Haines, E. B., Pace, M. L., and Wetzel, R. L.,** Aquatic Macroconsumers, in *The Ecology of a Salt Marsh, Ecological Studies,* Vol. 38, Pomeroy, L. R. and Wiegert, R. G., Eds., Springer-Verlag, New York, 1981, 69.
78. **Teal, J. M.,** Distribution of fiddler crabs in Georgia salt marshes, *Ecology,* 39, 185, 1958.
79. **Wolf, P. L., Shanholtzer, S. F., and Reimold, R. J.,** Population estimates for *Uca pugnax* (Smith, 1870) on the Duplin estuary marsh, Georgia, U. S. A. (Decapoda, Brachyura, Ocypodidae), *Crustaceana,* 29, 79, 1975.
80. **Odum, E. P. and Smalley, A. E.,** Comparison of population energy flow of a herbivorous and a deposit-feeding invertebrate in a salt-marsh ecosystem, *Proc. Natl. Acad. Sci. U.S.A.,* 45, 617, 1959.
81. **Kuenzler, E. J.,** Structure and energy flow of a mussel population in a Georgia salt marsh, *Limnol. Oceanogr.,* 6, 191, 1961.
82. **Pace, M. L., Shimmel, S., and Darley, W. M.,** The effect of grazing by a gastropod, *Nassarius obsoletus,* on the benthic microbial community of a salt marsh mud flat, *Estuarine Coastal Mar. Sci.,* 9, 121, 1979.
83. **Fry, B. and Sherr, E. B.,** Use of stable isotopes in marine ecology—a review, *Estuaries,* 6, 303, 1983.
84. **Knox, G. A.,** *Estuarine Ecosystems,* Vol. 1 and 2, CRC Press, Boca Raton, FL, 1986.
85. **Teal, J. M.,** Respiration of crabs in Georgia salt marshes and its relation to their ecology, *Physiol. Zool.,* 32, 1, 1959.
86. **Miller, D. C.,** Studies on the Systematics, Ecology, and Geographical Distribution of Certain Fiddler Crabs, Ph.D. thesis, Duke University, Durham, NC, 1965.
87. **Day, J., Hall, C., Kemp, M., and Yanez-Arancibia, A.,** Estuarine Ecology, John Wiley & Sons, New York, 1989.
88. **Johannes, R. E. and Satomi, M.,** Composition and nutritive value of fecal pellets of a marine crustacean, *Limnol. Oceanogr.,* 11, 191, 1966.

89. **Welsh, B.,** The role of grass shrimp, *Palaemonetes pugio,* in a tidal marsh ecosystem, *Ecology,* 56, 513, 1975.

90. **Odum, W. E.,** Utilization of the direct grazing and plant detritus food chains by the striped mullet, *Mugil cephalus,* in *Marine Food Chains,* Steele, J. H., Ed., University of California Press, Berkeley, 1970, 222.

91. **Darnell, R. M.,** Trophic spectrum of an estuarine community based on studies of Lake Pontchartrain, Louisiana, *Ecology,* 42, 553, 1961.

92. **Virnstein, R. W.,** The importance of predation by crabs and fishes on benthic infauna in Chesapeake Bay, *Ecology,* 58, 1199, 1977.

93. **Rickards, W. L.,** Ecology and growth of juvenile tarpon, *Megalops atlantis,* in a Georgia salt marsh, *Bull. Mar. Sci.,* 18, 220, 1968.

94. **Dahlberg, M. D. and Heard, R. W., III,** Observations on Elasmobranchs from Georgia, *Q. J. Fla. Acad. Sci.,* 32, 21, 1969.

95. **Dahlberg, M. D.,** *Fishes of Georgia and Nearby States,* University of Georgia Press, Athens, 1975.

96. **Heard, R. W.,** Feeding habits of white catfish from a Georgia estuary, *Fla. Sci.,* 38, 20, 1975.

97. **Emlen, J. M.,** *Ecology: An Evolutionary Approach,* Addison-Wesley, Reading, MA, 1973.

98. **Odum, E. P.,** *Fundamentals of Ecology,* 3rd ed., W. B. Saunders, Philadelphia, 1971.

99. **Clark, J.,** *Coastal Ecosystem Management,* Wiley-Interscience, New York, 1977.

100. **Thayer, G. W., Adams, S. M., and LaCroix, M. W.,** Structural and functional aspects of a recently established *Zostera marina* community, in *Estuarine Research,* Vol. 1, Cronin, L. E., Ed., Academic Press, New York, 1975, 518.

101. **Kikuchi, T.,** Faunal relationships in temperate seagrass beds, in *Handbook of Seagrass Biology: An Ecosystem Perspective,* Phillips, R. C. and McRoy, C. P., Eds., Garland STPM Press, New York, 1980, 153.

102. **Loveland, R. E., Brauner, J. F., Taylor, J. E., and Kennish, M. J.,** Macroflora, in *Ecology of Barnegat Bay, New Jersey,* Kennish, M. J. and Lutz, R. A., Eds., Springer-Verlag, New York, 1984, 78.

103. **Kennish, M. J. and Loveland, R. E.,** Trophic relationships, in *Ecology of Barnegat Bay, New Jersey,* Kennish, M. J. and Lutz, R. A., Eds., Springer-Verlag, New York, 1984, 302.

104. **Pfeiffer, W. J. and Wiegert, R. G.,** Grazers on *Spartina* and their predators, in *The Ecology of a Salt Marsh, Ecological Studies,* Vol. 38, Pomeroy, L. R. and Wiegert, R. G., Eds., Springer-Verlag, New York, 1981, 87.

105. **Reeve, M. R.,** The ecological significance of the zooplankton in the shallow subtropical waters of south Florida, in *Estuarine Research,* Vol. 1, Cronin, L. E., Ed., Academic Press, New York, 1975, 352.

106. **Thom, R. M.,** Composition, habitats, seasonal changes and productivity of macroalgae in Grays Harbor estuary, Washington, *Estuaries,* 7, 51, 1984.

107. **Howarth, R. W.,** Nutrient limitation of net primary production in marine ecosystems, *Annu. Rev. Ecol. Syst.,* 19, 89, 1988.

108. **Longhurst, A. R. and Pauly, D.,** *Ecology of Tropical Oceans,* Academic Press, Orlando, FL, 1987.

109. **Nybakken, J. W.,** *Marine Biology: An Ecological Approach,* Harper & Row, New York, 1982.

110. **de Sylva, D. P.,** Nektonic food webs in estuaries, in *Estuarine Research,* Vol. 1, Cronin, L. E., Ed., Academic Press, New York, 1975, 420.

111. **Wiegert, R. G., Pomeroy, L. R., and Wiebe, W. J.,** Ecology of salt marshes: an introduction, in *The Ecology of a Salt Marsh, Ecological Studies,* Vol. 38, Pomeroy, L. R. and Wiegert, R. G., Eds., Springer-Verlag, New York, 1981, 3.

112. **Marples, T. G.,** A radionuclide study of arthropod food chains in a *Spartina* salt marsh estuary, *Ecology,* 47, 270, 1966.

113. **Kale, H. W.,** Ecology and bioenergetics of the long-billed marsh wren, *Telmatodytes palustris griseus* (Brewster), in *Georgia Salt Marshes,* Publ. 5, Nuttall Ornithology Club, Cambridge, 1965.

114. **Williams, R. B.,** The Ecology of Diatom Populations in a Georgia Salt Marsh, Ph.D. thesis, Harvard University, Cambridge, 1962.

115. **Pomeroy, L. R.,** Algal productivity in salt marshes of Georgia, *Limnol. Oceanogr.,* 4, 386, 1959.

116. **Ragotzkie, R. A.,** Plankton productivity in estuarine waters of Georgia, *Publ. Inst. Mar. Sci. Univ. Tex.,* 6, 146, 1959.

117. **Brownlee, D. C. and Jacobs, R.,** Mesozooplankton and microzooplankton in the Chesapeake Bay, in *Contaminant Problems and Management of Living Chesapeake Bay Resources,* Majumdar, S. K., Hall, L. W., Jr., and Austin, H. M., Eds., Pennsylvania Academy of Science, Easton, 1987, 217.

118. **Sellner, K. G.,** Phytoplankton in Chesapeake Bay: role in carbon, oxygen and nutrient dynamics, in *Contaminant Problems and Management of Living Chesapeake Bay Resources,* Majumdar, S. K., Hall, L. W., Jr., and Austin, H. M., Eds., Pennsylvania Academy of Science, Easton, 1987, 134.

119. **Horwitz, R. J.,** Fish, in *Ecological Studies in the Middle Reach of Chesapeake Bay: Calvert Cliffs,* Heck, K. L., Jr., Ed., Springer-Verlag, Berlin, 1987, 167.

120. **Orth, R. J. and Moore, K. A.,** Chesapeake Bay: an unprecedented decline in submerged aquatic vegetation, *Science,* 222, 51, 1983.

121. **Kennish, M. J.,** Summary and conclusions, in *Ecology of Barnegat Bay, New Jersey,* Kennish, M. J. and Lutz, R. A., Eds., Springer-Verlag, New York, 1984, 339.

122. **Commito, J. A. and Ambrose, W. G., Jr.,** Multiple trophic levels in soft-bottom communities, *Mar. Ecol. Prog. Ser.,* 26, 289, 1985.

123. **Sprules, W. G. and Bowerman, J. E.,** Omnivory and food chain length in zooplankton food webs, *Ecology,* 69, 418, 1988.

124. **McVean, A. and Findlay, I.,** The incidence of autotomy in an estuarine population of the crab *Carcinus maenas, J. Mar. Biol. Assoc. U.K.,* 59, 341, 1979.

125. **Rasmussen, E.,** Systematics and ecology of the Isefjord marine fauna, *Ophelia,* 11, 1, 1973.

126. **Rulifson, R. A. and McKenna, S. A.,** Food of striped bass in the upper Bay of Fundy, Canada, *Trans. Am. Fish. Soc.,* 116, 119, 1987.

127. **Burger, J. and Olla, B. L., Eds.,** *Shorebirds: Breeding Behavior and Populations,* Vol. 5, Plenum Press, New York, 1984.

128. **Burger, J. and Olla, B. L., Eds.,** *Shorebirds: Migration and Foraging Behavior,* Vol. 6, Plenum Press, New York, 1984.

129. **Feare, C. J. and Summers, R.,** Birds as predators on rocky shores, in *The Ecology of Rocky Coasts,* Moore, P. G. and Seed, R., Eds., Hodder and Stoughton, Sevenoaks, England, 1985.

130. **Prater, A. J.,** Ecology of Morecambe Bay. III. The food and feeding habits of knot (*Calidris canutus*) in Morecambe Bay, *J. Appl. Ecol.,* 9, 179, 1972.

131. **Ferns, P. N.,** Birds of the Bristol Channel and Severn estuary, *Mar. Pollut. Bull.,* 15, 76, 1984.

132. **Prater, A. J.,** *Estuary Birds of Britain and Ireland, T. & A. D.,* Poyser, Calton, England, 1981.

133. **Ferns, P. N.,** Sediment mobility in the Severn estuary and its influence upon the distribution of shorebirds, *Can. J. Fish. Aquat. Sci.,* 40(Suppl. 1), 331, 1983.

134. **Perry, M. C.,** Waterfowl of Chesapeake Bay, in *Contaminant Problems and Management of Living Chesapeake Bay Resources,* Majumdar, S. K., Hall, L. W., Jr., and Austin, H. M., Eds., Pennsylvania Academy of Science, Easton, 1987, 94.

135. **Perry, M. C. and Uhler, F. M.,** Food habits and distribution of wintering canvasbacks, *Aythya valisineria,* on Chesapeake Bay, *Estuaries,* 11, 57, 1988.

136. **Milne, H. and Dunnet, G. M.,** Standing crop, productivity and trophic relations of the fauna of the Ythan estuary, in *The Estuarine Environment,* Barnes, R. S. K. and Green, J., Ed., Applied Science Publishers, London, 1972, 86.

137. **Wolff, W. J. and de Wolf, L.,** Biomass and production of zoobenthos in the Grevelingen estuary, the Netherlands, *Estuarine Coastal Mar. Sci.,* 5, 1, 1977.

138. **Burger, J.,** Effects of demolition and beach clean-up operations on birds on a coastal mudflat in New Jersey, *Estuarine Coastal Shelf Sci.,* 27, 95, 1988.

139. **Burger, J.,** The effect of human activity on birds at a coastal bay, *Biol. Conserv.,* 21, 231, 1981.

140. **Burger, J.,** The effects of human activity on shorebirds in two coastal bays in northeastern United States, *Environ. Conserv.,* 13, 123, 1986.

141. **Burger, J.,** Abiotic factors affecting migrant shorebirds, in *Behavior of Marine Animals,* Vol. 6, Burger, J. and Olla, B., Eds., Plenum Press, New York, 1984, 1.

142. **Burger, J.,** Effects of human disturbance on colonial species, particularly gulls, *Colon. Waterbirds,* 4, 28, 1981.

INDEX

A

Abundance
  of bacteria, 11—18, 210
  of benthic fauna, 239
  of benthic macroalgae, 167
  of benthos, 224
  of copepods, 133
  of diatoms, 74
  of dinoflagellates, 74
  of fishes, 291, 298—300, 319—320, 338
  of ichthyoplankton, 127
  of meiofauna, 210—215
  of microalgae, see under Microalgae
  of microzooplankton, 134
  of phytoplankton, 63, 75, 79, 98
  of silicoflagellates, 74
  of zooplankton, 98, 122, 130, 366
Acetate, 30
Acoustic current meters, 6, 253
Acoustic techniques, 253, see also specific types
Adenosine triphosphate (ATP), 10, 12, 18, 28, 58
Adenylases, 30
Aerobic environments, 24—32, see also specific
    types
  attached bacteria and, 24—25
  bacteria as decomposers in, 25—31
  bacterivory and, 31—32
  free-living bacteria and, 24—25
Agar, 173, 177
Agarase, 29
Alanine, 30
Algae, see also specific types
  benthic, see Benthic algae
  blue-green, 51, 52, 54, 57, 91, 156, 159
  brown, 51, 56, 165
  classification of, 166
  epibenthic, 160, 161
  epiphytic, 163, 164
  golden-brown, 160
  green, 51, 52, 57, 165
  macro-, see Benthic macroalgae
  micro-, see Microalgae
  red, 165, 177
Alginase, 29
Alginates, 173, 177
Amino acids, 93, see also specific types
Ammonia, 93, 95
Ammonification, 10
Amphipods, 113, 186, see also specific types
Amylase, 29
Anaerobic decomposition, 33
Anaerobic environments, 32—38, see also specific
    types
Anaerobic microbial metabolism, 33
Animal-sediment relationships, 251—253
Anoxia, 237, 300, 301
Anoxic zones, 32

Anoxyphotobacteria, 11
Anthropogenic factors, 1—3, 5, 374, see also specific
    types
Anthropogenic stress, 292
Anthropogenic wastes, 1
Aphids, 186
Arsenic, 1
Aspartic acid, 30
ATP, see Adenosine triphosphate
Attached bacteria, 12, 24—25, 29
Attached microalgae, 162

B

Bacillariophyceae, 4, 52, 54—55, 156, see also
    Diatoms
Bacteria, 3, 6, 9—41, 351, 353, see also
    Bacterioplankton; specific types
  abundance of, 11—18, 210
  attached, 12, 24—25, 29
  benthic, 10
  biomass of, 11, 18, 26
  chemoheterotrophic, 165
  chemolithotrophic, 34
  chemosynthetic, 11
  as decomposers, 25—31
  density of, 15
  detritus and, 359
  detritus food web and, 357
  distribution of, 12
  ecological roles of, 10
  endobiotic, 10
  epibiotic, 10
  extracellular production of, 15
  filamentous, 165
  free-living, 12, 24—25, 29
  grazing and, 140, 210
  growth of, 25
  habitat of, 10
  heterotrophic, 11, 25
  macrofauna and, 249
  methanogenic, 25, 32, 37
  microbial processes and, see Microbial processes
  neustonic, 10
  nitrate-respiring, 18
  nitrifying, 11, 18
  nutrient competition and, 26
  oxide-reducing, 25
  photosynthetic, 11
  phototrophic, 18
  physiological groups of, 18
  physiology of, 10
  planktonic, 10
  production of, 9, 15, 18—23, 31
  in salt marshes, 185
  saprophagic, 351
  sediment, 161
  sulfate-reducing, 18, 25, 32, 34

sulfur, 11, 209
types of, 10—11
Bacterioplankton, 3, 9—11, 25, 30, 31, 51, see also
      Bacteria
Bacterivores, 9, see also specific types
Bacterivory, 31—32
Barnegat Bay
   benthic macroalgae in, 167—170
   benthic macroinvertebrates in, 229, 236
   detritus food web in, 360—364
   fishes in, 320—326
   grazing food web in, 368—370
   macroalgae in, 167—170
   macrofauna in, 229, 236
   macroinvertebrates in, 229, 236
   microzooplankton in, 125
   phytoplankton in, 70—74
   zooplankton in, 125—129
Beer's law, 84
Beetles, 187
Belehradek equation, 135
Benthic algae, 159—177, see also specific types
   classification of, 166
   mangroves and, 206
   micro-, 159—165
   production of, 364—366
Benthic bacteria, 10
Benthic fauna, 117, 155, 158—159, 200—201, 209,
      see also Macrofauna; Meiofauna; Microfauna;
      specific types
   abundance of, 239
   biomass of, 207
   boring organisms and, 257—258
   classification of, 158—159
   distribution of, 223, 224
   mangroves and, 207
   salinity and distribution of, 224
Benthic flora, 156—209, see also Benthic algae;
      Mangroves; Vascular plants; specific types
Benthic infauna, 222, 251
Benthic invertebrates, see Benthic macroinvertebrates
Benthic macroalgae, 155, 165—177, 249
   abundance of, 167
   in Barnegat Bay, 167—170
   biomass of, 170
   distribution of, 166
   ecological significance of, 173
   economical value of, 173—177
   in Grays Harbor, 167—170
   production of, 167
   type examples of, 167—170
Benthic macrofauna, 9, 31, 197, 237
Benthic macroinvertebrates, 224, 229, 236—241, see
      also specific types
   in Barnegat Bay, 229, 236
   biomass of, 246—248, 251
   in Chesapeake Bay, 236—238
   in Corpus Christi Bay, 238—239
   dominance of, 239
   in Dutch Wadden Sea, 241
   production of, 246—248

in Puget Sound, 239
   salinity and, 236
   species richness of, 244, 246
   in Upper Clyde Estuary, 239—241
Benthic macrophytes, 155
Benthic microalgae, 159—165, 249
Benthos, see also entries beginning with Benthic;
      specific types
   abundance of, 224
   biomass of, 224
   defined, 155
   equilibrium-stage communities in, 252
   general features of, 156—159
   macro-, 222, 236—238
   meio-, 26, 210
   salinity and, 155
   species composition of, 224
   species diversity of, 224
Beverton and Holt model, 318
B-glucosidase, 29
Biodegradational process, 28
Bioenergetic model, 318
Biogenic silicon, 97
Biogenic structures, 252
Biogeochemical cycling of elements, 9
Bioluminescence, 56
Biomass
   of bacteria, 11, 18, 26
   of benthic macroalgae, 170
   of benthic macroinvertebrates, 246—248, 251
   of benthos, 224
   of ciliates, 210
   of fauna, 207
   of fishes, 291
   of macrofauna, 219, 241, 246—248
   of meiofauna, 210, 217
   of phytoplankton, 51, 58—61, 97, 98
   of seagrasses, 199
   of sea urchins, 173
   of toplankton, 69
   of zooflagellates, 210
   of zooplankton, 112, 124, 125, 132, 133, 143, 144
Biotic relationships, 10
Biotin, 92, 97
Bioturbation, 5, 32, 117, 207, 219, 220, 252, 354
Birds, 187, 197, 372—374, see also specific types
Blue crab, 138
Blue-green algae, 51, 52, 54, 57, 91, 156, 159
Blue mussel, 241
Boring organisms, 253, 257—258, see also specific
      types
Brown algae, 51, 56, 165
Brown shrimp, 186

C

Cadmium, 1
Caging experiments, 200, 250
Calanoids, 112, 115, 125, 139, see also specific types
Capitella capitata, 5
Carrageenan, 177

Carbon fixation, 10, 60
Carbon-14-glucose, 30
Carbon mineralization, 33
Carbon/nitrogen ratio, 29
Carotenoids, 57
Cell division, 89, 91, 97
Cell fission, 55
Celluloses, 173
Centrales, 54
Centricae, 54
Centric diatoms, 54
Chemoheterotrophic bacteria, 165
Chemolithotrophic bacteria, 34
Chemolithotrophs, 10, 11, see also specific types
Chemoreception, 139
Chemosynthesis, 9
Chemosynthetic bacteria, 11
Chemotaxis, 165
Chesapeake Bay
    benthic macroinvertebrates in, 236—238
    fishes in, 319, 326—329
    grazing food web in, 368
    ichthyofauna in, 368
    macrofauna in, 236—238
    meroplankton in, 129
    mesozooplankton in, 129
    microzooplankton in, 129, 368
    phytoplankton in, 63—70
    zooplankton in, 129—130
Chlorinated hydrocarbons, 1
Chlorophyceae, 52, 57
Chlorophyll, 55, 60, 68, 70, 75, 79, 84
    blue-green algae and, 57
    phytoplankton biomass and, 58
    synthesis of, 91
Chlorophytes, 91, 165
Chloroplasts, 55, 85
Choanoflagellates, 29
Chromatic adaptation, 88
Chrysolaminarin, 54
Chrysophyceae, 51, 52, 56
Ciliates, 29, 31, 210
Cistenides gouldii, 5
Cladocerans, 112, 116
Clams, 159
Classification, see also Taxonomy
    of algae, 166
    of benthic fauna, 158—159
    of benthic invertebrates, 209
    of fishes, 293—298
    of phytoplankton, 51—53
    of seawater, 83
    of zooplankton, 111—121, see also Zooplankton
Clostridia, 18
Cobalamine, 92, 97
Cobalt, 92, 97
Coccolithophores, 52, 56, 58, 91
Coenocline, 222, 223
Commensalism, 303
Coot clam, 159
Copepods, 98, 112, 115, 369, see also specific types

    abundance of, 133
    calanoid, 125, 139
    cyclopoid, 139
    density of, 130
    distribution of, 112, 115, 116, 133
    feeding of, 139—140
    harpacticoid, 139
    in Narragansett Bay, 122
    production of, 145
    salinity and, 137
    succession of, 137
Copper, 1, 92, 97
Coprophagy, 353
Corpus Christi Bay, 238—239
Crabs, 138, 186, 207, 357, 358, 367
Crustaceans, 113, 157, see also specific types
Cryptophyceae, 51
Current meters, 6, 253
Cyanophyceae, 57, 156
Cyclopoids, 112, 115, 139, see also specific types

D

Day classification, 293
DDT, 1
Decay phases, 26
Decomposition, 26
    anaerobic, 33, 37
    bacterial role in, 25—31
    organic detritus, 28, 29
    organic matter, 32
    seagrass, 12
Dentrification, 10, 35, 36
Destabilization, 252, 253
Desulfomaculum spp., 34
Desulfovibrio
    desulfuricans, 38
    spp., 34
Desulfuromonas spp., 34
Detrital food web, see Detritus food web
Detritivores, 26, 29, 30, 163, 202, 353, 359, see also
        specific types
Detritus, 197, 352—355
    concentration of, 353
    macrofauna and, 249
    mangroves and, 208—209
    production of, 184—185, 208, 356, 359
    radiolabeling of, 355
Detritus food web, 6, 173, 207, 351—364
    in Barnegat Bay, 360—364
    detritus and, 352—355
    in Georgia salt marsh systems, 356—359
    macroconsumers and, 357—359
    mangroves and, 359—360
    microbes and, 357
Diarrhetic shellfish poisoning, 56
Diatoms, 4, 5, 51, 52, 54—55, 156, 159, 161, see also
        Bacillariophyceae; specific types
    abundance of, 58, 74
    in Chesapeake Bay, 69, 70
    epipelic growth of, 161

epiphytic, 164
  in Long Island Sound, 75, 77
  in Potomac River, 162
  reproduction of, 55
  salinity and, 91
  succession of, 76
Dinoflagellates, 4, 5, 51, 52, 55—56, 58, 156, see also
      Dinophyceae; specific types
  abundance of, 74
  in Chesapeake Bay, 69, 70
  in Long Island Sound, 75
  in Potomac River, 162
  reproduction of, 55
  salinity and, 91
  succession of, 76
Dinophyceae, 4, 52, 55—56, 156, see also
      Dinoflagellates
Dissimilatory nitrogenous oxide reduction, 32, 35—
      37
Dissimilatory processes, 37, see also specific types
Dissimilatory sulfate-reducing bacteria, 25, 32
Dissimilatory sulfate reduction, 32, 34—35
Dissolved inorganic phosphorus, 96
Dissolved organic carbon, 29, 30
Dissolved organic matter (DOM), 9, 23—26, 29, 30
  concentration of, 31
  detritus food web and, 352
  metabolizers of, 31
  precipitation of, 354
  seagrasses and, 201
  uptake of by bacteria, 30
Dissolved organic phosphorus, 96
Dissolved oxygen, 155, 300—302
Distribution
  of bacteria, 12
  of benthic fauna, 223
  of benthic macroalgae, 166
  of benthic macroinvertebrates, 224
  of benthic organisms, 156
  of copepods, 112, 115, 116, 133
  of fauna, 207
  of fishes, 298, 299, 312, 319, 338
  of ichthyoplankton, 138
  of macrofauna, see under Macrofauna
  of macrofaunal larvae, 225
  of mangroves, 202, 204—205
  of meiofauna, 215—216
  of meroplankton, 138
  of microalgae, 161
  of microzooplankton, 134
  of phytoplankton, 58
  of salt marsh fauna, 185
  of seagrasses, 198, 199
  of toplankton, 69
  of zooplankton, 137, 138, 140
DNA, 58
DOM, see Dissolved organic matter
Dredge spoils, 1
Ducks, 187
Dutch Wadden Sea, 241

E

Econfina-Fenholloway system, 319
Effluents, 1
Electromagnetic current meters, 6, 253
Endobacteria, 10
Endobiotic bacteria, 10
Energy budget approach, 143
Energy flow diagrams, 24, 197
Environmental stress, 245—246
Enzymes, 29, 30, see also specific types
Epibacteria, 11
Epibenthic algae, 160, 161
Epibiotic bacteria, 10
Epifauna, 155, 209
Epiflora, 202
Epifluorescence microscopy, 10, 18
Epiphytes, 160, 164, 170, 200, 202
Epiphytic algae, 163, 164
Epiphytic diatoms, 164
Equilibrium-stage benthic communities, 252
Esterase, 29
Euglenoids, 53, 57, 162, see also specific types
Euglenophyceae, 53, 57
Eukaryotes, 69
Eutrophication, 2, 300
Extracellular enzymes, 29, 30, see also specific types
Extracellular production of bacteria, 15

F

Fauna, see also specific types
  benthic, see Benthic fauna
  distribution of, 207
  ichthyo-, 6, 302, 304, 360
  in-, see Infauna
  macro-, see Macrofauna
  mangroves and, 206—207
  meio-, see Meiofauna
  micro-, see Microfauna
  salt marsh, 185—187
  sessile, 200—201
  swimming, 201
Fermentation, 32, 33
Fiddler crabs, 186, 207, 357, 358, 367
Filamentous bacteria, 165
Fishes, 6, 197, 291—345, see also specific types
  abundance of, 291, 298—300, 319—320, 338
  adventitious visitor, 298
  age and, 312—313
  anadromous, 296
  in Barnegat Bay, 320—326
  biomass of, 291
  catadromous, 296
  in Chesapeake Bay, 319, 326—329
  classification of, 293—298
  competition and, 302—303
  diet of, 139, 304—306, 310—312
  dissolved oxygen and, 300—302
  distribution of, 298, 299, 312, 319, 338

diversity of, 200, 291, 319—320
in Econfina-Fenholloway system, 319
environmental adaptations of, 298—304
feeding behavior of, 300
food webs and, 370—372
freshwater, 295
in Galveston Bay, 319
growth of, 300, 312—315
interspecific interactions and, 302—304
major predatory, 364
mass mortality of, 301
migration of, 138
mortality of, 312, 315—316
in Narragansett Bay, 319
niche requirements of, 302—303
nursery habitats for, 186, 297—298
population dynamics of, 312—319
predation by, 238
recruitment of, 312, 316—319
reproduction of, 300, 316—319
salinity and, 298—300, 302
in San Francisco Bay, 331, 336—338, 340—342
seasonal alteration in, 318
seasonal migrant, 297
species abundance of, 319—320
species composition of, 298, 299, 318
species diversity of, 319—320
species richness of, 299
temperature and, 298—300, 302, 319
in Terminos Lagoon, 329—331
truly estuarine, 295—296, 299
Flagellates, 53, 58, see also specific types
    choano, 29
    dino-, see Dinoflagellates
    euglenoid, 57
    micro-, 9, 25, 29, 31, 70
    phyto-, 52
    silico-, 52, 56, 74
    zoo-, 210
Flicker effect, 85
Flies, 187
Flora, see also specific types
    benthic, see Benthic flora
    epi-, 202
    macro-, 165
    mangroves and, 206—207
    micro-, 156
    salt marsh, 179
Florida mangroves, 359—360
Fluid bioturbation, 220
Food chain, 351, 360
Food webs, 3, 9, 310, 352—374, see also Trophic
        relationships; specific types
    birds and, 372—374
    detritus, see Detritus food web
    fishes and, 370—372
    grazing, see Grazing food web
    meiobenthos-microbial, 26
    microbial, 3, 6, 352
    pelagic, 9, 351

seagrasses and, 201
    secondary consumers and, 370—374
    tertiary consumers and, 370—374
Founding organisms, 253—257, see also specific
        types
Free-living bacteria, 12, 24—25, 29
Freshwater fishes, 295
Freshwater marshes, 188—189, 197
Frustule, 54
Fulvic acid, 26
Fungi, 6, 28, 32, 353
    detritus and, 359
    detritus food web and, 357
    saprophagic, 351

G

Gadoid recruitment model, 318
Geese, 187
Geochemical cycles, 11, see also specific types
Georgia salt marshes, 356—359, 366—368
Ghost crabs, 207
Glucose, 30
B-Glucosidase, 29
Glutamic acid, 30
Glycine, 30
Golden-brown algae, 160
Gompertz equation, 313, 314
Graham-Schaefer model, 312
Grasshoppers, 186
Grays Harbor, 167—170
Grazing, 31, 161, 163
    bacteria and, 140
    microfauna, 210
    phytoplankton and, 83, 97—99, 140
    zonation and, 166
    zooplankton, 70, 90, 98, 124, 139, 140
Grazing food web, 6, 310, 351, 364—370
    in Barnegat Bay, 368—370
    benthic algae production and, 364—366
    in Chesapeake Bay, 368
    in Georgia salt marshes, 366—368
    phytoplankton and, 364—366
Green algae, 51, 52, 57, 165
Growth
    of bacteria, 25
    of fishes, 300, 312—315
    of meiofauna, 216—218
    of phytoplankton, 88—90, 92, 97
    of seagrasses, 199
Gymnodiniales, 55

H

Halocline, 90
Halophytes, 180
Haptophyceae, 52—53, 56
Harpacticoids, 113, 115, 139, see also specific types
Heat transfer-type current meters, 6, 253
Heavy metals, 1, see also specific types

Hermaphroditism, 217
Heterophic bacteria, 25
Heterotrophic bacteria, 11
Heterotrophic communities, 359
Heterotrophs, 351, see also specific types
Heterotrophy, 24
Holoplankton, 4, 51, 111, 115—117, 121, 136, 137
Homoihaline gradient, 223
Humic acid, 26
Hydraulic models, 2
Hydrocarbons, 1
Hydromedusae, 113
Hydroxycinnamic acids, 185
Hypoxia, 69

I

Ichthyofauna, 6, 302, 304, 368
Ichthyoplankton, 113, 114, 136, 305, 317, 329
  abundance of, 127
  development of, 137
  distribution of, 138
Impellors, 6, 253
Industrial effluents, 1
Infauna, 155, 209, 219, 222
  benthic, 251
  nonpredatory, 251
  predatory, 222, 251
Insects, 186, 187, see also specific types
Instantaneous mortality rates, 144
Invertebrates, 200, 207, 209, 225, see also specific
    types
  benthic, see Benthic macroinvertebrates
  macro-, see Macroinvertebrates
  as predators, 370
  salinity and, 236
  species richness of, 244
Ipifauns, 201
Iron, 38, 92, 97
Isopods, 113, 186, 258, see also specific types

J

Jellyfish, 113

K

Killifishes, 186, 291

L

Lactic acid, 30
Land crabs, 207
Laser-doppler current meters, 6, 253
Leaching, 26, 354
Lead, 1
Level-bottom community concept, 258—259
Light
  competition for, 166
  duration of, 95
  intensity of, 83—85, 88, 95
  phytoplankton production and, 79, 82—89

  seagrass distribution and, 198, 199
  spectral distribution of, 83
  zooplankton and, 134—136
Light-shade adaptation, 88
Lignins, 173, 185, see also specific types
Lignocellulose, 28
Lipids, 54, see also specific types
Lithophytes, 160
Lobsters, 186
Logistic equation, 313, 314
Long Island Sound, 74—77, 124—125

M

Macroalgae, see Benthic macroalgae
Macroalgal epiphytes, 202
Macrobenthic coenocline, 223
Macrobenthic species, 155, see also specific types
Macrobenthos, 222, 236—238
Macroconsumers, 357—359, see also specific types
Macrofauna, 9, 158, 164, 200, 209, 218—260, 354,
    see also Benthic fauna
  abundance of, 218, 222, 228—229, 236—241
    in Barnegat Bay, 229, 236
    in Chesapeake Bay, 236—238
    in Corpus Christi Bay, 238—239
    in Dutch Wadden Sea, 241
    in Puget Sound, 239
    in Upper Clyde Estuary, 239—241
  animal-sediment relationships and, 251—253
  benthic, 9, 31, 197, 237
  biomass of, 219, 241, 246—248
  boring organisms and, 253, 257—258
  density of, 218
  distribution of, 218—228
    on scale of estuary, 222—228
    small-scale patterns in, 218—222
    on world scale, 228
  feeding strategies in, 248—249
  founding organisms of, 253—257
  larval dispersal in, 224—228
  level-bottom community concept and, 258—259
  mortality of, 237
  production of, 246—248
  salinity and, 222, 224, 237, 244—246
  species composition of, 228—229, 236—241
  species diversity in, 241—246
  succession of, 218, 219, 221
  trophic interactions and, 250—251
  zonation and, 259—260
Macroflora, 165
Macroinvertebrates, 224, 229, 236—241, see also
    specific types
  in Barnegat Bay, 229, 236
  biomass of, 246—248, 251
  in Chesapeake Bay, 236—238
  in Corpus Christi Bay, 238—239
  predator-prey experiments and, 250
  production of, 246—248
  in Puget Sound, 239
  standing crops of, 251
  in Upper Clyde Estuary, 239—241

Macronutrients, 92, see also specific types
Macrophytes, 155
Macrophytoplankton, 51
Macrozooplankton, 4, 111—113, 125, see also
    Zooplankton
Mammals, 187, see also specific types
Mangals, see Mangroves
Manganese, 38, 92, 97
Mangroves, 155, 156, 158, 179, 202—209
    benthic fauna and, 207
    detritus and, 208—209
    detritus food web and, 359—360
    distribution of, 202, 204—205
    ecological importance of, 207—209
    fauna and, 206—207
    flora and, 206—207
    Florida, 359—360
    as habitat formeers, 209
    production and, 207—208
    as sediment stabilizers, 209
    species composition of, 202—204
    succession and, 206
    zonation and, 205, 260
Marsh crabs, 184
Marshes, 188—189, 197, see also Salt marshes
McHugh classification, 295—298
Mechanoreception, 139
Meiobenthos, 210
Meiobenthos-microbial food web, 26
Meiofauna, 5, 9, 23, 25, 31, 158, 209—218, see also
    Benthic fauna
    abundance of, 210—215
    biomass of, 210, 217
    density of, 210, 215
    detritus food web and, 354
    development of, 217
    dispersal of, 215
    distribution of, 215—216
    as food source, 249
    growth of, 216—218
    permanent, 217
    production of, 217—218
    reproduction of, 216—217
    salinity and distribution of, 215
    species composition of, 215
    species diversity of, 215
    taxonomy of, 210
    temperature and distribution of, 215
    turnover of, 218
    zonation of, 215, 216
Meroplankton, 4, 111, 112, 122, 124, 125, 130
    in Chesapeake Bay, 129
    development of, 137
    distribution of, 138
    length of life of, 113—115
    in North Inlet, 133
    phytoplankton and, 369
    salinity and, 136
Mesozooplankton, 4, 111—113, 129, see also
    Zooplankton
Metabolism, 30, 33, 94
Metals, 1, 97, 185, see also specific types

Metamorphosis, 226
Metazoan bacterivores, 9
Metazooplankton, 3
Methane, 37, 38
Methanogenesis, 32, 37—38
Methanogenic bacteria, 25, 32, 37
Methanogens, 18
Michaelis-Menten equation, 92
Microalgae, see also specific types
    attached, 162
    benthic, 159—165
    distribution of, 161
    production of, 162, 163
    types of, 164
Microbial decay, 26
Microbial food web, 3, 6, 26, 352
Microbial loop, 3, 6, 9, 351
Microbial processes, 23—38, see also specific types
    in aerobic environments, 24—32
    in anaerobic environments, 32—38
Microbial stripping, 26, 355
Microfauna, 5, 29, 158, 200—201, 209—210, see
    also Benthic fauna
    detritus food web and, 354
    as food source, 249
Microflagellates, 9, 25, 29, 31, 70
Microflora, 156
Microorganisms, 165, see also specific types
Microzooplankton, 4, 31, 98, 111, see also Zooplank-
    ton
    abundance of, 134
    in Barnegat Bay, 125
    in Chesapeake Bay, 129, 368
    distribution of, 134
    fishes and, 305
    salinity and, 129
    size of, 111—112
Mineralization, 9, 25, 33, 34, 36
Molecular nitrogen, 93
Mollusks, 185, 237
Molybdenum, 92, 93
Monad equation, 93
Mosquitos, 187
Moths, 186
Moyle and Cech classification, 293
Mucopolysaccharides, 29, 30, see also specific types
Mud snails, 186
Municipal effluents, 1
Muramic acid, 28
Muskrats, 187
Mussels, 183, 241
Mutualism, 303—304
Mysid shrimp, 113

N

Nanophytoplankton, 112
Nanoplankton, 51, 58, 69
Nanozooplankton, 31
Narragansett Bay
    copepods in, 122
    fishes in, 319

phytoplankton in, 77—79, 124
zooplankton in, 122—124
Nematodes, 31
Neustonic bacteria, 10
Nitrate, 36, 93, 95
Nitrate-respiring bacteria, 18
Nitrification, 10, 36
Nitrifying bacteria, 11, 18
Nitrite, 93, 95
Nitrobacteriaceae, 11
Nitrobacter spp., 11
Nitrogen, 92
  carbon ratio to, 29
  metabolism of, 94
  phosphorus ratio to, 96
  phytoplankton production and, 93—96
  zooplankton release of, 143
Nitrogen budgets, 95
Nitrogen cycle, 10, 11, 94, 95
Nitrogen fixation, 10
Nitrogenous oxide reduction, 32, 33, 35—37
Nitrosomonas spp., 11
Nitrous oxide, 93
Nonpredatory infauna, 251
North Inlet, 133
Nucleic acid synthesis, 10
Numerical models, 2
Nutrients, 1, see also specific types
  bacteria as competitors for, 26
  concentration of, 77, 90
  cycling of, 143, 202
  limitation of, 69
  macro-, 92
  phytoplankton production and, 79, 83, 92—97
  phytoplankton uptake of, 92
  remineralization of, 354
  in salt marshes, 185

O

Oil spills, 1
Organic loading, 300
Organic trace substances, 97, see also specific types
Orthochemical processes, 354, see also specific types
Orthophosphate, 96
Osmoconformers, 223
Osmoregulation, 91, 224, 298, 319
Osmotic pressure, 91
Osmotic shock, 91
Osmotic stress, 90, 91, 223
Oxidative metabolism, 30
Oxide-reducing bacteria, 25
Oxide reduction, 32, 33, 35—37
Oxygen demand, 35, 300
Oxygen production, 60
Oxyphotobacteria, 11

P

PAR, see Photosynthetically active radiation
Parallel bottom communities, 259

Paralytic shellfish poisoning, 1, 56
Parasites, 11, 159, see also specific types
Parasitism, 303, 304
Parthenogenesis, 217
Particulate organic matter (POM), 23, 24, 26, 29, 30
  detritus food web and, 352
  seagrasses and, 201
Pathogenic organisms, see also specific types
Pathogens, 1, 2, see also specific types
PCB, see Polychlorinated biphenyls
Pearcy and Richards classification, 293
Pelagic food webs, 9, 351
Penate diatoms, 54
Peptides, 93, see also specific types
Peridinales, 55
Periphyton, 163, 164
Periwinkles, 186
Perkins classification, 293
Permanent meiobenthos, 210
Permanent meiofauna, 217
Pheromones, 226
Phosphatase, 29
Phosphate, 96
Phosphorus, 92, 93
  nitrogen ratio to, 96
  phytoplankton production and, 96
  zooplankton release of, 143
Phosphorus cycle, 10
Photic zone, 94
Photographic techniques, 253, see also specific types
Photosynthesis, 9, 10, 33, 54, 55, 60, 83
  measurement of, 61
  phytoplankton growth and, 89
  rates of, 91
  salinity and, 91
Photosynthetically active radiation (PAR), 84
Photosynthetic bacteria, 11
Photosynthetic rate, 85
Phototrophic bacteria, 18
Phototrophs, 10, 11, see also specific types
Phycobilins, 57
Physiology, 10, 135
Phytoflagellates, 52
Phytoplankton, 1, 4, 9, 51—102, 352, see also
    specific types
  abundance of, 63, 75, 79, 98
  biomass of, 51, 58—61, 69, 97, 98
  blooms of, 77, 79, 88—91, 95, 98, 134
  cell characteristics of, 53
  classification of, 51—53
  defined, 51
  density of, 58, 63, 70
  distribution of, 58, 69, 98
  diversity of, 58
  grazing and, 140
  growth of, 88—90, 92, 97
  nutrient uptake by, 92
  production of, 19, 51, 60—62, 68, 69, 74, 75, 79—
    99
    biological factors affecting, 97—99
    chemical factors affecting, 91—97

grazing and, 83, 97—99
grazing food web and, 364—366
light and, 79, 82—89
in Narragansett Bay, 124
nutrients and, 79, 83, 92—97
physical factors affecting, 83—91
salinity and, 79, 83, 91
temperature and, 79, 82, 89—90
water circulation and, 82, 90—91
reproduction of, 92
in salt marshes, 185
sampling methods for, 57
seasonal dynamics of, 63, 74, 134
species composition of, 62—79
in Barnegat Bay, 70—74
in Chesapeake Bay, 63—70
in Long Island Sound, 74—77
in Narragansett Bay, 77—79
standing crop of, 79, 97
succession of, 62—79
in Barnegat Bay, 70—74
in Chesapeake Bay, 63—70
in Long Island Sound, 74—77
in Narragansett Bay, 77—79
taxonomy of, 52—57, 74, 79
turnover in, 90
Plankton, 30, see also specific types
bacterio-, see Bacterioplankton
defined, 51
holo-, see Holoplankton
ichthyo-, see Ichthyoplankton
larvae of, 224, 225
mero-, see Meroplankton
nano-, 51, 58, 69
phyto-, see Phytoplankton
ultra-, 51, 58
zoo-, see Zooplankton
Plankton cycles, 98
Planktonic bacteria, 10
Poikilohaline gradient, 223
Poikilotherms, 299
Pollutants, 1, 2, see also specific types
Polychaetes, 117, 237, 241
Polychlorinated biphenyls (PCBs), 1
Polydora ligni, 5
Polymeric substances, 165
Polysaccharides, 28, 54, 185, see also specific types
POM, see Particulate organic matter
Population dynamics approach, 143
Potomac River, 162
Predator-prey experiments, 250
Predator-prey relationships, 302
Predatory infauna, 222, 251
Production, see also Reproduction
of bacteria, 9, 15, 18—23, 31
of benthic algae, 364—366
of benthic macroalgae, 167
of benthic macroinvertebrates, 246—248
of benthic microalgae, 163
of copepods, 145
of detritus, 184—185, 208, 356

of macrofauna, 246—248
mangroves and, 207—208
of meiofauna, 217—218
of methane, 37
of microalgae, 162, 163
of phytoplankton, 75, see under Phytoplankton
of seagrasses, 199
of zooplankton, 143—145
Prohodiscus brevis, 1
Proteinase, 29
Protozoans, 9, 23, 25, 29, 31, 111, 118, see also
specific types
benthic algae and, 164
in salt marshes, 185
Prymnesiophyceae, 52
Puget Sound, 239
Pycnocline, 90, 301

R

Radionuclides, 1
Rats, 187
Red algae, 165, 177
Redox potential discontinuity (RPD), 210, 221, 252
Refractory phase, 26
Remineralization, 9, 354
REMOTS sediment profile camera, 6, 253
Reproduction, see also Production
of diatoms, 55
of dinoflagellates, 55
of fishes, 300, 316—319
of meiofauna, 216—217
of phytoplankton, 92
Resource stability, 246
Respiration, 33, 60
Ribbed mussel, 185
Ricker recruitment equation, 318
Rotifers, 112
RPD, see Redox potential discontinuity
Russell equation, 312

S

Salinity
benthic fauna distribution and, 224
benthic invertebrates and, 236
benthos and, 155
copepods and, 137
fishes and, 298—300, 302
holoplankton and, 136
macrofauna and, 222, 224, 237
macrofauna species diversity and, 244—246
mangrove distribution and, 204, 205
meiofauna distribution and, 215
meroplankton and, 136
microzooplankton and, 129
phytoplankton production and, 79, 83, 91
seagrass distribution and, 198
zooplankton and, 136—137
Salt marshes, 156, 161
climatic groups of, 179

ecological importance of, 182—185
fauna in, 185—187
flora of, 179
formation of, 180—182
Georgia, 356—359, 366—368
species composition of, 177
succession and, 180, 182
trophic relationships in, 356
types of, 177
zonation in, 260
Salt marsh grasses, 155, 177—187, see also specific
    types
ecological importance of, 182—185
fauna and, 185—187
salt marsh formation and, 180—182
species composition of, 177—180
San Francisco Bay
fishes in, 331, 336—338, 340—342
zooplankton in, 133—134
Saprophagic bacteria, 351
Saprophytes, 11
Saxitoxin, 1, 56
Scanning electron microscopy (SEM), 164
Seagrasses, 5, 30, 155—157, 163, 197—202
biomass of, 199
decomposition of, 12
density of, 200
distribution of, 198, 199
diversity of, 200
ecological significance of, 201—202
as food sources, 249
growth of, 199
production of, 199
species composition of, 197—200
stabilization and, 252
subhabitats associated with, 197
temperate, 360—364
Seasonal alteration in fishes, 318
Seasonal dynamics of phytoplankton, 63, 74, 134
Seasonal marine migrant fishes, 297
Sea urchins, 173
Seaweeds, 167, 173, see also Seagrasses
Sedimentation, 165, 180, 202
Sediment-profile cameras, 221
SEM, see Scanning electron microscopy
Sessile fauna, 200—201
Sewage sludge, 1
Shrews, 187
Shrimp, 113, 186
Side-scan sonography, 253
Silicic acid, 96, 97
Silicoflagellates, 52, 56, 74
Silicon, 92, 93, 96—97
Silicon budgets, 97
Silicon cycle, 97
Sludge, 1
Snails, 186, 367
Sonography, 253
Sorption, 354
Spartina spp., 5

Spotfin killifish, 186
Stabilization, 252
Stochastic events, 205
Stock-recruitment models, 318
Stress, see also specific types
anthropogenic, 292
environmental, 245—246
osmotic, 90, 91, 223
Substrates, 156
Succession, 51
of copepods, 137
of diatoms, 76
of dinoflagellates, 76
of macrofauna, 218, 219, 221
mangroves and, 206
of microorganisms, 165
of phytoplankton, see under Phytoplankton
salt marshes and, 180, 182
of zooplankton, 136
Succinate dehydrogenase, 29
Sulfate-reducing bacteria, 18, 25, 32, 34
Sulfate-reducing zone, 38
Sulfate reduction, 10, 32, 34—36
Sulfides, 10
Sulfur bacteria, 11, 209
Sulfur cycle, 10, 11
Sunlight, see Light
Swamps, 156, 158, 207
Swimming fauna, 201
Symbiosis, 159, 302, 303

T

Tagging experiments, 314
Taxonomy, see also Classification
of meiofauna, 210
of phytoplankton, 52—57, 74, 79
of zooplankton, 117—122
Teleosts, 291
Temperate seagrasses, 360—364
Temperature
benthos and, 155
fishes and, 298—300, 302, 319
mangrove distribution and, 204
meiofauna distribution and, 215
phytoplankton production and, 79, 82, 89—90
zooplankton and, 135—136
Temporary meiobenthos, 210
Terminos Lagoon, 329—331
Thermistors, 6, 253
Thiamin, 92, 97
Thiobacillus spp., 11
Thrips, 186
Tidal freshwater marshes, 188—189, 197
Tides, 204, 205
Trace elements, 92, see also specific types
Trace metals, 1, 97, 185, see also specific types
Trophic relationships, 6, 351—376, see also Food
    webs
of fishes, 306, 310, 331

food webs and, see Food webs
  macrofauna and, 250—251
  in salt marshes, 356
Trophodynamics, 9, 304, 351
Tropical land crabs, 207
Truly estuarine fishes, 295—296, 299
Turbidity, 90, 155, 165, 181, 198
Turbidity maximum zone, 25
Turbulance, 155
Tychoplankton, 4, 111, 117

U

Ultraplankton, 51, 58
Upper Clyde Estuary, 239—241
Urea, 30, 93

V

Vascular plants, 173, 177—202, 365, see also specific
    types
  detritus of, 359, 363
  salt marsh grasses and, see Salt marsh grasses
  seagrasses, see Seagrasses
  in tidal freshwater marshes, 189, 197
Vegetative cells, 54, 55
Venice system of classification, 166
Vitamins, 92, 97, see also specific types
Von Bertalanffy equation, 313, 314

W

Walford plot, 313—315
Waste disposal, 1
Water circulation
  phytoplankton production and, 82, 90—91
  zooplankton and, 138—139
Wave action, 155

X

Xenobiotic compounds, 1, 2, see also specific types

Z

Zinc, 1, 92, 97
Zonation, 155, 166—167, 179, 205
  macrofauna and, 259—260
  in mangroves, 260
  meiofauna and, 215, 216
  in salt marshes, 260
Zooflagellates, 210
Zooplankton, 4, 9, 54, 95, 111—148, 352, see also
    specific types
  abundance of, 98, 122, 130, 366
  in Barnegat Bay, 125—129
  biomass of, 112, 124, 125, 132, 133, 143, 144
  in Chesapeake Bay, 129—130
  classification of, 111—121, see also Classification;
    Taxonomy
  defined, 51
  density of, 132
  distribution of, 137, 138, 140
  ecology of, 135
  feeding of, 139—143
  length of life of, 113—117
  light and, 134—136
  in Long Island Sound, 124—125
  macro-, see Macrozooplankton
  meso-, see Mesozooplankton
  meta-, 3
  micro-, see Microzooplankton
  nano-, 31
  in Narragansett Bay, 122—124
  in North Inlet, 132—133
  physiology of, 135
  production of, 143—145
  salinity and, 136—137
  in San Francisco Bay, 133—134
  size of, 111—113
  standing crops of, 138
  succession of, 136
  taxonomy of, 117—122
  temperature and, 135—136
  water circulation and, 138—139
Zooplankton grazing, 70, 90, 98, 124, 139, 140
Zooxanthellae, 56